Springer Studium Mathematik (Master)

Die Buchreihe **Springer Studium Mathematik** orientiert sich am Aufbau der Bachelor- und Masterstudiengänge im deutschsprachigen Raum: Neben einer soliden Grundausbildung in Mathematik vermittelt sie fachübergreifende und anwendungsbezogene Kompetenzen.

Zielgruppe der Lehr- und Übungsbücher sind vor allem Studierende und Dozenten mathematischer Studiengänge, aber auch anderer Fachrichtungen.

Die Unterreihe **Springer Studium Mathematik (Bachelor)** bietet Orientierung insbesondere beim Übergang von der Schule zur Hochschule sowie zu Beginn des Studiums. Sie unterstützt durch ansprechend aufbereitete Grundlagen, Beispiele und Übungsaufgaben und möchte Studierende für die Prinzipien und Arbeitsweisen der Mathematik begeistern. Titel dieser Reihe finden Sie unter springer.com/series/ 16564.

Die Unterreihe **Springer Studium Mathematik (Master)** bietet studierendengerechte Darstellungen forschungsnaher Themen und unterstützt fortgeschrittene Studierende so bei der Vertiefung und Spezialisierung. Dem Wandel des Studienangebots entsprechend sind diese Bücher in deutscher oder englischer Sprache verfasst. Titel dieser Reihe finden Sie unter springer.com/series/16565.

Die Reihe bündelt verschiedene erfolgreiche Lehrbuchreihen und führt diese systematisch fort. Insbesondere werden hier Titel weitergeführt, die vorher veröffentlicht wurden unter:

- Springer Studium Mathematik – Bachelor (springer.com/series/13446)
- Springer Studium Mathematik – Master (springer.com/series/13893)
- Bachelorkurs Mathematik (springer.com/series/13643)
- Aufbaukurs Mathematik (springer.com/series/12357)
- Advanced Lectures in Mathematics (springer.com/series/12121)

More information about this series at http://www.springer.com/series/16565

Joel W. Robbin · Dietmar A. Salamon

Introduction to Differential Geometry

 Springer

Joel W. Robbin
University of Wisconsin–Madison
Madison, USA

Dietmar A. Salamon
ETH Zürich
Zürich, Switzerland

Springer Studium Mathematik (Master)
ISBN 978-3-662-64339-6   ISBN 978-3-662-64340-2 (eBook)
https://doi.org/10.1007/978-3-662-64340-2

This Springer Spektrum imprint is published by the registered company Springer-Verlag GmbH, DE part of Springer Nature.
The registered company address is: Heidelberger Platz 3, 14197 Berlin, Germany

Preface

These are notes for the lecture course *"Differential Geometry I"* given by the second author at ETH Zürich in the fall semester 2017. They are based on a lecture course[1] given by the first author at the University of Wisconsin–Madison in the fall semester 1983.

One can distinguish *extrinsic differential geometry* and *intrinsic differential geometry*. The former restricts attention to submanifolds of Euclidean space while the latter studies manifolds equipped with a Riemannian metric. The extrinsic theory is more accessible because we can visualize curves and surfaces in $\mathbb{R}^3$, but some topics can best be handled with the intrinsic theory. The definitions in Chap. 2 have been worded in such a way that it is easy to read them either extrinsically or intrinsically and the subsequent chapters are mostly (but not entirely) extrinsic. One can teach a self contained one semester course in extrinsic differential geometry by starting with Chap. 2 and skipping the sections marked with an asterisk such as Sect. 2.8.

Here is a description of the content of the book, chapter by chapter. Chapter 1 gives a brief historical introduction to differential geometry and explains the extrinsic versus the intrinsic viewpoint of the subject.[2] This chapter was not included in the lecture course at ETH.

The mathematical treatment of the field begins in earnest in Chap. 2, which introduces the foundational concepts used in differential geometry and topology. It begins by defining manifolds in the extrinsic setting as smooth submanifolds of Euclidean space, and then moves on to tangent spaces, submanifolds and embeddings, and vector fields and flows.[3] The chapter includes an introduction to Lie groups in the extrinsic setting and a proof of the Closed Subgroup Theorem. It then discusses vector bundles and submersions and proves the Theorem of Frobenius. The last two sections deal with the intrinsic setting and can be skipped at first reading.

---

[1] Extrinsic Differential Geometry.

[2] It is shown in Sect. 1.3 how any topological atlas on a set induces a topology.

[3] Our sign convention for the Lie bracket of vector fields is explained in Sect. 2.5.7.

Chapter 3 introduces the Levi-Civita connection as covariant derivatives of vector fields along curves.[4] It continues with parallel transport, introduces motions without sliding, twisting, and wobbling, and proves the Development Theorem. It also characterizes the Levi-Civita connection in terms of the Christoffel symbols. The last section introduces Riemannian metrics in the intrinsic setting, establishes their existence, and characterizes the Levi-Civita connection as the unique torsion-free Riemannian connection on the tangent bundle.

Chapter 4 defines geodesics as critical points of the energy functional and introduces the distance function defined in terms of the lengths of curves. It then examines the exponential map, establishes the local existence of minimal geodesics, and proves the existence of geodesically convex neighborhoods. A highlight of this chapter is the proof of the Hopf–Rinow Theorem and of the equivalence of geodesic and metric completeness. The last section shows how these concepts and results carry over to the intrinsic setting.

Chapter 5 introduces isometries and the Riemann curvature tensor and proves the Generalized Theorema Egregium, which asserts that isometries preserve geodesics, the covariant derivative, and the curvature.

Chapter 6 contains some answers to what can be viewed as the fundamental problem of differential geometry: When are two manifolds isometric? The central tool for answering this question is the Cartan–Ambrose–Hicks Theorem, which etablishes necessary and sufficient conditions for the existence of a (local) isometry between two Riemannian manifolds. The chapter then moves on to examine flat spaces, symmetric spaces, and constant sectional curvature manifolds. It also includes a discussion of manifolds with nonpositive sectional curvature, proofs of the Cartan–Hadamard Theorem and of Cartan's Fixed Point Theorem, and as the main example a discussion of the space of positive definite symmetric matrices equipped with a natural Riemannian metric of nonpositive sectional curvature.

This is the point at which the ETH lecture course ended. However, Chap. 6 contains some additional material such as a proof of the Bonnet–Myers Theorem about manifolds with positive Ricci curvature, and it ends with brief discussions of the scalar curvature and the Weyl tensor.

The logical progression of the book up to this point is linear in that every chapter builds on the material of the previous one, and so no chapter can be skipped except for the first. What can be skipped at first reading are only the sections labelled with an asterisk that carry over the various notions introduced in the extrinsic setting to the intrinsic setting.

Chapter 7 deals with various specific topics that are at the heart of the subject but go beyond the scope of a one semester lecture course. It begins with a section on conjugate points and the Morse Index Theorem, which follows on naturally from Chap. 4 about geodesics. These results in turn are used in the proof of continuity of the injectivity radius in the second section. The third section builds on Chap. 5 on isometries and the Riemann curvature tensor. It contains a proof of the Myers–Steenrod Theorem, which asserts that the group of isometries is always a finite-

---

[4] The covariant derivative of a global vector field is deferred to Sect. 5.2.2.

dimensional Lie group. The fourth section examines the special case of the isometry group of a compact Lie group equipped with a bi-invariant Riemannian metric. The last two sections are devoted to Donaldson's differential geometric approach to Lie algebra theory as explained in [17]. They build on all the previous chapters and especially on the material in Chap. 6 about manifolds with nonpositive sectional curvature. The fifth section establishes conditions under which a convex function on a Hadamard manifold has a critical point. The last section uses these results to show that the Killing form on a simple Lie algebra is nondegenerate, to establish uniqueness up to conjugation of maximal compact subgroups of the automorphism group of a semisimple Lie algebra, and to prove Cartan's theorem about the compact real form of a semisimple complex Lie algebra.

The appendix contains brief discussions of some fundamental notions of analysis such as maps and functions, normal forms, and Euclidean spaces, that play a central role throughout this book.

We thank everyone who pointed out errors or typos in earlier versions of this book. In particular, we thank Charel Antony and Samuel Trautwein for many helpful comments. We also thank Daniel Grieser for his constructive suggestions concerning the exposition.

28 August 2021   Joel W. Robbin
                                                                                                     Dietmar A. Salamon

Contents

What Is Differential Geometry?

1

This preparatory chapter contains a brief historical introduction to the subject of differential geometry (Sect. 1.1), explains the concept of a coordinate chart (Sect. 1.2), discusses topological manifolds and shows how an atlas on a set determines a topology (Sect. 1.3), introduces the notion of a smooth structure (Sect. 1.4), and outlines the master plan for this book (Sect. 1.5).

1.1 Cartography and Differential Geometry

Carl Friedrich Gauß (1777–1855) is the father of differential geometry. He was (among many other things) a cartographer and many terms in modern differential geometry (chart, atlas, map, coordinate system, geodesic, etc.) reflect these origins. He was led to his *Theorema Egregium* (see Theorem 5.3.1) by the question of whether it is possible to draw an accurate map of a portion of our planet. Let us begin by discussing a mathematical formulation of this problem.

Consider the two-dimensional sphere S^2 sitting in the three-dimensional Euclidean space $\mathbb{R}^3$. It is cut out by the equation

$$x^2 + y^2 + z^2 = 1.$$

A map of a small region $U \subset S^2$ is represented mathematically by a one-to-one correspondence with a small region in the plane $z = 0$. In this book we will represent this with the notation $\phi : U \to \phi(U) \subset \mathbb{R}^2$ and call such an object a *chart* or a *system of local coordinates* (see Fig. 1.1).

What does it mean that ϕ is an "accurate" map? Ideally the user would want to use the map to compute the length of a curve in S^2. The length of a curve γ connecting two points $p, q \in S^2$ is given by the formula

$$L(\gamma) = \int_0^1 |\dot{\gamma}(t)|\, dt, \qquad \gamma(0) = p, \ \gamma(1) = q,$$

© The Editor(s) (if applicable) and The Author(s), under exclusive license to
Springer-Verlag GmbH, DE, part of Springer Nature 2022
J.W. Robbin, D.A. Salamon, *Introduction to Differential Geometry*,
Springer Studium Mathematik (Master), https://doi.org/10.1007/978-3-662-64340-2_1

Fig. 1.1 A chart

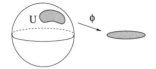

so the user will want the chart ϕ to satisfy $L(\gamma) = L(\phi \circ \gamma)$ for all curves γ. It is a consequence of the *Theorema Egregium* that there is no such chart.

Perhaps the user of such a map will be content to use the map to plot the shortest path between two points p and q in U. This path is called a *geodesic* and is denoted by γ_{pq}. It satisfies $L(\gamma_{pq}) = d_U(p,q)$, where

$$d_U(p,q) = \inf\{L(\gamma) \mid \gamma(t) \in U, \; \gamma(0) = p, \; \gamma(1) = q\}$$

so our less demanding user will be content if the chart ϕ satisfies

$$d_U(p,q) = d_E(\phi(p), \phi(q)),$$

where $d_E(\phi(p), \phi(q))$ is the length of the shortest path in the plane. It is also a consequence of the *Theorema Egregium* that there is no such chart.

Now suppose our user is content to have a map which makes it easy to navigate close to the shortest path connecting two points. Ideally the user would use a straight edge, magnetic compass, and protractor to do this. S/he would draw a straight line on the map connecting p and q and steer a course which maintains a constant angle (on the map) between the course and meridians. This *can* be done by the method of stereographic projection. This chart is *conformal* (which means that it preserves angles). According to Wikipedia stereographic projection was known to the ancient Greeks and a map using stereographic projection was constructed in the early 16th century. Exercises 3.7.5, 3.7.12, and 6.4.22 use stereographic projection; the latter exercise deals with the *Poincaré model* of the hyperbolic plane. The hyperbolic plane provides a counterexample to Euclid's Parallel Postulate.

Exercise 1.1.1 It is more or less obvious that for any surface $M \subset \mathbb{R}^3$ there is a unique shortest path in M connecting two points if they are sufficiently close. (This will be proved in Theorem 4.5.3.) This shortest path is called the *minimal geodesic* connecting p and q. Use this fact to prove that the minimal geodesic joining two points p and q in S^2 is an arc of the great circle through p and q. (This is the intersection of the sphere with the plane through p, q, and the center of the sphere.) Also prove that the minimal geodesic connecting two points in a plane is the straight line segment connecting them. **Hint:** Both a great circle in a sphere and a line in a plane are preserved by a reflection. (See also Exercise 4.2.5 below.)

Exercise 1.1.2 Stereographic projection is defined by the condition that for $p \in S^2 \setminus n$ the point $\phi(p)$ lies in the xy-plane $z = 0$ and the three points $n = (0, 0, 1)$, p, and $\phi(p)$ are collinear (see Fig. 1.2). Using the formula that the cosine of the

Fig. 1.2 Stereographic Projection

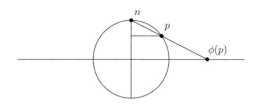

angle between two unit vectors is their inner product prove that ϕ is conformal. **Hint:** The plane through p, q, and n intersects the xy-plane in a straight line and the sphere in a circle through n. The plane through n, p, $\phi(p)$, and the center of the sphere intersects the sphere in a meridian. A proof that stereographic projection is conformal can be found in [27, page 248]. The proof is elementary in the sense that it doesn't use calculus. An elementary proof can also be found online at http://people.reed.edu/~jerry/311/stereo.pdf.

Exercise 1.1.3 It may seem fairly obvious that you can't draw an accurate map of a portion of the earth because the sphere is curved. However, the cylinder

$$C = \{(x, y, z) \in \mathbb{R}^3 \mid x^2 + y^2 = 1\}$$

is also curved, but the map $\psi : \mathbb{R}^2 \to C$ defined by $\psi(s, t) = (\cos(t), \sin(t), s)$ preserves lengths of curves, i.e. $L(\psi \circ \gamma) = L(\gamma)$ for any curve $\gamma : [a, b] \to \mathbb{R}^2$. Prove this.

Standard Notations The standard notations $\mathbb{N}$, $\mathbb{N}_0$, $\mathbb{Z}$, $\mathbb{Q}$, $\mathbb{R}$, $\mathbb{C}$ denote respectively the natural numbers (= positive integers), the non-negative integers, the integers, the rational numbers, the real numbers, and the complex numbers. We denote the identity map of a set X by id_X and the $n \times n$ identity matrix by $\mathbb{1}_n$ or simply $\mathbb{1}$. The notation V^* is used for the dual of a vector space V, but when $\mathbb{K}$ is a field such as $\mathbb{R}$ or $\mathbb{C}$ the notation $\mathbb{K}^*$ is sometimes used for the multiplicative group $\mathbb{K} \setminus \{0\}$. The terms *smooth*, *infinitely differentiable*, and C^∞ are all synonymous.

1.2 Coordinates

The rest of this chapter defines the category of smooth manifolds and smooth maps between them. Before giving the precise definitions we will introduce some terminology and give some examples.

Definition 1.2.1 A **chart** on a set M is a pair (ϕ, U) where U is a subset of M and $\phi : U \to \phi(U)$ is a bijection from U to an open set $\phi(U)$ in $\mathbb{R}^m$. An **atlas** on M is a collection

$$\mathcal{A} = \{(\phi_\alpha, U_\alpha)\}_{\alpha \in A}$$

of charts such that the domains U_α cover M, i.e. $M = \bigcup_{\alpha \in A} U_\alpha$.

The idea is that if $\phi(p) = (x_1(p), \ldots, x_m(p))$ for $p \in U$, then the functions x_i form a *system of local coordinates* defined on the subset U of M. The *dimension* of M should be m since it takes m numbers to uniquely specify a point of U. We will soon impose conditions on charts (ϕ, U), however *for the moment we are assuming nothing about the maps* ϕ (other than that they are bijective).

Example 1.2.2   Every open subset $U \subset \mathbb{R}^m$ has an atlas consisting of a single chart, namely $(\phi, U) = (\mathrm{id}_U, U)$ where id_U denotes the identity map of U.

Example 1.2.3   Assume that $\Omega \subset \mathbb{R}^m$ is an open set, that M is a subset of the product $\mathbb{R}^m \times \mathbb{R}^n = \mathbb{R}^{m+n}$, and that $h : \Omega \to \mathbb{R}^n$ is a continuous map whose graph is M, i.e.

$$\mathrm{graph}(h) := \{(x, y) \in \Omega \times \mathbb{R}^n \mid y = h(x)\} = M.$$

Let $U = \mathrm{graph}(h) = M$ and let $\phi(x, y) = x$ be the projection of U onto Ω. Then the pair (ϕ, U) is a chart on M. The inverse map is given by

$$\phi^{-1}(x) = (x, h(x))$$

for $x \in \Omega = \phi(U)$. Thus M has again an atlas consisting of a single chart.

Example 1.2.4   The m-sphere

$$S^m = \{p = (x_0, \ldots, x_m) \in \mathbb{R}^{m+1} \mid x_0^2 + \cdots + x_m^2 = 1\}$$

has an atlas consisting of the $2m + 2$ charts $\phi_{i\pm} : U_{i\pm} \to \mathbb{D}^m$ where $\mathbb{D}^m$ is the open unit disk in $\mathbb{R}^m$, $U_{i\pm} = \{p \in S^m \mid \pm x_i > 0\}$, and $\phi_{i\pm}$ is the projection which discards the ith coordinate. (See Example 2.1.14 below.)

Example 1.2.5   Let

$$A = A^\mathsf{T} \in \mathbb{R}^{(m+1)\times(m+1)}$$

be a symmetric matrix and define a quadratic form $F : \mathbb{R}^{m+1} \to \mathbb{R}$ by

$$F(p) := p^\mathsf{T} A p, \qquad p = (x_0, \ldots, x_m).$$

After a linear change of coordinates the function F has the form

$$F(p) = x_0^2 + \cdots + x_k^2 - x_{k+1}^2 - \cdots - x_r^2.$$

(Here $r + 1$ is the rank of the matrix A.) The set $M = F^{-1}(1)$ has an atlas of $2m + 2$ charts by the same construction as in Example 1.2.4, in fact S^m is the special case where $A = \mathbb{1}_{m+1}$, the $(m + 1) \times (m + 1)$ identity matrix. (See Example 2.1.13 below for another way to construct charts.)

Unit Sphere	$x^2 + y^2 + z^2 = 1$	
Ellipsoid	$\dfrac{x^2}{a^2} + \dfrac{y^2}{b^2} + \dfrac{z^2}{c^2} = 1$	
Cylinder	$x^2 + y^2 = 1$	
Elliptic Hyperboloid (of one sheet)	$\dfrac{x^2}{a^2} + \dfrac{y^2}{b^2} - \dfrac{z^2}{c^2} = 1$	
Elliptic Hyperboloid (of two sheets)	$\dfrac{x^2}{a^2} + \dfrac{y^2}{b^2} - \dfrac{z^2}{c^2} = -1$	
Hyperbolic Paraboloid	$z = \dfrac{x^2}{a^2} - \dfrac{y^2}{b^2}$	
Elliptic Paraboloid	$z = \dfrac{x^2}{a^2} + \dfrac{y^2}{b^2}$	

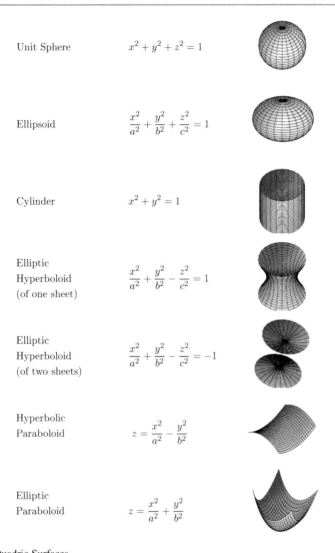

Fig. 1.3 Quadric Surfaces

Figure 1.3 enumerates the familiar **quadric surfaces** in $\mathbb{R}^3$. The *paraboloids* are examples of graphs as in Example 1.2.3 with $\Omega = \mathbb{R}^2$ and $n = 1$, and the *ellipsoid* and the two *hyperboloids* are instances of the *quadric hypersurfaces* defined in Example 1.2.5. The sphere is an instance of the ellipsoid (with $a = b = c = 1$) and the cylinder is a limit of the ellipsoid as well as of the elliptic hyperboloid of one sheet (as $a = b = 1$ and $c \to \infty$). The pictures were generated by computer using the parameterizations

$$x = a \cos(t) \cos(s), \qquad y = b \sin(t) \cos(s), \qquad z = c \sin(s)$$

for the ellipsoid,

$$x = a \cos(t) \cosh(s), \qquad y = b \sin(t) \cosh(s), \qquad z = c \sinh(s),$$

for the elliptic hyperboloid of one sheet, and

$$x = a \cos(t) \sinh(s), \qquad y = b \sin(t) \sinh(s), \qquad z = \pm c \cosh(s)$$

for the elliptic hyperboloid of two sheets. These quadric surfaces will be often used in the sequel to illustrate important concepts.

In the following two examples $\mathbb{K}$ denotes either the field $\mathbb{R}$ of real numbers or the field $\mathbb{C}$ of complex numbers, $\mathbb{K}^* := \{\lambda \in \mathbb{K} \mid \lambda \neq 0\}$ denotes the corresponding multiplicative group, and V denotes a vector space over $\mathbb{K}$.

Example 1.2.6   The **projective space** of V is the set of lines (through the origin) in V. In other words,

$$P(V) = \{\ell \subset V \mid \ell \text{ is a } 1\text{-dimensional } \mathbb{K}\text{-linear subspace}\}$$

When $\mathbb{K} = \mathbb{R}$ and $V = \mathbb{R}^{m+1}$ this is denoted by $\mathbb{R}P^m$ and when $\mathbb{K} = \mathbb{C}$ and $V = \mathbb{C}^{m+1}$ this is denoted by $\mathbb{C}P^m$. For our purposes we can identify the spaces $\mathbb{C}^{m+1}$ and $\mathbb{R}^{2m+2}$ but the projective spaces $\mathbb{C}P^m$ and $\mathbb{R}P^{2m}$ are very different. The various lines $\ell \in P(V)$ intersect in the origin, however, after the identification $P(V) = \{[v] \mid v \in V \setminus \{0\}\}$ with $[v] := \mathbb{K}^* v = \mathbb{K} v \setminus \{0\}$ the elements of $P(V)$ become disjoint, i.e. $P(V)$ is the set of equivalence classes of an equivalence relation on the open set $V \setminus \{0\}$. Assume that $V = \mathbb{K}^{m+1}$ and define an atlas on $P(V)$ as follows. For each integer $i = 0, 1, \ldots, m$ define $U_i := \{[v] \mid v = (x_0, \ldots, x_m), x_i \neq 0\}$ and define $\phi_i : U_i \to \mathbb{K}^m$ by

$$\phi_i([v]) = \left(\frac{x_0}{x_i}, \ldots, \frac{x_{i-1}}{x_i}, \frac{x_{i+1}}{x_i}, \ldots, \frac{x_m}{x_i} \right).$$

This atlas consists of $m + 1$ charts.

Example 1.2.7   For each positive integer k the set

$$G_k(V) := \{\ell \subset V \mid \ell \text{ is a } k\text{-dimensional } \mathbb{K}\text{-linear subspace}\}$$

is called the **Grassmann manifold** of k-planes in V. Thus $G_1(V) = P(V)$. Assume that $V = \mathbb{K}^n$ and define an atlas on $G_k(V)$ as follows. Let $e_1, \ldots, e_n$ be the standard basis for $\mathbb{K}^n$, i.e. e_i is the ith column of the $n \times n$ identity matrix $\mathbb{1}_n$. Each partition $\{1, 2, \ldots, n\} = I \cup J$, $I = \{i_1 < \cdots < i_k\}$, $J = \{j_1 < \cdots < j_{n-k}\}$ of the first n natural numbers determines a direct sum decomposition $\mathbb{K}^n = V = V_I \oplus V_J$ via the formulas

$$V_I = \mathbb{K} e_{i_1} + \cdots + \mathbb{K} e_{i_k}, \qquad V_J = \mathbb{K} e_{j_1} + \cdots + \mathbb{K} e_{j_{n-k}}.$$

Let U_I denote the set of all k-planes $\ell \in G_k(V)$ which are transverse to V_J, i.e. such that $\ell \cap V_J = \{0\}$. The elements of U_I are precisely those k-planes of the form $\ell = \mathrm{graph}(A)$, where $A : V_I \to V_J$ is a linear map. Define the map $\phi_I : U_I \to \mathbb{K}^{k \times (n-k)}$ by the formula

$$\phi_I(\ell) = (a_{rs}), \qquad Ae_{i_r} = \sum_{s=1}^{n-k} a_{rs} e_{j_s}, \qquad r = 1, \ldots, k.$$

Exercise: Prove that the set of all pairs (ϕ_I, U_I) as I ranges over the subsets of $\{1, \ldots, n\}$ of cardinality k form an atlas.

1.3 Topological Manifolds*

Definition 1.3.1 A **topological manifold** is a topological space M such that each point $p \in M$ has an open neighborhood U which is homeomorphic to an open subset of a Euclidean space.

Brouwer's *Invariance of Domain Theorem* asserts that, when $U \subset \mathbb{R}^m$ and $V \subset \mathbb{R}^n$ are nonempty open sets and $\phi : U \to V$ is a homeomorphism, then $m = n$. This means that if M is a connected topological manifold and some point of M has a neighborhood homeomorphic to an open subset of $\mathbb{R}^m$, then every point of M has a neighborhood homeomorphic to an open subset of that same $\mathbb{R}^m$. In this case we say that M has **dimension** m or is m-dimensional or is an m-manifold. Brouwer's theorem is fairly difficult (see [24, p. 126] for example) but if ϕ is a diffeomorphism, then the result is an easy consequence of the invariance of the rank in linear algebra and the chain rule. (See equation (1.4.1) below.)

By definition, a topological m-manifold M admits an atlas where every chart (ϕ, U) of the atlas is a homeomorphism $\phi : U \to \phi(U)$ from an open set $U \subset M$ to an open set $\phi(U) \subset \mathbb{R}^m$. The following definition and lemma explain when a given atlas determines a topology on M.

Definition 1.3.2 Let M be a set. Two charts (ϕ_1, U_1) and (ϕ_2, U_2) on M are said to be **topologically compatible** iff $\phi_1(U_1 \cap U_2)$ and $\phi_2(U_1 \cap U_2)$ are open subsets of $\mathbb{R}^m$ and the **transition map**

$$\phi_{21} := \phi_2 \circ \phi_1^{-1} : \phi_1(U_1 \cap U_2) \to \phi_2(U_1 \cap U_2)$$

is a homeomorphism. An atlas is said to be a **topological atlas** iff any two charts in this atlas are topologically compatible.

Lemma 1.3.3 *Let $\mathcal{A} = \{(\phi_\alpha, U_\alpha)\}_{\alpha \in A}$ be an atlas on a set M.*

(i) *If $\mathcal{A}$ is a topological atlas, then the collection*

$$\mathcal{U} := \left\{ U \subset M \;\middle|\; \begin{array}{l} \phi_\alpha(U \cap U_\alpha) \text{ is an open subset of } \mathbb{R}^m \\ \text{for every } \alpha \in A \end{array} \right\} \tag{1.3.1}$$

is a topology on M, and with this topology each U_α is an open subset of M and each ϕ_α is a homeomorphism. Thus M is a topological manifold with the topology (1.3.1).

(ii) If M is a topological manifold and each U_α is an open set and each ϕ_α is a homeomorphism, then $\mathcal{A}$ is a topological atlas and the topology (1.3.1) coincides with the topology of M.

Proof Exercise.   □

If M is a topological manifold, then the collection of all charts (U, ϕ) on M such that U is open and ϕ is a homeomorphism is a topological atlas. It is the unique **maximal topological atlas** in the sense that it contains every other topological atlas as in part (ii) of Lemma 1.3.3. However, we will often consider smaller, even finite, topological atlases, and by Lemma 1.3.3 each of these determines the topology of M.

Exercise 1.3.4 Show that the atlas in each example in Sect. 1.2 is a topological atlas. Conclude that each of these examples is a topological manifold.

Any subset $S \subset X$ of a topological space X inherits a topology from X, called the **relative topology** of S. A subset $U_0 \subset S$ is called **relatively open** in S (or S-**open**) iff there is an open set $U \subset X$ such that $U_0 = U \cap S$. A subset $A_0 \subset S$ is called **relatively closed** (or S-**closed**) iff there is a closed set $A \subset X$ such that $A_0 = A \cap S$. The relative topology on S is the coarsest topology such that the inclusion map $S \to X$ is continuous.

Exercise 1.3.5 Show that the relative topology satisfies the axioms of a topology (i.e. arbitrary unions and finite intersections of S-open sets are S-open, and the empty set and S itself are S-open). Show that the complement of an S-open set in S is S-closed and vice versa.

Exercise 1.3.6 Each of the sets defined in Examples 1.2.2, 1.2.3, 1.2.4, and 1.2.5 is a subset of some Euclidean space $\mathbb{R}^k$. Show that the topology in Exercise 1.3.4 is the relative topology inherited from the topology of $\mathbb{R}^k$. The topology on $\mathbb{R}^k$ is of course the metric topology defined by the distance function $d(x, y) = |x - y|$.

If $\sim$ is an equivalence relation on a topological space X, then the **quotient space** $Y := X/\sim := \{[x] \mid x \in X\}$ is the set of all equivalence classes $[x] := \{x' \in X \mid x' \sim x\}$. The map $\pi : X \to Y$ defined by $\pi(x) = [x]$ will be called the **obvious projection**. The quotient space inherits the **quotient topology** from Y. Namely, a set $V \subset Y$ is open in this topology iff the preimage $\pi^{-1}(V)$ is open in X. This topology is the finest topology on Y such that projection $\pi : X \to Y$ is continuous. Since the operation $V \mapsto \pi^{-1}(V)$ commutes with arbitrary unions and intersections the quotient topology obviously satisfies the axioms of a topology.

Exercise 1.3.7   Show that the topology on the projective space $P(V)$ in Example 1.2.6 determined by the atlas is the quotient topology inherited from the open set $V \setminus \{0\}$. Express the Grassmann manifold $G_k(V)$ in Example 1.2.7 as a quotient space and show that the topology determined by the atlas is the quotient topology. (Recall that in both examples $V = \mathbb{K}^n$ with $\mathbb{K} = \mathbb{R}$ or $\mathbb{K} = \mathbb{C}$.)

1.4   Smooth Manifolds Defined*

Let $U \subset \mathbb{R}^n$ and $V \subset \mathbb{R}^m$ be open sets. A map $f : U \to V$ is called **smooth** iff it is infinitely differentiable, i.e. iff all its partial derivatives

$$\partial^\alpha f = \frac{\partial^{\alpha_1 + \cdots + \alpha_n} f}{\partial x_1^{\alpha_1} \cdots \partial x_n^{\alpha_n}}, \qquad \alpha = (\alpha_1, \ldots, \alpha_n) \in \mathbb{N}_0^n,$$

exist and are continuous. In later chapters we will sometimes write $C^\infty(U, V)$ for the set of smooth maps from U to V.

Definition 1.4.1   Let $U \subset \mathbb{R}^n$ and $V \subset \mathbb{R}^m$ be open sets. For a smooth map $f = (f_1, \ldots, f_m) : U \to V$ and a point $x \in U$ the **derivative of** f **at** x is the linear map $df(x) : \mathbb{R}^n \to \mathbb{R}^m$ defined by

$$df(x)\xi := \frac{d}{dt}\bigg|_{t=0} f(x + t\xi) = \lim_{t \to 0} \frac{f(x + t\xi) - f(x)}{t}, \qquad \xi \in \mathbb{R}^n.$$

This linear map is represented by the **Jacobian matrix** of f at x which will also be denoted by

$$df(x) := \begin{pmatrix} \dfrac{\partial f_1}{\partial x_1}(x) & \cdots & \dfrac{\partial f_1}{\partial x_n}(x) \\ \vdots & & \vdots \\ \dfrac{\partial f_m}{\partial x_1}(x) & \cdots & \dfrac{\partial f_m}{\partial x_n}(x) \end{pmatrix} \in \mathbb{R}^{m \times n}.$$

Note that we use the same notation for the Jacobian matrix and the corresponding linear map from $\mathbb{R}^n$ to $\mathbb{R}^m$.

The derivative satisfies the **chain rule**. Namely, if $U \subset \mathbb{R}^n$, $V \subset \mathbb{R}^m$, and $W \subset \mathbb{R}^\ell$ are open sets and $f : U \to V$ and $g : V \to W$ are smooth map, then $g \circ f : U \to W$ is smooth and

$$d(g \circ f)(x) = dg(f(x)) \circ df(x) : \mathbb{R}^n \to \mathbb{R}^\ell \tag{1.4.1}$$

for every $x \in U$. Moreover the identity map $\mathrm{id}_U : U \to U$ is always smooth and its derivative at every point is the identity map of $\mathbb{R}^n$. This implies that, if $f : U \to V$

is a **diffeomorphism** (i.e. f is bijective and f and f^{-1} are both smooth), then its
derivative at every point is an invertible linear map. This is why the Invariance of
Domain Theorem (discussed after Definition 1.3.1) is easy for diffeomorphisms: if
$f : U \to V$ is a diffeomorphism, then the Jacobian matrix $df(x) \in \mathbb{R}^{m \times n}$ is invertible
for every $x \in U$ and so $m = n$. The Inverse Function Theorem (see Theorem A.2.2
in Appendix A.2) is a kind of converse.

Definition 1.4.2 (Smooth manifold)   Let M be a set. A **chart** on M is a pair
(ϕ, U) where $U \subset M$ and ϕ is a bijection from U to an open subset $\phi(U) \subset \mathbb{R}^m$
of some Euclidean space. Two charts (ϕ_1, U_1) and (ϕ_2, U_2) are said to be **smoothly
compatible** iff $\phi_1(U_1 \cap U_2)$ and $\phi_2(U_1 \cap U_2)$ are both open in $\mathbb{R}^m$ and the **transition
map**

$$\phi_{21} = \phi_2 \circ \phi_1^{-1} : \phi_1(U_1 \cap U_2) \to \phi_2(U_1 \cap U_2) \tag{1.4.2}$$

is a diffeomorphism. A **smooth atlas** on M is a collection $\mathcal{A}$ of charts on M any
two of which are smoothly compatible and such that the sets U, as (ϕ, U) ranges
over the elements of $\mathcal{A}$, cover M (i.e. for every $p \in M$ there is a chart $(\phi, U) \in \mathcal{A}$
with $p \in U$). A **maximal smooth atlas** is an atlas which contains every chart which
is smoothly compatible with each of its members. A **smooth manifold** is a pair
consisting of a set M and a maximal smooth atlas $\mathcal{A}$ on M.

Lemma 1.4.3   *If $\mathcal{A}$ is a smooth atlas, then so is the collection $\overline{\mathcal{A}}$ of all charts
smoothly compatible with each member of $\mathcal{A}$. The smooth atlas $\overline{\mathcal{A}}$ is obviously maximal. In other words, every smooth atlas extends uniquely to a maximal smooth
atlas.*

Proof   Let (ϕ_1, U_1) and (ϕ_2, U_2) be charts in $\overline{\mathcal{A}}$ and let $x \in \phi_1(U_1 \cap U_2)$. Choose a
chart $(\phi, U) \in \mathcal{A}$ such that $\phi_1^{-1}(x) \in U$. Then $\phi_1(U \cap U_1 \cap U_2)$ is an open neighborhood of x in $\mathbb{R}^m$ and the transition maps

$$\phi \circ \phi_1^{-1} : \phi_1(U \cap U_1 \cap U_2) \to \phi(U \cap U_1 \cap U_2),$$
$$\phi_2 \circ \phi^{-1} : \phi(U \cap U_1 \cap U_2) \to \phi_2(U \cap U_1 \cap U_2)$$

are smooth by definition of $\overline{\mathcal{A}}$. Hence so is their composition. This shows that the
map $\phi_2 \circ \phi_1^{-1} : \phi_1(U_1 \cap U_2) \to \phi_2(U_1 \cap U_2)$ is smooth near x. Since x was chosen
arbitrary, this map is smooth. Apply the same argument to its inverse to deduce that
it is a diffeomorphism. Thus $\overline{\mathcal{A}}$ is a smooth atlas. $\square$

Definitions 1.4.2 and 1.3.2 are *mutatis mutandis* the same, so every smooth atlas
on a set M is *a fortiori* a topological atlas, i.e. every smooth manifold is a topological manifold. (See Lemma 1.3.3.) Moreover the definitions are worded in such a
way that it is obvious that every smooth map is continuous.

Exercise 1.4.4 Show that each of the atlases from the examples in Sect. 1.2 is a smooth atlas. (You must show that the transition maps from Exercise 1.3.4 are smooth.)

When $\mathcal{A}$ is a smooth atlas on a topological manifold M one says that $\mathcal{A}$ is a **smooth structure** on the (topological) manifold M iff $\mathcal{A} \subset \mathcal{B}$, where $\mathcal{B}$ is the maximal topological atlas on M. *When no confusion can result we generally drop the notation for the maximal smooth atlas as in the following exercise.*

Exercise 1.4.5 Define the notion of a continuous map between topological manifolds and of a smooth map between smooth manifolds via continuity, respectively smoothness, in local coordinates. Let M, N, and P be smooth manifolds and $f : M \to N$ and $g : N \to P$ be smooth maps. Prove that the identity map id_M is smooth and that the composition $g \circ f : M \to P$ is a smooth map. (This is of course an easy consequence of the chain rule (1.4.1).)

Remark 1.4.6 It is easy to see that a topological manifold can have many distinct smooth structures. For example, $\{(\mathrm{id}_{\mathbb{R}}, \mathbb{R})\}$ and $\{(\phi, \mathbb{R})\}$ where $\phi(x) = x^3$ are atlases on the real numbers which extend to distinct smooth structures but determine the same topology. However these two manifolds are diffeomorphic via the map $x \mapsto x^{1/3}$. In the 1950's it was proved that there are smooth manifolds which are homeomorphic but not diffeomorphic and that there are topological manifolds which admit no smooth structure. In the 1980's it was proved in dimension $m = 4$ that there are uncountably many smooth manifolds that are all homeomorphic to $\mathbb{R}^4$ but no two of them are diffeomorphic to each other. These theorems are very surprising and very deep.

A collection of sets and maps between them is called a *category* iff the collection of maps contains the identity map of every set in the collection and the composition of any two maps in the collection is also in the collection. The sets are called the *objects* of the category and the maps are called the *morphisms* of the category. An invertible morphism whose inverse is also in the category is called an *isomorphism*. Some examples are the category of *all* sets and maps, the category of topological spaces and continuous maps (the isomorphisms are the homeomorphisms), the category of topological manifolds and continuous maps between them, and the category of smooth manifolds and smooth maps (the isomorphisms are the diffeomorphisms). Each of the last three categories is a subcategory of the preceding one.

Often categories are enlarged by a kind of "gluing process". For example, the "global" category of smooth manifolds and smooth maps was constructed from the "local" category of open sets in Euclidean space and smooth maps between them via the device of charts and atlases. (The chain rule shows that this local category is in fact a category.) The point of Definition 1.3.2 is to show (via Lemma 1.3.3) that topological manifolds can be defined in a manner analogous to the definition we gave for smooth manifolds in Definition 1.4.2.

Other kinds of manifolds (and hence other kinds of geometry) are defined by choosing other local categories, i.e. by imposing conditions on the transition maps in Equation (1.4.2). For example, a *real analytic manifold* is one where the transition maps are real analytic, a *complex manifold* is one whose coordinate charts take values in $\mathbb{C}^n$ and whose transition maps are holomorphic diffeomorphisms, and a *symplectic manifold* is one whose coordinate charts take values in $\mathbb{R}^{2n}$ and whose transition maps are *canonical transformations* in the sense of classical mechanics. Thus $\mathbb{C}P^n$ is a complex manifold and $\mathbb{R}P^n$ is a real analytic manifold.

1.5   The Master Plan

In studying differential geometry it is best to begin with *extrinsic differential geometry* which is the study of the geometry of submanifolds of Euclidean space as in Examples 1.2.3 and 1.2.5. This is because we can visualize curves and surfaces in $\mathbb{R}^3$. However, there are a few topics in the later chapters which require the more abstract Definition 1.4.2 even to say interesting things about extrinsic geometry. There is a generalization to these manifolds involving a structure called a *Riemannian metric*. We will call this generalization *intrinsic differential geometry*. Examples 1.2.6 and 1.2.7 fit into this more general definition so intrinsic differential geometry can be used to study them.

Since an open set in Euclidean space is a smooth manifold the definition of a submanifold of Euclidean space (see Sect. 2.1 below) is *mutatis mutandis* the same as the definition of a submanifold of a manifold. The definitions in Chap. 2 are worded in such a way that it is easy to read them either extrinsically or intrinsically and the subsequent chapters are mostly (but not entirely) extrinsic. Those sections which require intrinsic differential geometry (or which translate extrinsic concepts into intrinsic ones) are marked with a *.

Foundations

2

This chapter introduces various fundamental concepts that are central to the fields of differential geometry and differential topology. Both fields concern the study of smooth manifolds and their diffeomorphisms. The chapter begins with an introduction to submanifolds of Euclidean space and smooth maps (Sect. 2.1), to tangent spaces and derivatives (Sect. 2.2), and to submanifolds and embeddings (Sect. 2.3). In Sect. 2.4 we move on to vector fields and flows and introduce the Lie bracket of two vector fields. Lie groups and their Lie algebras, in the extrinsic setting, are the subject of Sect. 2.5, which includes a proof of the Closed Subgroup Theorem. In Sect. 2.6 we introduce vector bundles over a manifold as subbundles of a trivial bundle and in Sect. 2.7 we prove the theorem of Frobenius. The last two sections of this chapter are concerned with carrying over all these concepts from the extrinsic to the intrinsic setting and can be skipped at first reading (Sects. 2.8 and 2.9).

2.1 Submanifolds of Euclidean Space

To carry out the Master Plan Sect. 1.5 we must (as was done in [50]) extend the definition of smooth map to maps $f : X \to Y$ between subsets $X \subset \mathbb{R}^k$ and $Y \subset \mathbb{R}^\ell$ which are not necessarily open. In this case a map $f : X \to Y$ is called **smooth** iff for each $x_0 \in X$ there exists an open neighborhood $U \subset \mathbb{R}^k$ of x_0 and a smooth map $F : U \to \mathbb{R}^\ell$ that agrees with f on $U \cap X$. A map $f : X \to Y$ is called a **diffeomorphism** iff f is bijective and f and f^{-1} are smooth. When there exists a diffeomorphism $f : X \to Y$ then X and Y are called **diffeomorphic**. *When X and Y are open these definitions coincide with the usage in Sect. 1.4.*

Exercise 2.1.1 (Chain rule) Let $X \subset \mathbb{R}^k$, $Y \subset \mathbb{R}^\ell$, $Z \subset \mathbb{R}^m$ be arbitrary subsets. If $f : X \to Y$ and $g : Y \to Z$ are smooth maps, then so is the composition $g \circ f : X \to Z$. The identity map id : $X \to X$ is smooth.

© The Editor(s) (if applicable) and The Author(s), under exclusive license to
Springer-Verlag GmbH, DE, part of Springer Nature 2022
J.W. Robbin, D.A. Salamon, *Introduction to Differential Geometry*,
Springer Studium Mathematik (Master), https://doi.org/10.1007/978-3-662-64340-2_2

Fig. 2.1 A coordinate chart
$\phi : U \cap M \to \Omega$

Exercise 2.1.2 Let $E \subset \mathbb{R}^k$ be an m-dimensional linear subspace and let $v_1, \ldots, v_m$ be a basis of E. Then the map $f : \mathbb{R}^m \to E$ defined by $f(x) := \sum_{i=1}^{m} x_i v_i$ is a diffeomorphism.

Definition 2.1.3 Let $k, m \in \mathbb{N}_0$. A subset $M \subset \mathbb{R}^k$ is called a **smooth m-dimensional submanifold of $\mathbb{R}^k$** iff every point $p \in M$ has an open neighborhood $U \subset \mathbb{R}^k$ such that $U \cap M$ is diffeomorphic to an open subset $\Omega \subset \mathbb{R}^m$. A diffeomorphism

$$\phi : U \cap M \to \Omega$$

is called a **coordinate chart** of M and its inverse

$$\psi := \phi^{-1} : \Omega \to U \cap M$$

is called a **(smooth) parametrization** of $U \cap M$ (see Fig. 2.1).

In Definition 2.1.3 we have used the fact that the domain of a smooth map can be an arbitrary subset of Euclidean space and need not be open. The term *m-manifold in $\mathbb{R}^k$* is short for *m-dimensional submanifold of $\mathbb{R}^k$*. In keeping with the Master Plan Sect. 1.5 we will sometimes say *manifold* rather than *submanifold of $\mathbb{R}^k$* to indicate that the context holds in both the intrinsic and extrinsic settings.

Lemma 2.1.4 *If $M \subset \mathbb{R}^k$ is a nonempty smooth m-manifold, then $m \leq k$.*

Proof Fix an element $p_0 \in M$, choose a coordinate chart $\phi : U \cap M \to \Omega$ with $p_0 \in U$ and values in an open subset $\Omega \subset \mathbb{R}^m$, and denote its inverse by $\psi := \phi^{-1} : \Omega \to U \cap M$. Shrinking U, if necessary, we may assume that ϕ extends to a smooth map $\Phi : U \to \mathbb{R}^m$. This extension satisfies $\Phi(\psi(x)) = \phi(\psi(x)) = x$ and hence $d\Phi(\psi(x)) d\psi(x) = \mathrm{id} : \mathbb{R}^m \to \mathbb{R}^m$ for all $x \in \Omega$, by the chain rule. Hence the derivative $d\psi(x) : \mathbb{R}^m \to \mathbb{R}^k$ is injective for all $x \in \Omega$, and hence $m \leq k$ because Ω is nonempty. This proves Lemma 2.1.4. $\square$

Example 2.1.5 Consider the 2-sphere

$$M := S^2 = \{(x, y, z) \in \mathbb{R}^3 \,|\, x^2 + y^2 + z^2 = 1\}$$

depicted in Fig. 2.2 and let $U \subset \mathbb{R}^3$ and $\Omega \subset \mathbb{R}^2$ be the open sets

$$U := \{(x, y, z) \in \mathbb{R}^3 \,|\, z > 0\}, \qquad \Omega := \{(x, y) \in \mathbb{R}^2 \,|\, x^2 + y^2 < 1\}.$$

Fig. 2.2 The 2-sphere and the 2-torus

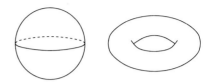

The map $\phi : U \cap M \to \Omega$ given by

$$\phi(x, y, z) := (x, y)$$

is bijective and its inverse $\psi := \phi^{-1} : \Omega \to U \cap M$ is given by

$$\psi(x, y) = (x, y, \sqrt{1 - x^2 - y^2}).$$

Since both ϕ and ψ are smooth, the map ϕ is a coordinate chart on S^2. Similarly, we can use the open sets $z < 0$, $y > 0$, $y < 0$, $x > 0$, $x < 0$ to cover S^2 by six coordinate charts. Hence S^2 is a manifold. A similar argument shows that the unit sphere $S^m \subset \mathbb{R}^{m+1}$ (see Example 2.1.14 below) is a manifold for every integer $m \geq 0$.

Example 2.1.6 Let $\Omega \subset \mathbb{R}^m$ be an open set and $h : \Omega \to \mathbb{R}^{k-m}$ be a smooth map. Then the graph of h is a smooth submanifold of $\mathbb{R}^m \times \mathbb{R}^{k-m} = \mathbb{R}^k$:

$$M := \text{graph}(h) := \{(x, y) \mid x \in \Omega, \ y = h(x)\}.$$

It can be covered by a single coordinate chart $\phi : U \cap M \to \Omega$, where $U := \Omega \times \mathbb{R}^{k-m}$, ϕ is the projection onto Ω, and $\psi := \phi^{-1} : \Omega \to U$ is given by $\psi(x) = (x, h(x))$ for $x \in \Omega$.

Exercise 2.1.7 (The case $m = 0$) Show that a subset $M \subset \mathbb{R}^k$ is a 0-dimensional submanifold if and only if M is discrete, i.e. for every $p \in M$ there exists an open set $U \subset \mathbb{R}^k$ such that $U \cap M = \{p\}$.

Exercise 2.1.8 (The case $m = k$) Show that a subset $M \subset \mathbb{R}^m$ is an m-dimensional submanifold if and only if M is open.

Exercise 2.1.9 (Products) If $M_i \subset \mathbb{R}^{k_i}$ is an m_i-manifold for $i = 1, 2$, show that $M_1 \times M_2$ is an $(m_1 + m_2)$-dimensional submanifold of $\mathbb{R}^{k_1 + k_2}$. Prove that the m-torus $\mathbb{T}^m := (S^1)^m$ is a smooth submanifold of $\mathbb{C}^m$.

The next theorem characterizes smooth submanifolds of Euclidean space. In particular condition (iii) will be useful in many cases for verifying the manifold condition. We emphasize that the sets $U_0 := U \cap M$ that appear in Definition 2.1.3 are open subsets of M with respect to the relative topology that M inherits from the ambient space $\mathbb{R}^k$ and that such relatively open sets are also called M-open (see Sect. 1.3).

Fig. 2.3 Submanifolds of
Euclidean space

Theorem 2.1.10 (Manifolds) *Let m and k be integers with $0 \le m \le k$. Let $M \subset \mathbb{R}^k$
be a set and $p_0 \in M$. Then the following are equivalent.*

(i) *There exists an M-open neighborhood $U_0 \subset M$ of p_0 and a diffeomorphism
$\phi_0 : U_0 \to \Omega_0$ onto an open set $\Omega_0 \subset \mathbb{R}^m$.*
(ii) *There exist open sets $U, \Omega \subset \mathbb{R}^k$ and a diffeomorphism $\phi : U \to \Omega$ such that
$p_0 \in U$ and*

$$\phi(U \cap M) = \Omega \cap (\mathbb{R}^m \times \{0\})$$

(see Fig. 2.3).
(iii) *There exists an open set $U \subset \mathbb{R}^k$ and a smooth map $f : U \to \mathbb{R}^{k-m}$ such
that $p_0 \in U$, the derivative $df(p) : \mathbb{R}^k \to \mathbb{R}^{k-m}$ is surjective for every point
$p \in U \cap M$, and*

$$U \cap M = f^{-1}(0) = \{p \in U \mid f(p) = 0\}.$$

*Moreover, if (i) holds, then the diffeomorphism $\phi : U \to \Omega$ in (ii) can be chosen
such that $U \cap M \subset U_0$ and $\phi(p) = (\phi_0(p), 0)$ for every $p \in U \cap M$.*

Proof First assume (ii) and denote the diffeomorphism in (ii) by

$$\phi = (\phi_1, \phi_2, \dots, \phi_k) : U \to \Omega \subset \mathbb{R}^k.$$

Then part (i) holds with $U_0 := U \cap M$, $\Omega_0 := \{x \in \mathbb{R}^m \mid (x, 0) \in \Omega\}$, and

$$\phi_0 := (\phi_1, \dots, \phi_m)|_{U_0} : U_0 \to \Omega_0,$$

and part (iii) holds with $f := (\phi_{m+1}, \dots, \phi_k) : U \to \mathbb{R}^{k-m}$. This shows that
part (ii) implies both (i) and (iii).

We prove that (i) implies (ii). Let $\phi_0 : U_0 \to \Omega_0$ be the coordinate chart in
part (i), let $\psi_0 := \phi_0^{-1} : \Omega_0 \to U_0$ be its inverse, and let $x_0 := \phi_0(p_0) \in \Omega_0$.
Then the derivative $d\psi_0(x_0) : \mathbb{R}^m \to \mathbb{R}^k$ is injective by Lemma 2.1.4. Hence
there exists a matrix $B \in \mathbb{R}^{k \times (k-m)}$ such that $\det([d\psi_0(x_0) \, B]) \ne 0$. Define the
map $\psi : \Omega_0 \times \mathbb{R}^{k-m} \to \mathbb{R}^k$ by

$$\psi(x, y) := \psi_0(x) + By.$$

Then the $k \times k$-matrix $d\psi(x_0, 0) = [d\psi_0(x_0) \, B] \in \mathbb{R}^{k \times k}$ is nonsingular, by choice
of B. Hence, by the Inverse Function Theorem A.2.2, there exists an open

neighborhood $\widetilde{\Omega} \subset \Omega_0 \times \mathbb{R}^{k-m}$ of $(x_0, 0)$ such that $\widetilde{U} := \psi(\widetilde{\Omega}) \subset \mathbb{R}^k$ is open and $\psi|_{\widetilde{\Omega}} : \widetilde{\Omega} \to \widetilde{U}$ is a diffeomorphism. In particular, the restriction of ψ to $\widetilde{\Omega}$ is injective. Now the set $\{x \in \Omega_0 \mid (x, 0) \in \widetilde{\Omega}\}$ is open and contains x_0. Hence the set

$$\widetilde{U}_0 := \{\psi_0(x) \mid x \in \Omega_0, \ (x, 0) \in \widetilde{\Omega}\} = \{p \in U_0 \mid (\phi_0(p), 0) \in \widetilde{\Omega}\} \subset M$$

is M-open and contains p_0. Hence, by the definition of the relative topology, there exists an open set $W \subset \mathbb{R}^k$ such that $\widetilde{U}_0 = W \cap M$. Define

$$U := \widetilde{U} \cap W, \qquad \Omega := \widetilde{\Omega} \cap \psi^{-1}(W).$$

Then $U \cap M = \widetilde{U}_0$ and ψ restricts to a diffeomorphism from Ω to U.

Now let $(x, y) \in \Omega$. We claim that

$$\psi(x, y) \in M \qquad \Longleftrightarrow \qquad y = 0. \tag{2.1.1}$$

If $y = 0$, then obviously $\psi(x, y) = \psi_0(x) \in M$. Conversely, let $(x, y) \in \Omega$ and suppose that $p := \psi(x, y) \in M$. Then $p \in U \cap M = \widetilde{U} \cap W \cap M = \widetilde{U}_0 \subset U_0$ and hence $(\phi_0(p), 0) \in \widetilde{\Omega}$, by definition of $\widetilde{U}_0$. This implies

$$\psi(\phi_0(p), 0) = \psi_0(\phi_0(p)) = p = \psi(x, y).$$

Since the pairs (x, y) and $(\phi_0(p), 0)$ both belong to the set $\widetilde{\Omega}$ and the restriction of ψ to $\widetilde{\Omega}$ is injective we obtain $x = \phi_0(p)$ and $y = 0$. This proves (2.1.1). It follows from (2.1.1) that the map $\phi := (\psi|_\Omega)^{-1} : U \to \Omega$ satisfies $\phi(U \cap M) = \{(x, y) \in \Omega \mid \psi(x, y) \in M\} = \Omega \cap (\mathbb{R}^m \times \{0\})$. Thus we have proved that (i) implies (ii).

We prove that (iii) implies (ii). Let $f : U \to \mathbb{R}^{k-m}$ be as in part (iii). Then $p_0 \in U$ and the derivative $df(p_0) : \mathbb{R}^k \to \mathbb{R}^{k-m}$ is a surjective linear map. Hence there exists a matrix $A \in \mathbb{R}^{m \times k}$ such that

$$\det \begin{pmatrix} A \\ df(p_0) \end{pmatrix} \neq 0.$$

Define the map $\phi : U \to \mathbb{R}^k$ by

$$\phi(p) := \begin{pmatrix} Ap \\ f(p) \end{pmatrix} \qquad \text{for } p \in U.$$

Then $\det(d\phi(p_0)) \neq 0$. Hence, by the Inverse Function Theorem A.2.2, there exists an open neighborhood $U' \subset U$ of p_0 such that $\Omega' := \phi(U')$ is an open subset of $\mathbb{R}^k$ and the restriction $\phi' := \phi|_{U'} : U' \to \Omega'$ is a diffeomorphism. In particular, the restriction $\phi|_{U'}$ is injective. Moreover, it follows from the assumptions on f and the definition of ϕ that

$$U' \cap M = \{p \in U' \mid f(p) = 0\} = \{p \in U' \mid \phi(p) \in \mathbb{R}^m \times \{0\}\}$$

and so $\phi'(U' \cap M) = \Omega' \cap (\mathbb{R}^m \times \{0\})$. Hence the diffeomorphism $\phi' : U' \to \Omega'$ satisfies the requirements of part (ii). This proves Theorem 2.1.10. $\qquad\square$

The next corollary relates the notion of a smooth map on a smooth submanifold as defined in the beginning of Sect. 2.1 to the standard notion of smoothness in local coordinates used in the intrinsic setting of Sect. 2.8 below.

Corollary 2.1.11 *Let $M \subset \mathbb{R}^k$ be a smooth m-dimensional submanifold and let $f : M \to \mathbb{R}^\ell$ be a map. Then the following are equivalent.*

(i) *For every $p_0 \in M$ there exists an open neighborhood $U \subset \mathbb{R}^k$ of p_0 and a smooth map $F : U \to \mathbb{R}^\ell$ that agrees with f on $U \cap M$.*
(ii) *If $U_0 \subset M$ is an M-open set and $\phi_0 : U_0 \to \Omega_0$ is a diffeomorhism onto an open set $\Omega_0 \subset \mathbb{R}^m$, then the composition $f \circ \phi_0^{-1} : \Omega_0 \to \mathbb{R}^\ell$ is smooth.*

Proof Assume (ii), let $p_0 \in M$, and choose $\phi = (\phi_1, \ldots, \phi_k) : U \to \Omega \subset \mathbb{R}^k$ as in part (ii) of Theorem 2.1.10. Shrinking U, if necessary, we may assume that $\Omega = \Omega_0 \times \Omega_1$, where $\Omega_0 \subset \mathbb{R}^m$ is an open set and $\Omega_1 \subset \mathbb{R}^{k-m}$ is an open neighborhood of the origin. Then the map $\Omega_0 \to \mathbb{R}^\ell : x \mapsto f \circ \phi^{-1}(x, 0)$ is smooth by part (ii). Define $F(p) := f \circ \phi^{-1}(\phi_1(p), \ldots, \phi_m(p), 0, \ldots, 0)$ for $p \in U$. Then the map $F : U \to \mathbb{R}^\ell$ is smooth and agrees with f on $U \cap M$. Thus f satisfies (i). That (i) implies (ii) follows from Exercise 2.1.1 and this proves Corollary 2.1.11. $\square$

Definition 2.1.12 (Regular value) Let $U \subset \mathbb{R}^k$ be an open set and let $f : U \to \mathbb{R}^\ell$ be a smooth map. An element $c \in \mathbb{R}^\ell$ is called a **regular value** of f iff, for all $p \in U$, we have

$$f(p) = c \qquad \Longrightarrow \qquad df(p) : \mathbb{R}^k \to \mathbb{R}^\ell \text{ is surjective.}$$

Otherwise c is called a **singular value** of f. Theorem 2.1.10 asserts that, if c is a regular value of f, then the preimage

$$M := f^{-1}(c) = \{p \in U \mid f(p) = c\}$$

is a smooth $(k - \ell)$-dimensional submanifold of $\mathbb{R}^k$.

Examples and Exercises

Example 2.1.13 Let $A = A^\mathsf{T} \in \mathbb{R}^{k \times k}$ be a symmetric matrix and define the function $f : \mathbb{R}^k \to \mathbb{R}$ by $f(x) := x^\mathsf{T} A x$. Then $df(x)\xi = 2x^\mathsf{T} A \xi$ for $x, \xi \in \mathbb{R}^k$ and hence the linear map $df(x) : \mathbb{R}^k \to \mathbb{R}$ is surjective if and only if $Ax \neq 0$. Thus $c = 0$ is the only singular value of f and hence, for every element $c \in \mathbb{R} \setminus \{0\}$, the set $M := f^{-1}(c) = \{x \in \mathbb{R}^k \mid x^\mathsf{T} A x = c\}$ is a smooth manifold of dimension $m = k - 1$.

Example 2.1.14 (The sphere) As a special case of Example 2.1.13 consider the case $k = m + 1$, $A = \mathbb{1}$, and $c = 1$. Then $f(x) = |x|^2$ and so we have another proof that the unit sphere

$$S^m = \left\{ x \in \mathbb{R}^{m+1} \mid |x|^2 = 1 \right\}$$

in $\mathbb{R}^{m+1}$ is a smooth m-manifold. (See Examples 1.2.4 and 2.1.5.)

Example 2.1.15 Define the map $f : \mathbb{R}^3 \times \mathbb{R}^3 \to \mathbb{R}$ by $f(x, y) := |x - y|^2$. This is another special case of Example 2.1.13 and so, for every $r > 0$, the set $M := \{(x, y) \in \mathbb{R}^3 \times \mathbb{R}^3 \mid |x - y| = r\}$ is a smooth 5-manifold.

Example 2.1.16 (The 2-torus) Let $0 < r < 1$ and define $f : \mathbb{R}^3 \to \mathbb{R}$ by

$$f(x, y, z) := (x^2 + y^2 + r^2 - z^2 - 1)^2 - 4(x^2 + y^2)(r^2 - z^2).$$

This map has zero as a regular value and $M := f^{-1}(0)$ is diffeomorphic to the 2-torus $\mathbb{T}^2 = S^1 \times S^1$. An explicit diffeomorphism is given by

$$(e^{is}, e^{it}) \mapsto \big((1 + r\cos(s))\cos(t), (1 + r\cos(s))\sin(t), r\sin(s)\big).$$

This example corresponds to the second surface in Fig. 2.2.

Exercise: Show that $f(x, y, z) = 0$ if and only if $(\sqrt{x^2 + y^2} - 1)^2 + z^2 = r^2$. Verify that zero is a regular value of f.

Example 2.1.17 The set

$$M := \{(x^2, y^2, z^2, yz, zx, xy) \mid x, y, z \in \mathbb{R}, \ x^2 + y^2 + z^2 = 1\}$$

is a smooth 2-manifold in $\mathbb{R}^6$. To see this, define an equivalence relation on the unit sphere $S^2 \subset \mathbb{R}^3$ by $p \sim q$ iff $q = \pm p$. The quotient space (the set of equivalence classes) is called the **real projective plane** and is denoted by

$$\mathbb{R}\mathrm{P}^2 := S^2/\{\pm 1\}.$$

(See Example 1.2.6.) It is equipped with the quotient topology, i.e. a subset $U \subset \mathbb{R}\mathrm{P}^2$ is open, by definition, iff its preimage under the obvious projection $S^2 \to \mathbb{R}\mathrm{P}^2$ is an open subset of S^2. Now the map $f : S^2 \to \mathbb{R}^6$ defined by

$$f(x, y, z) := (x^2, y^2, z^2, yz, zx, xy)$$

descends to a homeomorphism from $\mathbb{R}P^2$ onto M. The submanifold M is covered by the local smooth parameterizations

$$\Omega \to M : (x, y) \mapsto f(x, y, \sqrt{1 - x^2 - y^2}),$$
$$\Omega \to M : (x, z) \mapsto f(x, \sqrt{1 - x^2 - z^2}, z),$$
$$\Omega \to M : (y, z) \mapsto f(\sqrt{1 - y^2 - z^2}, y, z),$$

defined on the open unit disc $\Omega \subset \mathbb{R}^2$. We remark the following.

(a) M is *not* the preimage of a regular value under a smooth map $\mathbb{R}^6 \to \mathbb{R}^4$.
(b) M is *not* diffeomorphic to a submanifold of $\mathbb{R}^3$.
(c) The projection $\Sigma := \{(yz, zx, xy) \mid x, y, z \in \mathbb{R}, \ x^2 + y^2 + z^2 = 1\}$ of M onto the last three coordinates is called the **Roman surface** and was discovered by Jakob Steiner. The Roman surface can also be represented as the set of solutions $(\xi, \eta, \zeta) \in \mathbb{R}^3$ of the equation $\eta^2\zeta^2 + \zeta^2\xi^2 + \xi^2\eta^2 = \xi\eta\zeta$. It is not a submanifold of $\mathbb{R}^3$.

Exercise: Prove this. Show that M is diffeomorphic to a submanifold of $\mathbb{R}^4$. Show that M is diffeomorphic to $\mathbb{R}P^2$ as defined in Example 1.2.6.

Exercise 2.1.18 Let $V : \mathbb{R}^n \to \mathbb{R}$ be a smooth function and define the Hamiltonian function $H : \mathbb{R}^n \times \mathbb{R}^n \to \mathbb{R}$ (kinetic plus potential energy) by

$$H(x, y) := \frac{1}{2}|y|^2 + V(x).$$

Prove that c is a regular value of H if and only if it is a regular value of V.

Exercise 2.1.19 Consider the **general linear group**

$$GL(n, \mathbb{R}) = \{g \in \mathbb{R}^{n \times n} \mid \det(g) \neq 0\}$$

Prove that the derivative of the function $f = \det : \mathbb{R}^{n \times n} \to \mathbb{R}$ is given by

$$df(g)v = \det(g)\operatorname{trace}(g^{-1}v)$$

for all $g \in GL(n, \mathbb{R})$ and $v \in \mathbb{R}^{n \times n}$. Deduce that the **special linear group**

$$SL(n, \mathbb{R}) := \{g \in GL(n, \mathbb{R}) \mid \det(g) = 1\}$$

is a smooth submanifold of $\mathbb{R}^{n \times n}$.

Example 2.1.20 The **orthogonal group**

$$O(n) := \{g \in \mathbb{R}^{n \times n} \mid g^{\mathsf{T}}g = 1\}$$

is a smooth submanifold of $\mathbb{R}^{n \times n}$. To see this, denote by

$$S_n := \{S \in \mathbb{R}^{n \times n} \mid S^{\mathsf{T}} = S\}$$

the vector space of symmetric matrices and define $f : \mathbb{R}^{n \times n} \to S_n$ by

$$f(g) := g^{\mathsf{T}}g.$$

Its derivative $df(g) : \mathbb{R}^{n \times n} \to S_n$ is given by

$$df(g)v = g^{\mathsf{T}}v + v^{\mathsf{T}}g.$$

This map is surjective for every $g \in O(n)$: if $g^{\mathsf{T}}g = 1$ and $S = S^{\mathsf{T}} \in S_n$, then the matrix $v := \frac{1}{2}gS$ satisfies

$$df(g)v = \frac{1}{2}g^{\mathsf{T}}gS + \frac{1}{2}(gS)^{\mathsf{T}}g = \frac{1}{2}S + \frac{1}{2}S^{\mathsf{T}} = S.$$

Hence 1 is a regular value of f and so $O(n)$ is a smooth manifold. It has the dimension

$$\dim O(n) = n^2 - \dim S_n = n^2 - \frac{n(n+1)}{2} = \frac{n(n-1)}{2}.$$

Exercise 2.1.21 Prove that the set

$$M := \{(x, y) \in \mathbb{R}^2 \mid xy = 0\}$$

is not a submanifold of $\mathbb{R}^2$. **Hint:** If $U \subset \mathbb{R}^2$ is a neighborhood of the origin and $f : U \to \mathbb{R}$ is a smooth map such that $U \cap M = f^{-1}(0)$, then $df(0,0) = 0$.

2.2 Tangent Spaces and Derivatives

The main reason for first discussing the extrinsic notion of embedded manifolds in Euclidean space as explained in the Master Plan Sect. 1.5 is that the concept of a tangent vector is much easier to digest in the embedded case: it is simply the derivative of a curve in M, understood as a vector in the ambient Euclidean space in which M is embedded.

Fig. 2.4 The tangent space $T_p M$ and the translated tangent space $p + T_p M$

2.2.1 Tangent Space

Definition 2.2.1 (Tangent vector) Let $M \subset \mathbb{R}^k$ be a smooth m-dimensional manifold and fix a point $p \in M$. A vector $v \in \mathbb{R}^k$ is called a **tangent vector** of M at p iff there exists a smooth curve $\gamma : \mathbb{R} \to M$ such that

$$\gamma(0) = p, \qquad \dot{\gamma}(0) = v.$$

The set

$$T_p M := \{ \dot{\gamma}(0) \,|\, \gamma : \mathbb{R} \to M \text{ is smooth}, \gamma(0) = p \}$$

of tangent vectors of M at p is called the **tangent space** of M at p.

Theorem 2.2.3 below shows that $T_p M$ is a linear subspace of $\mathbb{R}^k$. As does any linear subspace it contains the origin; it need not actually intersect M. Its translate $p + T_p M$ touches M at p; this is what you should visualize for $T_p M$ (see Fig. 2.4).

Remark 2.2.2 Let $p \in M$ be as in Definition 2.2.1 and let $v \in \mathbb{R}^k$. Then

$$v \in T_p M \qquad \Longleftrightarrow \qquad \begin{cases} \exists \varepsilon > 0 \ \exists \gamma : (-\varepsilon, \varepsilon) \to M \text{ such that} \\ \gamma \text{ is smooth}, \gamma(0) = p, \dot{\gamma}(0) = v. \end{cases}$$

To see this suppose that $\gamma : (-\varepsilon, \varepsilon) \to M$ is a smooth curve with $\gamma(0) = p$ and $\dot{\gamma}(0) = v$. Define $\widetilde{\gamma} : \mathbb{R} \to M$ by

$$\widetilde{\gamma}(t) := \gamma \left(\frac{\varepsilon t}{\sqrt{\varepsilon^2 + t^2}} \right), \qquad t \in \mathbb{R}.$$

Then $\widetilde{\gamma}$ is smooth and satisfies $\widetilde{\gamma}(0) = p$ and $\dot{\widetilde{\gamma}}(0) = v$. Hence $v \in T_p M$.

Theorem 2.2.3 (Tangent spaces) *Let $M \subset \mathbb{R}^k$ be a smooth m-dimensional manifold and fix a point $p \in M$. Then the following holds.*

(i) *Let $U_0 \subset M$ be an M-open set with $p \in U_0$ and let $\phi_0 : U_0 \to \Omega_0$ be a diffeomorphism onto an open subset $\Omega_0 \subset \mathbb{R}^m$. Let $x_0 := \phi_0(p)$ and let $\psi_0 := \phi_0^{-1} : \Omega_0 \to U_0$ be the inverse map. Then*

$$T_p M = \mathrm{im} \left(d\psi_0(x_0) : \mathbb{R}^m \to \mathbb{R}^k \right).$$

(ii) *Let $U, \Omega \subset \mathbb{R}^k$ be open sets and $\phi : U \to \Omega$ be a diffeomorphism such that $p \in U$ and $\phi(U \cap M) = \Omega \cap (\mathbb{R}^m \times \{0\})$. Then*

$$T_p M = d\phi(p)^{-1}(\mathbb{R}^m \times \{0\}).$$

(iii) *Let $U \subset \mathbb{R}^k$ be an open neighborhood of p and $f : U \to \mathbb{R}^{k-m}$ be a smooth map such that 0 is a regular value of f and $U \cap M = f^{-1}(0)$. Then*

$$T_p M = \ker df(p).$$

(iv) *$T_p M$ is an m-dimensional linear subspace of $\mathbb{R}^k$.*

Proof Let $\psi_0 : \Omega_0 \to U_0$ and $x_0 \in \Omega_0$ be as in (i) and let $\phi : U \to \Omega$ be as in (ii). We prove that

$$\operatorname{im} d\psi_0(x_0) \subset T_p M \subset d\phi(p)^{-1}(\mathbb{R}^m \times \{0\}). \tag{2.2.1}$$

To prove the first inclusion in (2.2.1), choose a constant $r > 0$ such that

$$B_r(x_0) := \{x \in \mathbb{R}^m \mid |x - x_0| < r\} \subset \Omega_0.$$

Now let $\xi \in \mathbb{R}^m$ and choose $\varepsilon > 0$ so small that

$$\varepsilon|\xi| \le r.$$

Then $x_0 + t\xi \in \Omega_0$ for all $t \in \mathbb{R}$ with $|t| < \varepsilon$. Define $\gamma : (-\varepsilon, \varepsilon) \to M$ by

$$\gamma(t) := \psi_0(x_0 + t\xi) \qquad \text{for} \ -\varepsilon < t < \varepsilon.$$

Then γ is a smooth curve in M satisfying

$$\gamma(0) = \psi_0(x_0) = p, \qquad \dot{\gamma}(0) = \left.\frac{d}{dt}\right|_{t=0} \psi_0(x_0 + t\xi) = d\psi_0(x_0)\xi.$$

Hence it follows from Remark 2.2.2 that $d\psi_0(x_0)\xi \in T_p M$, as claimed.

To prove the second inclusion in (2.2.1) we fix a vector $v \in T_p M$. Then, by definition of the tangent space, there exists a smooth curve $\gamma : \mathbb{R} \to M$ such that $\gamma(0) = p$ and $\dot{\gamma}(0) = v$. Let $U \subset \mathbb{R}^k$ be as in (ii) and choose $\varepsilon > 0$ so small that $\gamma(t) \in U$ for $|t| < \varepsilon$. Then

$$\phi(\gamma(t)) \in \phi(U \cap M) \subset \mathbb{R}^m \times \{0\}$$

for $|t| < \varepsilon$ and hence

$$d\phi(p)v = d\phi(\gamma(0))\dot{\gamma}(0) = \left.\frac{d}{dt}\right|_{t=0} \phi(\gamma(t)) \in \mathbb{R}^m \times \{0\}.$$

This shows that $v \in d\phi(p)^{-1}(\mathbb{R}^m \times \{0\})$ and thus we have proved (2.2.1).

Now the sets $\operatorname{im} d\psi_0(x_0)$ and $d\phi(p)^{-1}(\mathbb{R}^m \times \{0\})$ are both m-dimensional linear subspaces of $\mathbb{R}^k$. Hence it follows from (2.2.1) that these subspaces agree and that they both agree with $T_p M$. Thus we have proved assertions (i), (ii), and (iv).

We prove (iii). Let $v \in T_p M$. Then there is a smooth curve $\gamma : \mathbb{R} \to M$ such that $\gamma(0) = p$ and $\dot{\gamma}(0) = v$. For t sufficiently small we have $\gamma(t) \in U$, where $U \subset \mathbb{R}^k$ is the open set in (iii), and $f(\gamma(t)) = 0$. Hence

$$df(p)v = df(\gamma(0))\dot{\gamma}(0) = \frac{d}{dt}\bigg|_{t=0} f(\gamma(t)) = 0.$$

This implies $T_p M \subset \ker df(p)$. Since $T_p M$ and the kernel of $df(p)$ are both m-dimensional linear subspaces of $\mathbb{R}^k$, we deduce that $T_p M = \ker df(p)$. This proves part (iii) and Theorem 2.2.3. $\qquad\square$

Exercise 2.2.4 Let $M \subset \mathbb{R}^k$ be a smooth m-dimensional manifold and let $p_i \in M$ be a sequence that converges to a point $p \in M$. Let τ_i be a sequence of nonzero real numbers and let $v \in \mathbb{R}^k$ such that

$$\lim_{i\to\infty} \tau_i = 0, \qquad \lim_{i\to 0} \frac{p_i - p}{\tau_i} = v.$$

Prove that $v \in T_p M$. **Hint:** Use part (iii) of Theorem 2.2.3.

Example 2.2.5 Let $A = A^\mathsf{T} \in \mathbb{R}^{k\times k}$ be a nonzero matrix as in Example 2.1.13 and let $c \neq 0$. Then part (iii) of Theorem 2.2.3 asserts that the tangent space of the manifold

$$M = \left\{ x \in \mathbb{R}^k \,\middle|\, x^\mathsf{T} A x = c \right\}$$

at a point $x \in M$ is the $(k-1)$-dimensional linear subspace

$$T_x M = \left\{ \xi \in \mathbb{R}^k \,\middle|\, x^\mathsf{T} A \xi = 0 \right\}.$$

Example 2.2.6 As a special case of Example 2.2.5 with $A = \mathbb{1}$ and $c = 1$ we find that the tangent space of the unit sphere $S^m \subset \mathbb{R}^{m+1}$ at a point $x \in S^m$ is the orthogonal complement of x. i.e.

$$T_x S^m = x^\perp = \{\xi \in \mathbb{R}^{m+1} \mid \langle x, \xi \rangle = 0\}.$$

Here $\langle x, \xi \rangle = \sum_{i=0}^m x_i \xi_i$ denotes the standard inner product on $\mathbb{R}^{m+1}$.

Exercise 2.2.7 What is the tangent space of the 5-manifold

$$M := \{(x, y) \in \mathbb{R}^3 \times \mathbb{R}^3 \mid |x - y| = r\}$$

at a point $(x, y) \in M$? (See Exercise 2.1.15.)

Example 2.2.8 Let $H(x, y) := \frac{1}{2}|y|^2 + V(x)$ be as in Exercise 2.1.18 and let c be a regular value of H. If $(x, y) \in M := H^{-1}(c)$, then

$$T_{(x,y)}M = \{(\xi, \eta) \in \mathbb{R}^n \times \mathbb{R}^n \mid \langle y, \eta \rangle + \langle \nabla V(x), \xi \rangle = 0\}.$$

Here $\nabla V := (\partial V/\partial x_1, \ldots, \partial V/\partial x_n) : \mathbb{R}^n \to \mathbb{R}^n$ denotes the gradient of V.

Exercise 2.2.9 The tangent space of $\mathrm{SL}(n, \mathbb{R})$ at the identity matrix is the space

$$\mathfrak{sl}(n, \mathbb{R}) := T_1 \mathrm{SL}(n, \mathbb{R}) = \{\xi \in \mathbb{R}^{n \times n} \mid \mathrm{trace}(\xi) = 0\}$$

of traceless matrices. (Prove this, using Exercise 2.1.19.)

Example 2.2.10 The tangent space of $O(n)$ at g is

$$T_g O(n) = \{v \in \mathbb{R}^{n \times n} \mid g^\mathsf{T} v + v^\mathsf{T} g = 0\}.$$

In particular, the tangent space of $O(n)$ at the identity matrix is the space of skew-symmetric matrices

$$\mathfrak{o}(n) := T_1 O(n) = \{\xi \in \mathbb{R}^{n \times n} \mid \xi^\mathsf{T} + \xi = 0\}$$

To see this, choose a smooth curve $\mathbb{R} \to O(n) : t \mapsto g(t)$. Then $g(t)^\mathsf{T} g(t) = 1$ for all $t \in \mathbb{R}$ and, differentiating this identity with respect to t, we obtain $g(t)^\mathsf{T} \dot{g}(t) + \dot{g}(t)^\mathsf{T} g(t) = 0$ for every t. Hence every matrix $v \in T_g O(n)$ satisfies the equation $g^\mathsf{T} v + v^\mathsf{T} g = 0$. With this understood, the claim follows from the fact that $g^\mathsf{T} v + v^\mathsf{T} g = 0$ if and only if the matrix $\xi := g^{-1} v$ is skew-symmetric and that the space of skew-symmetric matrices in $\mathbb{R}^{n \times n}$ has dimension $n(n-1)/2$.

Exercise 2.2.11 Let $\Omega \subset \mathbb{R}^m$ be an open set and $h : \Omega \to \mathbb{R}^{k-m}$ be a smooth map. Prove that the tangent space of the graph of h at a point $(x, h(x))$ is the graph of the derivative $dh(x) : \mathbb{R}^m \to \mathbb{R}^{k-m}$:

$$M = \{(x, h(x)) \mid x \in \Omega\}, \qquad T_{(x,h(x))}M = \{(\xi, dh(x)\xi) \mid \xi \in \mathbb{R}^m\}.$$

Exercise 2.2.12 (Monge coordinates) Let M be a smooth m-manifold in $\mathbb{R}^k$ and suppose that $p \in M$ is such that the projection $T_p M \to \mathbb{R}^m \times \{0\}$ is invertible. Prove that there exists an open set $\Omega \subset \mathbb{R}^m$ and a smooth map $h : \Omega \to \mathbb{R}^{k-m}$ such that the graph of h is an M-open neighborhood of p (see Example 2.1.6). Of course, the projection $T_p M \to \mathbb{R}^m \times \{0\}$ need not be invertible, but it must be invertible for at least one of the $\binom{k}{m}$ choices of the m-dimensional coordinate plane. Hence every point of M has an M-open neighborhood which may be expressed as a graph of a function of some of the coordinates in terms of the others as in e.g. Example 2.1.5.

2.2.2 Derivative

A key purpose behind the concept of a smooth manifold is to carry over the notion of a smooth map and its derivatives from the realm of first year analysis to the present geometric setting. Here is the basic definition. It appeals to the notion of a smooth map between arbitrary subsets of Euclidean spaces as introduced in the beginning of Sect. 2.1.

Definition 2.2.13 (Derivative) Let $M \subset \mathbb{R}^k$ be an m-dimensional smooth manifold and let

$$f : M \to \mathbb{R}^\ell$$

be a smooth map. The **derivative** of f at a point $p \in M$ is the map

$$df(p) : T_p M \to \mathbb{R}^\ell$$

defined as follows. Given a tangent vector $v \in T_p M$, choose a smooth curve

$$\gamma : \mathbb{R} \to M$$

satisfying

$$\gamma(0) = p, \qquad \dot{\gamma}(0) = v,$$

and define the vector $df(p)v \in \mathbb{R}^\ell$ by

$$df(p)v := \left. \frac{d}{dt} \right|_{t=0} f(\gamma(t)) = \lim_{h \to 0} \frac{f(\gamma(h)) - f(p)}{h}. \qquad (2.2.2)$$

That the limit on the right in equation (2.2.2) exists follows from our assumptions. We must prove, however, that the derivative is well defined, i.e. that the right hand side of (2.2.2) depends only on the tangent vector v and not on the choice of the curve γ used in the definition. This is the content of the first assertion in the next theorem.

Theorem 2.2.14 (Derivatives) *Let $M \subset \mathbb{R}^k$ be an m-dimensional smooth manifold and $f : M \to \mathbb{R}^\ell$ be a smooth map. Fix a point $p \in M$. Then the following holds.*

(i)   *The right hand side of (2.2.2) is independent of γ.*
(ii) *The map $df(p) : T_p M \to \mathbb{R}^\ell$ is linear.*
(iii) *If $N \subset \mathbb{R}^\ell$ is a smooth n-manifold and $f(M) \subset N$, then*

$$df(p)T_p M \subset T_{f(p)} N.$$

(iv) **(Chain Rule)** *Let N be as in (iii), suppose that $f(M) \subset N$, and let $g : N \to \mathbb{R}^d$ be a smooth map. Then*

$$d(g \circ f)(p) = dg(f(p)) \circ df(p) : T_p M \to \mathbb{R}^d.$$

(v) *If $f = \mathrm{id} : M \to M$, then $df(p) = \mathrm{id} : T_p M \to T_p M$.*

Proof We prove (i). Let $v \in T_p M$ and $\gamma : \mathbb{R} \to M$ be as in Definition 2.2.13. By definition there is an open neighborhood $U \subset \mathbb{R}^k$ of p and a smooth map $F : U \to \mathbb{R}^\ell$ such that

$$F(p') = f(p') \qquad \text{for all } p' \in U \cap M.$$

Let $dF(p) \in \mathbb{R}^{\ell \times k}$ denote the Jacobian matrix (i.e. the matrix of all first partial derivatives) of F at p. Then, since $\gamma(t) \in U \cap M$ for t sufficiently small, we have

$$
\begin{aligned}
dF(p)v &= dF(\gamma(0))\dot{\gamma}(0) \\
&= \frac{d}{dt}\bigg|_{t=0} F(\gamma(t)) \\
&= \frac{d}{dt}\bigg|_{t=0} f(\gamma(t)).
\end{aligned}
$$

The right hand side of this identity is independent of the choice of F while the left hand side is independent of the choice of γ. Hence the right hand side is also independent of the choice of γ and this proves (i). Assertion (ii) follows immediately from the identity

$$df(p)v = dF(p)v$$

just established.

Assertion (iii) follows directly from the definitions. Namely, if γ is as in Definition 2.2.13, then $\beta := f \circ \gamma : \mathbb{R} \to N$ is a smooth curve in N satisfying

$$\beta(0) = f(\gamma(0)) = f(p) =: q, \qquad \dot{\beta}(0) = df(p)v =: w.$$

Hence $w \in T_q N$. Assertion (iv) also follows directly from the definitions. If $g : N \to \mathbb{R}^d$ is a smooth map and β, q, w are as above, then

$$
\begin{aligned}
d(g \circ f)(p)v &= \frac{d}{dt}\bigg|_{t=0} g(f(\gamma(t))) \\
&= \frac{d}{dt}\bigg|_{t=0} g(\beta(t)) \\
&= dg(q)w \\
&= dg(f(p))df(p)v.
\end{aligned}
$$

This proves (iv). Assertion (v) follows again directly from the definitions and this proves Theorem 2.2.14. $\qquad \square$

Corollary 2.2.15 (Diffeomorphisms) *Let $M \subset \mathbb{R}^k$ be a nonempty smooth m-manifold and $N \subset \mathbb{R}^\ell$ be a smooth n-manifold and let $f : M \to N$ be a diffeomorphism. Then $m = n$ and the derivative $df(p) : T_pM \to T_{f(p)}N$ is a vector space isomorphism with inverse*

$$df(p)^{-1} = df^{-1}(f(p)) : T_{f(p)}N \to T_pM$$

for all $p \in M$.

Proof Define $g := f^{-1} : N \to M$ so that $g \circ f = \mathrm{id}_M$ and $f \circ g = \mathrm{id}_N$. Then it follows from Theorem 2.2.14 that, for $p \in M$ and $q := f(p) \in N$, we have $dg(q) \circ df(p) = \mathrm{id} : T_pM \to T_pM$ and $df(p) \circ dg(q) = \mathrm{id} : T_qN \to T_qN$. Hence $df(p) : T_pM \to T_qN$ is a vector space isomorphism and its inverse is given by $dg(q) = df(p)^{-1} : T_qN \to T_pM$. Hence $m = n$ and this proves Corollary 2.2.15. $\square$

Exercise 2.2.16 Let $M \subset \mathbb{R}^k$ be a smooth manifold and let $f : M \to \mathbb{R}^\ell$ be a smooth map. Let $p_i \in M$ be a sequence that converges to a point $p \in M$, let τ_i be a sequence of nonzero real numbers, and let $v \in T_pM$ such that

$$\lim_{i \to \infty} \tau_i = 0, \qquad \lim_{i \to \infty} \frac{p_i - p}{\tau_i} = v.$$

(See Exercise 2.2.4.) Prove that

$$\lim_{i \to \infty} \frac{f(p_i) - f(p)}{\tau_i} = df(p)v.$$

Hint: Use the local extension F of f in the proof of Theorem 2.2.14.

2.2.3   The Inverse Function Theorem

Corollary 2.2.15 is analogous to the corresponding assertion for smooth maps between open subsets of Euclidean space. Likewise, the inverse function theorem for manifolds is a partial converse of Corollary 2.2.15.

Theorem 2.2.17 (Inverse Function Theorem) *Assume that $M \subset \mathbb{R}^k$ and $N \subset \mathbb{R}^\ell$ are smooth m-manifolds and $f : M \to N$ is a smooth map. Let $p_0 \in M$ and suppose that the derivative*

$$df(p_0) : T_{p_0}M \to T_{f(p_0)}N$$

is a vector space isomorphism. Then there exists an M-open neighborhood $U \subset M$ of p_0 such that $V := f(U) \subset N$ is an N-open subset of N and the restriction $f|_U : U \to V$ is a diffeomorphism.

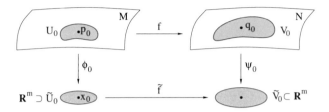

Fig. 2.5 The Inverse Function Theorem

Proof   Choose coordinate charts $\phi_0 : U_0 \to \widetilde{U}_0$, defined on an M-open neighborhood $U_0 \subset M$ of p_0 onto an open set $\widetilde{U}_0 \subset \mathbb{R}^m$, and $\psi_0 : V_0 \to \widetilde{V}_0$, defined on an N-open neighborhood $V_0 \subset N$ of $f(p_0)$ onto an open set $\widetilde{V}_0 \subset \mathbb{R}^m$. Shrinking U_0, if necessary, we may assume that $f(U_0) \subset V_0$. Then the map

$$\widetilde{f} := \psi_0 \circ f \circ \phi_0^{-1} : \widetilde{U}_0 \to \widetilde{V}_0$$

(see Fig. 2.5) is smooth and its derivative $d\widetilde{f}(x_0) : \mathbb{R}^m \to \mathbb{R}^m$ is bijective at $x_0 := \phi_0(p_0)$. Hence the Inverse Function Theorem A.2.2 asserts that there exists an open neighborhood $\widetilde{U} \subset \widetilde{U}_0$ of x_0 such that $\widetilde{V} := \widetilde{f}(\widetilde{U})$ is an open subset of $\widetilde{V}_0$ and the restriction of $\widetilde{f}$ to $\widetilde{U}$ is a diffeomorphism from $\widetilde{U}$ to $\widetilde{V}$. Hence the assertion holds with $U := \phi_0^{-1}(\widetilde{U})$ and $V := \psi_0^{-1}(\widetilde{V})$. This proves Theorem 2.2.17. $\square$

2.2.4   Regular Values

Definition 2.2.18 (Regular value)   Let $M \subset \mathbb{R}^k$ be a smooth m-manifold, let $N \subset \mathbb{R}^\ell$ be a smooth n-manifold, and let $f : M \to N$ be a smooth map. An element $q \in N$ is called a **regular value** of f iff, for every $p \in M$ with $f(p) = q$, the derivative $df(p) : T_p M \to T_{f(p)} N$ is surjective.

Theorem 2.2.19 (Regular values)   *Let $f : M \to N$ be as in Definition 2.2.18 and let $q \in N$ be a regular value of f. Then the set*

$$P := f^{-1}(q) = \{p \in M \mid f(p) = q\}$$

is a smooth submanifold of $\mathbb{R}^k$ of dimension $m - n$ and, for each point $p \in P$, its tangent space at p is given by

$$T_p P = \ker df(p) = \{v \in T_p M \mid df(p)v = 0\}.$$

Proof 1   Fix a point $p_0 \in P$ and choose a linear map $A : \mathbb{R}^k \to \mathbb{R}^{m-n}$ such that the restriction of A to the $(m - n)$-dimensional linear subspace

$$\ker df(p_0) = \{v \in T_{p_0} M \mid df(p_0) = 0\} \subset T_{p_0} M \subset \mathbb{R}^k$$

is a vector space isomorphism. Define the map $F : M \to N \times \mathbb{R}^{m-n}$ by

$$F(p) := (f(p), Ap)$$

for $p \in M$. The derivative of F at p_0 is given by $dF(p_0)v = (df(p_0)v, Av)$ for $v \in T_{p_0}M$ and is a vector space isomorphism. Hence, by Theorem 2.2.17 there exists an M-open neighborhood $U \subset M$ of p_0 such that $V := F(U)$ is an open subset of $N \times \mathbb{R}^{m-n}$ in the relative topology and the restriction $F|_U : U \to V$ is a diffeomorphism. Hence the P-open set $U \cap P$ is diffeomorphic to the open set $\Omega := \{y \in \mathbb{R}^{m-n} \mid (q, y) \in V\} \subset \mathbb{R}^{m-n}$ by the diffeomorphism $\phi : U \cap P \to \Omega$, defined by $\phi(p) := Ap$ for $p \in U \cap P$, whose inverse is the smooth map $\psi : \Omega \to U \cap P$ given by $\psi(y) = (F|_U)^{-1}(q, y)$ for $y \in \Omega$. This shows that P is a smooth $(m-n)$-manifold in $\mathbb{R}^k$.

Now let $p \in P$ and $v \in T_pP$. Then there exists a smooth curve $\gamma : \mathbb{R} \to P$ such that $\gamma(0) = p$ and $\dot\gamma(0) = v$. Since $f(\gamma(t)) = q$ for all t, we have

$$df(p)v = \frac{d}{dt}\bigg|_{t=0} f(\gamma(t)) = 0$$

and so $v \in \ker df(p)$. Hence $T_pP \subset \ker df(p)$ and equality holds because both T_pP and $\ker df(p)$ are $(m-n)$-dimensional linear subspaces of $\mathbb{R}^k$. This proves Theorem 2.2.19.   □

Proof 2   Here is another proof of Theorem 2.2.19 in local coordinates. Fix a point $p_0 \in P$ and choose a coordinate chart $\phi_0 : U_0 \to \phi_0(U_0) \subset \mathbb{R}^m$ on an M-open neighborhood $U_0 \subset M$ of p_0. Likewise, choose a coordinate chart $\psi_0 : V_0 \to \psi_0(V_0) \subset \mathbb{R}^n$ on an N-open neighborhood $V_0 \subset N$ of q. Shrinking U_0, if necessary, we may assume that $f(U_0) \subset V_0$. Then the point $c_0 := \psi_0(q)$ is a regular value of the map

$$f_0 := \psi_0 \circ f \circ \phi_0^{-1} : \phi_0(U_0) \to \mathbb{R}^n.$$

Namely, if $x \in \phi_0(U_0)$ satisfies $f_0(x) = c_0$, then $p := \phi_0^{-1}(x) \in U_0 \cap P$, so the maps $d\phi_0^{-1}(x) : \mathbb{R}^m \to T_pM$, $df(p) : T_pM \to T_qN$, and $d\psi_0(q) : T_qN \to \mathbb{R}^n$ are all surjective, hence so is their composition, and by the chain rule this composition is the derivative $df_0(x) : \mathbb{R}^m \to \mathbb{R}^n$. With this understood, it follows from Theorem 2.1.10 that the set

$$f_0^{-1}(c_0) = \{x \in \phi_0(U_0) \mid f(\phi_0^{-1}(x)) = q\} = \phi_0(U_0 \cap P)$$

is a manifold of of dimension $m - n$ contained in the open set $\phi_0(U_0) \subset \mathbb{R}^m$. Using Definition 2.1.3 and shrinking U_0 further, if necessary, we may assume that the set $\phi_0(U_0 \cap P)$ is diffeomorphic to an open subset of $\mathbb{R}^{m-n}$. Composing this diffeomorphism with ϕ_0 we find that $U_0 \cap P$ is diffeomorphic to the same open subset of $\mathbb{R}^{m-n}$. Since the set $U_0 \subset M$ is M-open, there exists an open set $U \subset \mathbb{R}^k$ such that $U \cap M = U_0$, hence $U \cap P = U_0 \cap P$, and so $U_0 \cap P$ is a P-open neigh-borhood of p_0. Thus we have proved that every element $p_0 \in P$ has a P-open neigborhood that is diffeomorphic to an open subset of $\mathbb{R}^{m-n}$. Thus $P \subset \mathbb{R}^k$ is a manifold of dimension $m - n$ (Definition 2.1.3). The proof that the tangent spaces of P are given by $T_p P = \ker df(p)$ remains unchanged and this completes the second proof of Theorem 2.2.19.   □

Definition 2.2.20 Let $M \subset \mathbb{R}^k$ and $N \subset \mathbb{R}^\ell$ be smooth m-manifolds. A smooth map $f : M \to N$ is called a **local diffeomorphism** iff its derivative $df(p) : T_p M \to T_{f(p)} N$ is a vector space isomorphism for every $p \in M$.

Example 2.2.21 The inclusion of an M-open subset $U \subset M$ into M and the map $\mathbb{R} \to S^1 : t \mapsto e^{it}$ are examples of local diffeomorphisms.

Exercise 2.2.22 Prove that the image of a local diffeomorphism is an open subset of the target manifold. **Hint:** Use the Inverse Function Theorem.

In the terminology introduced in Sect. 2.3 and Sect. 2.6.1 below, local diffeo-morphisms are both immersions and submersions. In particular, if $f : M \to N$ is a local diffeomorphism, then every element $q \in N$ is a regular value of f and its preimage $f^{-1}(q)$ is a discrete subset of M.

2.3   Submanifolds and Embeddings

This section deals with subsets of a manifold M that are themselves manifolds as in Definition 2.1.3. Such subsets are called submanifolds of M.

Definition 2.3.1 (Submanifold) Let $M \subset \mathbb{R}^k$ be an m-dimensional manifold. A subset $P \subset M$ is called a **submanifold** of M of dimension n, iff P itself is an n-manifold.

Definition 2.3.2 (Embedding) Let $M \subset \mathbb{R}^k$ be an m-dimensional manifold and $N \subset \mathbb{R}^\ell$ be an n-dimensional manifold. A smooth map $f : N \to M$ is called an **immersion** iff its derivatve $df(q) : T_q N \to T_{f(q)} M$ is injective for every $q \in N$. It is called **proper** iff, for every compact subset $K \subset f(N)$, the preimage $f^{-1}(K) = \{q \in N \mid f(q) \in K\}$ is compact. The map f is called an **embedding** iff it is a proper injective immersion.

Remark 2.3.3 In our definition of proper maps it is important that the compact set K is required to be contained in the image of f. The literature also contains a stronger definition of *proper* which requires that $f^{-1}(K)$ is a compact subset of N for every compact subset $K \subset M$, whether or not K is contained in the image of f. This holds if and only if the map f is proper in the sense of Definition 2.3.2 and has an M-closed image. (Exercise!)

Fig. 2.6 A coordinate chart
adapted to a submanifold

Theorem 2.3.4 (Submanifolds) *Let $M \subset \mathbb{R}^k$ be an m-dimensional manifold and $N \subset \mathbb{R}^\ell$ be an n-dimensional manifold.*

(i)   *If $f : N \to M$ is an embedding, then $f(N)$ is a submanifold of M.*
(ii) *If $P \subset M$ is a submanifold, then the inclusion $P \to M$ is an embedding.*
(iii) *A subset $P \subset M$ is a submanifold of dimension n if and only if, for every $p_0 \in P$, there exists a coordinate chart $\phi : U \to \mathbb{R}^m$ on an M-open neighborhood U of p_0 such that $\phi(U \cap P) = \phi(U) \cap (\mathbb{R}^n \times \{0\})$ (Fig. 2.6).*
(iv) *A subset $P \subset M$ is a submanifold of dimension n if and only if, for every $p_0 \in P$, there exists an M-open neighborhood $U \subset M$ of p_0 and a smooth map $g : U \to \mathbb{R}^{m-n}$ such that 0 is a regular value of g and $U \cap P = g^{-1}(0)$.*

The proof is based on the following lemma.

Lemma 2.3.5 (Embeddings) *Let M and N be as in Theorem 2.3.4, let $f : N \to M$ be an embedding, let $q_0 \in N$, and define*

$$P := f(N), \qquad p_0 := f(q_0) \in P.$$

Then there exists an M-open neighborhood $U \subset M$ of p_0, an N-open neighborhood $V \subset N$ of q_0, an open neighborhood $W \subset \mathbb{R}^{m-n}$ of the origin, and a diffeomorphism $F : V \times W \to U$ such that, for all $q \in V$ and all $z \in W$,

$$F(q,0) = f(q), \tag{2.3.1}$$
$$F(q,z) \in P \qquad \Longleftrightarrow \qquad z = 0. \tag{2.3.2}$$

Proof Choose any coordinate chart $\phi_0 : U_0 \to \mathbb{R}^m$ on an M-open neighborhood $U_0 \subset M$ of p_0. Then $d(\phi_0 \circ f)(q_0) = d\phi_0(f(q_0)) \circ df(q_0) : T_{q_0}N \to \mathbb{R}^m$ is injective. Hence there is a linear map $B : \mathbb{R}^{m-n} \to \mathbb{R}^m$ such that the map

$$T_{q_0}N \times \mathbb{R}^{m-n} \to \mathbb{R}^m : (w, \zeta) \mapsto d(\phi_0 \circ f)(q_0)w + B\zeta \tag{2.3.3}$$

is a vector space isomorphism. Define the set

$$\Omega := \{(q,z) \in N \times \mathbb{R}^{m-n} \mid f(q) \in U_0, \ \phi_0(f(q)) + Bz \in \phi_0(U_0)\}.$$

This is an open subset of $N \times \mathbb{R}^{m-n}$ and we define $F : \Omega \to M$ by

$$F(q,z) := \phi_0^{-1}(\phi_0(f(q)) + Bz).$$

This map is smooth, it satisfies $F(q,0) = f(q)$ for all $q \in f^{-1}(U_0)$, and the derivative $dF(q_0,0) : T_{q_0}N \times \mathbb{R}^{m-n} \to T_{p_0}M$ is the composition of the map (2.3.3) with $d\phi_0(p_0)^{-1} : \mathbb{R}^m \to T_{p_0}M$ and so is a vector space isomorphism. Thus the Inverse Function Theorem 2.2.17 asserts that there is an N-open neighborhood $V_0 \subset N$ of q_0 and an open neighborhood $W_0 \subset \mathbb{R}^{m-n}$ of the origin such that $V_0 \times W_0 \subset \Omega$, the set $U_0 := F(V_0 \times W_0)$ is M-open, and the restriction of F to $V_0 \times W_0$ is a diffeomorphism onto U_0. Thus we have constructed a diffeomorphism $F : V_0 \times W_0 \to U_0$ that satisfies (2.3.1).

We claim that the restriction of F to the product $V \times W$ of sufficiently small open neighborhoods $V \subset N$ of q_0 and $W \subset \mathbb{R}^{m-n}$ of the origin also satisfies (2.3.2). Otherwise, there exist sequences $q_i \in V_0$ converging to q_0 and $z_i \in W_0 \setminus \{0\}$ converging to zero such that $F(q_i, z_i) \in P$. Hence there exists a sequence $q_i' \in N$ such that $F(q_i, z_i) = f(q_i')$. This sequence converges to $f(q_0)$. Since f is proper we may assume, passing to a suitable subsequence if necessary, that q_i' converges to a point $q_0' \in N$. Then

$$f(q_0') = \lim_{i \to \infty} f(q_i') = \lim_{i \to \infty} F(q_i, z_i) = f(q_0).$$

Since f is injective, this implies $q_0' = q_0$. Hence $(q_i', 0) \in V_0 \times W_0$ for i sufficiently large and $F(q_i', 0) = f(q_i') = F(q_i, z_i)$. This contradicts the fact that the map $F : V_0 \times W_0 \to M$ is injective, and proves Lemma 2.3.5.   □

Proof of Theorem 2.3.4 We prove (i). Let $q_0 \in N$, denote $p_0 := f(q_0) \in P$, and choose a diffeomorphism $F : V \times W \to U$ as in Lemma 2.3.5. Then the set $V \subset N$ is diffeomorphic to an open subset of $\mathbb{R}^n$ (after schrinking V if necessary), the set $U \cap P$ is P-open because $U \subset M$ is M-open, and we have $U \cap P = \{F(q,0) \,|\, q \in V\} = f(V)$ by (2.3.1) and (2.3.2). Hence the map $f : V \to U \cap P$ is a diffeomorphism whose inverse is the composition of the smooth maps $F^{-1} : U \cap P \to V \times W$ and $V \times W \to V : (q,z) \mapsto q$. Hence a P-open neighborhood of p_0 is diffeomorphic to an open subset of $\mathbb{R}^n$. Since $p_0 \in P$ was chosen arbitrary, this shows that P is an n-dimensional submanifold of M.

We prove (ii). The inclusion $\iota : P \to M$ is obviously smooth and injective (it extends to the identity map on $\mathbb{R}^k$). Moreover, $T_pP \subset T_pM$ for every $p \in P$ and the derivative $d\iota(p) : T_pP \to T_pM$ is the obvious inclusion for every $p \in P$. That ι is proper follows immediately from the definition. Hence ι is an embedding.

We prove (iii). If a coordinate chart ϕ as in (iii) exists, then the set $U \cap P$ is P-open and is diffeomorphic to an open subset of $\mathbb{R}^n$. Since the point $p_0 \in P$ was chosen arbitrary this proves that P is an n-dimensional submanifold of M. Conversely, suppose that P is an n-dimensional submanifold of M and let $p_0 \in P$. Choose any coordinate chart $\phi_0 : U_0 \to \mathbb{R}^m$ of M defined on an M-open neighborhood $U_0 \subset M$ of p_0. Then $\phi_0(U_0 \cap P)$ is an n-dimensional submanifold of $\mathbb{R}^m$. Hence Theorem 2.1.10 asserts that there are open sets $V, W \subset \mathbb{R}^m$ with $p_0 \in V \subset \phi_0(U_0)$ and a diffeomorphism $\psi : V \to W$ such that

$$\phi_0(p_0) \in V, \qquad \psi(V \cap \phi_0(U_0 \cap P)) = W \cap (\mathbb{R}^n \times \{0\}).$$

Fig. 2.7 A proper immersion

Now define $U := \phi_0^{-1}(V) \subset U_0$. Then $p_0 \in U$, the chart ϕ_0 restricts to a diffeomorphism from U to V, the composition $\phi := \psi \circ \phi_0|_U : U \to W$ is a diffeomorphism, and $\phi(U \cap P) = \psi(V \cap \phi_0(U_0 \cap P)) = W \cap (\mathbb{R}^n \times \{0\})$.

We prove (iv). That the condition is sufficient follows directly from Theorem 2.2.19. To prove that it is necessary, assume that $P \subset M$ is a submanifold of dimension n, fix an element $p_0 \in P$, and choose a coordinate chart $\phi : U \to \mathbb{R}^m$ on an M-open neighborhood $U \subset M$ of p_0 as in part (iii). Define the map $g : U \to \mathbb{R}^{m-n}$ by $g(p) := (\phi_{n+1}(p), \ldots, \phi_m(p))$ for $p \in U$. Then 0 is a regular value of g and $g^{-1}(0) = U \cap P$. This proves Theorem 2.3.4. $\qquad\square$

Exercise 2.3.6 Let $M \subset \mathbb{R}^k$ be a smooth m-manifold and $\emptyset \neq P \subset M$.

(i)   If P is an n-dimensional submanifold of M, then $0 \leq n \leq m$.
(ii) P is a 0-dimensional submanifold of M if and only if P is discrete, i.e. every $p \in P$ has an M-open neighborhood U such that $U \cap P = \{p\}$.
(iii) P is an m-dimensional submanifold of M if and only if P is M-open.

Example 2.3.7 Let $S^1 \subset \mathbb{R}^2 \cong \mathbb{C}$ be the unit circle and consider the map $f : S^1 \to \mathbb{R}^2$ given by $f(x, y) := (x, xy)$. This map is a proper immersion but is not injective (the points $(0, 1)$ and $(0, -1)$ have the same image under f). The image $f(S^1)$ is a figure 8 in $\mathbb{R}^2$ and is not a submanifold (Fig. 2.7).

The restriction of f to the submanifold $N := S^1 \setminus \{(0, -1)\}$ is an injective immersion but it is not proper. It has the same image as before and hence $f(N)$ is not a manifold.

Example 2.3.8 The map $f : \mathbb{R} \to \mathbb{R}^2$ given by $f(t) := (t^2, t^3)$ is proper and injective, but is not an embedding (its derivatuve at $t = 0$ is not injective). The image of f is the set $f(\mathbb{R}) = C := \{(x, y) \in \mathbb{R}^2 \mid x^3 = y^2\}$ (see Fig. 2.8) and is not a submanifold. (Prove this!)

Fig. 2.8 A proper injection

Example 2.3.9 Define the map $f : \mathbb{R} \to \mathbb{R}^2$ by $f(t) := (\cos(t), \sin(t))$. This map is an immersion, but it is neither injective nor proper. However, its image is the unit circle in $\mathbb{R}^2$ and hence is a submanifold of $\mathbb{R}^2$. The map $\mathbb{R} \to \mathbb{R}^2 : t \mapsto f(t^3)$ is not an immersion and is neither injective nor proper, but its image is still the unit circle.

2.4 Vector Fields and Flows

This section introduces vector fields on manifolds (Sect. 2.4.1), explains the flow of a vector field and the group of diffeomorphisms (Sect. 2.4.2), and defines the Lie bracket of two vector fields (Sect. 2.4.4).

2.4.1 Vector Fields

Definition 2.4.1 (Vector field) Let $M \subset \mathbb{R}^k$ be a smooth m-manifold. A **(smooth) vector field** on M is a smooth map $X : M \to \mathbb{R}^k$ such that

$$X(p) \in T_p M$$

for every $p \in M$. The set of smooth vector fields on M will be denoted by

$$\text{Vect}(M) := \{X : M \to \mathbb{R}^k \mid X \text{ is smooth}, \ X(p) \in T_p M \text{ for all } p \in M\}.$$

Exercise 2.4.2 Prove that the set of smooth vector fields on M is a real vector space.

Example 2.4.3 Denote the standard cross product on $\mathbb{R}^3$ by

$$x \times y := \begin{pmatrix} x_2 y_3 - x_3 y_2 \\ x_3 y_1 - x_1 y_3 \\ x_1 y_2 - x_2 y_1 \end{pmatrix}$$

for $x, y \in \mathbb{R}^3$. Fix a vector $\xi \in S^2$ and define the maps $X, Y : S^2 \to \mathbb{R}^3$ by

$$X(p) := \xi \times p, \qquad Y(p) := (\xi \times p) \times p.$$

These are vector fields with zeros $\pm\xi$. Their integral curves (see Definition 2.4.6 below) are illustrated in Fig. 2.9.

Fig. 2.9 Two vector fields on the 2-sphere

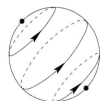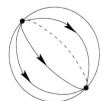

Fig. 2.10 A hyperbolic fixed
point

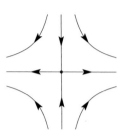

Example 2.4.4 Let $M := \mathbb{R}^2$. A vector field on M is then any smooth map $X :$
$\mathbb{R}^2 \to \mathbb{R}^2$. As an example consider the vector field

$$X(x, y) := (x, -y).$$

This vector field has a single zero at the origin and its integral curves are illustrated
in Fig. 2.10.

Example 2.4.5 Every smooth function $f : \mathbb{R}^m \to \mathbb{R}$ determines a gradient vector
field

$$X = \nabla f := \begin{pmatrix} \dfrac{\partial f}{\partial x_1} \\[4pt] \dfrac{\partial f}{\partial x_2} \\[4pt] \vdots \\[4pt] \dfrac{\partial f}{\partial x_m} \end{pmatrix} : \mathbb{R}^m \to \mathbb{R}^m.$$

Definition 2.4.6 (Integral curve) Let $M \subset \mathbb{R}^k$ be a smooth m-manifold, let X
be a smooth vector field on M, and let $I \subset \mathbb{R}$ be an open interval. A continuously
differentiable curve $\gamma : I \to M$ is called an **integral curve** of X iff it satisfies the
equation $\dot{\gamma}(t) = X(\gamma(t))$ for every $t \in I$. Note that every integral curve of X is
smooth.

Theorem 2.4.7 *Let $M \subset \mathbb{R}^k$ be a smooth m-manifold and $X \in \mathrm{Vect}(M)$ be a
smooth vector field. Fix a point $p_0 \in M$. Then the following holds.*

(i)   *There exists an open interval $I \subset \mathbb{R}$ containing 0 and a smooth curve
$\gamma : I \to M$ satisfying the equation*

$$\dot{\gamma}(t) = X(\gamma(t)), \qquad \gamma(0) = p_0 \tag{2.4.1}$$

     for every $t \in I$.
(ii) *If $\gamma_1 : I_1 \to M$ and $\gamma_2 : I_2 \to M$ are two solutions of (2.4.1) on open intervals
I_1 and I_2 containing 0, then $\gamma_1(t) = \gamma_2(t)$ for every $t \in I_1 \cap I_2$.*

Fig. 2.11 Vector fields in
local coordinates

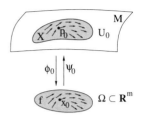

Proof We prove (i). Let $\phi_0 : U_0 \to \mathbb{R}^m$ be a coordinate chart on M, defined
on an M-open neighborhood $U_0 \subset M$ of p_0. The image of ϕ_0 is an open set
$\Omega := \phi_0(U_0) \subset \mathbb{R}^m$ and we denote the inverse map by $\psi_0 := \phi_0^{-1} : \Omega \to M$ (see
Fig. 2.11). Then, by Theorem 2.2.3, the derivative $d\psi_0(x) : \mathbb{R}^m \to \mathbb{R}^k$ is injective
and its image is the tangent space $T_{\psi_0(x)}M$ for every $x \in \Omega$. Define $f : \Omega \to \mathbb{R}^m$
by $f(x) := d\psi_0(x)^{-1} X(\psi_0(x))$ for $x \in \Omega$. This map is smooth and hence, by the
basic existence and uniqueness theorem for ordinary differential equations in $\mathbb{R}^m$
(see [63]), the equation

$$\dot{x}(t) = f(x(t)), \qquad x(0) = x_0 := \phi_0(p_0), \tag{2.4.2}$$

has a solution $x : I \to \Omega$ on some open interval $I \subset \mathbb{R}$ containing 0. Hence the
function $\gamma := \psi_0 \circ x : I \to U_0 \subset M$ is a smooth solution of (2.4.1). This proves (i).

   The local uniqueness theorem asserts that two solutions $\gamma_i : I_i \to M$ of (2.4.1)
for $i = 1, 2$ agree on the interval $(-\varepsilon, \varepsilon) \subset I_1 \cap I_2$ for $\varepsilon > 0$ sufficiently small.
This follows immediately from the standard uniqueness theorem for the solutions
of (2.4.2) in [63] and the fact that $x : I \to \Omega$ is a solution of (2.4.2) if and only if
$\gamma := \psi_0 \circ x : I \to U_0$ is a solution of (2.4.1).

   To prove (ii) we observe that the set $I := I_1 \cap I_2$ is an open interval containing
zero and hence is connected. Now consider the set

$$A := \{t \in I \mid \gamma_1(t) = \gamma_2(t)\}.$$

This set is nonempty, because $0 \in A$. It is closed, relative to I, because the maps $\gamma_1 :$
$I \to M$ and $\gamma_2 : I \to M$ are continuous. Namely, if $t_i \in I$ is a sequence converging to
$t \in I$, then $\gamma_1(t_i) = \gamma_2(t_i)$ for every i and, taking the limit $i \to \infty$, we obtain $\gamma_1(t) =$
$\gamma_2(t)$ and hence $t \in A$. The set A is also open by the local uniqueness theorem. Since
I is connected it follows that $A = I$. This proves (ii) and Theorem 2.4.7.   □

2.4.2   The Flow of a Vector Field

Definition 2.4.8 (The flow of a vector field) Let $M \subset \mathbb{R}^k$ be a smooth m-manifold
and $X \in \mathrm{Vect}(M)$ be a smooth vector field on M. For $p_0 \in M$ the **maximal existence interval** of p_0 is the open interval

$$I(p_0) := \bigcup \left\{ I \left| \begin{array}{l} I \subset \mathbb{R} \text{ is an open interval containing } 0 \\ \text{and there is a solution } \gamma : I \to M \text{ of (2.4.1)} \end{array} \right. \right\}.$$

By Theorem 2.4.7 equation (2.4.1) has a solution $\gamma : I(p_0) \to M$. The **flow of** X is the map $\phi : \mathcal{D} \to M$ defined by

$$\mathcal{D} := \{(t, p_0) \mid p_0 \in M, t \in I(p_0)\}$$

and $\phi(t, p_0) := \gamma(t)$, where $\gamma : I(p_0) \to M$ is the unique solution of (2.4.1).

Theorem 2.4.9 *Let* $M \subset \mathbb{R}^k$ *be a smooth* m-*manifold and* $X \in \text{Vect}(M)$ *be a smooth vector field on* M. *Let* $\phi : \mathcal{D} \to M$ *be the flow of* X. *Then the following holds.*

(i) $\mathcal{D}$ *is an open subset of* $\mathbb{R} \times M$.
(ii) *The map* $\phi : \mathcal{D} \to M$ *is smooth.*
(iii) *Let* $p_0 \in M$ *and* $s \in I(p_0)$. *Then*

$$I(\phi(s, p_0)) = I(p_0) - s \tag{2.4.3}$$

and, for every $t \in \mathbb{R}$ *with* $s + t \in I(p_0)$, *we have*

$$\phi(s + t, p_0) = \phi(t, \phi(s, p_0)). \tag{2.4.4}$$

The proof is based on the following lemma.

Lemma 2.4.10 *Let* M, X, $\mathcal{D}$, ϕ *be as in Theorem 2.4.9 and let* $K \subset M$ *be a compact set. Then there exists an* M-*open set* $U \subset M$ *and an* $\varepsilon > 0$ *such that* $K \subset U$, $(-\varepsilon, \varepsilon) \times U \subset \mathcal{D}$, *and* ϕ *is smooth on* $(-\varepsilon, \varepsilon) \times U$.

Proof In the case where $M = \Omega$ is an open subset of $\mathbb{R}^m$ this is proved in [64, Satz 4.1.4 & Satz 4.3.1 & Satz 4.4.1]. Using local coordinates (as in the proof of Theorem 2.4.7) we deduce that, for every $p \in M$, there exists an M-open neighborhood $U_p \subset M$ of p and a constant $\varepsilon_p > 0$ such that $(-\varepsilon_p, \varepsilon_p) \times U_p \subset \mathcal{D}$ and the restriction of ϕ to $(-\varepsilon_p, \varepsilon_p) \times U_p$ is smooth. Using this observation for every $p \in K$ (and the axiom of choice) we obtain an M-open cover $K \subset \bigcup_{p \in K} U_p$. Since the set K is compact there exists a finite subcover $K \subset U_{p_1} \cup \cdots \cup U_{p_N} =: U$. Now define $\varepsilon := \min\{\varepsilon_{p_1}, \ldots, \varepsilon_{p_N}\}$ to deduce that $(-\varepsilon, \varepsilon) \times U \subset \mathcal{D}$ and ϕ is smooth on $(-\varepsilon, \varepsilon) \times U$. This proves Lemma 2.4.10.   □

Proof of Theorem 2.4.9 We prove (iii). The map $\gamma : I(p_0) - s \to M$ defined by $\gamma(t) := \phi(s + t, p_0)$ is a solution of the initial value problem $\dot{\gamma}(t) = X(\gamma(t))$ with $\gamma(0) = \phi(s, p_0)$. Hence $I(p_0) - s \subset I(\phi(s, p_0))$ and equation (2.4.4) holds for every $t \in \mathbb{R}$ with $s + t \in I(p_0)$. In particular, with $t = -s$ we have $p_0 = \phi(-s, \phi(s, p_0))$. Thus we obtain equality in equation (2.4.3) by the same argument with the pair $(s. p_0)$ replaced by $(-s, \phi(s, p_0))$.

We prove (i) and (ii). Let $(t_0, p_0) \in \mathcal{D}$ so that $p_0 \in M$ and $t_0 \in I(p_0)$. Suppose $t_0 \geq 0$. Then $K := \{\phi(t, p_0) \mid 0 \leq t \leq t_0\}$ is a compact subset of M. (It is the image

of the compact interval $[0, t_0]$ under the unique solution $\gamma : I(p_0) \to M$ of (2.4.1).)
Hence, by Lemma 2.4.10, there is an M-open set $U \subset M$ and an $\varepsilon > 0$ such that

$$K \subset U, \qquad (-\varepsilon, \varepsilon) \times U \subset \mathcal{D},$$

and ϕ is smooth on $(-\varepsilon, \varepsilon) \times U$. Choose N so large that $t_0/N < \varepsilon$. Define $U_0 := U$
and, for $k = 1, \dots, N$, define the sets $U_k \subset M$ inductively by

$$U_k := \{p \in U \mid \phi(t_0/N, p) \in U_{k-1}\}.$$

These sets are open in the relative topology of M.

We prove by induction on k that $(-\varepsilon, k t_0/N + \varepsilon) \times U_k \subset \mathcal{D}$ and ϕ is smooth
on $(-\varepsilon, k t_0/N + \varepsilon) \times U_k$. For $k = 0$ this holds by definition of ε and U. If $k \in \{1, \dots, N\}$ and the assertion holds for $k - 1$, then we have

$$\begin{aligned}
p \in U_k &\implies p \in U, \, \phi(t_0/N, p) \in U_{k-1} \\
&\implies (-\varepsilon, \varepsilon) \subset I(p), \, (-\varepsilon, (k-1)t_0/N + \varepsilon) \subset I(\phi(t_0/N, p)) \\
&\implies (-\varepsilon, k t_0/N + \varepsilon) \subset I(p).
\end{aligned}$$

Here the last implication follows from (2.4.3). Moreover, for $p \in U_k$ and $t_0/N - \varepsilon < t < k t_0/N + \varepsilon$, we have, by (2.4.4), that

$$\phi(t, p) = \phi(t - t_0/N, \phi(t_0/N, p))$$

Since $\phi(t_0/N, p) \in U_{k-1}$ for $p \in U_k$ the right hand side is a smooth map on the
open set $(t_0/N - \varepsilon, k t_0/N + \varepsilon) \times U_k$. Since $U_k \subset U$, ϕ is also a smooth map on
$(-\varepsilon, \varepsilon) \times U_k$ and hence on $(-\varepsilon, k t_0/N + \varepsilon) \times U_k$. This completes the induction. With
$k = N$ we have found an open neighborhood of (t_0, p_0) contained in $\mathcal{D}$, namely the
set $(-\varepsilon, t_0 + \varepsilon) \times U_N$, on which ϕ is smooth. The case $t_0 \le 0$ is treated similarly.
This proves (i) and (ii) and Theorem 2.4.9.   □

Definition 2.4.11   A vector field $X \in \mathrm{Vect}(M)$ is called **complete** iff, for each
$p_0 \in M$, there exists an integral curve $\gamma : \mathbb{R} \to M$ of X with $\gamma(0) = p_0$.

Lemma 2.4.12   *Let $M \subset \mathbb{R}^k$ be a compact manifold. Then every vector field on M
is complete.*

Proof   Let $X \in \mathrm{Vect}(M)$. It follows from Lemma 2.4.10 with $K = M$ that there
exists an $\varepsilon > 0$ such that $(-\varepsilon, \varepsilon) \subset I(p)$ for all $p \in M$. By Theorem 2.4.9 this
implies $I(p) = \mathbb{R}$ for all $p \in M$. Hence X is complete.   □

Let $M \subset \mathbb{R}^k$ be a smooth manifold and $X \in \mathrm{Vect}(M)$. Then

$$X \text{ is complete} \quad \Longleftrightarrow \quad I(p) = \mathbb{R} \, \forall \, p \in M \quad \Longleftrightarrow \quad \mathcal{D} = \mathbb{R} \times M.$$

Assume X is complete, let $\phi : \mathbb{R} \times M \to M$ be the flow of X, and define the map $\phi^t : M \to M$ by $\phi^t(p) := \phi(t, p)$ for $t \in \mathbb{R}$ and $p \in M$. Then Theorem 2.4.9 asserts that ϕ^t is smooth for every $t \in \mathbb{R}$ and that

$$\phi^{s+t} = \phi^s \circ \phi^t, \qquad \phi^0 = \mathrm{id} \tag{2.4.5}$$

for all $s, t \in \mathbb{R}$. In particular, this implies that $\phi^t \circ \phi^{-t} = \phi^{-t} \circ \phi^t = \mathrm{id}$. Hence ϕ^t is bijective and $(\phi^t)^{-1} = \phi^{-t}$, so each ϕ^t is a diffeomorphism.

Exercise 2.4.13 Let $M \subset \mathbb{R}^k$ be a smooth manifold. A vector field X on M is said to have **compact support** iff there exists a compact subset $K \subset M$ such that $X(p) = 0$ for every $p \in M \setminus K$. Prove that every vector field with compact support is complete.

We close this subsection with an important observation about incomplete vector fields. The lemma asserts that an integral curve on a finite existence interval must leave every compact subset of M.

Lemma 2.4.14 *Let $M \subset \mathbb{R}^k$ be a smooth m-manifold, let $X \in \mathrm{Vect}(M)$, let $\phi : \mathcal{D} \to M$ be the flow of X, let $K \subset M$ be a compact set, and let $p_0 \in M$ be an element such that*

$$I(p_0) \cap [0, \infty) = [0, b), \qquad 0 < b < \infty.$$

Then there exists a number $0 < t_K < b$ such that

$$t_K < t < b \qquad \Longrightarrow \qquad \phi(t, p_0) \in M \setminus K$$

Proof By Lemma 2.4.10 there exists an $\varepsilon > 0$ such that $(-\varepsilon, \varepsilon) \subset I(p)$ for every $p \in K$. Choose ε so small that $\varepsilon < b$ and define

$$t_K := b - \varepsilon > 0.$$

Choose a real number $t_K < t < b$. Then $I(\phi(t, p_0)) \cap [0, \infty) = [0, b - t)$ by equation (2.4.3) in part (ii) of Theorem 2.4.9. Since $0 < b - t < b - t_K = \varepsilon$, this shows that $(-\varepsilon, \varepsilon) \not\subset I(\phi(t, p_0))$ and hence $\phi(t, p_0) \notin K$. This proves Lemma 2.4.14. $\square$

The next corollary is an immediate consequence of Lemma 2.4.14. In this formulation the result will be used in Sect. 4.6 and in Sect. 7.3.

Corollary 2.4.15 *Let $M \subset \mathbb{R}^k$ be a smooth m-manifold, let $X \in \mathrm{Vect}(M)$, and let $\gamma : (0, T) \to M$ be an integral curve of X. If there exists a compact set $K \subset M$ that contains the image of γ, then γ extends to an integral curve of X on the interval $(-\rho, T + \rho)$ for some $\rho > 0$.*

Proof   Here is another more direct proof that does not rely on Lemma 2.4.10. Since K is compact, there exists a constant $c > 0$ such that $|X(p)| \le c$ for all $p \in K$. Since $\gamma(t) \in K$ for $0 < t < T$, this implies

$$|\gamma(t) - \gamma(s)| = \left| \int_s^t \dot{\gamma}(r)\, dr \right| \le \int_s^t |\dot{\gamma}(r)|\, dr = \int_s^t |X(\gamma(r))|\, dr \le c(t - s)$$

for $0 < s < t < T$. Thus the limit $p_0 := \lim_{t \searrow 0} \gamma(t)$ exists in $\mathbb{R}^k$ and, since K is a closed subset of $\mathbb{R}^k$, we have $p_0 \in K \subset M$. Define $\gamma_0 : [0, T) \to M$ by

$$\gamma_0(t) := \begin{cases} p_0, & \text{for } t = 0, \\ \gamma(t), & \text{for } 0 < t < T. \end{cases}$$

We prove that γ_0 is differentiable at $t = 0$ and $\dot{\gamma}_0(0) = X(p_0)$. To see this, fix a constant $\varepsilon > 0$. Since the curve $[0, T) \to \mathbb{R}^k : t \mapsto X(\gamma(t))$ is continuous, there exists a constant $\delta > 0$ such that

$$0 < t \le \delta \qquad \Longrightarrow \qquad |X(\gamma(t)) - X(p_0)| \le \varepsilon.$$

Hence, for $0 < s < t \le \delta$, we have

$$|\gamma(t) - \gamma(s) - (t - s)X(p_0)| = \left| \int_s^t (\dot{\gamma}(r) - X(p_0))\, dr \right|$$

$$= \left| \int_s^t (X(\gamma(r)) - X(p_0))\, dr \right|$$

$$\le \int_s^t |X(\gamma(r)) - X(p_0)|\, dr$$

$$\le (t - s)\varepsilon.$$

Take the limit $s \to 0$ to obtain

$$\left| \frac{\gamma(t) - p_0}{t} - X(p_0) \right| = \lim_{s \to 0} \frac{|\gamma(t) - \gamma(s) - (t - s)X(p_0)|}{t - s} \le \varepsilon$$

for $0 < t \le \delta$. Thus γ_0 is differentiable at $t = 0$ with $\dot{\gamma}_0(0) = X(p_0)$, as claimed. Hence γ extends to an integral curve $\widetilde{\gamma} : (-\rho, T) \to M$ of X for some $\rho > 0$ via $\widetilde{\gamma}(t) := \phi(t, p_0)$ for $-\rho < t \le 0$ and $\widetilde{\gamma}(t) := \gamma(t)$ for $0 < t < T$. Here ϕ is the flow of X. That γ also extends beyond $t = T$, follows by replacing $\gamma(t)$ with $\gamma(T - t)$ and X with $-X$. This proves Corollary 2.4.15. $\square$

2.4.3   The Group of Diffeomorphisms

Let us denote the space of diffeomorphisms of M by

$$\text{Diff}(M) := \{\phi : M \to M \mid \phi \text{ is a diffeomorphism}\}.$$

This is a group. The group operation is composition and the neutral element is the identity. Now equation (2.4.5) asserts that the flow of a complete vector field $X \in \text{Vect}(M)$ is a group homomorphism

$$\mathbb{R} \to \text{Diff}(M) : t \mapsto \phi^t.$$

This homomorphism is smooth and is characterized by the equation

$$\frac{d}{dt}\phi^t(p) = X(\phi^t(p)), \qquad \phi^0(p) = p$$

for all $p \in M$ and $t \in \mathbb{R}$. We will often abbreviate this equation in the form

$$\frac{d}{dt}\phi^t = X \circ \phi^t, \qquad \phi^0 = \text{id}. \tag{2.4.6}$$

Exercise 2.4.16 (Isotopy) Let $M \subset \mathbb{R}^k$ be a compact manifold and $I \subset \mathbb{R}$ be an open interval containing 0. Let

$$I \times M \to \mathbb{R}^k : (t, p) \mapsto X_t(p)$$

be a smooth map such that $X_t \in \text{Vect}(M)$ for every t. Prove that there is a smooth family of diffeomorphisms $I \times M \to M : (t, p) \mapsto \phi_t(p)$ satisfying

$$\frac{d}{dt}\phi_t = X_t \circ \phi_t, \qquad \phi_0 = \text{id} \tag{2.4.7}$$

for every $t \in I$. Such a family of diffeomorphisms

$$I \to \text{Diff}(M) : t \mapsto \phi_t$$

is called an **isotopy** of M. Conversely prove that every smooth isotopy $I \to \text{Diff}(M) : t \mapsto \phi_t$ is generated (uniquely) by a smooth family of vector fields $I \to \text{Vect}(M) : t \mapsto X_t$.

2.4.4   The Lie Bracket

Let $M \subset \mathbb{R}^k$ and $N \subset \mathbb{R}^\ell$ be smooth m-manifolds and $X \in \text{Vect}(M)$ be smooth vector field on M. If $\psi : N \to M$ is a diffeomorphism, then the **pullback** of X under ψ is the vector field on N defined by

$$(\psi^* X)(q) := d\psi(q)^{-1} X(\psi(q)) \tag{2.4.8}$$

for $q \in N$. If $\phi : M \to N$ is a diffeomorphism, then the **pushforward** of X under ϕ is the vector field on N defined by

$$(\phi_* X)(q) := d\phi(\phi^{-1}(q)) X(\phi^{-1}(q)) \tag{2.4.9}$$

for $q \in N$.

Lemma 2.4.17   *Let $M \subset \mathbb{R}^k$, $N \subset \mathbb{R}^\ell$, and $P \subset \mathbb{R}^n$ be smooth m-dimensional submanifolds, let $\phi : M \to N$ and $\psi : N \to P$ be diffeomorphisms, and let $X \in \mathrm{Vect}(M)$ and $Z \in \mathrm{Vect}(P)$. Then*

$$\phi_* X = (\phi^{-1})^* X \tag{2.4.10}$$

and

$$(\psi \circ \phi)_* X = \psi_* \phi_* X, \qquad (\psi \circ \phi)^* Z = \phi^* \psi^* Z. \tag{2.4.11}$$

Proof   Equation (2.4.10) follows from the fact that

$$d\phi^{-1}(q) = d\phi(\phi^{-1}(q))^{-1} : T_q N \to T_{\phi^{-1}(q)} M$$

for all $q \in N$ (Corollary 2.2.15) and the equations in (2.4.11) follow directly from the chain rule (Theorem 2.2.14). This proves Lemma 2.4.17.   □

We think of a vector field on M as a smooth map

$$X : M \to \mathbb{R}^k$$

that satisfies the condition $X(p) \in T_p M$ for every $p \in M$. Ignoring this condition temporarily, we can differentiate X as a map from M to $\mathbb{R}^k$ and its derivative at p is then a linear map

$$dX(p) : T_p M \to \mathbb{R}^k.$$

In general, this derivative will not take values in the tangent space $T_p M$. However, if we have two vector fields X and Y on M, then the next lemma shows that the difference of the derivative of X in the direction Y and of Y in the direction X does take values in the tangent spaces of M.

Lemma 2.4.18 (Lie bracket)   *Let $M \subset \mathbb{R}^k$ be a smooth m-manifold and let $X, Y \in \mathrm{Vect}(M)$ be complete vector fields. Denote by*

$$\mathbb{R} \to \mathrm{Diff}(M) : t \mapsto \phi^t, \qquad \mathbb{R} \to \mathrm{Diff}(M) : t \mapsto \psi^t$$

the flows of X and Y, respectively. Fix a point $p \in M$ and define the smooth map $\gamma : \mathbb{R} \to M$ by

$$\gamma(t) := \phi^t \circ \psi^t \circ \phi^{-t} \circ \psi^{-t}(p). \tag{2.4.12}$$

Then $\dot{\gamma}(0) = 0$ and

$$
\begin{aligned}
\frac{d}{dt}\bigg|_{t=0} \gamma(\sqrt{t}) &= \frac{1}{2}\ddot{\gamma}(0) \\
&= \frac{d}{ds}\bigg|_{s=0} ((\phi^s)_* Y)(p) \\
&= \frac{d}{dt}\bigg|_{t=0} ((\psi^t)^* X)(p) \\
&= dX(p)Y(p) - dY(p)X(p) \qquad \in \quad T_p M.
\end{aligned}
\tag{2.4.13}
$$

Exercise 2.4.19 Let $\gamma : \mathbb{R} \to \mathbb{R}^k$ be a C^2-curve and assume $\dot{\gamma}(0) = 0$. Prove that $\frac{d}{dt}\big|_{t=0}\gamma(\sqrt{t}) = \lim_{t\to 0} t^{-2}(\gamma(t) - \gamma(0)) = \frac{1}{2}\ddot{\gamma}(0)$.

Proof of Lemma 2.4.18 Define the map $\beta : \mathbb{R}^2 \to M$ by

$$
\beta(s,t) := \phi^s \circ \psi^t \circ \phi^{-s} \circ \psi^{-t}(p)
$$

for $s, t \in \mathbb{R}$. Then $\gamma(t) = \beta(t, t)$ and

$$
\frac{\partial \beta}{\partial s}(0, t) = X(p) - d\psi^t(\psi^{-t}(p))X(\psi^{-t}(p)), \tag{2.4.14}
$$

$$
\frac{\partial \beta}{\partial t}(s, 0) = d\phi^s(\phi^{-s}(p))Y(\phi^{-s}(p)) - Y(p) \tag{2.4.15}
$$

for all $s, t \in \mathbb{R}$. Hence

$$
\dot{\gamma}(0) = \frac{\partial \beta}{\partial s}(0, 0) + \frac{\partial \beta}{\partial t}(0, 0) = 0.
$$

This implies the first equality in (2.4.13) by Exercise (2.4.19). To prove the remaining assertions, note that $\beta(s, 0) = \beta(0, t) = p$, hence the second derivatives $\partial^2 \beta/\partial s^2$ and $\partial^2 \beta/\partial t^2$ vanish at $s = t = 0$, and therefore

$$
\ddot{\gamma}(0) = 2\frac{\partial^2 \beta}{\partial s \partial t}(0, 0). \tag{2.4.16}
$$

Combining equations (2.4.15) and (2.4.16) we find

$$
\begin{aligned}
\frac{1}{2}\ddot{\gamma}(0) &= \frac{\partial}{\partial s}\bigg|_{s=0} \frac{\partial \beta}{\partial t}(s, 0) = \frac{d}{ds}\bigg|_{s=0} d\phi^s(\phi^{-s}(p))Y(\phi^{-s}(p)) \\
&= \frac{d}{ds}\bigg|_{s=0} ((\phi^s)_* Y)(p)).
\end{aligned}
$$

Likewise, combining equations (2.4.14) and (2.4.16) we find

$$
\begin{aligned}
\frac{1}{2}\ddot{\gamma}(0) &= \frac{\partial}{\partial t}\Big|_{t=0} \frac{\partial\beta}{\partial s}(0,t) = -\frac{d}{dt}\Big|_{t=0} d\psi^t(\psi^{-t}(p))X(\psi^{-t}(p)) \\
&= \frac{d}{dt}\Big|_{t=0} d\psi^{-t}(\psi^t(p))X(\psi^t(p)) \\
&= \frac{d}{dt}\Big|_{t=0} d\psi^t(p)^{-1}X(\psi^t(p)) \\
&= \frac{d}{dt}\Big|_{t=0} ((\psi^t)^*X)(p)).
\end{aligned}
$$

In both cases the right hand side is the derivative of a smooth curve in the tangent space T_pM and so is itself an element of T_pM. Moreover, we have

$$
\begin{aligned}
\frac{1}{2}\ddot{\gamma}(0) &= \frac{\partial}{\partial s}\Big|_{s=0} d\phi^s(\phi^{-s}(p))Y(\phi^{-s}(p)) \\
&= \frac{\partial}{\partial s}\Big|_{s=0} \frac{\partial}{\partial t}\Big|_{t=0} \phi^s \circ \psi^t \circ \phi^{-s}(p) \\
&= \frac{\partial}{\partial t}\Big|_{t=0} \frac{\partial}{\partial s}\Big|_{s=0} \phi^s \circ \psi^t \circ \phi^{-s}(p) \\
&= \frac{\partial}{\partial t}\Big|_{t=0} \big(X(\psi^t(p)) - d\psi^t(p)X(p)\big) \\
&= dX(p)Y(p) - \frac{\partial}{\partial t}\Big|_{t=0} \frac{\partial}{\partial s}\Big|_{s=0} \psi^t \circ \phi^s(p) \\
&= dX(p)Y(p) - \frac{\partial}{\partial s}\Big|_{s=0} \frac{\partial}{\partial t}\Big|_{t=0} \psi^t \circ \phi^s(p) \\
&= dX(p)Y(p) - \frac{\partial}{\partial s}\Big|_{s=0} Y(\phi^s(p)) \\
&= dX(p)Y(p) - dY(p)X(p).
\end{aligned}
$$

This proves Lemma 2.4.18.   □

Definition 2.4.20 (Lie bracket)   Let $M \subset \mathbb{R}^k$ be a smooth manifold and let $X, Y \in \mathrm{Vect}(M)$ be smooth vector fields on M. The **Lie bracket** of X and Y is the vector field $[X, Y] \in \mathrm{Vect}(M)$ defined by

$$
[X, Y](p) := dX(p)Y(p) - dY(p)X(p). \tag{2.4.17}
$$

Warning   In the literature on differential geometry the Lie bracket of two vector fields is often (but not always) defined with the opposite sign. The rationale behind the present choice of the sign will be explained in Sect. 2.5.7.

Lemma 2.4.21 *Let $M \subset \mathbb{R}^k$ and $N \subset \mathbb{R}^\ell$ be smooth manifolds, let X, Y, Z be smooth vector fields on M, and let*

$$\phi : N \to M$$

be a diffeomorphism. Then

$$\phi^*[X, Y] = [\phi^* X, \phi^* Y], \qquad (2.4.18)$$
$$[X, Y] + [Y, X] = 0, \qquad (2.4.19)$$
$$[X, [Y, Z]] + [Y, [Z, X]] + [Z, [X, Y]] = 0. \qquad (2.4.20)$$

*The last equation is called the **Jacobi identity**.*

Proof Let $\mathbb{R} \to \mathrm{Diff}(M) : t \mapsto \psi^t$ be the flow of Y. Then the map

$$\mathbb{R} \to \mathrm{Diff}(N) : t \mapsto \phi^{-1} \circ \psi^t \circ \phi$$

is the flow of the vector field $\phi^* Y$ on N. Hence, by Lemmas 2.4.17 and 2.4.18, we have

$$\begin{aligned}
[\phi^* X, \phi^* Y] &= \frac{d}{dt}\bigg|_{t=0} \left(\phi^{-1} \circ \psi^t \circ \phi\right)^* \phi^* X \\
&= \frac{d}{dt}\bigg|_{t=0} \phi^* \left(\psi^t\right)^* X \\
&= \phi^*[X, Y].
\end{aligned}$$

This proves (2.4.18). Equation (2.4.19) is obvious. To prove (2.4.20), let ϕ^t be the flow of X. Then by (2.4.18) and (2.4.19) and Lemma 2.4.18 we have

$$\begin{aligned}
[[Y, Z], X] &= \frac{d}{dt}\bigg|_{t=0} (\phi^t)^*[Y, Z] \\
&= \frac{d}{dt}\bigg|_{t=0} [(\phi^t)^* Y, (\phi^t)^* Z] \\
&= [[Y, X], Z] + [Y, [Z, X]] \\
&= [Z, [X, Y]] + [Y, [Z, X]].
\end{aligned}$$

This proves Lemma 2.4.21. □

Definition 2.4.22 A **Lie algebra** is a real vector space $\mathfrak{g}$ equipped with a skew-symmetric bilinear map $\mathfrak{g} \times \mathfrak{g} \to \mathfrak{g} : (\xi, \eta) \mapsto [\xi, \eta]$ that satisfies the Jacobi identity $[\xi, [\eta, \zeta]] + [\eta, [\zeta, \xi]] + [\zeta, [\xi, \eta]] = 0$ for all $\xi, \eta, \zeta \in \mathfrak{g}$.

Example 2.4.23 The Vector fields on a smooth manifold $M \subset \mathbb{R}^k$ form a Lie algebra with the Lie bracket (2.4.17). The space $\mathfrak{gl}(n, \mathbb{R}) = \mathbb{R}^{n \times n}$ of real $n \times n$-matrices is a Lie algebra with the Lie bracket

$$[\xi, \eta] := \xi\eta - \eta\xi.$$

It is also interesting to consider subspaces of $\mathfrak{gl}(n, \mathbb{R})$ that are invariant under this Lie bracket. An example is the space

$$\mathfrak{o}(n) := \{\xi \in \mathfrak{gl}(n, \mathbb{R}) \,|\, \xi^\mathsf{T} + \xi = 0\}$$

of skew-symmetric $n \times n$-matrices. It is a nontrivial fact that every finite-dimensional Lie algebra is isomorphic to a Lie subalgebra of $\mathfrak{gl}(n, \mathbb{R})$ for some n. For example, the cross product defines a Lie algebra structure on $\mathbb{R}^3$ and the resulting Lie algebra is isomorphic to $\mathfrak{o}(3)$.

Remark 2.4.24 There is a linear map $\mathbb{R}^{m \times m} \to \mathrm{Vect}(\mathbb{R}^m) : \xi \mapsto X_\xi$ which assigns to a matrix $\xi \in \mathfrak{gl}(m, \mathbb{R})$ the linear vector field $X_\xi : \mathbb{R}^m \to \mathbb{R}^m$ given by $X_\xi(x) := \xi x$ for $x \in \mathbb{R}^m$. This map preserves the Lie bracket, i.e. $[X_\xi, X_\eta] = X_{[\xi,\eta]}$, and hence is a **Lie algebra homomorphism**.

To understand the Lie bracket geometrically, consider again the curve

$$\gamma(t) := \phi^t \circ \psi^t \circ \phi^{-t} \circ \psi^{-t}(p)$$

in Lemma 2.4.18, where ϕ^t and ψ^t are the flows of the vector fields X and Y, respectively. Since $\dot\gamma(0) = 0$, Exercise 2.4.19 asserts that

$$[X, Y](p) = \frac{1}{2}\ddot\gamma(0) = \left.\frac{d}{dt}\right|_{t=0} \phi^{\sqrt{t}} \circ \psi^{\sqrt{t}} \circ \phi^{-\sqrt{t}} \circ \psi^{-\sqrt{t}}(p). \qquad (2.4.21)$$

Geometrically this means that by following first the backward flow of Y for time ε, then the backward flow of X for time ε, then the forward flow of Y for time ε, and finally the forward flow of X for time ε, we will not, in general, get back to the original point p where we started but approximately obtain an *"error"* $\varepsilon^2[X, Y](p)$. An example of this (which we learned from Donaldson) is the mathematical formulation of parking a car.

Example 2.4.25 (Parking a car) The configuration space for driving a car in the plane is the manifold $M := \mathbb{C} \times S^1$, where $S^1 \subset \mathbb{C}$ denotes the unit circle. Thus a point in M is a pair $p = (z, \lambda) \in \mathbb{C} \times \mathbb{C}$ with $|\lambda| = 1$. The point $z \in \mathbb{C}$ represents the position of the car and the unit vector $\lambda \in S^1$ represents the direction in which it is pointing. The *left turn* is represented by a vector field X and the *right turn* by a vector field Y on M. These vector field are given by $X(z, \lambda) := (\lambda, i\lambda)$ and $Y(z, \lambda) := (\lambda, -i\lambda)$. Their Lie bracket is the vector field $[X, Y](z, \lambda) = (-2i\lambda, 0)$.

This vector field represents a sideways move of the car to the right. And a sideways move by $2\varepsilon^2$ can be achieved by following a backward right turn for time ε, then a backward left turn for time ε, then a forward right turn for time ε, and finally a forward left turn for time ε.

This example can be reformulated by identifying $\mathbb{C}$ with $\mathbb{R}^2$ via $z = x + iy$ and representing a point in the unit circle by the angle $\theta \in \mathbb{R}/2\pi\mathbb{Z}$ via $\lambda = e^{i\theta}$. In this formulation the manifold is $M = \mathbb{R}^2 \times \mathbb{R}/2\pi\mathbb{Z}$, a point in M is represented by a triple $(x, y, \theta) \in \mathbb{R}^3$, the vector fields X and Y are

$$X(x, y, \theta) := (\cos(\theta), \sin(\theta), 1), \qquad Y(x, y, \theta) := (\cos(\theta), \sin(\theta), -1),$$

and their Lie bracket is $[X, Y](x, y, \theta) = 2(\sin(\theta), -\cos(\theta), 0)$.

Lemma 2.4.26 *Let $X, Y \in \mathrm{Vect}(M)$ be complete vector fields on a manifold M and $\phi^t, \psi^t \in \mathrm{Diff}(M)$ be the flows of X and Y, respectively. Then the Lie bracket $[X, Y]$ vanishes if and only if the flows of X and Y commute, i.e. $\phi^s \circ \psi^t = \psi^t \circ \phi^s$ for all $s, t \in \mathbb{R}$.*

Proof If the flows of X and Y commute, then the Lie bracket $[X, Y]$ vanishes by Lemma 2.4.18. Conversely, suppose that $[X, Y] = 0$. Then we have

$$\frac{d}{ds}(\phi^s)_* Y = (\phi^s)_* \frac{d}{dr}\bigg|_{r=0} (\phi^r)_* Y = (\phi^s)_*[X, Y] = 0$$

for every $s \in \mathbb{R}$ and hence

$$(\phi^s)_* Y = Y. \tag{2.4.22}$$

Fix a real number s and define the curve $\gamma : \mathbb{R} \to M$ by $\gamma(t) := \phi^s(\psi^t(p))$ for $t \in \mathbb{R}$. Then $\gamma(0) = \phi^s(p)$ and

$$\dot{\gamma}(t) = d\phi^s(\psi^t(p))Y(\psi^t(p)) = ((\phi^s)_* Y)(\gamma(t)) = Y(\gamma(t))$$

for all t. Here the last equation follows from (2.4.22). Since ψ^t is the flow of Y we obtain $\gamma(t) = \psi^t(\phi^s(p))$ for all $t \in \mathbb{R}$ and this proves Lemma 2.4.26. $\qquad\square$

Exercise 2.4.27 In the situation of Lemma 2.4.26 prove that $\{\phi^t \circ \psi^t\}_{t \in \mathbb{R}}$ is the flow of the vector field $X + Y$.

2.5 Lie Groups

Combining the concept of a group and a manifold, it is interesting to consider groups which are also manifolds and have the property that the group operation and the inverse define smooth maps. We shall only consider groups of matrices.

2.5.1 Definition and Examples

Definition 2.5.1 (Lie group) A nonempty subset $G \subset \mathbb{R}^{n \times n}$ is called a **Lie group** iff it is a submanifold of $\mathbb{R}^{n \times n}$ and a subgroup of $GL(n, \mathbb{R})$, i.e.

$$g, h \in G \quad \Longrightarrow \quad gh \in G$$

(where gh denotes the product of the matrices g and h) and

$$g \in G \quad \Longrightarrow \quad \det(g) \neq 0 \text{ and } g^{-1} \in G.$$

(Since $G \neq \emptyset$ it follows from these conditions that the identity matrix $\mathbb{1}$ is an element of G.)

Example 2.5.2 The general linear group $G = GL(n, \mathbb{R})$ is an open subset of $\mathbb{R}^{n \times n}$ and hence is a Lie group. By Exercise 2.1.19 the special linear group

$$SL(n, \mathbb{R}) = \left\{ g \in GL(n, \mathbb{R}) \,\middle|\, \det(g) = 1 \right\}$$

is a Lie group and, by Example 2.1.20, the special orthogonal group

$$SO(n) := \left\{ g \in GL(n, \mathbb{R}) \,\middle|\, g^{\mathsf{T}} g = \mathbb{1}, \det(g) = 1 \right\}$$

is a Lie group. In fact every orthogonal matrix has determinant ± 1 and so $SO(n)$ is an open subset of $O(n)$ (in the relative topology).

In a similar vein the group $GL(n, \mathbb{C}) := \{ g \in \mathbb{C}^{n \times n} \mid \det(g) \neq 0 \}$ of complex matrices with nonzero (complex) determinant is an open subset of $\mathbb{C}^{n \times n}$ and hence is a Lie group. As in the real case, the subgroups

$$SL(n, \mathbb{C}) := \left\{ g \in GL(n, \mathbb{C}) \,\middle|\, \det(g) = 1 \right\},$$
$$U(n) := \left\{ g \in GL(n, \mathbb{C}) \,\middle|\, g^* g = \mathbb{1} \right\},$$
$$SU(n) := \left\{ g \in GL(n, \mathbb{C}) \,\middle|\, g^* g = \mathbb{1}, \det(g) = 1 \right\}$$

are submanifolds of $GL(n, \mathbb{C})$ and hence are Lie groups. Here $g^* := \bar{g}^{\mathsf{T}}$ denotes the conjugate transpose of a complex matrix.

Exercise 2.5.3 Prove that $SL(n, \mathbb{C})$, $U(n)$, and $SU(n)$ are Lie groups. Prove that $SO(n)$ is connected and that $O(n)$ has two connected components.

Exercise 2.5.4 Prove that $GL(n, \mathbb{C})$ can be identified with the group

$$G := \{ \Phi \in GL(2n, \mathbb{R}) \mid \Phi J_0 = J_0 \Phi \}, \qquad J_0 := \begin{pmatrix} 0 & -\mathbb{1} \\ \mathbb{1} & 0 \end{pmatrix}.$$

Hint: Use the isomorphism $\mathbb{R}^n \times \mathbb{R}^n \to \mathbb{C}^n : (x, y) \mapsto x + iy$. Show that a matrix $\Phi \in \mathbb{R}^{2n \times 2n}$ commutes with J_0 if and only if it has the form

$$\Phi = \begin{pmatrix} X & -Y \\ Y & X \end{pmatrix}, \qquad X, Y \in \mathbb{R}^{n \times n}.$$

What is the relation between the real determinant of Φ and the complex determinant of $X + iY$?

Exercise 2.5.5 Let J_0 be as in Exercise 2.5.4 and define

$$\mathrm{Sp}(2n) := \left\{ \Psi \in \mathrm{GL}(2n, \mathbb{R}) \,\middle|\, \Psi^{\mathsf{T}} J_0 \Psi = J_0 \right\}.$$

This is the **symplectic linear group**. Prove that $\mathrm{Sp}(2n)$ is a Lie group. **Hint:** See [49, Lemma 1.1.12].

Example 2.5.6 (Unit quaternions) The **quaternions** form a four-dimensional associative unital algebra $\mathbb{H}$, equipped with a basis $1, i, j, k$. The elements of $\mathbb{H}$ are vectors of the form

$$x = x_0 + ix_1 + jx_2 + kx_3 \qquad x_0, x_1, x_2, x_3 \in \mathbb{R}. \tag{2.5.1}$$

The product structure is the bilinear map $\mathbb{H} \times \mathbb{H} \to \mathbb{H} : (x, y) \mapsto xy$, determined by the relations

$$i^2 = j^2 = k^2 = -1, \quad ij = -ji = k, \quad jk = -kj = i, \quad ki = -ik = j.$$

This product structure is associative but not commutative. The quaternions are equipped with an involution $\mathbb{H} \to \mathbb{H} : x \mapsto \bar{x}$, which assigns to a quaternion x of the form (2.5.1) its **conjugate** $\bar{x} := x_0 - ix_1 - jx_2 - kx_3$. This involution satisfies the conditions

$$\overline{x + y} = \bar{x} + \bar{y}, \qquad \overline{xy} = \bar{y}\bar{x}, \qquad x\bar{x} = |x|^2, \qquad |xy| = |x||y|$$

for $x, y \in \mathbb{H}$, where $|x| := \sqrt{x_0^2 + x_2^2 + x_2^2 + x_3^2}$ denotes the Euclidean norm of the quaternion (2.5.1). Thus the **unit quaternions** form a group

$$\mathrm{Sp}(1) := \left\{ x \in \mathbb{H} \,\middle|\, |x| = 1 \right\}$$

with the inverse map $x \mapsto \bar{x}$. Note that the group $\mathrm{Sp}(1)$ is diffeomorphic to the 3-sphere $S^3 \subset \mathbb{R}^4$ under the isomorphism $\mathbb{H} \cong \mathbb{R}^4$. **Warning:** The unit quaternions (a compact Lie group) are not to be confused with the symplectic linear group in Exercise 2.5.5 (a noncompact Lie group) despite the similarity in notation.

Let $G \subset GL(n, \mathbb{R})$ be a Lie group. Then the maps

$$G \times G \to G : (g, h) \mapsto gh, \qquad G \to G : g \mapsto g^{-1}$$

are smooth (see [64]). Fixing an element $h \in G$ we find that the derivative of the map $G \to G : g \mapsto gh$ at $g \in G$ is given by the linear map

$$T_g G \to T_{gh} G : \widehat{g} \mapsto \widehat{g}h. \tag{2.5.2}$$

Here $\widehat{g}$ and h are both matrices in $\mathbb{R}^{n \times n}$ and $\widehat{g}h$ denotes the matrix product. In fact, if $\widehat{g} \in T_g G$, then, since G is a manifold, there exists a smooth curve $\gamma : \mathbb{R} \to G$ with $\gamma(0) = g$ and $\dot{\gamma}(0) = \widehat{g}$. Since G is a group we obtain a smooth curve $\beta : \mathbb{R} \to G$ given by $\beta(t) := \gamma(t)h$. It satisfies $\beta(0) = gh$ and so $\widehat{g}h = \dot{\beta}(0) \in T_{gh} G$.

The linear map (2.5.2) is obviously a vector space isomorphism whose inverse is given by right multiplication with h^{-1}. It is sometimes convenient to define the map $R_h : G \to G$ by

$$R_h(g) := gh$$

for $g \in G$ (*right multiplication* by h). This is a diffeomorphism and the linear map (2.5.2) is the derivative of R_h at g, so

$$dR_h(g)\widehat{g} = \widehat{g}h \qquad \text{for } \widehat{g} \in T_g G.$$

Similarly, each element $g \in G$ determines a diffeomorphism $L_g : G \to G$, given by

$$L_g(h) := gh$$

for $h \in G$ (*left multiplication* by g). Its derivative at $h \in G$ is again given by matrix multiplication, i.e. the linear map $dL_g(h) : T_h G \to T_{gh} G$ is given by

$$dL_g(h)\widehat{h} = g\widehat{h} \qquad \text{for } \widehat{h} \in T_h G. \tag{2.5.3}$$

Since L_g is a diffeomorphism its derivative $dL_g(h) : T_h G \to T_{gh} G$ is again a vector space isomorphism for every $h \in G$.

Exercise 2.5.7 Prove that the map $G \to G : g \mapsto g^{-1}$ is a diffeomorphism and that its derivative at $g \in G$ is the vector space isomorphism

$$T_g G \to T_{g^{-1}} G : v \mapsto -g^{-1}vg^{-1}.$$

Hint: Use [64] or any textbook on first year analysis.

2.5.2   The Lie Algebra of a Lie Group

Let

$$G \subset GL(n, \mathbb{R})$$

be a Lie group. Its tangent space at the identity matrix $\mathbb{1} \in G$ is called the **Lie algebra** of G and will be denoted by

$$\mathfrak{g} = \text{Lie}(G) := T_{\mathbb{1}} G.$$

This terminology is justified by the fact that $\mathfrak{g}$ is in fact a Lie algebra, i.e. it is invariant under the standard Lie bracket operation

$$[\xi, \eta] := \xi\eta - \eta\xi$$

on the space $\mathbb{R}^{n \times n}$ of square matrices (see Lemma 2.5.9 below). The proof requires the notion of the **exponential matrix**. For $\xi \in \mathbb{R}^{n \times n}$ and $t \in \mathbb{R}$ we define

$$\exp(t\xi) := \sum_{k=0}^{\infty} \frac{t^k \xi^k}{k!}. \tag{2.5.4}$$

A standard result in first year analysis asserts that this series converges absolutely (and uniformly on compact t-intervals), that the map

$$\mathbb{R} \to \mathbb{R}^{n \times n} : t \mapsto \exp(t\xi)$$

is smooth and satisfies the differential equation

$$\frac{d}{dt} \exp(t\xi) = \xi \exp(t\xi) = \exp(t\xi)\xi, \tag{2.5.5}$$

and that

$$\exp((s + t)\xi) = \exp(s\xi) \exp(t\xi), \qquad \exp(0\xi) = \mathbb{1} \tag{2.5.6}$$

for all $s, t \in \mathbb{R}$. This shows that the matrix $\exp(t\xi)$ is invertible for each t and that the map $\mathbb{R} \to GL(n, \mathbb{R}) : t \mapsto \exp(t\xi)$ is a group homomorphism.

Exercise 2.5.8   Prove the following analogue of (2.4.12). For $\xi, \eta \in \mathfrak{g}$

$$\frac{d}{dt}\Big|_{t=0} \exp(\sqrt{t}\xi) \exp(\sqrt{t}\eta) \exp(-\sqrt{t}\xi) \exp(-\sqrt{t}\eta) = [\xi, \eta]. \tag{2.5.7}$$

In other words, the infinitesimal Lie group commutator is the matrix commutator. (Compare Equations (2.5.7) and (2.4.21).)

Lemma 2.5.9 *Let* $G \subset GL(n, \mathbb{R})$ *be a Lie group and denote by* $\mathfrak{g} := \mathrm{Lie}(G)$ *its Lie algebra. Then the following holds.*

(i)   *If* $\xi \in \mathfrak{g}$*, then* $\exp(t\xi) \in G$ *for every* $t \in \mathbb{R}$.
(ii) *If* $g \in G$ *and* $\eta \in \mathfrak{g}$*, then* $g\eta g^{-1} \in \mathfrak{g}$.
(iii) *If* $\xi, \eta \in \mathfrak{g}$*, then* $[\xi, \eta] = \xi\eta - \eta\xi \in \mathfrak{g}$.

Proof We prove (i). For every $g \in G$ we have a vector space isomorphism $\mathfrak{g} = T_1 G \to T_g G : \xi \mapsto \xi g$ as in (2.5.2). Hence each element $\xi \in \mathfrak{g}$ determines a vector field $X_\xi \in \mathrm{Vect}(G)$, defined by

$$X_\xi(g) := \xi g \in T_g G, \qquad g \in G. \tag{2.5.8}$$

By Theorem 2.4.7 there is an integral curve $\gamma : (-\varepsilon, \varepsilon) \to G$ satisfying

$$\dot{\gamma}(t) = X_\xi(\gamma(t)) = \xi\gamma(t), \qquad \gamma(0) = 1.$$

By (2.5.5), the curve $(-\varepsilon, \varepsilon) \to \mathbb{R}^{n \times n} : t \mapsto \exp(t\xi)$ satisfies the same initial value problem and hence, by uniqueness, we have $\exp(t\xi) = \gamma(t) \in G$ for all $t \in \mathbb{R}$ with $|t| < \varepsilon$. Now let $t \in \mathbb{R}$ and choose $N \in \mathbb{N}$ such that $\left|\frac{t}{N}\right| < \varepsilon$. Then $\exp(\frac{t}{N}\xi) \in G$ and hence it follows from (2.5.6) that

$$\exp(t\xi) = \exp\left(\frac{t}{N}\xi\right)^N \in G.$$

This proves (i).

We prove (ii). Consider the smooth curve $\gamma : \mathbb{R} \to \mathbb{R}^{n \times n}$ defined by

$$\gamma(t) := g \exp(t\eta) g^{-1}.$$

By (i) we have $\gamma(t) \in G$ for every $t \in \mathbb{R}$. Since $\gamma(0) = 1$ we have

$$g\eta g^{-1} = \dot{\gamma}(0) \in \mathfrak{g}.$$

This proves (ii).

We prove (iii). Define the smooth map $\eta : \mathbb{R} \to \mathbb{R}^{n \times n}$ by

$$\eta(t) := \exp(t\xi)\eta \exp(-t\xi).$$

By (i) we have $\exp(t\xi) \in G$ and, by (ii), we have $\eta(t) \in \mathfrak{g}$ for every $t \in \mathbb{R}$. Hence $[\xi, \eta] = \dot{\eta}(0) \in \mathfrak{g}$. This proves (iii) and Lemma 2.5.9. $\square$

By Lemma 2.5.9 the curve $\gamma : \mathbb{R} \to G$ defined by $\gamma(t) := \exp(t\xi)g$ is the integral curve of the vector field X_ξ in (2.5.8) with initial condition $\gamma(0) = g$. Thus X_ξ is complete for every $\xi \in \mathfrak{g}$.

Lemma 2.5.10 *If $\xi \in \mathfrak{g}$ and $\gamma : \mathbb{R} \to G$ is a smooth curve satisfying*

$$\gamma(s + t) = \gamma(s)\gamma(t), \qquad \gamma(0) = \mathbb{1}, \qquad \dot{\gamma}(0) = \xi, \qquad (2.5.9)$$

then $\gamma(t) = \exp(t\xi)$ for every $t \in \mathbb{R}$.

Proof For every $t \in \mathbb{R}$ we have

$$\dot{\gamma}(t) = \frac{d}{ds}\bigg|_{s=0} \gamma(s + t) = \frac{d}{ds}\bigg|_{s=0} \gamma(s)\gamma(t) = \dot{\gamma}(0)\gamma(t) = \xi\gamma(t).$$

Hence γ is the integral curve of the vector field X_ξ in (2.5.8) with $\gamma(0) = \mathbb{1}$. This implies $\gamma(t) = \exp(t\xi)$ for every $t \in \mathbb{R}$, as claimed. $\square$

Example 2.5.11 Since the general linear group $GL(n, \mathbb{R})$ is an open subset of $\mathbb{R}^{n \times n}$ its Lie algebra is the space of all real $n \times n$-matrices

$$\mathfrak{gl}(n, \mathbb{R}) := \text{Lie}(GL(n, \mathbb{R})) = \mathbb{R}^{n \times n}.$$

The Lie algebra of the special linear group is

$$\mathfrak{sl}(n, \mathbb{R}) := \text{Lie}(SL(n, \mathbb{R})) = \{\xi \in \mathfrak{gl}(n, \mathbb{R}) \,|\, \text{trace}(\xi) = 0\}$$

(see Exercise 2.2.9) and the Lie algebra of the special orthogonal group is

$$\mathfrak{so}(n) := \text{Lie}(SO(n)) = \{\xi \in \mathfrak{gl}(n, \mathbb{R}) \,|\, \xi^\mathsf{T} + \xi = 0\} = \mathfrak{o}(n)$$

(see Example 2.2.10).

Exercise 2.5.12 Prove that the Lie algebras of the general linear group over $\mathbb{C}$, the special linear group over $\mathbb{C}$, the unitary group, and the special unitary group are given by

$$\mathfrak{gl}(n, \mathbb{C}) := \text{Lie}(GL(n, \mathbb{C})) = \mathbb{C}^{n \times n},$$
$$\mathfrak{sl}(n, \mathbb{C}) := \text{Lie}(SL(n, \mathbb{C})) = \{\xi \in \mathfrak{gl}(n, \mathbb{C}) \,|\, \text{trace}(\xi) = 0\},$$
$$\mathfrak{u}(n) := \text{Lie}(U(n)) = \{\xi \in \mathfrak{gl}(n, \mathbb{R}) \,|\, \xi^* + \xi = 0\},$$
$$\mathfrak{su}(n) := \text{Lie}(SU(n)) = \{\xi \in \mathfrak{gl}(n, \mathbb{C}) \,|\, \xi^* + \xi = 0, \text{ trace}(\xi) = 0\}.$$

These are vector spaces over the reals. Determine their real dimensions. Which of these are also complex vector spaces?

Remark 2.5.13 Let $G \subset GL(n, \mathbb{R})$ be a subgroup. In Theorem 2.5.27 below it is shown that G is a Lie group if and only if it is a closed subset of $GL(n, \mathbb{R})$ in the relative topology. This observation can be used in many of the examples and exercises of the present section.

Exercise 2.5.14 Let V be a finite-dimensional vector space. Prove that the vector space $\mathfrak{g} := V \times \mathrm{End}(V)$ is a Lie algebra with the Lie bracket

$$[(u, A), (v, B)] := (Av - Bu, AB - BA) \qquad (2.5.10)$$

for $u, v \in V$ and $A, B \in \mathrm{End}(V)$. Find the corresponding Lie group. Find an embedding of $\mathfrak{g}$ into $\mathrm{End}(\mathbb{R} \times V)$ as a Lie subalgebra.

Exercise 2.5.15 Let (V, ω) be a $2n$-dimensional symplectic vector space, so $\omega : V \times V \to \mathbb{R}$ is a nondegenerate skew-symmetric bilinear form. The **Heisenberg algebra** of (V, ω) is the Lie algebra $\mathfrak{h} := V \times \mathbb{R}$ with the Lie bracket of two elements $(v, t), (v', t') \in V \times \mathbb{R}$ defined by

$$\big[(v, t), (v', t')\big] := \big(0, \omega(v, v')\big). \qquad (2.5.11)$$

Find a corresponding Lie group structure on $H = V \times \mathbb{R}$. Embed H as a Lie subgroup into $\mathrm{GL}(n + 2, \mathbb{R})$ and find a formula for the exponential map. **Hint:** Take $V = \mathbb{R}^n \times \mathbb{R}^n$ and $\omega\big((x, y), (x', y')\big) = \langle x, y' \rangle - \langle y, x' \rangle$ and define $(x, y, t) \cdot (x', y', t') := (x + x', y + y', t + t' + \langle x, y' \rangle)$.

2.5.3 Lie Group Homomorphisms

Let G, H be Lie groups and $\mathfrak{g}, \mathfrak{h}$ be Lie algebras. A **Lie group homomorphism** from G to H is a smooth map $\rho : G \to H$ that is a group homomorphism. A **Lie group isomorphism** is a bijective Lie group homomorphism whose inverse is also a Lie group homomorphism. A **Lie group automorphism** is a Lie group isomorphism from a Lie group to itself. A **Lie algebra homomorphism** from $\mathfrak{g}$ to $\mathfrak{h}$ is a linear map $\Phi : \mathfrak{g} \to \mathfrak{h}$ that preserves the Lie bracket. A **Lie algebra isomorphism** is a bijective Lie algebra homomorphism whose inverse is also a Lie algebra homomorphism. A **Lie algebra automorphism** is a Lie algebra isomorphism from a Lie algebra to itself.

Lemma 2.5.16 *Let G and H be Lie groups and denote their Lie algebras by $\mathfrak{g} := \mathrm{Lie}(G)$ and $\mathfrak{h} := \mathrm{Lie}(H)$. Let $\rho : G \to H$ be a Lie group homomorphism and denote its derivative at $\mathbb{1} \in G$ by*

$$\dot{\rho} := d\rho(\mathbb{1}) : \mathfrak{g} \to \mathfrak{h}.$$

Then $\dot{\rho}$ is a Lie algebra homomorphism. Moreover,

$$\rho(\exp(\xi)) = \exp(\dot{\rho}(\xi)), \qquad \rho(g\xi g^{-1}) = \rho(g)\dot{\rho}(\xi)\rho(g)^{-1}$$

for all $\xi \in \mathfrak{g}$ and all $g \in G$.

Proof The proof has three steps.

Step 1 *For all $\xi \in \mathfrak{g}$ and $t \in \mathbb{R}$ we have $\rho(\exp(t\xi)) = \exp(t\dot\rho(\xi))$.*

Fix an element $\xi \in \mathfrak{g}$. Then $\exp(t\xi) \in G$ for every $t \in \mathbb{R}$ by Lemma 2.5.9. Thus we can define a curve $\gamma : \mathbb{R} \to H$ by $\gamma(t) := \rho(\exp(t\xi))$. Since ρ is smooth, this is a smooth curve in H and, since ρ is a group homomorphism and the exponential map satisfies (2.5.6), our curve γ satisfies the conditions

$$\gamma(s+t) = \gamma(s)\gamma(t), \qquad \gamma(0) = \mathbb{1}, \qquad \dot\gamma(0) = d\rho(\mathbb{1})\xi = \dot\rho(\xi).$$

Hence $\gamma(t) = \exp(t\dot\rho(\xi))$ by Lemma 2.5.10. This proves Step 1.

Step 2 *For all $g \in G$ and $\eta \in \mathfrak{g}$ we have $\dot\rho(g\eta g^{-1}) = \rho(g)\dot\rho(\eta)\rho(g)^{-1}$.*

Define the smooth curve $\gamma : \mathbb{R} \to G$ by $\gamma(t) := g\exp(t\eta)g^{-1}$. It takes values in G by Lemma 2.5.9. By Step 1 we have

$$\rho(\gamma(t)) = \rho(g)\rho(\exp(t\eta))\rho(g)^{-1} = \rho(g)\exp(t\dot\rho(\eta))\rho(g)^{-1}$$

for every t. Since $\gamma(0) = \mathbb{1}$ and $\dot\gamma(0) = g\eta g^{-1}$ we obtain

$$\begin{aligned}
\dot\rho(g\eta g^{-1}) &= d\rho(\gamma(0))\dot\gamma(0) \\
&= \frac{d}{dt}\bigg|_{t=0} \rho(\gamma(t)) \\
&= \frac{d}{dt}\bigg|_{t=0} \rho(g)\exp(t\dot\rho(\eta))\rho(g)^{-1} \\
&= \rho(g)\dot\rho(\eta)\rho(g)^{-1}.
\end{aligned}$$

This proves Step 2.

Step 3 *For all $\xi, \eta \in \mathfrak{g}$ we have $\dot\rho([\xi,\eta]) = [\dot\rho(\xi), \dot\rho(\eta)]$.*

Define the curve $\eta : \mathbb{R} \to \mathfrak{g}$ by $\eta(t) := \exp(t\xi)\eta\exp(-t\xi)$ for $t \in \mathbb{R}$. It takes values in the Lie algebra of G by Lemma 2.5.9 and $\dot\eta(0) = [\xi, \eta]$. Hence

$$\begin{aligned}
\dot\rho([\xi,\eta]) &= \frac{d}{dt}\bigg|_{t=0} \dot\rho(\exp(t\xi)\eta\exp(-t\xi)) \\
&= \frac{d}{dt}\bigg|_{t=0} \rho(\exp(t\xi))\dot\rho(\eta)\rho(\exp(-t\xi)) \\
&= \frac{d}{dt}\bigg|_{t=0} \exp(t\dot\rho(\xi))\dot\rho(\eta)\exp(-t\dot\rho(\xi)) \\
&= [\dot\rho(\xi), \dot\rho(\eta)].
\end{aligned}$$

Here the first equality follows from the fact that $\dot{\rho}$ is linear, the second equality follows from Step 2 with $g = \exp(t\xi)$, and the third equality follows from Step 1. This proves Step 3 and Lemma 2.5.16. □

Exercise 2.5.17 A Lie group homomorphism $\rho : G \to H$ is uniquely determined by the Lie algebra homomorphism $\dot{\rho}$ whenever G is connected. **Hint:** If $\rho_1, \rho_2 : G \to H$ are Lie group homomorphisms such that $\dot{\rho}_1 = \dot{\rho}_2$, prove that the set $A := \{g \in G \mid \rho_1(g) = \rho_2(g)\}$ is both open and closed.

Exercise 2.5.18 If $\dot{\rho} : \mathfrak{g} \to \mathfrak{h}$ is a bijective Lie algebra homomorphism, then its inverse is also a Lie algebra homomorphism.

Exercise 2.5.19 If $\rho : G \to H$ is a bijective Lie group homomorphism, then $\rho^{-1} : H \to G$ is smooth and hence ρ is a Lie group isomorphism. **Hint:** Use Lemma 2.5.16 to prove that $\dot{\rho} : \mathfrak{g} \to \mathfrak{h}$ is injective. If $\dot{\rho}$ is not surjective, show that ρ has no regular value in contradiction to Sard's theorem.

Example 2.5.20 The complex determinant defines a Lie group homomorphism $\det : U(n) \to S^1$. The associated Lie algebra homomorphism is

$$\text{trace} = \dot{\det} : \mathfrak{u}(n) \to i\mathbb{R} = \text{Lie}(S^1).$$

Example 2.5.21 (Unit quaternions and SU(2)) The Lie group SU(2) is diffeomorphic to the 3-sphere. Every matrix in SU(2) can be written as

$$g = \begin{pmatrix} x_0 + ix_1 & x_2 + ix_3 \\ -x_2 + ix_3 & x_0 - ix_1 \end{pmatrix}, \qquad x_0^2 + x_1^2 + x_2^2 + x_3^2 = 1. \qquad (2.5.12)$$

Here the x_i are real numbers. They can be interpreted as the coordinates of a unit quaternion $x = x_0 + ix_1 + jx_2 + kx_3 \in Sp(1)$ (see Example 2.5.6). The reader may verify that the map $Sp(1) \to SU(2) : x \mapsto g$ in (2.5.12) is a Lie group isomorphism.

Exercise 2.5.22 (The double cover of SO(3)) Identify the imaginary part of $\mathbb{H}$ with $\mathbb{R}^3$ and write a vector $\xi \in \mathbb{R}^3 = \text{Im}(\mathbb{H})$ as a purely imaginary quaternion $\xi = i\xi_1 + j\xi_1 + k\xi_3$. Prove that if $\xi \in \text{Im}(\mathbb{H})$ and $x \in Sp(1)$, then $x\xi\bar{x} \in \text{Im}(\mathbb{H})$. Define the map $\rho : Sp(1) \to SO(3)$ by $\rho(x)\xi := x\xi\bar{x}$ for $x \in Sp(1)$ and $\xi \in \text{Im}(\mathbb{H})$. Prove that the linear map $\rho(x) : \mathbb{R}^3 \to \mathbb{R}^3$ is represented by the 3×3-matrix

$$\rho(x) = \begin{pmatrix} x_0^2 + x_1^2 - x_2^2 - x_3^2 & 2(x_1x_2 - x_0x_3) & 2(x_1x_3 + x_0x_2) \\ 2(x_1x_2 + x_0x_3) & x_0^2 + x_2^2 - x_3^2 - x_1^2 & 2(x_2x_3 - x_0x_1) \\ 2(x_1x_3 - x_0x_2) & 2(x_2x_3 + x_0x_1) & x_0^2 + x_3^2 - x_1^2 - x_2^2 \end{pmatrix}.$$

Show that ρ is a Lie group homomorphism. Find a formula for the map

$$\dot{\rho} := d\rho(\mathbb{1}) : \mathfrak{sp}(1) \to \mathfrak{so}(3)$$

and show that it is a Lie algebra isomorphism. For $x, y \in \mathrm{Sp}(1)$ prove that $\rho(x) = \rho(y)$ if and only if $y = \pm x$.

Example 2.5.23 Let $\mathfrak{g}$ be a finite-dimensional Lie algebra. Then the set

$$\mathrm{Aut}(\mathfrak{g}) := \left\{ \Phi : \mathfrak{g} \to \mathfrak{g} \,\middle|\, \begin{array}{l} \Phi \text{ is a bijective linear map and} \\ \Phi[\xi, \eta] = [\Phi\xi, \Phi\eta] \\ \text{for all } \xi, \eta \in \mathfrak{g} \end{array} \right\} \tag{2.5.13}$$

of **Lie algebra automorphisms** of $\mathfrak{g}$ is a Lie group. Its Lie algebra is the space of **derivations** on $\mathfrak{g}$ denoted by

$$\mathrm{Der}(\mathfrak{g}) := \left\{ \delta : \mathfrak{g} \to \mathfrak{g} \,\middle|\, \begin{array}{l} \delta \text{ is a linear map and} \\ \delta[\xi, \eta] = [\delta\xi, \eta] + [\xi, \delta\eta] \\ \text{for all } \xi, \eta \in \mathfrak{g} \end{array} \right\}. \tag{2.5.14}$$

Now suppose that $\mathfrak{g} = \mathrm{Lie}(G)$ is the Lie algebra of a Lie group G. Then there is a map $\mathrm{Ad} : G \to \mathrm{Aut}(\mathfrak{g})$ defined by

$$\mathrm{Ad}(g)\eta := g\eta g^{-1} \tag{2.5.15}$$

for $g \in G$ and $\eta \in \mathfrak{g}$. Part (ii) of Lemma 2.5.9 asserts that $\mathrm{Ad}(g)$ maps $\mathfrak{g}$ to itself for every $g \in G$. It follows directly from the definitions that the map $\mathrm{Ad}(g) : \mathfrak{g} \to \mathfrak{g}$ is a Lie algebra automorphism for every $g \in G$ and that the map $\mathrm{Ad} : G \to \mathrm{Aut}(\mathfrak{g})$ is a Lie group homomorphism. The associated Lie algebra homomorphism is the linear map $\mathrm{ad} : \mathfrak{g} \to \mathrm{Der}(\mathfrak{g})$ defined by

$$\mathrm{ad}(\xi)\eta := [\xi, \eta] \tag{2.5.16}$$

for $\xi, \eta \in \mathfrak{g}$. To verify the equation $\mathrm{ad} = \dot{\mathrm{Ad}}$ we compute

$$\dot{\mathrm{Ad}}(\xi)\eta = \left.\frac{d}{dt}\right|_{t=0} \mathrm{Ad}(\exp(t\xi))\eta = \left.\frac{d}{dt}\right|_{t=0} \exp(t\xi)\eta\exp(-t\xi) = [\xi, \eta].$$

Exercise 2.5.24 Let $\mathfrak{g}$ be any Lie algebra. Define the map $\mathrm{ad} : \mathfrak{g} \to \mathrm{End}(\mathfrak{g})$ by (2.5.16) and prove that the endomorphism $\mathrm{ad}(\xi) : \mathfrak{g} \to \mathfrak{g}$ is a derivation for every $\xi \in \mathfrak{g}$. Prove that $\mathrm{ad} : \mathfrak{g} \to \mathrm{Der}(\mathfrak{g})$ is a Lie algebra homomorphism.

Exercise 2.5.25 Let $\mathfrak{g}$ be any finite-dimensional Lie algebra. Prove that the group $\mathrm{Aut}(\mathfrak{g})$ in (2.5.13) is a Lie subgroup of $\mathrm{GL}(\mathfrak{g})$ with the Lie algebra $\mathrm{Lie}(\mathrm{Aut}(\mathfrak{g})) = \mathrm{Der}(\mathfrak{g})$. **Hint:** Show that a linear map $\delta : \mathfrak{g} \to \mathfrak{g}$ is a derivation if and only if the linear map $\exp(t\delta) : \mathfrak{g} \to \mathfrak{g}$ is a Lie algebra automorphism for every $t \in \mathbb{R}$. Use the Closed Subgroup Theorem 2.5.27.

2.5.4 Closed Subgroups

This section deals with subgroups of a Lie group G that are also submanifolds of G. Such subgroups are called Lie subgroups. We assume throughout that $G \subset GL(n, \mathbb{R})$ is a Lie group with the Lie algebra $\mathfrak{g} := \mathrm{Lie}(G) = T_1 G$.

Definition 2.5.26 (Lie subgroup) A subset $H \subset G$ is called a **Lie subgroup of** G iff it is both a subgroup and a smooth submanifold of G.

A useful general criterion is the Closed Subgroup Theorem which asserts that a subgroup $H \subset G$ is a Lie subgroup if and only if it is a closed subset of G. This was first proved in 1929 by John von Neumann [54] for the special case $G = GL(n, \mathbb{R})$ and then in 1930 by Élie Cartan [15] in full generality.

Theorem 2.5.27 (Closed Subgroup Theorem) *Let* H *be a subgroup of* G. *Then the following are equivalent.*

(i) H *is a smooth submanifold (and hence a Lie subgroup) of* G.
(ii) H *is a closed subset of* G.

If (i) holds, then the Lie algebra of H *is the space*

$$\mathfrak{h} := \{\eta \in \mathfrak{g} \mid \exp(t\eta) \in H \text{ for all } t \in \mathbb{R}\}. \tag{2.5.17}$$

Proof of Theorem 2.5.27 (i) $\Longrightarrow$ ***(ii) and*** *(2.5.17)* Assume that H is a Lie subgroup of G and let $\mathfrak{h} \subset \mathfrak{g}$ be defined by (2.5.17). We prove that $\mathfrak{h}$ is the Lie algebra of H. Assume first that $\eta \in \mathfrak{h}$. Then the curve $\gamma : \mathbb{R} \to G$ defined by $\gamma(t) := \exp(t\eta)$ for $t \in \mathbb{R}$ takes values in H and satisfies $\gamma(0) = 1$ and $\dot{\gamma}(0) = \eta$, and this implies $\eta \in T_1 H = \mathrm{Lie}(H)$. Conversely, if $\eta \in \mathrm{Lie}(H)$, then Lemma 2.5.9 asserts that $\exp(t\eta) \in H$ for all $t \in \mathbb{R}$ and hence $\eta \in \mathfrak{h}$. This shows that $\mathfrak{h} = \mathrm{Lie}(H)$.

Next we prove in three steps that H is a closed subset of G. Choose any inner product on $\mathfrak{g}$, denote by $|\cdot|$ the associated norm, and denote by $\mathfrak{h}^\perp \subset \mathfrak{g}$ the orthogonal complement of $\mathfrak{h}$ with respect to this inner product.

Step 1 *There exist open neighborhoods $V \subset H$ of 1 and $W \subset \mathfrak{h}^\perp$ of the origin such that the map $\phi : V \times W \to G$, defined by*

$$\phi(h, \xi) := h \exp(\xi)$$

for $h \in V$ and $\xi \in W$, is a diffeomorphism from $V \times W$ onto an open neighborhood $U = \phi(V \times W) \subset G$ of 1.

The derivative of the map $H \times \mathfrak{h}^\perp \to G : (h, \xi) \mapsto h \exp(\xi)$ at the point $(1, 0)$ is bijective. Hence Step 1 follows from the Inverse Function Theorem.

Step 2 *There exists a $\delta > 0$ such that, if $\xi, \xi' \in \mathfrak{h}^{\perp}$ satisfy $|\xi|, |\xi'| < \delta$ and* $\exp(\xi') \exp(-\xi) \in H$, *then $\xi = \xi'$.*

Let V, W, ϕ be as in Step 1, choose an open neigborhood $V' \subset G$ of $\mathbb{1}$ such that $V' \cap H = V$, and choose a constant $\delta > 0$ such that the following holds.

(a) If $\xi \in \mathfrak{h}^{\perp}$ satisfies $|\xi| < \delta$, then $\xi \in W$.
(b) If $\xi, \xi' \in \mathfrak{g}$ satisfy $|\xi|, |\xi'| < \delta$, then $\exp(\xi') \exp(-\xi) \in V'$.

Let $\xi, \xi' \in \mathfrak{h}^{\perp}$ such that $|\xi|, |\xi'| < \delta$ and $h := \exp(\xi') \exp(-\xi) \in H$. Then we have $\xi, \xi' \in W$ by (a) and $h \in V' \cap H = V$ by (b). Also $\phi(h, \xi) = \phi(\mathbb{1}, \xi')$ and so $\xi = \xi'$, because ϕ is injective on $V \times W$. This proves Step 2.

Step 3 *Let h_i be a sequence in H that converges to an element $g \in G$. Then $g \in H$.*

Let $\phi : V \times W \to U$ be as in Step 1 and let $\delta > 0$ be as in Step 2. Since the sequence $h_i^{-1} g$ converges to $\mathbb{1}$, there exists an $i_0 \in \mathbb{N}$ such that $h_i^{-1} g \in U$ for all $i \geq i_0$. Hence, for each $i \geq i_0$, there exists a unique pair $(h_i', \xi_i) \in V \times W$ such that $h_i^{-1} g = h_i' \exp(\xi_i)$. This sequence satisfies $\lim_{i \to \infty} \xi_i = 0$. Hence there exists an integer $i_1 \geq i_0$ such that $|\xi_i| < \delta$ for all $i \geq i_1$. Since

$$h_i h_i' \exp(\xi_i) = g = h_j h_j' \exp(\xi_j),$$

we also have $\exp(\xi_i) \exp(-\xi_j) = (h_i h_i')^{-1} h_j h_j' \in H$ for all $i, j \geq i_1$. By Step 3, this implies $\xi_i = \xi_j$ for all $i, j \geq i_1$. Hence $\xi_i = \lim_{j \to \infty} \xi_j = 0$ for all $i \geq i_1$ and so $g = h_i h_i' \in H$. This proves Step 3.

By Step 3 the Lie subgroup H is a closed subset of G. Thus we have proved that (i) implies (ii) and (2.5.17) in Theorem 2.5.27. $\square$

The proof of the converse implication requires three preparatory lemmas.

Lemma 2.5.28 *Let $\xi \in \mathfrak{g}$ and let $\gamma : \mathbb{R} \to G$ be a curve that is differentiable at $t = 0$ and satisfies $\gamma(0) = \mathbb{1}$ and $\dot{\gamma}(0) = \xi$. Then*

$$\exp(t\xi) = \lim_{k \to \infty} \gamma(t/k)^k \qquad (2.5.18)$$

for every $t \in \mathbb{R}$.

Proof Fix a nonzero real number t and define $\xi_k := k\big(\gamma(t/k) - \mathbb{1}\big) \in \mathbb{R}^{n \times n}$ for $k \in \mathbb{N}$. Then

$$\lim_{k \to \infty} \xi_k = t \lim_{k \to \infty} \frac{\gamma(t/k) - \gamma(0)}{t/k} = t\dot{\gamma}(0) = t\xi$$

and hence

$$\exp(t\xi) = \lim_{k\to\infty} \left(1 + \frac{\xi_k}{k}\right)^k = \lim_{k\to\infty} \gamma(t/k)^k.$$

(See [64, Satz 1.5.2].) This proves Lemma 2.5.28.   □

Lemma 2.5.29 *Let* $H \subset G$ *be a closed subgroup. Then the set* $\mathfrak{h}$ *in* (2.5.17) *is a Lie subalgebra of* $\mathfrak{g}$

Proof Let $\xi, \eta \in \mathfrak{h}$ and define the curve $\gamma : \mathbb{R} \to H$ by

$$\gamma(t) := \exp(t\xi)\exp(t\eta)$$

for $t \in \mathbb{R}$. This curve is smooth and satisfies $\gamma(0) = \mathbb{1}$ and $\dot\gamma(0) = \xi + \eta$. Since H is closed, it follows from Lemma 2.5.28 that

$$\exp(t(\xi + \eta)) = \lim_{k\to\infty} \gamma(t/k)^k \in H$$

for all $t \in \mathbb{R}$ and so $\xi + \eta \in \mathfrak{h}$ by definition. Thus $\mathfrak{h}$ is a linear subspace of $\mathfrak{g}$.
  Now fix an element $\xi \in \mathfrak{h}$. If $h \in H$, then

$$\exp(sh^{-1}\xi h) = h^{-1}\exp(s\xi)h \in H$$

for all $s \in \mathbb{R}$ and hence $h^{-1}\xi h \in \mathfrak{h}$ by definition. Take $h = \exp(t\eta)$ with $\eta \in \mathfrak{h}$ to obtain $\exp(-t\eta)\xi \exp(t\eta) \in \mathfrak{h}$ for all $t \in \mathbb{R}$. Differentiating this curve at $t = 0$ gives $[\xi, \eta] \in \mathfrak{h}$ and this proves Lemma 2.5.29.   □

Lemma 2.5.30 *Let* $H \subset G$ *be a closed subgroup and let* $\mathfrak{h} \subset \mathfrak{g}$ *be the Lie subalgebra in* (2.5.17). *Let* $\xi \in \mathfrak{g}$, *let* $(\xi_i)_{i\in\mathbb{N}}$ *be a sequence in* $\mathfrak{g}$, *and let* $(\tau_i)_{i\in\mathbb{N}}$ *be a sequence of positive real numbers such that*

$$\exp(\xi_i) \in H, \qquad \xi_i \neq 0$$

for all $i \in \mathbb{N}$ *and*

$$\lim_{i\to\infty} \tau_i = 0, \qquad \lim_{i\to\infty} \xi_i = 0, \qquad \lim_{i\to\infty} \frac{\xi_i}{\tau_i} = \xi.$$

Then $\xi \in \mathfrak{h}$.

Proof Fix a real number t. Then, for each $i \in \mathbb{N}$, there exists a unique integer $m_i \in \mathbb{Z}$ such that $m_i\tau_i \leq t < (m_i + 1)\tau_i$. The sequence m_i satisfies

$$\lim_{i\to\infty} m_i\tau_i = t, \qquad \lim_{i\to\infty} m_i\xi_i = \lim_{i\to\infty} m_i\tau_i\frac{\xi_i}{\tau_i} = t\xi$$

and hence

$$\exp(t\xi) = \lim_{i\to\infty} \exp(m_i\xi_i) = \lim_{i\to\infty} \exp(\xi_i)^{m_i} \in H.$$

Thus $\exp(t\xi) \in H$ for every $t \in \mathbb{R}$ and so $\xi \in \mathfrak{h}$ by (2.5.17). This proves Lemma 2.5.30.

$\square$

Proof of Theorem 2.5.27 (ii) $\Longrightarrow$ ***(i)***   Choose any inner product on $\mathfrak{g}$. Let $H \subset G$ be a closed subgroup of G and define the set $\mathfrak{h} \subset \mathfrak{g}$ by (2.5.17). Then $\mathfrak{h}$ is a Lie subalgebra of $\mathfrak{g}$ by Lemma 2.5.29. Define

$$k := \dim(\mathfrak{h}), \qquad \ell := \dim(\mathfrak{g}) \geq k,$$

and choose a basis $\eta_1, \ldots, \eta_\ell$ of $\mathfrak{g}$ such that the vectors $\eta_1, \ldots, \eta_k$ form a basis of $\mathfrak{h}$ and $\eta_\nu \in \mathfrak{h}^\perp$ for $\nu > k$. Let $h_0 \in H$ and define the map $\Theta : \mathbb{R}^\ell \to G$ by

$$\Theta(t^1, \ldots, t^\ell) := h_0 \exp(t^1\eta_1 + \cdots + t^k\eta_k)\exp(t^{k+1}\eta_{k+1} + \cdots + t^\ell\eta_\ell).$$

Then $\Theta(0) = h_0$, $\Theta(\mathbb{R}^k \times \{0\}) \subset H$, and the derivative $d\Theta(0) : \mathbb{R}^\ell \to T_{h_0}G$ is bijective. Hence the inverse function theorem asserts that Θ restricts to a diffeomorphism from an open neighborhood $\Omega \subset \mathbb{R}^\ell$ of the origin to an open neighborhood $U := \Theta(\Omega) \subset G$ of h_0 that satisfies

$$\Theta(0) = h_0, \qquad \Theta\big(\Omega \cap (\mathbb{R}^k \times \{0\})\big) \subset U \cap H.$$

We prove the following.

Claim *There exists an open set $\Omega_0 \subset \mathbb{R}^\ell$ such that*

$$0 \in \Omega_0 \subset \Omega, \qquad \Theta\big(\Omega_0 \cap (\mathbb{R}^k \times \{0\})\big) = U_0 \cap H, \qquad U_0 := \Theta(\Omega_0). \quad (2.5.19)$$

Assume, by contradiction, that such an open set Ω_0 does not exist. Then there exists a sequence $t_i = (t_i^1, \ldots, t_i^\ell) \in \mathbb{R}^\ell$ such that

$$\lim_{i\to\infty} t_i = 0, \qquad t_i \in \Omega \setminus (\mathbb{R}^k \times \{0\}), \qquad \Theta(t_i) \in H.$$

Define

$$h_i := h_0 \exp\left(\sum_{\nu=1}^k t_i^\nu \eta_\nu\right) \in H, \qquad \xi_i := \sum_{\nu=k+1}^\ell t_i^\nu \eta_\nu \in \mathfrak{h}^\perp \setminus \{0\}.$$

Then $h_i \exp(\xi_i) = \Theta(t_i) \in H$ and hence

$$\lim_{i\to\infty} \xi_i = 0, \qquad \xi_i \neq 0, \qquad \exp(\xi_i) = h_i^{-1}\Theta(t_i) \in H.$$

Passing to a subsequence, if necessary, we may assume that the sequence $\xi_i/|\xi_i|$ converges. Denote its limit by $\xi := \lim_{i\to\infty} \xi_i/|\xi_i|$. Then $\xi \in \mathfrak{h}$ by Lemma 2.5.30 and $\xi \in \mathfrak{h}^\perp$ by definition. Since $|\xi| = 1$, this is a contradiction. This contradiction proves the Claim. Thus there does, after all, exist an open set $\Omega_0 \subset \mathbb{R}^\ell$ that satisfies (2.5.19), and the map $\Theta^{-1} : U_0 \to \Omega_0$ is then a coordinate chart on G which satisfies $\Theta^{-1}(U_0 \cap H) = \Omega_0 \cap (\mathbb{R}^k \times \{0\})$. Hence H is a submanifold of G and this proves Theorem 2.5.27. □

Exercise 2.5.31   The subgroup $\{\exp(it) \mid t \in \mathbb{Q}\} \subset S^1$ is not closed.

Exercise 2.5.32   Choose a nonzero vector $(\omega_1, \dots, \omega_n) \in \mathbb{R}^n$ such that at least one of the ratios ω_i/ω_j is irrational. Prove that the subgroup

$$S_\omega := \{(e^{2\pi it\omega_1}, e^{2\pi it\omega_2}, \dots, e^{2\pi it\omega_n}) \mid t \in \mathbb{R}\} \subset (S^1)^n \cong \mathbb{T}^n$$

of the torus is not closed. Similar examples exist in any Lie group that contains a torus of dimension at least two.

Exercise 2.5.33   Let G_0 and G_1 be Lie subgroups of $GL(n, \mathbb{R})$ with the Lie algebras $\mathfrak{g}_0 := \mathrm{Lie}(G_0)$ and $\mathfrak{g}_1 := \mathrm{Lie}(G_1)$. Prove that $G := G_0 \cap G_1$ is a Lie subgroup of $GL(n, \mathbb{R})$ with the Lie algebra $\mathfrak{g} = \mathfrak{g}_0 \cap \mathfrak{g}_1$.

Exercise 2.5.34 (Center)   The **center of a group** G is the subgroup

$$Z(G) := \{g \in G \mid gh = hg \text{ for all } h \in G\}. \tag{2.5.20}$$

Let $G \subset GL(n, \mathbb{R})$ be a Lie group. Prove that its center $Z(G)$ is a Lie subgroup of G. If G is connected, prove that the Lie algebra of the center $Z(G)$ is the **center of the Lie algebra** $\mathfrak{g} = \mathrm{Lie}(G)$, defined by

$$Z(\mathfrak{g}) := \{\xi \in \mathfrak{g} \mid [\xi, \eta] = 0 \text{ for all } \eta \in \mathfrak{g}\}. \tag{2.5.21}$$

Hint: If G is connected, prove that an element $\xi \in \mathfrak{g}$ satisfies $[\xi, \eta] = 0$ for all $\eta \in \mathfrak{g}$ if and only if $\exp(t\xi)h = h\exp(t\xi)$ for all $t \in \mathbb{R}$ and all $h \in G$.

Exercise 2.5.35   Let $G \subset GL(n, \mathbb{R})$ be a compact Lie group with the Lie algebra $\mathfrak{g} := \mathrm{Lie}(G)$ and let $\xi \in \mathfrak{g}$. Prove that the set $T_\xi := \{\exp(t\xi) \mid t \in \mathbb{R}\}$ is a closed, connected, abelian subgroup of G and deduce that it is a Lie subgroup of G (called the **torus generated by** ξ).

Exercise 2.5.36   Let $G \subset GL(n, \mathbb{R})$ be a Lie group with the Lie algebra $\mathfrak{g}$ and let $\xi : \mathbb{R} \to \mathfrak{g}$ be a smooth function. Prove that the differential equation $\dot{\gamma}(t) = \xi(t)\gamma(t)$, $\gamma(0) = \mathbb{1}$, has a unique solution $\gamma : \mathbb{R} \to G$. **Hint:** Prove the existence of a solution $\gamma : \mathbb{R} \to GL(n, \mathbb{R})$ and show that the set $\{t \in \mathbb{R} \mid \gamma(t) \in G\}$ is open and closed.

Remark 2.5.37 (Malcev's Theorem)   Let $G \subset GL(n, \mathbb{R})$ be a Lie group with the Lie algebra $\mathfrak{g}$ and let $\mathfrak{h} \subset \mathfrak{g}$ be a Lie subalgebra. Then the set

$$H := \left\{ h(1) \,\middle|\, \begin{array}{l} h : [0, 1] \to G \text{ is a smooth path} \\ \text{such that } h(0) = \mathbb{1} \text{ and} \\ \dot{h}(t)h(t)^{-1} \in \mathfrak{h} \text{ for all } t \in [0, 1] \end{array} \right\} \qquad (2.5.22)$$

is a subgroup of G, called the **integral subgroup of** $\mathfrak{h}$. A theorem by Anatolij Ivanovich Malcev [47] (see also [28, Corollary 13.4.6]) asserts that H is a Lie subgroup of G if and only if $T_\eta := \overline{\{\exp(t\eta) \,|\, t \in \mathbb{R}\}} \subset H$ for all $\eta \in \mathfrak{h}$.

2.5.5   Lie Groups and Diffeomorphisms

There is a natural correspondence between Lie groups and Lie algebras on the one hand and diffeomorphisms and vector fields on the other hand. We summarize this correspondence in the following table.

Lie groups	**Diffeomorphisms**
$G \subset GL(n, \mathbb{R})$	$\mathrm{Diff}(M)$
$\mathfrak{g} = \mathrm{Lie}(G) = T_\mathbb{1}G$	$\mathrm{Vect}(M) = T_{\mathrm{id}}\mathrm{Diff}(M)$
exponential map	flow of a vector field
$t \mapsto \exp(t\xi)$	$t \mapsto \phi^t = \text{``}\exp(tX)\text{''}$
adjoint representation	pushforward
$\xi \mapsto g\xi g^{-1}$	$X \mapsto \phi_* X$
Lie bracket on $\mathfrak{g}$	Lie bracket of vector fields
$[\xi, \eta] = \xi\eta - \eta\xi$	$[X, Y] = dX \cdot Y - dY \cdot X$

To understand the correspondence between the exponential map and the flow of a vector field compare equation (2.4.6) with equation (2.5.5). To understand the correspondence between the adjoint representation and pushforward observe that

$$\phi_* Y = \frac{d}{dt}\bigg|_{t=0} \phi \circ \psi^t \circ \phi^{-1}, \qquad g\eta g^{-1} = \frac{d}{dt}\bigg|_{t=0} g\exp(t\eta)g^{-1},$$

where ψ^t denotes the flow of Y. To understand the correspondence between the Lie brackets recall that

$$[X, Y] = \frac{d}{dt}\bigg|_{t=0} (\phi^t)_* Y, \qquad [\xi, \eta] = \frac{d}{dt}\bigg|_{t=0} \exp(t\xi)\eta \exp(-t\xi),$$

where ϕ^t denotes the flow of X. We emphasize that the analogy between Lie groups and Diffeomorphisms only works well when the manifold M is compact so that every vector field on M is complete. The next exercise gives another parallel between the Lie bracket on the Lie algebra of a Lie group and the Lie bracket of two vector fields.

Exercise 2.5.38 Let $G \subset GL(n, \mathbb{R})$ be a Lie group with Lie algebra $\mathfrak{g}$ and let $\xi, \eta \in \mathfrak{g}$. Define the smooth curve $\gamma : \mathbb{R} \to G$ by

$$\gamma(t) := \exp(t\xi) \exp(t\eta) \exp(-t\xi) \exp(-t\eta).$$

Show that $\dot{\gamma}(0) = 0$ and $\frac{1}{2}\ddot{\gamma}(0) = [\xi, \eta]$ (cf. Exercise 2.5.8 and Lemma 2.4.18).

Exercise 2.5.39 Let $G \subset GL(n, \mathbb{R})$ be a Lie group with Lie algebra $\mathfrak{g}$ and let $\xi, \eta \in \mathfrak{g}$. Show that $[\xi, \eta] = 0$ if and only if the exponential maps commute, i.e. $\exp(s\xi) \exp(t\eta) = \exp(t\eta) \exp(s\xi) = \exp(s\xi + t\eta)$ for all $s, t \in \mathbb{R}$. How can this observation be deduced from Lemma 2.4.26?

Definition 2.5.40 Let $M \subset \mathbb{R}^k$ be a smooth manifold and let $G \subset GL(n, \mathbb{R})$ be a Lie group. A **(smooth) group action** of G on M is a smooth map

$$G \times M \to M : (g, p) \mapsto \phi_g(p) \tag{2.5.23}$$

that for each pair $g, h \in G$ satisfies the condition

$$\phi_g \circ \phi_h = \phi_{gh}, \qquad \phi_{\mathbb{1}} = \mathrm{id}. \tag{2.5.24}$$

If (2.5.23) is a smooth group action, then the **infinitesimal action** of the Lie algebra $\mathfrak{g} := \mathrm{Lie}(G)$ on M is the map $\mathfrak{g} \to \mathrm{Vect}(M) : \xi \mapsto X_\xi$ defined by

$$X_\xi(p) := \left.\frac{d}{dt}\right|_{t=0} \phi_{\exp(y\xi)}(p) \tag{2.5.25}$$

for $\xi \in \mathfrak{g}$ and $p \in M$.

Exercise 2.5.41 Let (2.5.23) be a smooth group action of a Lie group G on a manifold M. Prove that

$$X_{g^{-1}\xi g} = \phi_g^* X_\xi, \qquad X_{[\xi,\eta]} = [X_\xi, X_\eta] \tag{2.5.26}$$

for all $g \in G$ and all $\xi, \eta \in \mathfrak{g} = \mathrm{Lie}(G)$.

Exercise 2.5.42 Show that the maps $GL(m, \mathbb{R}) \times \mathbb{R}^m \to \mathbb{R}^m : (gx) \mapsto gx$, $SO(m + 1) \times S^m \to S^m : (g, x) \mapsto gx$, and $\mathbb{R} \times S^1 \to S^1 : (\theta, z) \mapsto e^{i\theta}z$ are smooth group actions. Verify the formulas in (2.5.26) in these examples.

A smooth group action of a Lie group G on a manifold M can the thought of as a *"Lie group homomorphism"*

$$G \to \mathrm{Diff}(M) : g \mapsto \phi_g. \tag{2.5.27}$$

While the group $\mathrm{Diff}(M)$ is infinite-dimensional, and so cannot cannot be a Lie group in the formal sense, it has many properties in common with Lie groups as explained above. For example, one can define what is meant by a smooth path in $\mathrm{Diff}(M)$ and extend formally the notion of a tangent vector (as the derivative of a path through a given element of $\mathrm{Diff}(M)$) to this setting. In particular, the tangent space of $\mathrm{Diff}(M)$ at the identity can then be identified with the space of vector fields $T_{\mathrm{id}}\mathrm{Diff}(M) = \mathrm{Vect}(M)$, and the infinitesimal action in (2.5.25) is the Lie algebra homomorphism associated to the *"Lie group homomorphism"* (2.5.27). In fact, we have chosen the sign in the definition of the Lie bracket of vector fields so that the map $\xi \mapsto X_\xi$ is a Lie algebra homomorphism and not a Lie algebra anti-homomorphism. This will be discussed further in Sect. 2.5.7.

2.5.6　Smooth Maps and Algebra Homomorphisms

Let M be a smooth submanifold of $\mathbb{R}^k$. Denote by $\mathcal{F}(M) := C^\infty(M, \mathbb{R})$ the space of smooth real valued functions $f : M \to \mathbb{R}$. Then $\mathcal{F}(M)$ is a commutative unital algebra. Each $p \in M$ determines a unital algebra homomorphism $\varepsilon_p : \mathcal{F}(M) \to \mathbb{R}$ defined by $\varepsilon_p(f) = f(p)$ for $p \in M$.

Theorem 2.5.43 *Every unital algebra homomorphism $\varepsilon : \mathcal{F}(M) \to \mathbb{R}$ has the form $\varepsilon = \varepsilon_p$ for some $p \in M$.*

Proof Assume that $\varepsilon : \mathcal{F}(M) \to \mathbb{R}$ is an algebra homomorphism.

Claim *For all $f, g \in \mathcal{F}(M)$ we have $\varepsilon(g) = 0 \implies \varepsilon(f) \in f(g^{-1}(0))$.*

Indeed, the function $f - \varepsilon(f) \cdot 1$ lies in the kernel of ε and so the function $h := (f - \varepsilon(f) \cdot 1)^2 + g^2$ also lies in the kernel of ε. There must be at least one point $p \in M$ where $h(p) = 0$ for otherwise $1 = \varepsilon(h)\varepsilon(1/h) = 0$. For this point p we have $f(p) = \varepsilon(f)$ and $g(p) = 0$, hence $p \in g^{-1}(0)$, and therefore $\varepsilon(f) = f(p) \in f(g^{-1}(0))$. This proves the claim.

The theorem asserts that there exists a $p \in M$ such that every $f \in \mathcal{F}(M)$ satisfies $\varepsilon(f) = f(p)$. Assume, by contradiction, that this is false. Then for every $p \in M$ there exists a function $f \in \mathcal{F}(M)$ such that $f(p) \neq \varepsilon(f)$. Replace f by $f - \varepsilon(f)$ to obtain $f(p) \neq 0 = \varepsilon(f)$. Now use the axiom of choice to obtain a family of functions $f_p \in \mathcal{F}(M)$, one for every $p \in M$, such that $f_p(p) \neq 0 = \varepsilon(f_p)$ for all $p \in M$. Then the set $U_p := f_p^{-1}(\mathbb{R} \setminus \{0\})$ is an M-open neighborhood of p for every $p \in M$. Choose a sequence of compact sets $K_n \subset M$ such that $K_n \subset \mathrm{int}_M(K_{n+1})$ for all n and $M = \bigcup_n K_n$. Then, for each n, there is a $g_n \in \mathcal{F}(M)$ (a finite sum of the form $\sum_i f_{p_i}^2$) such that $\varepsilon(g_n) = 0$ and $g_n(q) > 0$ for all $q \in K_n$. If M is compact, this is already a contradiction because a positive function cannot belong to the kernel of ε. Otherwise, choose $f \in \mathcal{F}(M)$ such that $f(q) \geq n$ for all $q \in M \setminus K_n$ and all $n \in \mathbb{N}$. Then $\varepsilon(f) \in f(g_n^{-1}(0)) \subset f(M \setminus K_n) \subset [n, \infty)$ by the claim and so $\varepsilon(f) \geq n$ for all n. This is a contradiction and proves Theorem 2.5.43.　　　　　□

Now let N be another smooth submanifold (say of $\mathbb{R}^\ell$) and let $C^\infty(M, N)$ denote the space of smooth maps from M to N. A **homomorphism** from $\mathcal{F}(N)$ to $\mathcal{F}(M)$ is a (real) linear map $\Phi : \mathcal{F}(N) \to \mathcal{F}(M)$ that satisfies

$$\Phi(fg) = \Phi(f)\Phi(g), \qquad \Phi(1) = 1.$$

Let $\mathrm{Hom}(\mathcal{F}(N), \mathcal{F}(M))$ denote the space of homomorphisms from $\mathcal{F}(N)$ to $\mathcal{F}(M)$. An **automorphism** of the algebra $\mathcal{F}(M)$ is a bijective homomorphism $\Phi : \mathcal{F}(M) \to \mathcal{F}(M)$. The automorphisms of $\mathcal{F}(M)$ form a group denoted by $\mathrm{Aut}(\mathcal{F}(M))$.

Corollary 2.5.44 *The pullback operation*

$$C^\infty(M, N) \to \mathrm{Hom}(\mathcal{F}(N), \mathcal{F}(M)) : \phi \mapsto \phi^*$$

is bijective. In particular, the map $\mathrm{Diff}(M) \to \mathrm{Aut}(\mathcal{F}(M)) : \phi \mapsto \phi^*$ *is an anti-isomorphism of groups.*

Proof This is an exercise with hint. Let $\Phi : \mathcal{F}(N) \to \mathcal{F}(M)$ be a unital algebra homomorphism. By Theorem 2.5.43 there exists a map $\phi : M \to N$ such that $\varepsilon_p \circ \Phi = \varepsilon_{\phi(p)}$ for all $p \in M$. Prove that $f \circ \phi : M \to \mathbb{R}$ is smooth for every smooth map $f : N \to \mathbb{R}$ and deduce that ϕ is smooth. $\square$

Remark 2.5.45 The pullback operation is **functorial**, i.e.

$$(\psi \circ \phi)^* = \phi^* \circ \psi^*, \qquad \mathrm{id}_M^* = \mathrm{id}_{\mathcal{F}(M)}.$$

for $\phi \in C^\infty(M, N)$ and $\psi \in C^\infty(N, P)$. Here id denotes the identity map of the space indicated in the subscript. Hence Corollary 2.5.44 may be summarized by saying that the category of smooth manifolds and smooth maps is anti-isomorphic to a subcategory of the category of commutative unital algebras and unital algebra homomorphisms.

Exercise 2.5.46 If M is compact, then there is a slightly different way to prove Theorem 2.5.43. An **ideal** in $\mathcal{F}(M)$ is a linear subspace $\mathcal{J} \subset \mathcal{F}(M)$ satisfying the condition $f \in \mathcal{F}(M)$, $g \in \mathcal{J} \implies fg \in \mathcal{J}$. A **maximal ideal** in $\mathcal{F}(M)$ is an ideal $\mathcal{J} \subsetneq \mathcal{F}(M)$ such that every ideal $\mathcal{J}' \subsetneq \mathcal{F}(M)$ containing $\mathcal{J}$ is equal to $\mathcal{J}$. Prove that, if M is compact and $\mathcal{J} \subset \mathcal{F}(M)$ is an ideal with the property that for every $p \in M$ there is an $f \in \mathcal{J}$ with $f(p) \neq 0$, then $\mathcal{J} = \mathcal{F}(M)$. Deduce that each maximal ideal in $\mathcal{F}(M)$ has the form $\mathcal{J}_p := \{f \in \mathcal{F}(M) \mid f(p) = 0\}$ for some $p \in M$.

Exercise 2.5.47 If M is compact, give another proof of Corollary 2.5.44 as follows. The set $\Phi^{-1}(\mathcal{J}_p)$ is a maximal ideal in $\mathcal{F}(N)$ for each $p \in M$. Use Exercise 2.5.46 to deduce that there is a unique map $\phi : M \to N$ such that $\Phi^{-1}(\mathcal{J}_p) = \mathcal{J}_{\phi(p)}$ for all $p \in M$. Show that ϕ is smooth and $\phi^* = \Phi$.

Exercise 2.5.48 It is a theorem of ring theory that, when $I \subset R$ is an ideal in a ring R, the quotient ring R/I is a field if and only if the ideal I is maximal. Show that the kernel of the ring homomorphism $\varepsilon_p : \mathcal{F}(M) \to \mathbb{R}$ of Theorem 2.5.43 is the ideal $\mathfrak{d}_p$ of Exercise 2.5.46. Conclude that M is compact if and only if every maximal ideal $\mathfrak{d}$ in $\mathcal{F}(M)$ is of the form $\mathfrak{d} = \mathfrak{d}_p$ for some $p \in M$. **Hint:** The functions of compact support form an ideal. It can be shown that if M is not compact and $\mathfrak{d}$ is a maximal ideal containing all functions of compact support, then the quotient field $\mathcal{F}(M)/\mathfrak{d}$ is a non-Archimedean ordered field which properly contains $\mathbb{R}$.

2.5.7 Vector Fields and Derivations

A **derivation** of $\mathcal{F}(M)$ is a linear map $\delta : \mathcal{F}(M) \to \mathcal{F}(M)$ that satisfies

$$\delta(fg) = \delta(f)g + f\delta(g).$$

and the derivations form a Lie algebra denoted by $\mathrm{Der}(\mathcal{F}(M))$. We may think of $\mathrm{Der}(\mathcal{F}(M))$ as the Lie algebra of $\mathrm{Aut}(\mathcal{F}(M))$ with the Lie bracket given by the commutator. By Theorem 2.5.43 the pullback operation

$$\mathrm{Diff}(M) \to \mathrm{Aut}(\mathcal{F}(M)) : \phi \mapsto \phi^* \tag{2.5.28}$$

can be thought of as a Lie group anti-isomorphism. Differentiating it at the identity $\phi = \mathrm{id}$ gives a linear map

$$\mathrm{Vect}(M) \to \mathrm{Der}(\mathcal{F}(M)) : X \mapsto \mathcal{L}_X. \tag{2.5.29}$$

Here the operator $\mathcal{L}_X : \mathcal{F}(M) \to \mathcal{F}(M)$ is given by the derivative of a function f in the direction of the vector field X, i.e.

$$\mathcal{L}_X f := df \cdot X = \left.\frac{d}{dt}\right|_{t=0} f \circ \phi^t,$$

where ϕ^t denotes the flow of X. Since the map (2.5.29) is the derivative of the "Lie group" anti-homomorphism (2.5.28) we expect it to be a Lie algebra anti-homomorphism. Indeed, one can show that

$$\mathcal{L}_{[X,Y]} = \mathcal{L}_Y \mathcal{L}_X - \mathcal{L}_X \mathcal{L}_Y = -[\mathcal{L}_X, \mathcal{L}_Y] \tag{2.5.30}$$

for $X, Y \in \mathrm{Vect}(M)$. This confirms that our sign in the definition of the Lie bracket in Sect. 2.4.4 is consistent with the standard conventions in the theory of Lie groups. In the literature the difference between a vector field and the associated derivation $\mathcal{L}_X$ is sometimes neglected in the notation and many authors write $Xf := df \cdot X = \mathcal{L}_X f$, thus thinking of a vector field on a manifold M as an operator on the space of functions. With this notation one obtains the equation $[X, Y]f = Y(Xf) - X(Yf)$ and here lies the origin for the use of the opposite sign for the Lie bracket in many books on differential geometry.

Exercise 2.5.49   Prove that the map (2.5.29) is bijective. **Hint:** Fix a derivation $\delta \in \mathrm{Der}(\mathcal{F}(M))$ and prove the following. **Fact 1:** If $U \subset M$ is an open set and $f \in \mathcal{F}(M)$ vanishes on U, then $\delta(f)$ vanishes on U. **Fact 2:** If $p \in M$ and the derivative $df(p) : T_p M \to \mathbb{R}$ is zero, then $(\delta(f))(p) = 0$. (By Fact 1, the proof of Fact 2 can be reduced to an argument in local coordinates.)

Exercise 2.5.50   Verify the formula (2.5.30).

2.6   Vector Bundles and Submersions

This section characterizes submersions (Sect. 2.6.1) and introduces the concept of a smooth vector bundle in the extrinsic setting (Sect. 2.6.2).

2.6.1   Submersions

Let $M \subset \mathbb{R}^k$ be a smooth m-manifold and $N \subset \mathbb{R}^\ell$ be a smooth n-manifold. A smooth map $f : N \to M$ is called a **submersion** iff its derivative

$$df(q) : T_q N \to T_{f(q)} M$$

is surjective for every $q \in N$.

Lemma 2.6.1   *Let $M \subset \mathbb{R}^k$ be a smooth m-manifold, $N \subset \mathbb{R}^\ell$ be a smooth n-manifold, and $f : N \to M$ be a smooth map. The following are equivalent.*

(i)   *f is a submersion.*
(ii)   *For every $q_0 \in N$ there is an M-open neighborhood U of $p_0 := f(q_0)$ and a smooth map $g : U \to N$ such that $g(f(q_0)) = q_0$ and $f \circ g = \mathrm{id} : U \to U$. Thus f has a local right inverse near every point in N (see Fig. 2.12).*

Proof   We prove that (i) implies (ii). Since the derivative

$$df(q_0) : T_{q_0} N \to T_{p_0} M$$

is surjective we have $n \geq m$ and

$$\dim \ker df(q_0) = n - m.$$

Fig. 2.12 A local right inverse of a submersion

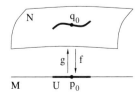

Hence there is a linear map $A : \mathbb{R}^{\ell} \to \mathbb{R}^{n-m}$ whose restriction to the kernel of $df(q_0)$ is bijective. Now define the map $\psi : N \to M \times \mathbb{R}^{n-m}$ by

$$\psi(q) := (f(q), A(q - q_0))$$

for $q \in N$. Then $\psi(q_0) = (p_0, 0)$ and its derivative

$$d\psi(q_0) : T_{q_0} N \to T_{p_0} M \times \mathbb{R}^{n-m}$$

sends $w \in T_{q_0} N$ to $(df(q_0)w, Aw)$ and is therefore bijective. Hence it follows from the inverse function theorem for manifolds (Theorem 2.2.17) that there exists an N-open neighborhood $V \subset N$ of q_0 such that the set

$$W := \psi(N) \subset M \times \mathbb{R}^{n-m}$$

is an open neighborhood of $(p_0, 0)$ and $\psi|_V : V \to W$ is a diffeomorphism. Let

$$U := \{ p \in M \mid (p, 0) \in W \}$$

and define the map $g : U \to N$ by

$$g(p) := \psi^{-1}(p, 0).$$

Then $p_0 \in U$, g is smooth and

$$(p, 0) = \psi(g(p)) = (f(g(p)), A(g(p) - q_0)).$$

Hence $f(g(p)) = p$ for all $p \in U$ and

$$g(p_0) = \psi^{-1}(p_0, 0) = q_0.$$

This shows that (i) implies (ii). The converse is an easy consequence of the chain rule and is left to the reader. This proves Lemma 2.6.1 $\square$

Corollary 2.6.2 *The image of a submersion* $f : N \to M$ *is open.*

Proof If $p_0 = f(q_0) \in f(N)$, then the neighborhood $U \subset M$ of p_0 in Lemma 2.6.1 (ii) is contained in the image of f. $\square$

Corollary 2.6.3 *If N is a nonempty compact manifold, M is a connected manifold, and $f : N \to M$ is a submersion, then f is surjective and M is compact.*

Proof The image $f(N)$ is an open subset of M by Corollary 2.6.2, it is a relatively closed subset of M because N is compact, and it is nonempty because N is nonempty. Since M is connected this implies that $f(N) = M$. In particular, M is compact. $\square$

Exercise 2.6.4   Let $f : N \to M$ be a smooth map. Prove that the sets $\{q \in N \mid df(q) \text{ is injective}\}$ and $\{q \in N \mid df(q) \text{ is surjective}\}$ are open (in the relative topology of N).

2.6.2   Vector Bundles

Let $M \subset \mathbb{R}^k$ be an m-dimensional smooth manifold.

Definition 2.6.5   A **(smooth) vector bundle (over M of rank n)** is a smooth submanifold $E \subset M \times \mathbb{R}^\ell$ of dimension $m + n$ such that, for every point $p \in M$, the set

$$E_p := \{v \in \mathbb{R}^\ell \mid (p, v) \in E\}$$

is an n-dimensional linear subspace of $\mathbb{R}^\ell$ (called the **fiber of E over p**). A vector bundle E over M is equipped with a smooth map

$$\pi : E \to M$$

defined by $\pi(p, v) := p$. This map is called the **canonical projection** of E. If $E \subset M \times \mathbb{R}^\ell$ is a vector bundle, then a **(smooth) section** of E is a smooth map $s : M \to \mathbb{R}^\ell$ such that $s(p) \in E_p$ for every $p \in M$.

A section $s : M \to \mathbb{R}^\ell$ of a vector bundle E over M determines a smooth map $\sigma : M \to E$ which sends the point $p \in M$ to the pair $(p, s(p)) \in E$. This map satisfies $\pi \circ \sigma = \text{id}$. It is sometimes convenient to abuse notation and eliminate the distinction between s and σ. Thus we will sometimes use the same letter s for the map from M to $\mathbb{R}^\ell$ and the map from M to E.

Definition 2.6.6   Let $M \subset \mathbb{R}^k$ be a smooth m-manifold. The set

$$TM := \{(p, v) \mid p \in M, v \in T_p M\}$$

is called the **tangent bundle** of M.

The tangent bundle is a subset of $M \times \mathbb{R}^k$ and, for every $p \in M$, its fiber $T_p M$ is an m-dimensional linear subspace of $\mathbb{R}^k$ by Theorem 2.2.3. However, it is not immediately obvious from the definition that TM is a submanifold of $M \times \mathbb{R}^k$. This will be proved below. The sections of TM are the vector fields on M.

Exercise 2.6.7   Let $f : M \to N$ be a smooth map between manifolds. Prove that the tangent map $TM \to TN : (p, v) \mapsto (f(p), df(p)v)$ is smooth.

Exercise 2.6.8   Let $M \subset \mathbb{R}^k$ be a smooth m-manifold and let $\phi : U \to \Omega$ be a smooth coordinate chart on an M-open set $U \subset M$ with values in an open set $\Omega = \phi(U) \subset \mathbb{R}^m$. Prove that the map $\widetilde{\phi} : TU \to \Omega \times \mathbb{R}^m$ defined by $\widetilde{\phi}(p, v) := (\phi(p), d\phi(p))$ for $p \in U$ and $v \in T_p M$ is a diffeomorphism. It is called a **standard coordinate chart on** TM. Deduce that TM is a smooth $2m$-dimensional submanifold of $M \times \mathbb{R}^k$ and hence is a smooth vector bundle over M. (See also Corollary 2.6.12 below.)

Exercise 2.6.9   Let $V \subset \mathbb{R}^\ell$ be an n-dimensional linear subspace. The **orthogonal projection** of $\mathbb{R}^\ell$ onto V is the matrix $\Pi \in \mathbb{R}^{\ell \times \ell}$ that satisfies

$$\Pi = \Pi^2 = \Pi^\mathsf{T}, \qquad \operatorname{im} \Pi = V. \tag{2.6.1}$$

Prove that there is a unique matrix $\Pi \in \mathbb{R}^{\ell \times \ell}$ satisfying (2.6.1). Prove that, for every symmetric matrix $S = S^\mathsf{T} \in \mathbb{R}^{\ell \times \ell}$, the kernel of S is the orthogonal complement of the image of S. If $D \in \mathbb{R}^{\ell \times n}$ is any injective matrix whose image is V, prove that $\det(D^\mathsf{T} D) \neq 0$ and

$$\Pi = D(D^\mathsf{T} D)^{-1} D^\mathsf{T}. \tag{2.6.2}$$

Theorem 2.6.10 (Vector bundles)   *Let $M \subset \mathbb{R}^k$ be a smooth m-manifold and let $E \subset M \times \mathbb{R}^\ell$ be a subset. Assume that, for every $p \in M$, the set*

$$E_p := \left\{ v \in \mathbb{R}^\ell \,|\, (p, v) \in E \right\} \tag{2.6.3}$$

is an n-dimensional linear subspace of $\mathbb{R}^\ell$. Let $\Pi : M \to \mathbb{R}^{\ell \times \ell}$ be the map that assigns to each $p \in M$ the orthogonal projection of $\mathbb{R}^\ell$ onto E_p, i.e.

$$\Pi(p) = \Pi(p)^2 = \Pi(p)^\mathsf{T}, \qquad \operatorname{im} \Pi(p) = E_p. \tag{2.6.4}$$

Then the following are equivalent.

(i)   *E is a vector bundle.*
(ii)   *For every $p_0 \in M$ and every $v_0 \in E_{p_0}$ there is a smooth map $s : M \to \mathbb{R}^\ell$ such that $s(p_0) = v_0$ and $s(p) \in E_p$ for all $p \in M$.*
(iii)   *The map $\Pi : M \to \mathbb{R}^{\ell \times \ell}$ is smooth.*
(iv)   *For every $p_0 \in M$ there is an open neighborhood $U \subset M$ of p_0 and a diffeomorphism $\pi^{-1}(U) \to U \times \mathbb{R}^n : (p, v) \mapsto \Phi(p, v) = (p, \Phi_p(v))$ such that the map $\Phi_p : E_p \to \mathbb{R}^n$ is an isometric isomorphism for all $p \in U$.*
(v)   *For every $p_0 \in M$ there is an open neighborhood $U \subset M$ of p_0 and a diffeomorphism $\pi^{-1}(U) \to U \times \mathbb{R}^n : (p, v) \mapsto \Phi(p, v) = (p, \Phi_p(v))$ such that the map $\Phi_p : E_p \to \mathbb{R}^n$ is a vector space isomorphism for all $p \in U$.*

Condition (i) implies that the projection $\pi : E \to M$ is a submersion. In (ii) the section s can be chosen to have compact support, i.e. there exists a compact subset $K \subset M$ such that $s(p) = 0$ for all $p \in M \setminus K$.

Before giving the proof of Theorem 2.6.10 we explain some of its consequences.

Definition 2.6.11 The maps $\Phi : \pi^{-1}(U) \to U \times \mathbb{R}^n$ in Theorem 2.6.10 are called **local trivializations** of E. They fit into commutative diagrams

Corollary 2.6.12 *Let $M \subset \mathbb{R}^k$ be a smooth m-manifold. Then TM is a vector bundle over M and hence is a smooth 2m-manifold in $\mathbb{R}^k \times \mathbb{R}^k$.*

Proof Let $\phi : U \to \Omega$ be a coordinate chart on an M-open set $U \subset M$ with values in an open subset $\Omega \subset \mathbb{R}^m$. Denote its inverse by $\psi := \phi^{-1} : \Omega \to M$. By Theorem 2.2.3 the linear map $d\psi(x) : \mathbb{R}^m \to \mathbb{R}^k$ is injective and its image is $T_{\psi(x)}M$ for every $x \in \Omega$. Hence the map $D : U \to \mathbb{R}^{k \times m}$ defined by

$$D(p) := d\psi(\phi(p)) \in \mathbb{R}^{k \times m}$$

is smooth and, for every $p \in U$, the linear map $D(p) : \mathbb{R}^m \to \mathbb{R}^k$ is injective and its image is T_pM. Thus the function $\Pi^{TM} : M \to \mathbb{R}^{k \times k}$ defined by (2.6.4) with $E_p = T_pM$ is given by

$$\Pi^{TM}(p) = D(p)\big(D(p)^\mathsf{T} D(p)\big)^{-1} D(p)^\mathsf{T} \qquad \text{for } p \in U.$$

Hence Π^{TM} is smooth and so TM is a vector bundle by Theorem 2.6.10. $\qquad \square$

Let $M \subset \mathbb{R}^k$ be an m-manifold, $N \subset \mathbb{R}^\ell$ be an n-manifold, $f : N \to M$ be a smooth map, and $E \subset M \times \mathbb{R}^d$ be a vector bundle. The **pullback bundle** is the vector bundle $f^*E \to N$ defined by

$$f^*E := \big\{(q, v) \in N \times \mathbb{R}^d \mid v \in E_{f(q)}\big\}$$

and the **normal bundle of E** is the vector bundle $E^\perp \to M$ defined by

$$E^\perp := \big\{(p, w) \in M \times \mathbb{R}^d \mid \langle v, w \rangle = 0 \ \forall \ v \in E_p\big\}.$$

Corollary 2.6.13 *The pullback and normal bundles are vector bundles.*

Proof Let $\Pi = \Pi^E : M \to \mathbb{R}^{d \times d}$ be the map defined by (2.6.4). This map is smooth by Theorem 2.6.10. Moreover, the corresponding maps for f^*E and $E^\perp$ are given by

$$\Pi^{f^*E} = \Pi^E \circ f : N \to \mathbb{R}^{d \times d}, \qquad \Pi^{E^\perp} = \mathbb{1} - \Pi^E : M \to \mathbb{R}^{d \times d}.$$

These maps are smooth and hence it follows from Theorem 2.6.10 that f^*E and $E^\perp$ are vector bundles. $\qquad \square$

Proof of Theorem 2.6.10 We first assume that E is a vector bundle and prove that $\pi : E \to M$ is a submersion. Let $\sigma : M \to E$ denote the zero section given by $\sigma(p) :=(p, 0)$. Then $\pi \circ \sigma = \mathrm{id}$ and hence it follows from the chain rule that the derivative $d\pi(p, 0) : T_{(p,0)}E \to T_p M$ is surjective. Now it follows from Exercise 2.6.4 that for every $p \in M$ there is an $\varepsilon > 0$ such that the derivative $d\pi(p, v) : T_{(p,v)}E \to T_p M$ is surjective for every $v \in E_p$ with $|v| < \varepsilon$. Consider the map $f_\lambda : E \to E$ defined by

$$f_\lambda(p, v) := (p, \lambda v).$$

This map is a diffeomorphism for every $\lambda > 0$. It satisfies

$$\pi = \pi \circ f_\lambda$$

and hence

$$d\pi(p, v) = d\pi(p, \lambda v) \circ df_\lambda(p, v) : T_{(p,v)}E \to T_p M.$$

Since $df_\lambda(p, v)$ is bijective and $d\pi(p, \lambda v)$ is surjective for $\lambda < \varepsilon/|v|$ it follows that $d\pi(p, v)$ is surjective for every $p \in M$ and every $v \in E_p$. Thus the projection $\pi : E \to M$ is a submersion for every vector bundle E over M.

We prove that (i) implies (ii). Let $p_0 \in M$ and $v_0 \in E_{p_0}$. We have already proved that π is a submersion. Hence it follows from Lemma 2.6.1 that there exists an M-open neighborhood $U \subset M$ of p_0 and a smooth map

$$\sigma_0 : U \to E$$

such that

$$\pi \circ \sigma_0 = \mathrm{id} : U \to U, \qquad \sigma_0(p_0) = (p_0, v_0).$$

Define the map $s_0 : U \to \mathbb{R}^\ell$ by

$$(p, s_0(p)) := \sigma_0(p) \qquad \text{for } p \in U.$$

Then $s_0(p_0) = v_0$ and $s_0(p) \in E_p$ for all $p \in U$. Now choose $\varepsilon > 0$ such that

$$\{p \in M \mid |p - p_0| < \varepsilon\} \subset U$$

and choose a smooth cutoff function $\beta : \mathbb{R}^k \to [0, 1]$ such that $\beta(p_0) = 1$ and $\beta(p) = 0$ for $|p - p_0| \geq \varepsilon$. Define $s : M \to \mathbb{R}^\ell$ by

$$s(p) := \begin{cases} \beta(p)s_0(p), & \text{if } p \in U, \\ 0, & \text{if } p \notin U. \end{cases}$$

This map satisfies the requirements of (ii).

We prove that (ii) implies (iii). Thus we assume that E satisfies (ii). Choose $p_0 \in M$ and a basis $v_1, \ldots, v_n$ of E_{p_0}. By (ii) there exists smooth sections $s_1, \ldots, s_n :$ $M \to \mathbb{R}^\ell$ of E such that $s_i(p_0) = v_i$ for $i = 1, \ldots, n$. Now there exists an M-open neighborhood $U \subset M$ of p_0 such that the vectors $s_1(p), \ldots, s_n(p)$ are linearly independent, and hence form a basis of E_p for every $p \in U$. Hence, for every $p \in U$, we have

$$E_p = \mathrm{im} D(p), \qquad D(p) := [s_1(p) \cdots s_n(p)] \in \mathbb{R}^{\ell \times n}.$$

By Exercise 2.6.9, this implies $\Pi(p) = D(p)(D(p)^\mathsf{T} D(p))^{-1} D(p)^\mathsf{T}$ for every $p \in U$. Thus every $p_0 \in M$ has a neighborhood U such that the restriction of Π to U is smooth. This shows that (ii) implies (iii).

We prove that (iii) implies (iv). Fix a point $p_0 \in M$ and choose a basis $v_1, \ldots, v_n$ of E_{p_0}. For $p \in M$ define

$$D(p) := [\Pi(p)v_1 \cdots \Pi(p)v_n] \in \mathbb{R}^{\ell \times n}$$

Then $D : M \to \mathbb{R}^{\ell \times n}$ is a smooth map and $D(p_0)$ has rank n. Hence the set

$$U := \{p \in M \mid \mathrm{rank} D(p) = n\} \subset M$$

is an open neighborhood of p_0 and $E_p = \mathrm{im} D(p)$ for all $p \in U$. Thus

$$\pi^{-1}(U) = \{(p, v) \in E \mid p \in U\} \subset E$$

is an open set containing $\pi^{-1}(p_0)$. Define the map $\Phi : \pi^{-1}(U) \to U \times \mathbb{R}^n$ by

$$\Phi(p, v) := (p, \Phi_p(v)), \qquad \Phi_p(v) := (D(p)^\mathsf{T} D(p))^{-1/2} D(p)^\mathsf{T} v$$

for $p \in U$ and $v \in E_p$. This map is bijective and its inverse is given by

$$\Phi^{-1}(p, \xi) = (p, \Phi_p^{-1}(\xi)), \qquad \Phi_p^{-1}(\xi) = D(p)(D(p)^\mathsf{T} D(p))^{-1/2} \xi$$

for $p \in U$ and $\xi \in \mathbb{R}^n$. Thus Φ is a diffeomorphism and $|\Phi_p(v)| = |v|$ for all $p \in U$ and all $v \in E_p$. This shows that (iii) implies (iv).

That (iv) implies (v) is obvious.

We prove that (v) implies (i). Shrinking U if necessary, we may assume that there exists a coordinate chart $\phi : U \to \Omega$ with values in an open set $\Omega \subset \mathbb{R}^m$. Then the composition $(\phi \times \mathrm{id}) \circ \Phi : \pi^{-1}(U) \to \Omega \times \mathbb{R}^n$ is a diffeomorphism. Thus $E \subset \mathbb{R}^k \times \mathbb{R}^\ell$ is a manifold of dimension $m + n$ and this proves Theorem 2.6.10. $\square$

Exercise 2.6.14 Define the notion of an isomorphism between two vector bundles E and F over M. Construct a vector bundle $E \subset S^1 \times \mathbb{R}^2$ of rank 1 that does not admit a *global trivialization*, i.e. that is not isomorphic to the trivial bundle $S^1 \times \mathbb{R}$. Such a vector bundle is called a **Möbius strip**.

2.6.3   The Implicit Function Theorem

Next we carry over the Implicit Function Theorem in Corollary A.2.6 to smooth maps on vector bundles.

Theorem 2.6.15 (Implicit Function Theorem) *Let $M \subset \mathbb{R}^k$ be a smooth m-manifold, let $N \subset \mathbb{R}^k$ be a smooth n-manifold, let $E \subset M \times \mathbb{R}^\ell$ be a smooth vector bundle of rank n, let $W \subset E$ be open, and let $f : W \to N$ be a smooth map. For $p \in M$ define $f_p : W_p \to N$ by*

$$W_p := \{v \in E_p \,|\, (p,v) \in W\}, \qquad f_p(v) := f(p,v).$$

Let $p_0 \in M$ such that $0 \in W_{p_0}$ and $df_{p_0}(0) : E_{p_0} \to T_{q_0}N$ is bijective, where $q_0 := f(p_0, 0) \in N$. Then there exists a constant $\varepsilon > 0$, open neighborhoods $U_0 \subset M$ of p_0 and $V_0 \subset N$ of q_0, and a smooth map $h : U_0 \times V_0 \to \mathbb{R}^\ell$ such that $\{(p,v) \in E \,|\, p \in U_0, |v| < \varepsilon\} \subset W$ and

$$h(p,q) \in E_p, \qquad |h(p,q)| < \varepsilon \tag{2.6.5}$$

for all $(p,q) \in U_0 \times V_0$ and

$$f_p(v) = 0 \qquad \Longleftrightarrow \qquad v = h(p,q) \tag{2.6.6}$$

for all $(p,q) \in U_0 \times V_0$, and all $v \in E_p$ with $|v| < \varepsilon$.

Proof   Choose a coordinate chart $\psi : V \to \mathbb{R}^n$ on an open set $V \subset N$ containing q_0. Choose an open neighborhood $U \subset M$ of p_0 such that $(p,0) \in W$ and $f(p,0) \in V$ for all $p \in \overline{U}$, there is a coordinate chart $\phi : U \to \Omega \subset \mathbb{R}^m$, and there is a local trivialization $\Phi : \pi^{-1}(U) \to U \times \mathbb{R}^n$ as in Theorem 2.6.10 with $|\Phi_p(v)| = |v|$ for $p \in U$ and $v \in E_p$. Define $B_r := \{\xi \in \mathbb{R}^n \,|\, |\xi| < r\}$ and choose $r > 0$ so small that $\Phi^{-1}(U \times B_r) \subset W$ and $f \circ \Phi^{-1}(U \times B_r) \subset V$. Define the map $F : \Omega \times \mathbb{R}^n \times B_r \to \mathbb{R}^n$ by

$$F(x,y,\xi) := \psi \circ f \circ \Phi^{-1}(\phi^{-1}(x), \xi) - y$$

for $(x,y) \in \Omega \times \mathbb{R}^n$ and $\xi \in B_r$. Let $x_0 := \phi(p_0)$ and $y_0 := \psi(q_0)$. Then we have $F(x_0, y_0, 0) = 0$ and the derivative $d_3 F(x_0, y_0, 0) : \mathbb{R}^n \to \mathbb{R}^n$ of F with respect to ξ at $(x_0, y_0, 0)$ is bijective. Hence Corollary A.2.6 asserts that there exist open neighborhoods $U_0 \subset U$ of p_0 and $V_0 \subset V$ of q_0, a constant $0 < \varepsilon < r$, and a smooth map $g : \phi(U_0) \times \psi(V_0) \to B_\varepsilon$ such that

$$F(x,y,\xi) = 0 \qquad \Longleftrightarrow \qquad g(x,y) = \xi$$

for all $(x,y) \in \phi(U_0) \times \psi(V_0)$ and all $\xi \in B_\varepsilon$. Thus the map

$$h : U_0 \times V_0 \to \mathbb{R}^\ell, \qquad h(p,q) := \Phi_p^{-1}\big(g(\phi(p), \psi(q))\big),$$

satisfies the requirements of Theorem 2.6.15. $\square$

2.7 The Theorem of Frobenius

Let $M \subset \mathbb{R}^k$ be an m-dimensional manifold and n be a nonnegative integer. A **subbundle of rank** n of the tangent bundle TM is a subset $E \subset TM$ that is itself a vector bundle of rank n over M, i.e. it is a submanifold of TM and the fiber $E_p = \{v \in T_pM \mid (p, v) \in E\}$ is an n-dimensional linear subspace of T_pM for every $p \in M$. Note that the rank n of a subbundle is necessarily less than or equal to m. In the literature a subbundle of the tangent bundle is sometimes called a *distribution* on M. We shall, however, not use this terminology in order to avoid confusion with the concept of a distribution in the functional analytic setting.

Definition 2.7.1 Let $M \subset \mathbb{R}^k$ be an m-dimensional manifold and $E \subset TM$ be a subbundle of rank n. The subbundle E is called **involutive** if, for any two vector fields $X, Y \in \mathrm{Vect}(M)$, we have

$$X(p), Y(p) \in E_p \ \forall \ p \in M \qquad \Longrightarrow \qquad [X, Y](p) \in E_p \ \forall \ p \in M. \quad (2.7.1)$$

The subbundle E is called **integrable** if, for every $p_0 \in M$, there exists a submanifold $N \subset M$ such that $p_0 \in N$ and $T_pN = E_p$ for every $p \in N$. A **foliation box for** E (see Fig. 2.13) is a coordinate chart $\phi : U \to \Omega$ on an M-open subset $U \subset M$ with values in an open set $\Omega \subset \mathbb{R}^n \times \mathbb{R}^{m-n}$ such that the set $\Omega \cap (\mathbb{R}^n \times \{y\})$ is connected for every $y \in \mathbb{R}^{m-n}$ and, for every $p \in U$ and every $v \in T_pM$, we have

$$v \in E_p \qquad \Longleftrightarrow \qquad d\phi(p)v \in \mathbb{R}^n \times \{0\}.$$

Theorem 2.7.2 (Frobenius) *Let $M \subset \mathbb{R}^k$ be an m-dimensional manifold and $E \subset TM$ be a subbundle of rank n. Then the following are equivalent.*

(i) *E is involutive.*
(ii) *E is integrable.*
(iii) *For every $p_0 \in M$ there exists a foliation box $\phi : U \to \Omega$ with $p_0 \in U$.*

It is easy to show that (iii) $\Longrightarrow$ (ii) $\Longrightarrow$ (i) (see below). The hard part of the theorem is to prove that (i) $\Longrightarrow$ (iii). This requires the following lemma.

Lemma 2.7.3 *Let $E \subset TM$ be an involutive subbundle and $X \in \mathrm{Vect}(M)$ be a complete vector field such that $X(p) \in E_p$ for every $p \in M$. Denote by*

$$\mathbb{R} \to \mathrm{Diff}(M) : t \mapsto \phi^t$$

the flow of X. Then, for all $t \in \mathbb{R}$ and all $p \in M$, we have

$$d\phi^t(p)E_p = E_{\phi^t(p)}. \quad (2.7.2)$$

Fig. 2.13 A foliation box

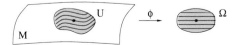

We show first how Theorem 2.7.2 follows from Lemma 2.7.3.

Lemma 2.7.3 implies Theorem 2.7.2   We prove first that (iii) implies (ii). Let $p_0 \in M$, choose a foliation box $\phi : U \to \Omega$ for E with $p_0 \in U$, and define

$$N := (p \in U \mid \phi(p) \in \mathbb{R}^n \times \{y_0\}\}$$

where $(x_0, y_0) := \phi(p_0) \in \Omega$. Then N satisfies the requirements of (ii).

We prove that (ii) implies (i). Choose two vector fields $X, Y \in \text{Vect}(M)$ that satisfy $X(p), Y(p) \in E_p$ for all $p \in M$ and fix a point $p_0 \in M$. Then, by (ii), there exists a submanifold $N \subset M$ containing p_0 such that $T_p N = E_p$ for every $p \in N$. Hence the restrictions $X|_N$ and $Y|_N$ are vector fields on N and so is the restriction of the Lie bracket $[X, Y]$ to N. Thus we have $[X, Y](p_0) \in T_{p_0} N = E_{p_0}$ as claimed.

We prove that (i) implies (iii). Thus we assume that E is an involutive sub-bundle of TM and fix a point $p_0 \in M$. By Theorem 2.6.10 there exist vector fields $X_1, \ldots, X_n \in \text{Vect}(M)$ such that $X_i(p) \in E_p$ for all i and p and the vectors $X_1(p_0), \ldots, X_n(p_0)$ form a basis of E_{p_0}. Using Theorem 2.6.10 again we find vector fields $Y_1, \ldots, Y_{m-n} \in \text{Vect}(M)$ such that the vectors

$$X_1(p_0), \ldots, X_n(p_0), Y_1(p_0), \ldots, Y_{m-n}(p_0)$$

form a basis of $T_{p_0} M$. Using cutoff functions as in the proof of Theorem 2.6.10 we may assume without loss of generality that the vector fields X_i and Y_j have compact support and hence are complete (see Exercise 2.4.13). Denote by $\phi_1^t, \ldots, \phi_n^t$ the flows of the vector fields $X_1, \ldots, X_n$, respectively, and by $\psi_1^t, \ldots, \psi_{m-n}^t$ the flows of the vector fields $Y_1, \ldots, Y_{m-n}$. Define the map

$$\psi : \mathbb{R}^n \times \mathbb{R}^{m-n} \to M$$

by

$$\psi(x, y) := \phi_1^{x_1} \circ \cdots \circ \phi_n^{x_n} \circ \psi_1^{y_1} \circ \cdots \circ \psi_{m-n}^{y_{m-n}}(p_0).$$

By Lemma 2.7.3, this map satisfies

$$\frac{\partial \psi}{\partial x_i}(x, y) \in E_{\psi(x,y)} \tag{2.7.3}$$

for all $x \in \mathbb{R}^n$ and $y \in \mathbb{R}^{m-n}$. Moreover,

$$\frac{\partial \psi}{\partial x_i}(0, 0) = X_i(p_0), \qquad \frac{\partial \psi}{\partial y_j}(0, 0) = Y_j(p_0),$$

and so the derivative

$$d\psi(0, 0) : \mathbb{R}^n \times \mathbb{R}^{m-n} \to T_{p_0} M$$

is bijective. Hence, by the Inverse Function Theorem 2.2.17, there exists an open neighborhood $\Omega \subset \mathbb{R}^n \times \mathbb{R}^{m-n}$ of the origin such that the set

$$U := \psi(\Omega) \subset M$$

is an M-open neighborhood of p_0 and $\psi|_\Omega : \Omega \to U$ is a diffeomorphism. Thus the vectors $\partial \psi / \partial x_i(x, y)$ are linearly independent for every $(x, y) \in \Omega$ and, by (2.7.3), form a basis of $E_{\psi(x,y)}$. Hence

$$\phi := (\psi|_\Omega)^{-1} : U \to \Omega$$

is a foliation box and this proves Theorem 2.7.2, assuming Lemma 2.7.3. $\qquad\square$

To complete the proof of the Frobenius theorem it remains to prove Lemma 2.7.3. This requires the following result.

Lemma 2.7.4 *Let $E \subset TM$ be an involutive subbundle. If $\beta : \mathbb{R}^2 \to M$ is a smooth map such that*

$$\frac{\partial \beta}{\partial s}(s, 0) \in E_{\beta(s,0)}, \qquad \frac{\partial \beta}{\partial t}(s, t) \in E_{\beta(s,t)}, \tag{2.7.4}$$

for all $s, t \in \mathbb{R}$, then

$$\frac{\partial \beta}{\partial s}(s, t) \in E_{\beta(s,t)}, \tag{2.7.5}$$

for all $s, t \in \mathbb{R}$.

We first show how Lemma 2.7.3 follows from Lemma 2.7.4.

Lemma 2.7.4 implies Lemma 2.7.3 Let $X \in \mathrm{Vect}(M)$ be a complete vector field satisfying $X(p) \in E_p$ for every $p \in M$ and let ϕ^t be the flow of X. Choose a point $p_0 \in M$ and a vector $v_0 \in E_{p_0}$. By Theorem 2.6.10 there is a vector field $Y \in \mathrm{Vect}(M)$ with values in E such that $Y(p_0) = v_0$. Moreover this vector field may be chosen to have compact support and hence it is complete (see Exercise 2.4.13). Thus there is a solution $\gamma : \mathbb{R} \to M$ of the initial value problem

$$\dot{\gamma}(s) = Y(\gamma(s)), \qquad \gamma(0) = p_0.$$

Define $\beta : \mathbb{R}^2 \to M$ by

$$\beta(s, t) := \phi^t(\gamma(s))$$

for $s, t \in \mathbb{R}$. Then

$$\frac{\partial \beta}{\partial s}(s, 0) = \dot{\gamma}(s) = Y(\gamma(s)) \in E_{\beta(s,0)},$$

$$\frac{\partial \beta}{\partial t}(s, t) = X(\beta(s, t)) \in E_{\beta(s,t)}$$

for all $s, t \in \mathbb{R}$. Hence it follows from Lemma 2.7.4 that

$$d\phi^t(p_0)v_0 = d\phi^t(\gamma(0))\dot{\gamma}(0) = \frac{\partial\beta}{\partial s}(0, t) \in E_{\phi^t(p_0)}$$

for every $t \in \mathbb{R}$. This proves Lemma 2.7.3, assuming Lemma 2.7.4. $\qquad\square$

Proof of Lemma 2.7.4 Given any point $p_0 \in M$ we choose a coordinate chart $\phi : U \to \Omega$, defined on an M-open set $U \subset M$ with values in an open set $\Omega \subset \mathbb{R}^n \times \mathbb{R}^{m-n}$, such that $p_0 \in U$ and $d\phi(p_0)E_{p_0} = \mathbb{R}^n \times \{0\}$. Shrinking U, if necessary, we find that for every $p \in U$ the linear subspace $d\phi(p)E_p \subset \mathbb{R}^n \times \mathbb{R}^{m-n}$ is the graph of a matrix $A \in \mathbb{R}^{(m-n)\times n}$. Thus there exists a smooth map $A : \Omega \to \mathbb{R}^{(m-n)\times n}$ such that, for every $p \in U$,

$$d\phi(p)E_p = \{(\xi, A(x, y)\xi) \mid \xi \in \mathbb{R}^n\}, \qquad (x, y) := \phi(p) \in \Omega. \qquad (2.7.6)$$

For $(x, y) \in \Omega$ define the linear maps

$$\frac{\partial A}{\partial x}(x, y) : \mathbb{R}^n \to \mathbb{R}^{(m-n)\times n}, \qquad \frac{\partial A}{\partial y}(x, y) : \mathbb{R}^{m-n} \to \mathbb{R}^{(m-n)\times n}$$

by

$$\frac{\partial A}{\partial x}(x, y) \cdot \xi := \sum_{i=1}^{n} \xi_i \frac{\partial A}{\partial x_i}(x, y), \qquad \frac{\partial A}{\partial y}(x, y) \cdot \eta := \sum_{j=1}^{m-n} \eta_j \frac{\partial A}{\partial y_j}(x, y),$$

for $\xi = (\xi_1, \dots, \xi_n) \in \mathbb{R}^n$ and $\eta = (\eta_1, \dots, \eta_{m-n}) \in \mathbb{R}^{m-n}$. We prove the following.

Claim 1 *Let $(x, y) \in \Omega$, $\xi, \xi' \in \mathbb{R}^n$ and define $\eta, \eta' \in \mathbb{R}^{m-n}$ by $\eta := A(x, y)\xi$ and $\eta' := A(x, y)\xi'$. Then*

$$\left(\frac{\partial A}{\partial x}(x, y) \cdot \xi + \frac{\partial A}{\partial y}(x, y) \cdot \eta\right)\xi' = \left(\frac{\partial A}{\partial x}(x, y) \cdot \xi' + \frac{\partial A}{\partial y}(x, y) \cdot \eta'\right)\xi.$$

The graphs of the matrices $A(z)$ determine a subbundle $\widetilde{E} \subset \Omega \times \mathbb{R}^m$ with the fibers

$$\widetilde{E}_z := \{(\xi, \eta) \in \mathbb{R}^n \times \mathbb{R}^{m-n} \mid \eta = A(x, y)\xi\}$$

for $z = (x, y) \in \Omega$. This subbundle is the image of the restriction

$$E|_U := \{(p, v) \mid p \in U, v \in E_p\}$$

under the diffeomorphism $TM|_U \to \Omega \times \mathbb{R}^m : (p, v) \mapsto (\phi(p), d\phi(p)v)$ and hence it is involutive. Now fix two elements $\xi, \xi' \in \mathbb{R}^n$ and define the vector fields $\zeta, \zeta' : \Omega \to \mathbb{R}^m$ by

$$\zeta(z) := (\xi, A(z)\xi), \qquad \zeta'(z) := (\xi', A(z)\xi'), \qquad z \in \Omega.$$

Then ζ and ζ' are sections of $\widetilde{E}$ and their Lie bracket $[\zeta, \zeta']$ is given by

$$[\zeta, \zeta'](z) = \left(0, \left(dA(z)\zeta'(z)\right)\xi - \left(dA(z)\zeta(z)\right)\xi'\right).$$

Since $\widetilde{E}$ is involutive the Lie bracket $[\zeta, \zeta']$ must take values in the graph of A. Hence the right hand side vanishes and this proves Claim 1.

Claim 2 *Let $I, J \subset \mathbb{R}$ be open intervals and let $z = (x, y) : I \times J \to \Omega$ be a smooth map. Fix two points $s_0 \in I$ and $t_0 \in J$ and assume that*

$$\frac{\partial y}{\partial s}(s_0, t_0) = A\left(x(s_0, t_0), y(s_0, t_0)\right)\frac{\partial x}{\partial s}(s_0, t_0), \tag{2.7.7}$$

$$\frac{\partial y}{\partial t}(s, t) = A\left(x(s, t), y(s, t)\right)\frac{\partial x}{\partial t}(s, t) \tag{2.7.8}$$

for all $s \in I$ and $t \in J$. Then

$$\frac{\partial y}{\partial s}(s_0, t) = A\left(x(s_0, t), y(s_0, t)\right)\frac{\partial x}{\partial s}(s_0, t) \tag{2.7.9}$$

for all $t \in J$.

Equation (2.7.9) holds by assumption for $t = t_0$. Moreover, dropping the argument $z(s_0, t) = z = (x, y)$ for notational convenience we obtain

$$\begin{aligned}
\frac{\partial}{\partial t}\left(\frac{\partial y}{\partial s} - A \cdot \frac{\partial x}{\partial s}\right) &= \frac{\partial^2 y}{\partial s \partial t} - A\frac{\partial^2 x}{\partial s \partial t} - \left(\frac{\partial A}{\partial x} \cdot \frac{\partial x}{\partial t} + \frac{\partial A}{\partial y} \cdot \frac{\partial y}{\partial t}\right)\frac{\partial x}{\partial s} \\
&= \frac{\partial^2 y}{\partial s \partial t} - A\frac{\partial^2 x}{\partial s \partial t} - \left(\frac{\partial A}{\partial x} \cdot \frac{\partial x}{\partial t} + \frac{\partial A}{\partial y} \cdot \left(A\frac{\partial x}{\partial t}\right)\right)\frac{\partial x}{\partial s} \\
&= \frac{\partial^2 y}{\partial s \partial t} - A\frac{\partial^2 x}{\partial s \partial t} - \left(\frac{\partial A}{\partial x} \cdot \frac{\partial x}{\partial s} + \frac{\partial A}{\partial y} \cdot \left(A\frac{\partial x}{\partial s}\right)\right)\frac{\partial x}{\partial t} \\
&= \frac{\partial^2 y}{\partial s \partial t} - A\frac{\partial^2 x}{\partial s \partial t} - \left(\frac{\partial A}{\partial x} \cdot \frac{\partial x}{\partial s} + \frac{\partial A}{\partial y} \cdot \frac{\partial y}{\partial s}\right)\frac{\partial x}{\partial t} \\
&\quad + \left(\frac{\partial A}{\partial y} \cdot \left(\frac{\partial y}{\partial s} - A\frac{\partial x}{\partial s}\right)\right)\frac{\partial x}{\partial t} \\
&= \left(\frac{\partial A}{\partial y} \cdot \left(\frac{\partial y}{\partial s} - A\frac{\partial x}{\partial s}\right)\right)\frac{\partial x}{\partial t}.
\end{aligned}$$

Here the second step follows from (2.7.8), the third step follows from Claim 1, and the last step follows by differentiating equation (2.7.8) with respect to s. Define the curve $\eta : J \to \mathbb{R}^{m-n}$ by

$$\eta(t) := \frac{\partial y}{\partial s}(s_0, t) - A\left(x(s_0, t), y(s_0, t)\right)\frac{\partial x}{\partial s}(s_0, t).$$

By (2.7.7) and what we have just proved, the curve η satisfies the linear differential equation

$$\dot{\eta}(t) = \left(\frac{\partial A}{\partial y} (x(s_0, t), y(s_0, t)) \cdot \eta(t) \right) \frac{\partial x}{\partial t} (s_0, t), \qquad \eta(t_0) = 0.$$

Hence $\eta(t) = 0$ for all $t \in J$. This proves (2.7.9) and Claim 2.

Now let $\beta : \mathbb{R}^2 \to M$ be a smooth map satisfying (2.7.4) and fix a real number s_0. Consider the set $W := \{ t \in \mathbb{R} \mid \partial_s \beta(s_0, t) \in E_{\beta(s_0,t)} \}$. By going to local coordinates, we obtain from Claim 2 that W is open. Moreover, W is obviously closed, and $W \neq \emptyset$ because $0 \in W$ by (2.7.4). Hence $W = \mathbb{R}$. Since $s_0 \in \mathbb{R}$ was chosen arbitrarily, this proves (2.7.5) and Lemma 2.7.4.   □

Any subbundle $E \subset TM$ determines an equivalence relation on M via

$$p_0 \sim p_1 \quad \Longleftrightarrow \quad \begin{array}{l} \text{there is a smooth curve } \gamma : [0, 1] \to M \\ \text{such that } \gamma(0) = p_0, \ \gamma(1) = p_1, \ \dot{\gamma}(t) \in E_{\gamma(t)} \ \forall \, t. \end{array} \quad (2.7.10)$$

If E is integrable, this equivalence relation is called a **foliation** and the equivalence class of $p_0 \in M$ is called the **leaf** of the foliation through p_0. The next example shows that the leaves do not need to be submanifolds.

Example 2.7.5 Consider the torus $M := S^1 \times S^1 \subset \mathbb{C}^2$ with the tangent bundle

$$TM = \{ (z_1, z_2, i\lambda_1 z_1, i\lambda_2 z_2) \in \mathbb{C}^4 \mid |z_1| = |z_2| = 1, \ \lambda_1, \lambda_2 \in \mathbb{R} \}.$$

Let ω_1, ω_2 be real numbers and consider the subbundle

$$E := \{ (z_1, z_2, it\omega_1 z_1, it\omega_2 z_2) \in \mathbb{C}^4 \mid |z_1| = |z_2| = 1, \ t \in \mathbb{R} \}.$$

The leaf of this subbundle through $z = (z_1, z_2) \in \mathbb{T}^2$ is given by

$$L = \left\{ \left(e^{it\omega_1} z_1, e^{it\omega_2} z_2 \right) \mid t \in \mathbb{R} \right\}.$$

It is a submanifold if and only if the quotient ω_1/ω_2 is a rational number (or $\omega_2 = 0$). Otherwise each leaf is a dense subset of $\mathbb{T}^2$.

Exercise 2.7.6 Prove that (2.7.10) defines an equivalence relation for every subbundle $E \subset TM$.

Exercise 2.7.7 Each subbundle $E \subset TM$ of rank 1 is integrable.

Exercise 2.7.8 Consider the manifold $M = \mathbb{R}^3$. Prove that the subbundle $E \subset TM = \mathbb{R}^3 \times \mathbb{R}^3$ with fiber $E_p = \{ (\xi, \eta, \zeta) \in \mathbb{R}^3 \mid \zeta - y\xi = 0 \}$ over $p = (x, y, z) \in \mathbb{R}^3$ is not integrable and that any two points in $\mathbb{R}^3$ can be joined by a path tangent to E.

Exercise 2.7.9   Consider the manifold $M = S^3 \subset \mathbb{R}^4 = \mathbb{C}^2$ and define

$$E := \{(z, \zeta) \in \mathbb{C}^2 \times \mathbb{C}^2 \,|\, |z| = 1, \, \zeta \perp z, \, i\zeta \perp z\}.$$

Thus the fiber

$$E_z \subset T_z S^3 = z^\perp$$

is the maximal complex linear subspace of $T_z S^3$. Prove that E has real rank 2 and is not integrable.

Exercise 2.7.10   Let $E \subset TM$ be an involutive subbundle of rank n and let $L \subset M$ be a leaf of the foliation determined by E. A subset $V \subset L$ is called L-**open** iff it can be written as a union of submanifolds N of M with tangent spaces $T_p N = E_p$ for $p \in N$. Prove that the L-open sets form a topology on L (called the **intrinsic topology**). Prove that the obvious inclusion $\iota : L \to M$ is continuous with respect to the intrinsic topology on L. Prove that the inclusion $\iota : L \to M$ is proper if and only if the intrinsic topology on L agrees with the relative topology inherited from M (called the **extrinsic topology**).

Remark 2.7.11   It is surprisingly difficult to prove that each closed leaf L of a foliation is a submanifold of M. A proof due to David Epstein [19] is sketched in Sect. 2.9.4 below.

2.8   The Intrinsic Definition of a Manifold*

It is somewhat restrictive to only consider manifolds that are embedded in some Euclidean space. Although we shall see that (at least) every compact manifold admits an embedding into a Euclidean space, such an embedding is in many cases not a natural part of the structure of a manifold. In particular, we encounter manifolds that are described as quotient spaces and there are manifolds that are embedded in certain infinite-dimensional Hilbert spaces. For this reason it is convenient, at this point, to introduce a more general *intrinsic* definition of a manifold. (See Chap. 1 for an overview.) This requires some background from point set topology that is not covered in the first year analysis courses. We shall then see that all the definitions and results of this chapter carry over in a natural manner to the intrinsic setting. We begin by recalling the intrinsing definition of a smooth manifold in Sect. 1.4.

2.8.1   Definition and Examples

Definition 2.8.1 (Smooth m-manifold)   Let $m \in \mathbb{N}_0$ and M be a set. A **chart** on M is a pair (ϕ, U) where $U \subset M$ and ϕ is a bijection from U to an open

Fig. 2.14 Coordinate charts
and transition maps

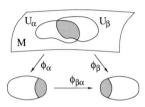

set $\phi(U) \subset \mathbb{R}^m$. Two charts (ϕ_1, U_1), (ϕ_2, U_2) are called **compatible** iff $\phi_1(U_1 \cap U_2)$
and $\phi_2(U_1 \cap U_2)$ are open and the **transition map**

$$\phi_{21} = \phi_2 \circ \phi_1^{-1} : \phi_1(U_1 \cap U_2) \to \phi_2(U_1 \cap U_2) \tag{2.8.1}$$

is a diffeomorphism (see Fig. 2.14). A **smooth atlas** on M is a collection $\mathcal{A}$ of
charts on M any two of which are compatible and such that the sets U, as (ϕ, U)
ranges over $\mathcal{A}$, cover M (i.e. for every $p \in M$ there exists a chart $(\phi, U) \in \mathcal{A}$ with
$p \in U$). A **maximal smooth atlas** is an atlas which contains every chart which is
compatible with each of its members. A **smooth m-manifold** is a pair consisting of
a set M and a maximal atlas $\mathcal{A}$ on M.

In Lemma 1.4.3 it was shown that, if $\mathcal{A}$ is an atlas, then so is the collection $\overline{\mathcal{A}}$
of all charts compatible with each member of $\mathcal{A}$. Moreover, the atlas $\overline{\mathcal{A}}$ is maximal,
so every atlas extends uniquely to a maximal atlas. For this reason, a manifold is
usually specified by giving its underlying set M and some atlas on M. Generally,
the notation for the atlas is suppressed and the manifold is denoted simply by M.
The members of the atlas are called **coordinate charts** or simply **charts** on M. By
Lemma 1.3.3 a smooth m-manifold admits a unique topology such that, for each
chart (ϕ, U) of the smooth atlas, the set $U \subset M$ is open and the bijection

$$\phi : U \to \phi(U)$$

is a homeomorphism onto the open set $\phi(U) \subset \mathbb{R}^m$. This topology is called the
intrinsic topology of M and is described in the following definition.

Definition 2.8.2 Let M be a smooth m-manifold. The **intrinsic topology** on the
set M is the topology induced by the charts, i.e. a subset

$$W \subset M$$

is open in the intrinsic topology iff $\phi(U \cap W)$ is an open subset of $\mathbb{R}^m$ for every
chart (ϕ, U) on M.[1]

---

[1] At this point we do not assume that the intrinsic topology on the manifold M is Hausdorff or
second countable. These hypotheses will be imposed after the end of the present chapter. For
explanations see the comments at the end of Sect. 2.8.1 and of Sect. 2.9.5.

Remark 2.8.3   Let $M \subset \mathbb{R}^k$ be smooth m-dimensional submanifold of $\mathbb{R}^k$ as in Definition 2.1.3. Then the set of all diffeomorphisms $(\phi, U \cap M)$ as in Definition 2.1.3 form a smooth atlas as in Definition 2.8.1. The intrinsic topology on the resulting smooth manifold is the same as the relative topology defined in Sect. 1.3.

Remark 2.8.4   A **topological manifold** is a topological space such that each point has a neighborhood U homeomorphic to an open subset of $\mathbb{R}^m$. Thus a smooth manifold (with the intrinsic topology) is a topological manifold and its maximal smooth atlas $\mathcal{A}$ is a subset of the set $\mathcal{A}_0$ of all pairs (ϕ, U) where $U \subset M$ is an open set and ϕ is a homeomorphism from U to an open subset of $\mathbb{R}^m$. One says that the maximal smooth atlas $\mathcal{A}$ is a **smooth structure** on the topological manifold M iff the topology of M is the intrinsic topology of the smooth structure and every chart of the smooth structure is a homeomorphism. As explained in Sect. 1.4 a topological manifold can have many distinct smooth structures (see Remark 1.4.6). However, it is a deep theorem beyond the scope of this book that there are topological manifolds which do not admit any smooth structure.

Example 2.8.5   The **complex projective space** $\mathbb{C}P^n$ is the set

$$\mathbb{C}P^n = \{\ell \subset \mathbb{C}^{n+1} \,|\, \ell \text{ is a 1-dimensional complex subspace}\}$$

of complex lines in $\mathbb{C}^{n+1}$. It can be identified with the quotient space

$$\mathbb{C}P^n = \big(\mathbb{C}^{n+1} \setminus \{0\}\big)/\mathbb{C}^*$$

of nonzero vectors in $\mathbb{C}^{n+1}$ modulo the action of the multiplicative group $\mathbb{C}^* = \mathbb{C} \setminus \{0\}$ of nonzero complex numbers. The equivalence class of a nonzero vector $z = (z_0, \dots, z_n) \in \mathbb{C}^{n+1}$ will be denoted by

$$[z] = [z_0 : z_1 : \cdots : z_n] := \{\lambda z \,|\, \lambda \in \mathbb{C}^*\}$$

and the associated line is $\ell = \mathbb{C}z$. An atlas on $\mathbb{C}P^n$ is given by the open cover $U_i := \{[z_0 : \cdots : z_n] \,|\, z_i \neq 0\}$ for $i = 0, 1, \dots, n$ and the coordinate charts $\phi_i : U_i \to \mathbb{C}^n$ are

$$\phi_i([z_0 : \cdots : z_n]) := \left(\frac{z_0}{z_i}, \dots, \frac{z_{i-1}}{z_i}, \frac{z_{i+1}}{z_i}, \dots, \frac{z_n}{z_i}\right). \qquad (2.8.2)$$

Exercise: Prove that each ϕ_i is a homeomorphism and the transition maps are holomorphic. Prove that the manifold topology is the quotient topology, i.e. if $\pi : \mathbb{C}^{n+1} \setminus \{0\} \to \mathbb{C}P^n$ denotes the obvious projection, then a subset $U \subset \mathbb{C}P^n$ is open if and only if $\pi^{-1}(U)$ is an open subset of $\mathbb{C}^{n+1} \setminus \{0\}$.

Example 2.8.6   The **real projective space** $\mathbb{R}P^n$ is the set

$$\mathbb{R}P^n = \{\ell \subset \mathbb{R}^{n+1} \,|\, \ell \text{ is a 1-dimensional linear subspace}\}$$

of real lines in $\mathbb{R}^{n+1}$. It can again be identified with the quotient space

$$\mathbb{R}\mathrm{P}^n = \left(\mathbb{R}^{n+1} \setminus \{0\}\right)/\mathbb{R}^*$$

of nonzero vectors in $\mathbb{R}^{n+1}$ modulo the action of the multiplicative group $\mathbb{R}^* = \mathbb{R} \setminus \{0\}$ of nonzero real numbers, and the equivalence class of a nonzero vector $x = (x_0, \dots, x_n) \in \mathbb{R}^{n+1}$ will be denoted by

$$[x] = [x_0 : x_1 : \cdots : x_n] := \{\lambda x \mid \lambda \in \mathbb{R}^*\}.$$

An atlas on $\mathbb{R}\mathrm{P}^n$ is given by the open cover

$$U_i := \{[x_0 : \cdots : x_n] \mid x_i \neq 0\}$$

and the coordinate charts $\phi_i : U_i \to \mathbb{R}^n$ are again given by (2.8.2), with z_j replaced by x_j. The arguments in Example 2.8.5 show that these coordinate charts form an atlas and the manifold topology is the quotient topology. The transition maps are real analytic diffeomorphisms.

Example 2.8.7 The **real n-torus** is the topological space

$$\mathbb{T}^n := \mathbb{R}^n / \mathbb{Z}^n$$

equipped with the quotient topology. Thus two vectors $x, y \in \mathbb{R}^n$ are equivalent iff their difference $x - y \in \mathbb{Z}^n$ is an integer vector and we denote by $\pi : \mathbb{R}^n \to \mathbb{T}^n$ the obvious projection which assigns to each vector $x \in \mathbb{R}^n$ its equivalence class

$$\pi(x) := [x] := x + \mathbb{Z}^n.$$

Then a set $U \subset \mathbb{T}^n$ is open if and only if the set $\pi^{-1}(U)$ is an open subset of $\mathbb{R}^n$. An atlas on $\mathbb{T}^n$ is given by the open cover

$$U_\alpha := \{[x] \mid x \in \mathbb{R}^n, \ |x - \alpha| < 1/2\},$$

parametrized by vectors $\alpha \in \mathbb{R}^n$, and the coordinate charts $\phi_\alpha : U_\alpha \to \mathbb{R}^n$ defined by $\phi_\alpha([x]) := x$ for $x \in \mathbb{R}^n$ with $|x - \alpha| < 1/2$. **Exercise:** Show that each transition map for this atlas is a translation by an integer vector.

Example 2.8.8 Consider the **complex Grassmannian**

$$\mathrm{G}_k(\mathbb{C}^n) := \{V \subset \mathbb{C}^n \mid v \text{ is a } k\text{-dimensional complex linear subspace}\}.$$

This set can again be described as a quotient space $\mathrm{G}_k(\mathbb{C}^n) \cong \mathcal{F}_k(\mathbb{C}^n)/\mathrm{U}(k)$. Here

$$\mathcal{F}_k(\mathbb{C}^n) := \{D \in \mathbb{C}^{n \times k} \mid D^* D = \mathbb{1}\}$$

denotes the set of unitary k-frames in $\mathbb{C}^n$ and the group $U(k)$ acts on $\mathcal{F}_k(\mathbb{C}^n)$ contravariantly by $D \mapsto Dg$ for $g \in U(k)$. The projection

$$\pi : \mathcal{F}_k(\mathbb{C}^n) \to G_k(\mathbb{C}^n)$$

sends a matrix $D \in \mathcal{F}_k(\mathbb{C}^n)$ to its image $V := \pi(D) := \operatorname{im} D$. A subset $U \subset G_k(\mathbb{C}^n)$ is open if and only if $\pi^{-1}(U)$ is an open subset of $\mathcal{F}_k(\mathbb{C}^n)$. Given a k-dimensional subspace $V \subset \mathbb{C}^n$ we can define an open set $U_V \subset G_k(\mathbb{C}^n)$ as the set of all k-dimensional subspaces of $\mathbb{C}^n$ that can be represented as graphs of linear maps from V to $V^\perp$. This set of graphs can be identified with the complex vector space $\operatorname{Hom}^{\mathbb{C}}(V, V^\perp)$ of complex linear maps from V to $V^\perp$ and hence with $\mathbb{C}^{(n-k) \times k}$. This leads to an atlas on $G_k(\mathbb{C}^n)$ with holomorphic transition maps and shows that $G_k(\mathbb{C}^n)$ is a manifold of complex dimension $kn - k^2$. **Exercise:** Verify the details of this construction. Find explicit formulas for the coordinate charts and their transition maps. Carry this over to the real setting. Show that $\mathbb{C}P^n$ and $\mathbb{R}P^n$ are special cases.

Example 2.8.9 (The real line with two zeros)   A topological space M is called **Hausdorff** iff any two points in M can be separated by disjoint open neighborhoods. This example shows that a manifold need not be a Hausdorff space. Consider the quotient space

$$M := \mathbb{R} \times \{0, 1\}/ \equiv$$

where $[x, 0] \equiv [x, 1]$ for $x \neq 0$. An atlas on M consists of two coordinate charts $\phi_0 : U_0 \to \mathbb{R}$ and $\phi_1 : U_1 \to \mathbb{R}$ where

$$U_i := \{[x, i] \mid x \in \mathbb{R}\}, \qquad \phi_i([x, i]) := x$$

for $i = 0, 1$. Thus M is a 1-manifold. But the topology on M is not Hausdorff, because the points $[0, 0]$ and $[0, 1]$ cannot be separated by disjoint open neighborhoods.

Example 2.8.10 (A 2-manifold without a countable atlas)   Consider the vector space $X = \mathbb{R} \times \mathbb{R}^2$ with the equivalence relation

$$[t_1, x_1, y_1] \equiv [t_2, x_2, y_2] \iff \begin{array}{l} \text{either } y_1 = y_2 \neq 0, \ t_1 + x_1 y_1 = t_2 + x_2 y_2 \\ \text{or } y_1 = y_2 = 0, \ t_1 = t_2, \ x_1 = x_2. \end{array}$$

For $y \neq 0$ we have $[0, x, y] \equiv [t, x - t/y, y]$, however, each point $(x, 0)$ on the x-axis gets replaced by the uncountable set $\mathbb{R} \times \{(x, 0)\}$. Our manifold is the quotient space $M := X/\equiv$. This time we do not use the quotient topology but the topology induced by our atlas (see Definition 2.8.2). The coordinate charts are parametrized by the reals: for $t \in \mathbb{R}$ the set $U_t \subset M$ and the coordinate chart $\phi_t : U_t \to \mathbb{R}^2$ are given by

$$U_t := \{[t, x, y] \mid x, y \in \mathbb{R}\}, \qquad \phi_t([t, x, y]) := (x, y).$$

A subset $U \subset M$ is open, by definition, iff $\phi_t(U \cap U_t)$ is an open subset of $\mathbb{R}^2$ for every $t \in \mathbb{R}$. With this topology each ϕ_t is a homeomorphism from U_t onto $\mathbb{R}^2$ and M admits a countable dense subset $S := \{[0, x, y] \mid x, y \in \mathbb{Q}\}$. However, there is no atlas on M consisting of countably many charts. (Each coordinate chart can contain at most countably many of the points $[t, 0, 0]$.) The function $f : M \to \mathbb{R}$ given by $f([t, x, y]) := t + xy$ is smooth and each point $[t, 0, 0]$ is a critical point of f with value t. Thus f has no regular value. **Exercise:** Show that M is a path-connected Hausdorff space.

In Theorem 2.9.12 we will show that smooth manifolds whose topology is Hausdorff and second countable are precisely those that can be embedded in Euclidean space. Most authors tacitly assume that manifolds are Hausdorff and second countable and so will we after the end of the present chapter. However before Sect. 2.9.1 there is no need to impose these hypotheses.

2.8.2   Smooth Maps and Diffeomorphisms

Our next goal is to carry over all the definitions from embedded manifolds in Euclidean space to the intrinsic setting.

Definition 2.8.11 (Smooth map)   Let

$$(M, \{(\phi_\alpha, U_\alpha)\}_{\alpha \in A}), \qquad (N, \{(\psi_\beta, V_\beta)\}_{\beta \in B})$$

be smooth manifolds. A map $f : M \to N$ is called **smooth** iff it is continuous and the map

$$f_{\beta\alpha} := \psi_\beta \circ f \circ \phi_\alpha^{-1} : \phi_\alpha(U_\alpha \cap f^{-1}(V_\beta)) \to \psi_\beta(V_\beta) \qquad (2.8.3)$$

is smooth for every $\alpha \in A$ and every $\beta \in B$. It is called a **diffeomorphism** iff it is bijective and f and f^{-1} are smooth. The manifolds M and N are called **diffeomorphic** iff there exists a diffeomorphism $f : M \to N$.

The reader may check that the notion of a smooth map is independent of the atlas used in the definition, that compositions of smooth maps are smooth, and that sums and products of smooth maps from M to $\mathbb{R}$ are smooth.

Exercise 2.8.12   Let M be a smooth m-dimensional manifold with an atlas

$$\mathcal{A} = \{(\phi_\alpha, U_\alpha)\}_{\alpha \in A}.$$

Consider the quotient space

$$\widetilde{M} := \bigcup_{\alpha \in A} \{\alpha\} \times \phi_\alpha(U_\alpha) \Big/ \sim,$$

where

$$(\alpha, x) \sim (\beta, y) \quad \overset{\text{def}}{\Longleftrightarrow} \quad \phi_\alpha^{-1}(x) = \phi_\beta^{-1}(y).$$

for $\alpha, \beta \in A$, $x \in \phi_\alpha(U_\alpha)$, and $y \in \phi_\beta(U_\beta)$. Define an atlas on $\widetilde{M}$ by

$$\widetilde{U}_\alpha := \{[\alpha, x] \mid x \in \phi_\alpha(U_\alpha)\}, \qquad \widetilde{\phi}_\alpha([\alpha, x]) := x.$$

Prove that $\widetilde{M}$ is a smooth m-manifold and that it is diffeomorphic to M.

Exercise 2.8.13   Prove that $\mathbb{C}P^1$ is diffeomorphic to S^2. **Hint:** Stereographic projection.

2.8.3   Tangent Spaces and Derivatives

In the situation where M is a submanifold of Euclidean space and $p \in M$ we have defined the tangent space of M at p as the set of all derivatives $\dot{\gamma}(0)$ of smooth curves $\gamma : \mathbb{R} \to M$ that pass through $p = \gamma(0)$. We cannot do this for manifolds in the intrinsic sense, as the derivative of a curve has yet to be defined. In fact, the purpose of introducing a tangent space of M is precisely to allow us to define what we mean by the derivative of a smooth map. There are two approaches. One is to introduce an appropriate equivalence relation on the set of curves through p and the other is to use local coordinates.

Definition 2.8.14   Let M be a manifold with an atlas $\mathcal{A} = \{(\phi_\alpha, U_\alpha)\}_{\alpha \in A}$ and let $p \in M$. Two smooth curves $\gamma_0, \gamma_1 : \mathbb{R} \to M$ with $\gamma_0(0) = \gamma_1(0) = p$ are called p-**equivalent** iff for some (and hence every) $\alpha \in A$ with $p \in U_\alpha$ we have

$$\frac{d}{dt}\bigg|_{t=0} \phi_\alpha(\gamma_0(t)) = \frac{d}{dt}\bigg|_{t=0} \phi_\alpha(\gamma_1(t)).$$

We write $\gamma_0 \overset{p}{\sim} \gamma_1$ iff γ_0 is p-equivalent to γ_1 and denote the equivalence class of a smooth curve $\gamma : \mathbb{R} \to M$ with $\gamma(0) = p$ by $[\gamma]_p$. Every such equivalence class is called a **tangent vector** of M at p. The **tangent space** of M at p is the set of equivalence classes

$$T_p M := \{[\gamma]_p \mid \gamma : \mathbb{R} \to M \text{ is smooth and } \gamma(0) = p\}. \qquad (2.8.4)$$

Definition 2.8.15   Let M be a manifold with an atlas $\mathcal{A} = \{(\phi_\alpha, U_\alpha)\}_{\alpha \in A}$ and let $p \in M$. The $\mathcal{A}$-**tangent space** of M at p is the quotient space

$$T_p^{\mathcal{A}} M := \bigcup_{p \in U_\alpha} \{\alpha\} \times \mathbb{R}^m \bigg/ \overset{p}{\sim}, \qquad (2.8.5)$$

where the union runs over all $\alpha \in A$ with $p \in U_\alpha$ and

$$(\alpha, \xi) \overset{p}{\sim} (\beta, \eta) \qquad \Longleftrightarrow \qquad d\left(\phi_\beta \circ \phi_\alpha^{-1}\right)(x)\xi = \eta, \quad x := \phi_\alpha(p).$$

Each equivalence class $[\alpha, \xi]_p$ is called a **tangent vector** of M at p.

In Definition 2.8.14 it is not immediately obvious that the set $T_p M$ in (2.8.4) is a vector space. However, the quotient space $T_p^{\mathcal{A}} M$ in (2.8.5) is obviously a vector space of dimension m and there is a natural bijection

$$T_p M \to T_p^{\mathcal{A}} M : [\gamma]_p \mapsto \left[\alpha, \frac{d}{dt}\bigg|_{t=0} \phi_\alpha(\gamma(t))\right]_p. \tag{2.8.6}$$

This bijection induces a vector space structure on the set $T_p M$. In other words, the set $T_p M$ in (2.8.4) admits a unique vector space structure such that the map $T_p M \to T_p^{\mathcal{A}} M$ in (2.8.6) is a vector space isomorphism.

Exercise 2.8.16 Verify the phrase *"and hence every"* in Definition 2.8.14 and deduce that the map $T_p M \to T_p^{\mathcal{A}} M$ in (2.8.6) is well defined. Show that it is bijective.

From now on we will use either Definition 2.8.14 or Definition 2.8.15 or both, whichever way is most convenient, and drop the superscript $\mathcal{A}$.

Definition 2.8.17 (Derivative of a smooth curve) For each smooth curve $\gamma : \mathbb{R} \to M$ with $\gamma(0) = p$ we define the derivative $\dot{\gamma}(0) \in T_p M$ as the equivalence class

$$\dot{\gamma}(0) := [\gamma]_p \cong \left[\alpha, \frac{d}{dt}\bigg|_{t=0} \phi_\alpha(\gamma(t))\right]_p \in T_p M.$$

Definition 2.8.18 (Derivative of a smooth map) Let $f : M \to N$ be a smooth map between two smooth manifolds $(M, \{(\phi_\alpha, U_\alpha)\}_{\alpha \in A})$ and $(N, \{(\psi_\beta, V_\beta)\}_{\beta \in B})$ and let $p \in M$. The derivative of f at p is the map

$$df(p) : T_p M \to T_{f(p)} N$$

defined by the formula

$$df(p)[\gamma]_p := [f \circ \gamma]_{f(p)} \tag{2.8.7}$$

for each smooth curve $\gamma : \mathbb{R} \to M$ with $\gamma(0) = p$. Here we use (2.8.4). Under the isomorphism (2.8.6) this corresponds to the linear map

$$df(p)[\alpha, \xi]_p := [\beta, df_{\beta\alpha}(x)\xi]_{f(p)}, \qquad x := \phi_\alpha(p), \tag{2.8.8}$$

for $\alpha \in A$ with $p \in U_\alpha$ and $\beta \in B$ with $f(p) \in V_\beta$, where $f_{\beta\alpha}$ is given by (2.8.3).

Remark 2.8.19 Think of $N = \mathbb{R}^n$ as a manifold with a single coordinate chart, namely the identity map $\psi_\beta = \mathrm{id} : \mathbb{R}^n \to \mathbb{R}^n$. For every $q \in N = \mathbb{R}^n$ the tangent space $T_q N$ is then canonically isomorphic to $\mathbb{R}^n$ via (2.8.5). Thus for every smooth map $f : M \to \mathbb{R}^n$ the derivative of f at $p \in M$ is a linear map $df(p) : T_p M \to \mathbb{R}^n$, and the formula (2.8.8) reads

$$df(p)[\alpha, \xi]_p = d(f \circ \phi_\alpha^{-1})(x)\xi, \qquad x := \phi_\alpha(p).$$

This formula also applies to maps defined on some open subset of M. In particular, with $f = \phi_\alpha : U_\alpha \to \mathbb{R}^m$ we have

$$d\phi_\alpha(p)[\alpha, \xi]_p = \xi.$$

Thus the map $d\phi_\alpha(p) : T_p M \to \mathbb{R}^m$ is the canonical vector space isomorphism determined by α.

With these definitions the derivative of f at p is a linear map and we have the chain rule for the composition of two smooth maps as in Theorem 2.2.14.

2.8.4 Submanifolds and Embeddings

Definition 2.8.20 (Submanifold) Let M be a smooth m-manifold and let $n \in \{0, 1, \ldots, m\}$. A subset $N \subset M$ is called an n-dimensional **submanifold** of M iff, for every $p \in N$, there exists a local coordinate chart $\phi : U \to \Omega$ for M, defined on an an open neighborhood $U \subset M$ of p and with values in an open set $\Omega \subset \mathbb{R}^n \times \mathbb{R}^{m-n}$, such that $\phi(U \cap N) = \Omega \cap (\mathbb{R}^n \times \{0\})$.

By Theorem 2.1.10 an m-manifold $M \subset \mathbb{R}^k$ in the sense of Definition 2.1.3 is a submanifold of $\mathbb{R}^k$ in the sense of Definition 2.8.20. By Theorem 2.3.4 the notion of a submanifold $N \subset M$ of a manifold $M \subset \mathbb{R}^k$ in Definition 2.3.1 agrees with the notion of a submanifold in Definition 2.8.20.

Exercise 2.8.21 Let N be a submanifold of M. Show that if M is Hausdorff, so is N, and if M is paracompact, so is N.

Exercise 2.8.22 Let N be a submanifold of M and let P be a submanifold of N. Prove that P is a submanifold of M. **Hint:** Use Theorem 2.1.10.

Exercise 2.8.23 Let N be a submanifold of M. Prove that there exists an open set $U \subset M$ such that $N \subset U$ and N is closed in the relative topology of U.

All the theorems we have proved for embedded manifolds and their proofs carry over almost word for word to the present setting. For example we have the inverse function theorem, the notion of a regular value, the notions of a submersion and of an immersion, the notion of an embedding as a proper injective immersion, and the fact from Theorem 2.3.4 that a subset $P \subset M$ is a submanifold if and only if it is the image of an embedding.

Exercise 2.8.24 (Lines in Euclidean space)   The tangent bundle of the 2-sphere is the 4-manifold

$$TS^2 = \{(x, y) \in \mathbb{R}^3 \times \mathbb{R}^3 \mid |x| = 1, \langle x, y \rangle = 0\}$$

(see Example 2.2.6). Define an equivalence relation on TS^2 by

$$(x, y) \sim (x', y') \qquad \overset{\text{def}}{\Longleftrightarrow} \qquad x' = \pm x, \quad y' = y$$

for $(x, y), (x', y') \in TS^2$. Show that the quotient space $TS^2/\sim$ can be identified with the set L of all lines in $\mathbb{R}^3$, by assigning to each pair $(x, y) \in TS^2$ the line $\ell_{x,y} := \{y + tx \mid t \in \mathbb{R}\} \subset \mathbb{R}^3$. Show that the space L of lines in $\mathbb{R}^3$ admits the unique structure of a smooth manifold such that the canonical projection $TS^2 \to L : (x, y) \mapsto \ell_{x,y}$ is a submersion. Show that the manifold topology on L agrees with the quotient topology on $TS^2/\sim$. Show that the map $L \to \mathbb{RP}^2 \times \mathbb{R}^3 : \ell_{x,y} \mapsto ([x], y)$ is an embedding.

Example 2.8.25 (Veronese embedding)   The map

$$\mathbb{CP}^2 \to \mathbb{CP}^5 : [z_0 : z_1 : z_2] \mapsto [z_0^2 : z_1^2 : z_2^2 : z_1 z_2 : z_2 z_0 : z_0 z_1]$$

is an embedding. (**Exercise:** Prove this.) It restricts to an embedding of the real projective plane into $\mathbb{RP}^5$ and also gives rise to embeddings of $\mathbb{RP}^2$ into $\mathbb{R}^4$ as well as to the Roman surface: an immersion of $\mathbb{RP}^2$ into $\mathbb{R}^3$. (See Example 2.1.17.) There are similar embeddings

$$\mathbb{CP}^n \to \mathbb{CP}^{N-1}, \qquad N := \binom{n + d}{d},$$

for all n and d, defined in terms of monomials of degree d in $n + 1$ variables. These are the **Veronese embeddings**.

Example 2.8.26 (Plücker embedding)   The Grassmannian $G_2(\mathbb{R}^4)$ of 2-planes in $\mathbb{R}^4$ is a smooth 4-manifold and can be expressed as the quotient of the space $\mathcal{F}_2(\mathbb{R}^4)$ of orthonormal 2-frames in $\mathbb{R}^4$ by the orthogonal group $O(2)$. (See Example 2.8.8.) Write an orthonormal 2-frame in $\mathbb{R}^4$ as a matrix

$$D = \begin{pmatrix} x_0 & y_0 \\ x_1 & y_1 \\ x_2 & y_2 \\ x_3 & y_3 \end{pmatrix}, \qquad D^\mathsf{T} D = \mathbb{1}.$$

Then the map $f : G_2(\mathbb{R}^4) \to \mathbb{R}P^5$, defined by

$$f([D]) := [p_{01} : p_{02} : p_{03} : p_{23} : p_{31} : p_{12}], \qquad p_{ij} := x_i y_j - x_j y_i,$$

is an embedding and its image is the quadric

$$X := f(G_2(\mathbb{R}^4)) = \{p \in \mathbb{R}P^5 \mid p_{01} p_{23} + p_{02} p_{31} + p_{03} p_{12} = 0\}.$$

(**Exercise:** Prove this.) There are analogous embeddings

$$f : G_k(\mathbb{R}^n) \to \mathbb{R}P^{N-1}, \qquad N := \binom{n}{k},$$

for all k and n, defined in terms of the $k \times k$-minors of the (orthonormal) frames. These are the **Plücker embeddings**.

2.8.5   Tangent Bundle and Vector Fields

Let M be a m-manifold with an atlas $\mathcal{A} = \{(\phi_\alpha, U_\alpha)\}_{\alpha \in A}$. The **tangent bundle** of M is defined as the disjoint union of the tangent spaces, i.e.

$$TM := \bigcup_{p \in M} \{p\} \times T_p M = \{(p, v) \mid p \in M, \ v \in T_p M\}.$$

Denote by $\pi : TM \to M$ the projection given by $\pi(p, v) := p$. Recall the notion of a submersion as a smooth map between smooth manifolds, whose derivative is surjective at each point.

Lemma 2.8.27 *The tangent bundle of M is a smooth $2m$-manifold with coordinate charts*

$$\widetilde{\phi}_\alpha : \widetilde{U}_\alpha := \pi^{-1}(U_\alpha) \to \phi_\alpha(U_\alpha) \times \mathbb{R}^m, \qquad \widetilde{\phi}_\alpha(p, v) := (\phi_\alpha(p), d\phi_\alpha(p)v).$$

The projection $\pi : TM \to M$ is a surjective submersion. If M is second countable and Hausdorff, so is TM.

Proof   For each pair $\alpha, \beta \in A$ the set

$$\widetilde{\phi}_\alpha(\widetilde{U}_\alpha \cap \widetilde{U}_\beta) = \phi_\alpha(U_\alpha \cap U_\beta) \times \mathbb{R}^m$$

is open in $\mathbb{R}^m \times \mathbb{R}^m$ and the transition map

$$\widetilde{\phi}_{\beta\alpha} := \widetilde{\phi}_\beta \circ \widetilde{\phi}_\alpha^{-1} : \widetilde{\phi}_\alpha(\widetilde{U}_\alpha \cap \widetilde{U}_\beta) \to \widetilde{\phi}_\beta(\widetilde{U}_\alpha \cap \widetilde{U}_\beta)$$

is given by

$$\widetilde{\phi}_{\beta\alpha}(x, \xi) = \left(\phi_{\beta\alpha}(x), d\phi_{\beta\alpha}(x)\xi\right)$$

for $x \in \phi_\alpha(U_\alpha \cap U_\beta)$ and $\xi \in \mathbb{R}^m$ where

$$\phi_{\beta\alpha} := \phi_\beta \circ \phi_\alpha^{-1}.$$

Thus the transition maps are all diffeomorphisms and so the coordinate charts $\widetilde{\phi}_\alpha$ define an atlas on TM. The topology on TM is determined by this atlas via Definition 2.8.2. If M has a countable atlas, so does TM. The remaining assertions are easy exercises.   □

Definition 2.8.28 Let M be a smooth m-manifold. A **(smooth) vector field** on M is a collection of tangent vectors $X(p) \in T_pM$, one for each point $p \in M$, such that the map $M \to TM : p \mapsto (p, X(p))$ is smooth. The set of all smooth vector fields on M will be denoted by Vect(M).

Associated to a vector field is a smooth map $M \to TM$ whose composition with the projection $\pi : TM \to M$ is the identity map on M. Strictly speaking this map should be denoted by a symbol other than X, for example by $\widetilde{X}$. However, it is convenient at this point, and common practice, to slightly abuse notation and denote the map from M to TM also by X. Thus a vector field can be defined as a smooth map

$$X : M \to TM$$

such that

$$\pi \circ X = \text{id} : M \to M.$$

Such a map is also called a *section of the tangent bundle*.

Now suppose $\mathcal{A} = \{(\phi_\alpha, U_\alpha)\}_{\alpha \in A}$ is an atlas on M and $X : M \to TM$ is a vector field on M. Then X determines a collection of smooth maps

$$X_\alpha : \phi_\alpha(U_\alpha) \to \mathbb{R}^m$$

given by

$$X_\alpha(x) := d\phi_\alpha(p)X(p), \qquad p := \phi_\alpha^{-1}(x), \qquad (2.8.9)$$

for $x \in \phi_\alpha(U_\alpha)$. We can think of each X_α as a vector field on the open set $\phi_\alpha(U_\alpha) \subset \mathbb{R}^m$, representing the vector field X on the coordinate patch U_α. These local vector fields X_α satisfy the condition

$$X_\beta(\phi_{\beta\alpha}(x)) = d\phi_{\beta\alpha}(x)X_\alpha(x) \qquad (2.8.10)$$

for $x \in \phi_\alpha(U_\alpha \cap U_\beta)$. This equation can also be expressed in the form

$$X_\alpha|_{\phi_\alpha(U_\alpha \cap U_\beta)} = \phi^*_{\beta\alpha} X_\beta|_{\phi_\beta(U_\alpha \cap U_\beta)}. \qquad (2.8.11)$$

Conversely, any collection of smooth maps $X_\alpha : \phi_\alpha(U_\alpha) \to \mathbb{R}^m$ satisfying (2.8.10) determines a unique vectorfield X on M via (2.8.9). Thus we can define the Lie bracket of two vector fields $X, Y \in \mathrm{Vect}(M)$ by

$$[X, Y]_\alpha(x) := [X_\alpha, Y_\alpha](x) = dX_\alpha(x)Y_\alpha(x) - dY_\alpha(x)X_\alpha(x) \qquad (2.8.12)$$

for $\alpha \in A$ and $x \in \phi_\alpha(U_\alpha)$. It follows from equation (2.4.18) in Lemma 2.4.21 that the local vector fields

$$[X, Y]_\alpha : \phi_\alpha(U_\alpha) \to \mathbb{R}^m$$

satisfy (2.8.11) and hence determine a unique vector field $[X, Y]$ on M via

$$[X, Y](p) := d\phi_\alpha(p)^{-1}[X_\alpha, Y_\alpha](\phi_\alpha(p)), \qquad p \in U_\alpha. \qquad (2.8.13)$$

Thus the **Lie bracket** of X and Y is defined on U_α as the pullback of the Lie bracket of the vector fields X_α and Y_α under the coordinate chart ϕ_α. With this understood all the results in Sect. 2.4 about vector fields and flows along with their proofs carry over word for word to the intrinsic setting whenever M is a Hausdorff space. This includes the existence and uniquess result for integral curves in Theorem 2.4.7, the concept of the flow of a vector field in Definition 2.4.8 and its properties in Theorem 2.4.9, the notion of completeness of a vector field (that the integral curves exist for all time), and the various properties of the Lie bracket such as the Jacobi identity (2.4.20), the formulas in Lemma 2.4.18, and the fact that the Lie bracket of two vector fields vanishes if and only if the corresponding flows commute (see Lemma 2.4.26). One can also carry over the notion of a **subbundle** $E \subset TM$ **of rank** n to the intrinsic setting by the condition that E is a smooth submanifold of TM and intersects each fiber T_pM in an n-dimensional linear subspace

$$E_p := \{v \in T_pM \mid (p, v) \in E\}.$$

Then the characterization of subbundles in Theorem 2.6.10 and the theorem of Frobenius 2.7.2 including their proofs also carry over to the intrinsic setting whenever M is a Hausdorff space.

2.8.6 Coordinate Notation

Fix a coordinate chart $\phi_\alpha : U_\alpha \to \mathbb{R}^m$ on an m-manifold M. The components of ϕ_α are smooth real valued functions on the open subset U_α of M and it is customary to denote them by

$$x^1, \ldots, x^m : U_\alpha \to \mathbb{R}.$$

The derivatives of these functions at $p \in U_\alpha$ are linear functionals

$$dx^i(p) : T_p M \to \mathbb{R}, \qquad i = 1, \ldots, m. \qquad (2.8.14)$$

They form a basis of the dual space

$$T_p^* M := \text{Hom}(T_p M, \mathbb{R}).$$

(A coordinate chart on M can in fact be characterized as an m-tuple of real valued functions on an open subset of M whose derivatives are everywhere linearly independent and which, taken together, form an injective map.) The dual basis of $T_p M$ will be denoted by

$$\frac{\partial}{\partial x^1}(p), \ldots, \frac{\partial}{\partial x^m}(p) \in T_p M. \qquad (2.8.15)$$

Thus

$$dx^i(p) \frac{\partial}{\partial x^j}(p) = \delta_j^i := \begin{cases} 1, & \text{if } i = j, \\ 0, & \text{if } i \neq j, \end{cases}$$

for $i, j = 1, \ldots, m$ and $\partial/\partial x^i$ is a vector field on the coordinate patch U_α. For each $p \in U_\alpha$ it is the canonical basis of $T_p M$ determined by ϕ_α. In the notation of (2.8.5) and Remark 2.8.19 we have

$$\frac{\partial}{\partial x^i}(p) = [\alpha, e_i]_p = d\phi_\alpha(p)^{-1} e_i,$$

where $e_i = (0, \ldots, 0, 1, 0, \ldots, 0)$ (with 1 in the ith place) denotes the standard basis vector of $\mathbb{R}^m$. In other words, for all $\xi = (\xi^1, \ldots, \xi^m) \in \mathbb{R}^m$ and all $p \in U_\alpha$, the tangent vector

$$v := d\phi_\alpha(p)^{-1}\xi \in T_p M$$

is given by

$$v = [\alpha, \xi]_p = \sum_{i=1}^m \xi^i \frac{\partial}{\partial x^i}(p). \qquad (2.8.16)$$

Thus the restriction of a vector field $X \in \text{Vect}(M)$ to U_α has the form

$$X|_{U_\alpha} = \sum_{i=1}^m \xi^i \frac{\partial}{\partial x^i},$$

where $\xi^1, \ldots, \xi^m : U_\alpha \to \mathbb{R}$ are smooth real valued functions. If the map

$$X_\alpha : \phi_\alpha(U_\alpha) \to \mathbb{R}^m$$

is defined by (2.8.9), then

$$X_\alpha \circ \phi_\alpha^{-1} = (\xi^1, \ldots, \xi^m).$$

The above notation is motivated by the observation that the derivative of a smooth function $f : M \to \mathbb{R}$ in the direction of a vector field X on a coordinate patch U_α is given by

$$\mathcal{L}_X f|_{U_\alpha} = \sum_{i=1}^{m} \xi^i \frac{\partial f}{\partial x^i}.$$

Here the term $\partial f / \partial x^i$ is understood as first writing f as a function of $x^1, \ldots, x^m$, then taking the partial derivative, and afterwards expressing this partial derivative again as a function of p. Thus $\partial f / \partial x^i$ is the shorthand notation for the function $\left(\frac{\partial}{\partial x^i}(f \circ \phi_\alpha^{-1})\right) \circ \phi_\alpha : U_\alpha \to \mathbb{R}$.

2.9 Consequences of Paracompactness*

In geometry it is often necessary to turn a construction in local coordinates into a global geometric object. A key technical tool for such *"local to global"* constructions is an existence theorem for partitions of unity.

2.9.1 Paracompactness

The existence of a countable atlas is of fundamental importance for almost everything we will prove about manifolds. The next two remarks describe several equivalent conditions.

Remark 2.9.1 Let M be a smooth manifold and denote by

$$\mathcal{U} \subset 2^M$$

the topology induced by the atlas as in Definition 2.8.2. Then the following are equivalent.

(a) M admits a countable atlas.
(b) M is σ-**compact**, i.e. there is a sequence of compact subsets $K_i \subset M$ such that $K_i \subset \mathrm{int}(K_{i+1})$ for every $i \in \mathbb{N}$ and $M = \bigcup_{i=1}^{\infty} K_i$.
(c) Every open cover of M has a countable subcover.
(d) M is **second countable**, i.e. there is a countable collection of open sets $\mathcal{B} \subset \mathcal{U}$ such that every open set $U \in \mathcal{U}$ is a union of open sets from the collection $\mathcal{B}$. ($\mathcal{B}$ is then called a **countable base** for the topology of M.)

That (a) $\Longrightarrow$ (b) $\Longrightarrow$ (c) $\Longrightarrow$ (a) and (a) $\Longrightarrow$ (d) follows directly from the definitions. The proof that (d) implies (a) requires the construction of a countable refinement and the axiom of choice. (A **refinement** of an open cover $\{U_i\}_{i \in I}$ is an open cover $\{V_j\}_{j \in J}$ such that each set V_j is contained in one of the sets U_i.)

Remark 2.9.2 Let M and $\mathcal{U}$ be as in Remark 2.9.1 and suppose in addition that M is a connected Hausdorff space. Then the existence of a countable atlas is also equivalent to each of the following conditions.

(e) M is **metrizable**, i.e. there is a distance function $d : M \times M \to [0, \infty)$ such that $\mathcal{U}$ is the topology induced by d.
(f) M is **paracompact**, i.e. every open cover of M has a locally finite refinement. (An open cover $\{V_j\}_{j \in J}$ is called **locally finite** iff every $p \in M$ has a neighborhood that intersects only finitely many V_j.)

That (a) implies (e) follows from the **Urysohn Metrization Theorem** which asserts (in its original form) that every normal second countable topological space is metrizable [51, Theorem 34.1]. A topological space M is called **normal** iff points are closed and, for any two disjoint closed sets $A, B \subset M$, there are disjoint open sets $U, V \subset M$ such that $A \subset U$ and $B \subset V$. It is called **regular** iff points are closed and, for every closed set $A \subset M$ and every $b \in M \setminus A$, there are disjoint open sets $U, V \subset M$ such that $A \subset U$ and $b \in V$. It is called **locally compact** iff, for every open set $U \subset M$ and every $p \in U$, there is a compact neighborhood of p contained in U. It is easy to show that every manifold is locally compact and every locally compact Hausdorff space is regular. **Tychonoff's Lemma** asserts that a regular topological space with a countable base is normal [51, Theorem 32.1]. Hence it follows from the Urysohn Metrization Theorem that every Hausdorff manifold with a countable base is metrizable. That (e) implies (f) follows from a more general theorem which asserts that every metric space is paracompact (see [51, Theorem 41.4] and [62]). Conversely, the **Smirnov Metrization Theorem** asserts that a paracompact Hausdorff space is metrizable if and only if it is locally metrizable, i.e. every point has a metrizable neighborhood (see [51, Theorem 42.1]). Since every manifold is locally metrizable this shows that (f) implies (e). Thus we have (a) $\Longrightarrow$ (e) $\Longleftrightarrow$ (f) for every Hausdorff manifold.

The proof that (f) implies (a) does not require the Hausdorff property but we do need the assumption that M is connected. (A manifold with uncountably many connected components, each of which is paracompact, is itself paracompact but does not admit a countable atlas.) Here is a sketch. If M is a paracompact manifold, then there is a locally finite open cover $\{U_\alpha\}_{\alpha \in A}$ by coordinate charts. Since each set U_α has a countable dense subset, the set $\{\alpha \in A \mid U_\alpha \cap U_{\alpha_0} \neq \emptyset\}$ is at most countable for each $\alpha_0 \in A$. Since M is connected we can reach each point from U_{α_0} through a finite sequence of sets $U_{\alpha_1}, \dots, U_{\alpha_\ell}$ with $U_{\alpha_{i-1}} \cap U_{\alpha_i} \neq \emptyset$. This implies that the index set A is countable and hence M admits a countable atlas.

Remark 2.9.3 A **Riemann surface** is a 1-dimensional complex manifold (i.e. the coordinate charts take values in $\mathbb{C}$ and the transition maps are holomorphic) with a Hausdorff topology. It is a deep theorem in the theory of Riemann surfaces that every connected Riemann surface is necessarily second countable (see [2]). Thus pathological examples of the type discussed in Example 2.8.10 cannot be constructed with holomorphic transition maps.

Exercise 2.9.4 Prove that every manifold is locally compact. Find an example of a manifold M and a point $p_0 \in M$ such that every closed neighborhood of p_0 is non-compact. **Hint:** The example is necessarly non-Hausdorff.

Exercise 2.9.5 Prove that a manifold M admits a countable atlas if and only if it is σ-compact if and only if every open cover of M has a countable subcover if and only if it is second countable. **Hint:** The topology of $\mathbb{R}^m$ is second countable and every open subset of $\mathbb{R}^m$ is σ-compact.

Exercise 2.9.6 Prove that every submanifold $M \subset \mathbb{R}^k$ (Definition 2.1.3) is second countable.

Exercise 2.9.7 Prove that every connected component of a manifold M is an open subset of M and is path-connected.

2.9.2 Partitions of Unity

Definition 2.9.8 (Partition of unity) Let M be a smooth manifold. A **partition of unity** on M is a collection of smooth functions

$$\theta_\alpha : M \to [0, 1], \qquad \alpha \in A,$$

such that each point $p \in M$ has an open neighborhood $V \subset M$ on which only finitely many θ_α do not vanish, i.e.

$$\#\{\alpha \in A \mid \theta_\alpha|_V \not\equiv 0\} < \infty, \tag{2.9.1}$$

and, for every $p \in M$, we have

$$\sum_{\alpha \in A} \theta_\alpha(p) = 1. \tag{2.9.2}$$

If $\{U_\alpha\}_{\alpha \in A}$ is an open cover of M, then a partition of unity $\{\theta_\alpha\}_{\alpha \in A}$ (indexed by the same set A) is called **subordinate to the cover** iff each θ_α is supported in U_α, i.e.

$$\mathrm{supp}(\theta_\alpha) := \overline{\{p \in M \mid \theta_\alpha(p) \neq 0\}} \subset U_\alpha.$$

Theorem 2.9.9 (Partitions of unity) *Let M be a smooth manifold whose topology is paracompact and Hausdorff. Then, for every open cover of M, there exists a partition of unity subordinate to that cover.*

The proof requires two preparatory lemmas.

Lemma 2.9.10 *Let M be a smooth manifold with a Hausdorff topology. Then, for every open set $V \subset M$ and every compact set $K \subset V$, there exists a smooth function $\kappa : M \to [0, \infty)$ with compact support such that $\kappa(p) > 0$ for every $p \in K$ and $\mathrm{supp}(\kappa) \subset V$.*

Proof Assume first that $K = \{p_0\}$ is a single point. Since M is a manifold it is locally compact. Hence there is a compact neighborhood $C \subset V$ of p_0. Since M is Hausdorff C is closed and hence the set $U := \mathrm{int}(C)$ is a neighborhood of p_0 whose closure $\overline{U} \subset C$ is compact and contained in V. Shrinking U, if necessary, we may assume that there is a coordinate chart $\phi : U \to \Omega$ with values in some open neighborhood $\Omega \subset \mathbb{R}^m$ of the origin such that $\phi(p_0) = 0$. (Here m is the dimension of M.) Now choose a smooth function $\kappa_0 : \Omega \to [0, \infty)$ with compact support such that $\kappa_0(0) > 0$. Then the function $\kappa : M \to [0, 1]$, defined by $\kappa|_U := \kappa_0 \circ \phi$ and $\kappa(p) := 0$ for $p \in M \setminus U$ is supported in V and satisfies $\kappa(p_0) > 0$. This proves the lemma in the case where K is a point.

Now let K be any compact subset of V. Then, by the first part of the proof, there is a collection of smooth functions $\kappa_p : M \to [0, \infty)$, one for every $p \in K$, such that $\kappa_p(p) > 0$ and $\mathrm{supp}(\kappa_p) \subset V$. Since K is compact there are finitely many points $p_1, \ldots, p_k \in K$ such that the sets $\{p \in M \mid \kappa_{p_j}(p) > 0\}$ cover K. Hence the function $\kappa := \sum_j \kappa_{p_j}$ is supported in V and is everywhere positive on K. This proves Lemma 2.9.10. $\qquad\square$

Lemma 2.9.11 *Let M be a topological space. If $\{V_i\}_{i \in I}$ is a locally finite collection of open sets in M, then*

$$\overline{\bigcup_{i \in I_0} V_i} = \bigcup_{i \in I_0} \overline{V}_i$$

for every subset $I_0 \subset I$.

Proof The set $\bigcup_{i \in I_0} \overline{V}_i$ is obviously contained in the closure of $\bigcup_{i \in I_0} V_i$. To prove the converse choose a point $p_0 \in M \setminus \bigcup_{i \in I_0} \overline{V}_i$. Since the collection $\{V_i\}_{i \in I}$ is locally finite, there exists an open neighborhood U of p_0 such that the set $I_1 := \{i \in I \mid V_i \cap U \neq \emptyset\}$ is finite. Hence the set

$$U_0 := U \setminus \bigcup_{i \in I_0 \cap I_1} \overline{V}_i$$

is an open neighborhood of p_0 and we have $U_0 \cap V_i = \emptyset$ for every $i \in I_0$. Hence $p_0 \notin \overline{\bigcup_{i \in I_0} V_i}$. This proves Lemma 2.9.11. $\qquad\square$

Proof of Theorem 2.9.9   Let $\{U_\alpha\}_{\alpha \in A}$ be an open cover of M. We prove in four steps that there is a partition of unity subordinate to this cover. The proofs of steps one and two are taken from [51, Lemma 41.6].

Step 1 *There is a locally finite open cover $\{V_i\}_{i \in I}$ of M such that, for every $i \in I$, the closure $\overline{V}_i$ is compact and contained in one of the sets U_α.*

Denote by $\mathcal{V} \subset 2^M$ the set of all open sets $V \subset M$ such that $\overline{V}$ is compact and $\overline{V} \subset U_\alpha$ for some $\alpha \in A$. Since M is a locally compact Hausdorff space the collection $\mathcal{V}$ is an open cover of M. (If $p \in M$, then there is an $\alpha \in A$ such that $p \in U_\alpha$; since M is locally compact there is a compact neighborhood $K \subset U_\alpha$ of p; since M is Hausdorff K is closed and thus $V := \text{int}(K)$ is an open neighborhood of p with $\overline{V} \subset K \subset U_\alpha$.) Since M is paracompact the open cover $\mathcal{V}$ has a locally finite refinement $\{V_i\}_{i \in I}$. This cover satisfies the requirements of Step 1.

Step 2 *There is a collection of compact sets $K_i \subset V_i$, one for each $i \in I$, such that $M = \bigcup_{i \in I} K_i$.*

Denote by $\mathcal{W} \subset 2^M$ the set of all open sets $W \subset M$ such that $\overline{W} \subset V_i$ for some i. Since M is a locally compact Hausdorff space, the collection $\mathcal{W}$ is an open cover of M. Since M is paracompact $\mathcal{W}$ has a locally finite refinement $\{W_j\}_{j \in J}$. By the axiom of choice there is a map

$$J \to I : j \mapsto i_j$$

such that

$$\overline{W}_j \subset V_{i_j} \qquad \forall \, j \in J.$$

Since the collection $\{W_j\}_{j \in J}$ is locally finite, we have

$$K_i := \overline{\bigcup_{i_j = i} W_j} = \bigcup_{i_j = i} \overline{W}_j \subset V_i$$

by Lemma 2.9.11. Since $\overline{V}_i$ is compact so is K_i.

Step 3 *There is a partition of unity subordinate to the cover $\{V_i\}_{i \in I}$.*

Choose a collection of compact sets $K_i \subset V_i$ for $i \in I$ as in Step 2. Then, by Lemma 2.9.10 and the axiom of choice, there is a collection of smooth functions $\kappa_i : M \to [0, \infty)$ with compact support such that

$$\text{supp}(\kappa_i) \subset V_i, \qquad \kappa_i|_{K_i} > 0 \qquad \forall \, i \in I.$$

Since the cover $\{V_i\}_{i \in I}$ is locally finite the sum

$$\kappa := \sum_{i \in I} \kappa_i : M \to \mathbb{R}$$

is **locally finite** (i.e. each point in M has a neighborhood in which only finitely many terms do not vanish) and thus defines a smooth function on M. This function is everywhere positive, because each summand is nonnegative and, for each $p \in M$, there is an $i \in I$ with $p \in K_i$ so that $\kappa_i(p) > 0$. Thus the funtions $\chi_i := \kappa_i/\kappa$ define a partition of unity satisfying $\text{supp}(\chi_i) \subset V_i$ for every $i \in I$ as required.

Step 4 *There is a partition of unity subordinate to the cover $\{U_\alpha\}_{\alpha \in A}$.*

Let $\{\chi_i\}_{i \in I}$ be the partition of unity constructed in Step 3. By the axiom of choice there is a map $I \to A : i \mapsto \alpha_i$ such that $V_i \subset U_{\alpha_i}$ for every $i \in I$. For $\alpha \in A$ define $\theta_\alpha : M \to [0, 1]$ by

$$\theta_\alpha := \sum_{\alpha_i = \alpha} \chi_i.$$

Here the sum runs over all indices $i \in I$ with $\alpha_i = \alpha$. This sum is locally finite and hence is a smooth function on M. Moreover, each point in M has an open neighborhood in which only finitely many of the θ_α do not vanish. Hence the sum of the θ_α is a well defined function on M and

$$\sum_{\alpha \in A} \theta_\alpha = \sum_{\alpha \in A} \sum_{\alpha_i = \alpha} \chi_i = \sum_{i \in I} \chi_i \equiv 1.$$

This shows that the functions θ_α form a partition of unity. To prove the inclusion $\text{supp}(\theta_\alpha) \subset U_\alpha$ we consider the open sets

$$W_i := \{p \in M \mid \chi_i(p) > 0\}$$

for $i \in I$. Since $W_i \subset V_i$ this collection is locally finite. Hence, by Lemma 2.9.11, we have

$$\text{supp}(\theta_\alpha) = \overline{\bigcup_{\alpha_i = \alpha} W_i} = \bigcup_{\alpha_i = \alpha} \overline{W_i} = \bigcup_{\alpha_i = \alpha} \text{supp}(\chi_i) \subset \bigcup_{\alpha_i = \alpha} V_i \subset U_\alpha.$$

This proves Theorem 2.9.9. $\qquad\qquad\qquad\qquad\qquad\qquad\qquad\qquad\qquad\qquad\qquad\qquad$ $\square$

2.9.3 Embedding in Euclidean Space

Theorem 2.9.12 *Let M be a second countable smooth m-manifold with a Hausdorff topology. Then there exists an embedding $f : M \to \mathbb{R}^{2m+1}$ with a closed image.*

Proof The proof has five steps.

Step 1 *Let $U \subset M$ be an open set and let $K \subset U$ be a compact set. Then there exists an integer $k \in \mathbb{N}$, a smooth map $f : M \to \mathbb{R}^k$, and an open set $V \subset M$, such that $K \subset V \subset U$, the restriction $f|_V : V \to \mathbb{R}^k$ is an injective immersion, and $f(p) = 0$ for all $p \in M \setminus U$.*

Choose a smooth atlas $\mathcal{A} = \{(\phi_\alpha, U_\alpha)\}_{\alpha \in A}$ on M such that, for each $\alpha \in A$, either $U_\alpha \subset U$ or $U_\alpha \cap K = \emptyset$. Since M is a paracompact Hausdorff manifold, Theorem 2.9.9 asserts that there exists a partition of unity $\{\theta_\alpha\}_{\alpha \in A}$ subordinate to the open cover $\{U_\alpha\}_{\alpha \in A}$ of M. Since the sets $\{p \in U_\alpha \mid \theta_\alpha(p) > 0\}$ with $U_\alpha \subset U$ form an open cover of K and K is a compact subset of M, there exist finitely many indices $\alpha_1, \dots, \alpha_\ell \in A$ such that

$$K \subset \{p \in M \mid \theta_{\alpha_1}(p) + \dots + \theta_{\alpha_\ell}(p)) > 0\} =: V \subset U.$$

Let $k := \ell(m + 1)$ and, for $i = 1, \dots, \ell$, abbreviate

$$\phi_i := \phi_{\alpha_i}, \qquad \theta_i := \theta_{\alpha_i}.$$

Define the smooth map $f : M \to \mathbb{R}^k$ by

$$f(p) := \begin{pmatrix} \theta_1(p) \\ \theta_1(p)\phi_1(p) \\ \vdots \\ \theta_\ell(p) \\ \theta_\ell(p)\phi_\ell(p) \end{pmatrix} \qquad \text{for } p \in M.$$

Then the restriction $f|_V : V \to \mathbb{R}^k$ is injective. Namely, if p_0, $p_1 \in V$ satisfy

$$f(p_0) = f(p_1),$$

then

$$I := \{i \mid \theta_i(p_0) > 0\} = \{i \mid \theta_i(p_1) > 0\} \neq \emptyset$$

and, for $i \in I$, we have $\theta_i(p_0) = \theta_i(p_1)$, hence $\phi_i(p_0) = \phi_i(p_1)$, and so $p_0 = p_1$. Moreover, for every $p \in V$ the derivative $df(p) : T_pM \to \mathbb{R}^k$ is injective, and this proves Step 1.

Step 2 *Let $f : M \to \mathbb{R}^k$ be an injective immersion and let $\mathcal{A} \subset \mathbb{R}^{(2m+1) \times k}$ be a nonempty open set. Then there exists a matrix $A \in \mathcal{A}$ such that the map $Af : M \to \mathbb{R}^{2m+1}$ is an injective immersion.*

The proof of Step 2 uses the Theorem of Sard (see [1, 50]). The sets

$$W_0 := \{(p, q) \in M \times M \mid p \neq q\},$$
$$W_1 := \{(p, v) \in TM \mid v \neq 0\}$$

are open subsets of smooth second countable Hausdorff $2m$-manifolds and the maps

$$F_0 : \mathcal{A} \times W_0 \to \mathbb{R}^{2m+1}, \qquad F_1 : \mathcal{A} \times W_1 \to \mathbb{R}^{2m+1},$$

defined by

$$F_0(A, p, q) := A(f(p) - f(q)), \qquad F_1(A, p, v) := Adf(p)v$$

for $A \in \mathcal{A}$, $(p, q) \in W_0$, and $(p, v) \in W_1$, are smooth. Moreover, the zero vector in $\mathbb{R}^{2m+1}$ is a regular value of F_0 because f is injective and of F_1 because f is an immersion. Hence it follows from the intrinsic analogue of Theorem 2.2.19 that the sets

$$\mathcal{M}_0 := F_0^{-1}(0) = \{(A, p, q) \in \mathcal{A} \times W_0 \mid Af(p) = Af(q)\},$$
$$\mathcal{M}_1 := F_1^{-1}(0) = \{(A, p, v) \in \mathcal{A} \times W_1 \mid Adf(p)v = 0\}$$

are smooth manifolds of dimension

$$\dim \mathcal{M}_0 = \dim \mathcal{M}_1 = (2m + 1)k - 1.$$

Since M is a second countable Hausdorff manifold, so are $\mathcal{M}_0$ and $\mathcal{M}_1$. Hence the Theorem of Sard asserts that the canonical projections

$$\mathcal{M}_0 \to \mathcal{A} : (A, p, q) \mapsto A =: \pi_0(A, p, q),$$
$$\mathcal{M}_1 \to \mathcal{A} : (A, p, v) \mapsto A =: \pi_1(A, p, v),$$

have a common regular value $A \in \mathcal{A}$. Since

$$\dim \mathcal{M}_0 = \dim \mathcal{M}_1 < \dim \mathcal{A},$$

this implies

$$A \in \mathcal{A} \setminus (\pi_0(\mathcal{M}_0) \cup \pi_1(\mathcal{M}_1)).$$

Hence $Af : M \to \mathbb{R}^{2m+1}$ is an injective immersion and this proves Step 2.

If M is compact, the result follows from Steps 1 and 2 with $K = U = M$. In the noncompact case the proof requires two more steps to construct an embedding into $\mathbb{R}^{4m+4}$ and a further step to reduce the dimension to $2m + 1$.

Step 3 *Assume M is not compact. Then there exists a sequence of open sets $U_i \subset M$, a sequence of smooth functions $\rho_i : M \to [0, 1]$, and a sequence of compact sets $K_i \subset U_i$ such that*

$$\mathrm{supp}(\rho_i) \subset U_i, \qquad K_i = \rho_i^{-1}(1) \subset U_i, \qquad U_i \cap U_j = \emptyset$$

for all $i, j \in \mathbb{N}$ with $|i - j| \geq 2$ and $M = \bigcup_{i=1}^{\infty} K_i$.

Since M is second countable, there exists a sequence of compact sets $C_i \subset M$ such that $C_i \subset \mathrm{int}(C_{i+1})$ for all $i \in \mathbb{N}$ and $M = \bigcup_{i \in \mathbb{N}} C_i$ (Remark 2.9.1). Define the compact sets $B_i \subset M$ by $C_0 := \emptyset$ and

$$B_i := \overline{C_i \setminus C_{i-1}} \qquad \text{for } i \in \mathbb{N}.$$

Then $M = \bigcup_{i \in \mathbb{N}} B_i$ and, for all $i, j \in \mathbb{N}$ with $j \geq i + 2$, we have

$$B_i \subset C_i \subset \mathrm{int}(C_{j-1}), \qquad B_j \subset C_j \setminus \mathrm{int}(C_{j-1})$$

and so $B_i \cap B_j = \emptyset$. Since M is metrizable by Remark 2.9.2, there exists a distance function $d : M \times M \to [0, \infty)$ that induces the intrinsic topology on M. Define

$$A_i := \bigcup_{j \in \mathbb{N} \setminus \{i-1, i, i+1\}} B_j, \qquad \varepsilon_i := d(A_i, B_i) = \inf_{p \in A_i, q \in B_i} d(p, q).$$

Then A_i is a closed subset of M, because any convergent sequence in M must belong to a finite union of the B_j. Since $A_i \cap B_i = \emptyset$, this implies $\varepsilon_i > 0$. For $i \in \mathbb{N}$ define the set $U_i \subset M$ by

$$U_i := \{p \in M \mid \text{there exists a } q \in B_i \text{ with } d(p, q) < \varepsilon_i / 3\}.$$

Then $\{U_i\}_{i \in \mathbb{N}}$ is a sequence of open subsets of M such that $B_i \subset U_i \subset C_{i+1}$ for all $i \in \mathbb{N}$ and $U_i \cap U_j = \emptyset$ for $|i - j| \geq 2$. In particular, each set U_i has a compact closure.

For each i there exists of a partition of unity subordinate to the open cover $M = U_i \cup (M \setminus B_i)$ and hence a smooth function $\rho_i : M \to [0, 1]$ such that $\mathrm{supp}(\rho_i) \subset U_i$ and $\rho_i|_{B_i} \equiv 1$. Define $K_i := \rho_i^{-1}(1) = \{p \in U_i \mid \rho_i(p) = 1\}$ for $i \in \mathbb{N}$. Then K_i is a compact set and $B_i \subset K_i \subset U_i$ for each $i \in \mathbb{N}$. Hence $M = \bigcup_{i \in \mathbb{N}} K_i$ and this proves Step 3.

Step 4 *Assume M is not compact. Then there exists an embedding*

$$f : M \to \mathbb{R}^{4m+4}$$

with a closed image and a pair of orthonormal vectors $x, y \in \mathbb{R}^{4m+4}$ such that, for every $\varepsilon > 0$, there exists a compact set $K \subset M$ with

$$\sup_{p \in M \setminus K} \inf_{s, t \in \mathbb{R}} \left| \frac{f(p)}{|f(p)|} - sx - ty \right| < \varepsilon. \tag{2.9.3}$$

Assume M is not compact and let K_i, U_i, ρ_i be as in Step 3. Then, by Steps 1 and 2, there exists a sequence of smooth maps $g_i : M \to \mathbb{R}^{2m+1}$ such that $g_i|_{M \setminus U_i} \equiv 0$, the restriction $g_i|_{K_i} : K_i \to \mathbb{R}^{2m+1}$ is injective, and the derivative $dg_i(p) : T_p M \to \mathbb{R}^{2m+1}$ is injective for all $p \in K_i$ and all $i \in \mathbb{N}$. Let $\xi \in \mathbb{R}^{2m+1}$ be a unit vector and define the maps $f_i : M \to \mathbb{R}^{2m+1}$ by

$$f_i(p) := \rho_i(p) \left(i\xi + \frac{g_i(p)}{\sqrt{1 + |g_i(p)|^2}} \right) \tag{2.9.4}$$

for $p \in M$ and $i \in \mathbb{N}$. Then the restriction $f_i|_{K_i} : K_i \to \mathbb{R}^{2m+1}$ is injective, the derivative $df_i(p) : T_p M \to \mathbb{R}^{2m+1}$ is injective for all $p \in K_i$, and

$$\mathrm{supp}(f_i) \subset U_i, \qquad f_i(K_i) \subset B_1(i\xi), \qquad f_i(M) \subset B_{i+1}(0).$$

Define the maps $f^{\mathrm{odd}}, f^{\mathrm{ev}} : M \to \mathbb{R}^{2m+1}$ and $\rho^{\mathrm{odd}}, \rho^{\mathrm{ev}} : M \to \mathbb{R}$ by

$$\rho^{\mathrm{odd}}(p) := \begin{cases} \rho_{2i-1}(p), & \text{if } i \in \mathbb{N} \text{ and } p \in U_{2i-1}, \\ 0, & \text{if } p \in M \setminus \bigcup_{i \in \mathbb{N}} U_{2i-1}, \end{cases}$$

$$f^{\mathrm{odd}}(p) := \begin{cases} f_{2i-1}(p), & \text{if } i \in \mathbb{N} \text{ and } p \in U_{2i-1}, \\ 0, & \text{if } p \in M \setminus \bigcup_{i \in \mathbb{N}} U_{2i-1}, \end{cases}$$

$$\rho^{\mathrm{ev}}(p) := \begin{cases} \rho_{2i}(p), & \text{if } i \in \mathbb{N} \text{ and } p \in U_{2i}, \\ 0, & \text{if } p \in M \setminus \bigcup_{i \in \mathbb{N}} U_{2i}, \end{cases}$$

$$f^{\mathrm{ev}}(p) := \begin{cases} f_{2i}(p), & \text{if } i \in \mathbb{N} \text{ and } p \in U_{2i}, \\ 0, & \text{if } p \in M \setminus \bigcup_{i \in \mathbb{N}} U_{2i}, \end{cases}$$

and define the map $f : M \to \mathbb{R}^{4m+4}$ by

$$f(p) := \left(\rho^{\mathrm{odd}}(p), f^{\mathrm{odd}}(p), \rho^{\mathrm{ev}}(p), f^{\mathrm{ev}}(p) \right)$$

for $p \in M$.

We prove that f is injective. To see this, note that

$$\begin{aligned} p \in K_{2i-1} &\implies \begin{cases} 2i - 2 < |f^{\mathrm{odd}}(p)| < 2i, \\ |f^{\mathrm{ev}}(p)| < 2i + 1, \end{cases} \\ p \in K_{2i} &\implies \begin{cases} 2i - 1 < |f^{\mathrm{ev}}(p)| < 2i + 1, \\ |f^{\mathrm{odd}}(p)| < 2i + 2, \end{cases} \end{aligned} \tag{2.9.5}$$

Now let $p_0, p_1 \in M$ such that $f(p_0) = f(p_1)$. Assume first that $p_0 \in K_{2i-1}$. Then $\rho^{\mathrm{odd}}(p_1) = \rho^{\mathrm{odd}}(p_0) = 1$ and hence $p_1 \in \bigcup_{j \in \mathbb{N}} K_{2j-1}$. By (2.9.5), we also have $2i - 2 < |f^{odd}(p_1)| = |f^{odd}(p_0)| < 2i$ and hence $p_1 \in K_{2i-1}$. This implies $f_{2i-1}(p_1) = f^{odd}(p_1) = f^{odd}(p_0) = f_{2i-1}(p_0)$ and so $p_0 = p_1$. Now assume $p_0 \in K_{2i}$. Then $\rho^{ev}(p_1) = \rho^{ev}(p_0) = 1$ and hence $p_1 \in \bigcup_{j \in \mathbb{N}} K_{2j}$. By (2.9.5), we also have $2i - 1 < |f^{ev}(p_1)| = |f^{ev}(p_0)| < 2i + 1$, so $p_1 \in K_{2i}$, which implies $f_{2i}(p_1) = f^{ev}(p_1) = f^{ev}(p_0) = f_{2i}(p_0)$, and so again $p_0 = p_1$. This shows that f is injective. That f is an immersion follows from the fact that the derivative $df_i(p)$ is injective for all $p \in K_i$ and all $i \in \mathbb{N}$.

We prove that f is proper and has a closed image. Let $(p_\nu)_{\nu \in \mathbb{N}}$ be a sequence in M such that the sequence $(f(p_\nu))_{\nu \in \mathbb{N}}$ in $\mathbb{R}^{4m+4}$ is bounded. Choose $i \in \mathbb{N}$ such that $|f^{odd}(p_\nu)| < 2i$ and $|f^{ev}(p_\nu)| < 2i + 1$ for all $\nu \in \mathbb{N}$. Then $p_\nu \in \bigcup_{j=1}^{2i} K_j$ for all $\nu \in \mathbb{N}$ by (2.9.5). Hence $(p_\nu)_{\nu \in \mathbb{N}}$ has a convergent subsequence. Thus $f : M \to \mathbb{R}^{4m+4}$ is an embedding with a closed image.

Next consider the pair of orthonormal vectors

$$x := (0, \xi, 0, 0), \qquad y := (0, 0, 0, \xi)$$

in $\mathbb{R}^{4m+4} = \mathbb{R} \times \mathbb{R}^{2m+1} \times \mathbb{R} \times \mathbb{R}^{2m+1}$. Let $(p_\nu)_{\nu \in \mathbb{N}}$ be a sequence in M that does not have a convergent subsequence and choose a sequence $i_\nu \in \mathbb{N}$ such that $p_\nu \in K_{2i_\nu-1} \cup K_{2i_\nu}$ for all $\nu \in \mathbb{N}$. Then i_ν tends to infinity. If $p_\nu \in K_{2i_\nu-1}$ for all ν, then we have $\limsup_{\nu \to \infty} |f^{odd}(p_\nu)|^{-1}|f^{ev}(p_\nu)| \le 1$ by (2.9.5). Passing to a subsequence, still denoted by $(p_\nu)_{\nu \in \mathbb{N}}$, we may assume that the limit $\lambda := \lim_{\nu \to \infty} |f^{odd}(f_\nu)|^{-1}|f^{ev}(p_\nu)|$ exists. Then

$$0 \le \lambda \le 1, \quad \lim_{\nu \to \infty} \frac{|f^{odd}(p_\nu)|}{|f(p_\nu)|} = \frac{1}{\sqrt{1+\lambda^2}}, \quad \lim_{\nu \to \infty} \frac{|f^{ev}(p_\nu)|}{|f(p_\nu)|} = \frac{\lambda}{\sqrt{1+\lambda^2}},$$

and it follows from (2.9.4) that

$$\lim_{\nu \to \infty} \frac{f^{odd}(p_\nu)}{|f^{odd}(p_\nu)|} = \xi, \quad \lim_{\nu \to \infty} \frac{f^{ev}(p_\nu)}{|f^{odd}(p_\nu)|} = \lambda \xi.$$

This implies

$$\lim_{\nu \to \infty} \frac{f(p_\nu)}{|f(p_\nu)|} = \left(0, \frac{\xi}{\sqrt{1+\lambda^2}}, 0, \frac{\lambda \xi}{\sqrt{1+\lambda^2}}\right) = \frac{1}{\sqrt{1+\lambda^2}}x + \frac{\lambda}{\sqrt{1+\lambda^2}}y.$$

Similarly, if $p_\nu \in K_{2i_\nu}$ for all ν, there exists a subsequence such that the limit $\lambda := \lim_{\nu \to \infty} |f^{ev}(p_\nu)|^{-1}|f^{odd}(p_\nu)|$ exists and, by (2.9.4), this implies

$$\lim_{\nu \to \infty} \frac{f(p_\nu)}{|f(p_\nu)|} = \left(0, \frac{\lambda \xi}{\sqrt{1+\lambda^2}}, 0, \frac{\xi}{\sqrt{1+\lambda^2}}\right) = \frac{\lambda}{\sqrt{1+\lambda^2}}x + \frac{1}{\sqrt{1+\lambda^2}}y.$$

This shows that the vectors x and y satisfy the requirements of Step 4.

Step 5 *There exists an embedding* $f : M \to \mathbb{R}^{2m+1}$ *with a closed image.*

For compact manifolds the result was proved in Steps 1 and 2 and for $m = 0$ the assertion is obvious, because then M is a finite or countable set with the discrete topology. Thus assume that M is not compact and $m \geq 1$. Choose $f : M \to \mathbb{R}^{4m+4}$ and $x, y \in \mathbb{R}^{4m+4}$ as in Step 4 and define

$$\mathcal{A} := \left\{ A \in \mathbb{R}^{(2m+1)\times(4m+4)} \;\middle|\; \begin{array}{l} \text{the vectors } Ax \text{ and } Ay \\ \text{are linearly independent} \end{array} \right\}.$$

Since $m \geq 1$, this is a nonempty open subset of $\mathbb{R}^{(2m+1)\times(4m+4)}$. We prove that the map $Af : M \to \mathbb{R}^{2m+1}$ is proper and has a closed image for every $A \in \mathcal{A}$. To see this, fix a matrix $A \in \mathcal{A}$. Let $(p_\nu)_{\nu\in\mathbb{N}}$ be a sequence in M that does not have a convergent subsequence. Then by Step 4 there exists a subsequence, still denoted by $(p_\nu)_{\nu\in\mathbb{N}}$, and real numbers $s, t \in \mathbb{R}$ such that

$$s^2 + t^2 = 1, \qquad \lim_{\nu\to\infty} \frac{f(p_\nu)}{|f(p_\nu)|} = sx + ty, \qquad \lim_{\nu\to\infty} |f(p_\nu)| = \infty.$$

This implies

$$\lim_{\nu\to\infty} \frac{Af(p_\nu)}{|f(p_\nu)|} = sAx + tAy \neq 0$$

and hence $\lim_{\nu\to\infty} |Af(p_\nu)| = \infty$. Thus the preimage of every compact subset of $\mathbb{R}^{2m+1}$ under the map $Af : M \to \mathbb{R}^{2m+1}$ is a compact subset of M, and hence Af is proper and has a closed image (Remark 2.3.3).

Now it follows from Step 2 that there exists a matrix $A \in \mathcal{A}$ such that the map $Af : M \to \mathbb{R}^{2m+1}$ is an injective immersion. Hence it is an embedding with a closed image. This proves Step 5 and Theorem 2.9.12. $\qquad\qquad\square$

The **Whitney Embedding Theorem** asserts that every second countable Hausdorff m-manifold M admits an embedding $f : M \to \mathbb{R}^{2m}$. The proof is based on the *Whitney Trick* and goes beyond the scope of this book. The next exercise shows that Whitney's theorem is sharp.

Remark 2.9.13 The manifold $\mathbb{R}P^2$ cannot be embedded into $\mathbb{R}^3$. The same is true for the **Klein bottle** $K := \mathbb{R}^2/\equiv$ where the equivalence relation is given by $[x, y] \equiv [x + k, \ell - y]$ for $x, y \in \mathbb{R}$ and $k, \ell \in \mathbb{Z}$.

2.9.4 Leaves of a Foliation

Let M be an m-dimensional paracompact Hausdorff manifold and $E \subset TM$ be an integrable subbundle of rank n. Let $L \subset M$ be a **closed leaf** of the foliation determined by E. Then L is a smooth n-dimensional submanifold of M. Here is a sketch of David Epstein's proof of this fact in [19].

(a) *The space L with the intrinsic topology admits the structure of a manifold such that the obvious inclusion $\iota : L \to M$ is an injective immersion.* This is an easy exercise. For the definition of the intrinsic topology see Exercise 2.7.10. The dimension of L is n.

(b) *If $f : X \to Y$ is a continuous map between topological spaces such that Y is paracompact and there is an open cover $\{V_j\}_{j \in J}$ of Y such that $f^{-1}(V_j)$ is paracompact for each j, then X is paracompact.* To see this, we may assume that the cover $\{V_j\}_{j \in J}$ is locally finite. Now let $\{U_\alpha\}_{\alpha \in A}$ be an open cover of X. Then the sets $U_\alpha \cap f^{-1}(V_j)$ define an open cover of $f^{-1}(V_j)$. Choose a locally finite refinement $\{W_{ij}\}_{i \in I_j}$ of this cover. Then the open cover $\{W_{ij}\}_{j \in J, i \in I_j}$ of M is a locally finite refinement of $\{U_\alpha\}_{\alpha \in A}$.

(c) *The intrinsic topology of L is paracompact.* This follows from (b) and the fact that the intersection of L with every foliation box is paracompact in the intrinsic topology.

(d) *The intrinsic topology of L is second countable.* This follows from (a) and (c) and the fact that every connected paracompact manifold is second countable (see Remark 2.9.2).

(e) *The intersection of L with a foliation box consists of at most countably many connected components.* This follows immediately from (d).

(f) *If L is a closed subset of M, then the intersection of L with a foliation box has only finitely many connected components.* To see this, we choose a transverse slice of the foliation at $p_0 \in L$, i.e. a connected submanifold $T \subset M$ through p_0, diffeomorphic to an open ball in $\mathbb{R}^{m-n}$, whose tangent space at each point $p \in T$ is a complement of E_p. By (d) we have that $T \cap L$ is at most countable. If this set is not finite, even after shrinking T, there must be a sequence $p_i \in (T \cap L) \setminus \{p_0\}$ converging to p_0. Using the holonomy of the leaf (obtained by transporting transverse slices along a curve via a lifting argument) we find that every point $p \in T \cap L$ is the limit point of a sequence in $(T \cap L) \setminus \{p\}$. Hence the one-point set $\{p\}$ has empty interior in the relative topology of $T \cap L$ for each $p \in T \cap L$. Thus $T \cap L$ is a countable union of closed subsets with empty interior. Since $T \cap L$ admits the structure of a complete metric space, this contradicts the Baire category theorem.

(g) It follows immediately from (f) that L is a submanifold of M.

2.9.5 Principal Bundles

An interesting class of foliations arises from smooth Lie group actions. Let $G \subset \mathrm{GL}(N, \mathbb{R})$ be a compact Lie group and let P be a smooth m-manifold whose topology is Hausdorff and second countable. A **smooth (contravariant) G-action**

on P is a smooth map

$$P \times G \to P : (p, g) \mapsto pg \tag{2.9.6}$$

that satisfies the conditions

$$(pg)h = p(gh), \qquad p\mathbb{1} = p \tag{2.9.7}$$

for all $p \in P$ and all $g, h \in G$. Fix any such group action. Then every group element $g \in G$ determines a diffeomorphism $P \to P : p \mapsto pg$, whose derivative at $p \in P$ is denoted by $T_p P \to T_{pg} P : v \mapsto vg$. Every Lie algebra element $\xi \in \mathfrak{g} := \mathrm{Lie}(G) = T_{\mathbb{1}} G$ determines a vector field $X_\xi \in \mathrm{Vect}(P)$ which assigns to each $p \in P$ the tangent vector

$$X_\xi(p) := p\xi := \frac{d}{dt}\bigg|_{t=0} p\exp(t\xi) \in T_p P. \tag{2.9.8}$$

The linear map $\mathfrak{g} \to \mathrm{Vect}(P) : \xi \mapsto X_\xi$ is called the **infinitesimal action**. It is a Lie algebra anti-homomorphism because the group action is contravariant. (Exercise: Prove that $[X_\xi, X_\eta] = -X_{[\xi,\eta]}$ for $\xi, \eta \in \mathfrak{g}$.) The group action (2.9.6) is said to be with **finite isotropy** iff the **isotropy subgroup**

$$G_p := \{g \in G \mid pg = p\}$$

is finite for all $p \in P$. The isotropy subgroup G_p is a Lie subgroup of G with Lie algebra $\mathfrak{g}_p := \mathrm{Lie}(G_p) = \{\xi \in \mathfrak{g} \mid X_\xi(p) = 0\}$. Since G is compact, this shows that G_p is a finite subgroup of G if and only if $\mathfrak{g}_p = \{0\}$ or, equivalently, the map $\mathfrak{g} \to T_p P : \xi \mapsto X_\xi(p) = p\xi$ is injective. Thus, in the finite isotropy case, the group action determines an involutive subbundle $E \subset TP$ with the fibers $E_p := p\mathfrak{g} = \{X_\xi(p) \mid \xi \in \mathfrak{g}\}$ for $p \in P$. When G is connected, the leaves of the corresponding foliation are the group orbits $pG := \{pg \mid g \in G\}$. These are the elements of the **orbit space**

$$P/G := \{pG \mid p \in P\}.$$

There is a natural projection $\pi : P \to P/G$ defined by $\pi(p) := pG$ for $p \in P$ and the orbit space P/G is equipped with the quotient topology (a subset $U \subset P/G$ is open if and only if $\pi^{-1}(U)$ is an open subset of P). The group action is called **free** iff $G_p = \{\mathbb{1}\}$ for all $p \in P$. The next theorem shows that, in the case of a free action, the quotient space admits a unique smooth structure such that the projection $\pi : P \to P/G$ is a submersion.

Theorem 2.9.14 (Principal Bundle) *Let P be a smooth m-manifold whose topology is Hausdorff and second countable. Suppose P is equipped with a smooth contravariant action of a compact Lie group G and assume the group action is free. Then $\dim(G) \leq m$ and $B := P/G$ admits a unique smooth structure such that the projection $\pi : P \to B$ is a submersion. The intrinsic topology of B, induced by the smooth structure, agrees with the quotient topology, and it is Hausdorff and second countable.*

Proof For each $p \in P$ the map $G \to P : g \mapsto pg$ is an embedding and this implies $k := \dim(G) \leq \dim(P) = m$. Define $n := m - k$. A **local slice** of the group action is a smooth map $\iota : \Omega \to P$, defined on an open set $\Omega \subset \mathbb{R}^n$, such that the map $\Omega \times G \to P : (x, g) \mapsto \iota(x)g$ is an embedding. With this understood, we prove the assertions in five steps.

Step 1 *For every $p_0 \in P$ there exists a local slice $\iota_0 : \Omega_0 \to P$, defined on an open neighborhood $\Omega_0 \subset \mathbb{R}^n$ of the origin, such that $\iota_0(0) = p_0$.*

Choose a coordinate chart $\phi : V \to \mathbb{R}^m$ on an open neighborhood $V \subset P$ of p_0 such that $\phi(p_0) = 0$ and $\phi(V) = \mathbb{R}^m$. Define $v_1, \ldots, v_m \in T_{p_0} P$ by

$$d\phi(p_0)v_i := e_i \qquad \text{for } i = 1, \ldots, m,$$

where $e_1, \ldots, e_m$ is the standard basis of $\mathbb{R}^m$. Reorder the coordinates on $\mathbb{R}^m$, if necessary, such that the vectors $v_1, \ldots, v_n$ project to a basis of the quotient space $T_{p_0} P / p_0 \mathfrak{g}$. Define $\iota : \mathbb{R}^n \to P$ by

$$\iota(x_1, \ldots, x_n) := \phi^{-1}(x_1, \ldots, x_n, 0, \ldots, 0)$$

and define the map $\psi : \mathbb{R}^n \times G \to P$ by

$$\psi(x, g) := \iota(x)g \qquad \text{for } x \in \mathbb{R}^n \text{ and } g \in G.$$

Then ψ is smooth and its derivative $d\psi(0, \mathbb{1}) : \mathbb{R}^n \times \mathfrak{g} \to T_{p_0} P$ is given by

$$d\psi(0, \mathbb{1})(\widehat{x}, \xi) = \sum_{i=1}^{n} \widehat{x}_i v_i + p_0 \xi$$

for $\widehat{x} = (\widehat{x}_1, \ldots, \widehat{x}_n) \in \mathbb{R}^n$ and $\xi \in \mathfrak{g}$. Hence $d\psi(0, \mathbb{1})$ is bijective and so it follows from the Inverse Function Theorem 2.2.17 that there exist open neighborhoods $\Omega \subset \mathbb{R}^n$ of 0, $\Omega_1 \subset G$ of $\mathbb{1}$, and $W \subset P$ of p_0 such that the restricted map

$$\psi_1 := \psi|_{\Omega \times \Omega_1} : \Omega \times \Omega_1 \to W$$

is a diffeomorphism.

Next we prove that there exists an open neigborhood $\Omega_0 \subset \Omega$ of the origin such that the restricted map

$$\psi_0 := \psi|_{\Omega_0 \times G} : \Omega_0 \times G \to P$$

is injective. Suppose otherwise that no such neighborhood Ω_0 exists. Then there exist sequences $(x_i, g_i), (x'_i, g'_i) \in \Omega \times G$ such that $(x_i, g_i) \neq (x'_i, g'_i)$ and $\psi(x_i, g_i) = \psi(x'_i, g'_i)$ for all i and the sequences $(x_i)_{i \in \mathbb{N}}$ and $(x'_i)_{i \in \mathbb{N}}$ in Ω converge to the origin. Since G is compact we may assume, by passing to a subsequence if necessary, that the sequences $(g_i)_{i\in\mathbb{N}}$ and $(g_i')_{i\in\mathbb{N}}$ converge. Denote the limits by

$$g := \lim_{i\to\infty} g_i \in G, \qquad g' := \lim_{i\to\infty} g_i' \in G.$$

Then

$$p_0 g = \lim_{i\to\infty} \iota(x_i)g_i = \lim_{i\to\infty} \iota(x_i')g_i' = p_0 g'$$

and so $g = g'$ because the group action is free. Thus the sequence $(g_i'g_i^{-1})_{i\in\mathbb{N}}$ in G converges to $\mathbb{1}$ and hence belongs to the set Ω_1 for i sufficiently large. Since

$$\psi_1(x_i, \mathbb{1}) = \iota(x_i) = \iota(x_i')g_i'g_i^{-1} = \psi_1(x_i', g_i'g_i^{-1})$$

for all i, this contradicts the injectivity of ψ_1. Thus we have proved that the map $\psi_0 : \Omega_0 \times G \to P$ is injective for a suitable neighborhood $\Omega_0 \subset \Omega$ of the origin. That it is an immersion is a direct consequence of the formula

$$d\psi_0(x,g)(\widehat{x},\widehat{g}) = \big(d\iota(x)\widehat{x} + \iota(x)(\widehat{g}g^{-1})\big)g = \big(d\psi_0(x,\mathbb{1})(\widehat{x},\widehat{g}g^{-1})\big)g$$

for all $x \in \Omega_0$, $\widehat{x} \in \mathbb{R}^n$, $g \in G$, and $\widehat{g} \in T_g G$, and the fact that the derivative $d\psi_0(x,\mathbb{1})$ is bijective for all $x \in \Omega_0$ (even for all $x \in \Omega$).

Thus we have proved that $\psi_0 : \Omega_0 \times G \to P$ is an injective immersion. Shrinking Ω_0 further, if necessary, we may assume that Ω_0 has a compact closure and that ψ is injective on $\overline{\Omega}_0 \times G$. This implies that ψ_0 is proper. Namely, if $(x_i, g_i)_{i\in\mathbb{N}}$ is a sequence in $\Omega_0 \times G$ and $(x, g) \in \Omega_0 \times G$ such that $\psi_0(x, g) = \lim_{i\to\infty} \psi_0(x_i, g_i)$, then there is a subsequence $(x_{i_v}, g_{i_v})_{v\in\mathbb{N}}$ that converges to a pair $(x', g') \in \overline{\Omega}_0 \times G$. This subsequence satisfies

$$\psi(x', g') = \lim_{v\to\infty} \psi_0(x_{i_v}, g_{i_v}) = \psi(x, g).$$

Since ψ is injective on $\overline{\Omega}_0 \times G$, this implies $x = x'$ and $g = g'$. Thus every subsequence of $(x_i, g_i)_{i\in\mathbb{N}}$ has a further subsequence that converges to (x, g) and so the sequence $(x_i, g_i)_{i\in\mathbb{N}}$ itself converges to (x, g). Thus the map $\psi_0 : \Omega_0 \times G \to P$ is a proper injective immersion and this proves Step 1.

Step 2 *Let $\iota : \Omega \to P$ be a local slice. Then the set $U := \pi(\iota(\Omega)) \subset B$ is open in the quotient topology and the map $\pi \circ \iota : \Omega \to U$ is a homeomorphism with respect to the quotient topology on U.*

The map $\psi : \Omega \times G \to P$, defined by $\psi(x, g) := \iota(x)g$ for $x \in \Omega$ and $g \in G$, is an embedding. Hence $W := \psi(\Omega \times G)$ is an open G-invariant subset of P and $\psi : \Omega \times G \to W$ is a G-equivariant homeomorphism. Moreover, for every element $p \in P$, we have $\pi(p) \in U$ if and only if there exists an element $x \in \Omega$ and an element $g \in G$ such that $p = \iota(x)g = \psi(x, g)$. Thus $\pi^{-1}(U) = \psi(\Omega \times G) = W$ is an open subset of P, and so U is an open subset of $B = P/G$ with respect to the quotient topology. The continuity of $\pi \circ \iota : \Omega \to U$ follows directly from the definition. Moreover, if $\Omega' \subset \Omega$ is an open set and $U' := \pi(\iota(\Omega'))$, then $\pi^{-1}(U') = \psi(\Omega' \times G)$ is open by the same argument, and so $U' \subset B$ is open with respect to the quotient topology. Thus $\pi \circ \iota : \Omega \to U$ is a homeomorphism and this proves Step 2.

Step 3 *By Step 1 there exists a collection* $\iota_\alpha : \Omega_\alpha \to P, \alpha \in A$, *of local slices such that the sets* $U_\alpha := \pi(\iota_\alpha(\Omega_\alpha))$ *cover the orbit space* $B = P/G$. *For* $\alpha \in A$ *define*

$$\phi_\alpha := (\pi \circ \iota_\alpha)^{-1} : U_\alpha \to \Omega_\alpha.$$

Then $\mathcal{A} = \{(\phi_\alpha, U_\alpha)\}_{\alpha \in A}$ *is a smooth structure on* B *which renders the canonical projection* $\pi : P \to B$ *into a submersion. Moreover, this smooth structure is compatible with the quotient topology on* B.

For $\alpha, \beta \in A$ define $\Omega_{\alpha\beta} := \phi_\alpha(U_\alpha \cap U_\beta)$ and $\phi_{\beta\alpha} := \phi_\beta \circ \phi_\alpha^{-1} : \Omega_{\alpha\beta} \to \Omega_{\beta\alpha}$. We must prove that $\phi_{\beta\alpha}$ is smooth. To see this, define $\psi_\alpha : \Omega_\alpha \times G \to P$ by $\psi_\alpha(x, g) := \iota_\alpha(x)g$ for $\alpha \in A$, $x \in \Omega_\alpha$, and $g \in G$. Then ψ_α is a diffeomorphism onto its image and $\psi_\alpha(\Omega_{\alpha\beta} \times G) = \psi_\beta(\Omega_{\beta\alpha} \times G) = \pi^{-1}(U_\alpha \cap U_\beta)$. For $x \in \Omega_{\alpha\beta}$ the element $\phi_{\beta\alpha}(x) \in \Omega_{\beta\alpha}$ is the projection of $\psi_\beta^{-1} \circ \psi_\alpha(x, \mathbb{1})$ onto the first factor. Thus $\phi_{\beta\alpha}$ is smooth and so is its inverse $\phi_{\alpha\beta}$. This shows that $\{(U_\alpha, \phi_\alpha)\}_{\alpha \in A}$ is a smooth structure on B. Second, π is a submersion with respect to this smooth structure, because $\phi_\alpha \circ \pi \circ \psi_\alpha(x, g) = x$ for all $\alpha \in A$, all $x \in \Omega_\alpha$, and all $g \in G$. Third, this smooth structure is compatible with the quotient topology by Step 2. This proves Step 3.

Step 4 *There is only one smooth structure on* B *with respect to which the projection* $\pi : P \to B$ *is a submersion.*

Fix any smooth structure on B for which the projection $\pi : P \to B$ is a submersion. Then the dimension of B is $n = \dim(P) - \dim(G)$, and so the smooth structure consists of bijections $\phi_\alpha : U_\alpha \to \Omega_\alpha$ from subsets $U_\alpha \subset B$ onto open sets $\Omega_\alpha \subset \mathbb{R}^n$ such that the sets U_α cover B and the transition maps are diffeomorphisms between open subsets of $\mathbb{R}^n$.

We prove that the intrinsic topology on B agrees with the quotient topology. To see this, fix a subset $U \subset B$. Then the following are equivalent.

(a) U is open with respect to the intrinsic topology on B.
(b) $\phi_\alpha(U \cap U_\alpha)$ is open in $\mathbb{R}^n$ for all $\alpha \in A$.
(c) $\pi^{-1}(U \cap U_\alpha)$ is open in P for all $\alpha \in A$.
(d) $\pi^{-1}(U)$ is open in P.
(e) U is open with respect to the quotient topology on B.

The equivalence of (a) and (b) follows from the definition of the intrinsic topology. That (b) implies (c) follows from the three observations that the set $\pi^{-1}(U_\alpha)$ is open in P, the map $\phi_\alpha \circ \pi : \pi^{-1}(U_\alpha) \to \Omega_\alpha$ is continuous, and $(\phi_\alpha \circ \pi)^{-1}(\phi_\alpha(U \cap U_\alpha)) = \pi^{-1}(U \cap U_\alpha)$. That (c) implies (b) follows from the fact that the map $\phi_\alpha \circ \pi : \pi^{-1}(U_\alpha) \to \Omega_\alpha$ is a submersion and hence maps the open set $\pi^{-1}(U \cap U_\alpha)$ onto an open subset of Ω_α (Corollary 2.6.2). The equivalence of (c) and (d) follows from the fact that the map $\pi : P \to B$ is continuous and $U_\alpha \subset B$ is open (both with respect to the intrinsic topology on B) and so $\pi^{-1}(U_\alpha)$ is open in P for all $\alpha \in A$. The equivalence of (d) and (e) follows from the definition of the quotient topology on B.

Now let $\iota : \Omega \to P$ be a local slice and define the set $U := \pi(\iota(\Omega)) \subset B$ and the map $\phi := (\pi \circ \iota)^{-1} : U \to \Omega$. Then the composition

$$\phi_\alpha \circ \phi^{-1} = \phi_\alpha \circ \pi \circ \iota : \phi(U \cap U_\alpha) \to \phi_\alpha(U \cap U_\alpha)$$

is a homeomorphism between open subsets of $\mathbb{R}^n$. Moreover, $\phi_\alpha \circ \phi^{-1}$ is the composition of the smooth maps $\iota : \{x \in \Omega \mid \pi(\iota(x)) \in U_\alpha\} \to \pi^{-1}(U \cap U_\alpha)$, $\pi : \pi^{-1}(U \cap U_\alpha) \to U \cap U_\alpha$, and $\phi_\alpha : U \cap U_\alpha \to \phi_\alpha(U \cap U_\alpha)$. So $\phi_\alpha \circ \phi^{-1}$ is smooth and its derivative is everywhere bijective because π is a submersion and the kernel of $d\pi(\iota(x))$ is transverse to the image of $d\iota(x)$. Thus $\phi_\alpha \circ \phi^{-1}$ is a diffeomorphism by the Inverse Function Theorem and this proves Step 4.

Step 5 *The quotient topology on B is a Hausdorff and second countable.*

Let $\iota_\alpha : \Omega_\alpha \to P$ for $\alpha \in A$ be a collection of local slices such that the sets $U_\alpha := \pi(\iota_\alpha(\Omega_\alpha))$ cover B. Then the open sets $\pi^{-1}(U_\alpha)$ form an open cover of P and so there is a countable subcover. Thus B is second countable. To prove that B is Hausdorff, fix two distinct elements $b_0, b_1 \in B$ and choose $p_0, p_1 \in P$ such that $\pi(p_0) = b_0$ and $\pi(p_1) = b_1$. Then $p_0 G$ and $p_1 G$ are disjoint compact subsets of P and hence can be separated by disjoint open subsets $U_0, U_1 \subset P$, because P is a Hausdorff space. Now for $i = 0, 1$ the set $V_i := \{p \in P \mid pG \subset U_i\}$ is open (exercise) and contains the orbit $p_i G$. Hence $W_0 := \pi(V_0)$ and $W_1 := \pi(V_1)$ are disjoint open subsets of B such that $b_0 \in W_0$ and $b_1 \in W_1$. This proves Step 5 and Theorem 2.9.14. $\square$

Example 2.9.15 There are many important examples of free group actions and principal bundles. A class of examples arises from orthonormal frame bundles (Sect. 3.4). The complex projective space $B = \mathbb{C}P^n$ arises from the action of the circle $G = S^1$ on the unit sphere $P = S^{2n+1} \subset \mathbb{C}^{n+1}$ (Example 2.8.5). The real projective space $B = \mathbb{R}P^n$ arises from the action of the finite group $G = \mathbb{Z}/2\mathbb{Z}$ on the unit sphere $P = S^n \subset \mathbb{R}^{n+1}$ (Example 2.8.6). The complex Grassmannian $B = G_k(\mathbb{C}^n)$ arises from the action of $G = U(k)$ on the space $P = \mathcal{F}_k(\mathbb{C}^n)$ of unitary k-frames in $\mathbb{C}^n$ (Example 3.7.6). If G is a Lie group and $K \subset G$ is a compact subgroup, then by Theorem 2.9.14 the **homogeneous space** G/K admits a unique smooth structure such that the projection $\pi : G \to G/K$ is a submersion. The example $SL(2, \mathbb{C})/SU(2)$ can be identified with hyperbolic 3-space (Sect. 6.4.4), the example $U(n)/O(n)$ can be identified with the space of Lagrangian subspaces of a symplectic vector space ([49, Lemma 2.3.2]), the example $Sp(2n)/U(n)$ can be identified with Siegel upper half space or the space of compatible linear complex structures on a symplectic vector space (Exercise 6.5.24 and [49, Lemma 2.5.12]), and the example $G_2/SO(4)$ can be identified with the associative Grassmannian ([68, Remark 8.4]).

2.9.6 Standing Assumption

We have seen that all the results in the present chapter carry over to the intrinsic setting, assuming that the topology of M is Hausdorff and paracompact. In fact, in many cases it is enough to assume the Hausdorff property. However, these results mainly deal with introducing the basic concepts such as smooth maps, embeddings, submersions, vector fields, flows, and verifying their elementary properties, i.e. with setting up the language for differential geometry and topology. When it comes to the substance of the subject we shall deal with Riemannian metrics and they only exist on paracompact Hausdorff manifolds. Another central ingredient in differential topology is the theorem of Sard and that requires second countability. To quote Moe Hirsch [29]: "Manifolds that are not paracompact are amusing, but they never occur naturally and it is difficult to prove anything about them." Thus we will set the following convention for the remaining chapters.

We assume from now on that each intrinsic manifold M is Hausdorff and second countable and hence is also paracompact.

For most of this text we will in fact continue to develop the theory for submanifolds of Euclidean space and indicate, wherever necessary, how to extend the definitions, theorems, and proofs to the intrinsic setting.

The Levi-Civita Connection

3

For a submanifold of Euclidean space the inner product on the ambient space determines an inner product on each tangent space, the *first fundamental form*. The *second fundamental form* is obtained by differentiating the map which assigns to each point in $M \subset \mathbb{R}^n$ the orthogonal projection onto the tangent space (Sect. 3.1). The covariant derivative of a vector field along a curve is the orthogonal projection of the derivative in the ambient space onto the tangent space (Sect. 3.2). We will show how the covariant derivative gives rise to parallel transport (Sect. 3.3), examine the frame bundle (Sect. 3.4), discuss motions without "sliding, twisting, and wobbling", and prove the development theorem (Sect. 3.5).

In Sect. 3.6 we will see that the covariant derivative is determined by the Christoffel symbols in local coordinates and thus carries over to the intrinsic setting. The intrinsic setting of Riemannian manifolds is explained in Sect. 3.7. The covariant derivative takes the form of a family of linear operators $\nabla : \mathrm{Vect}(\gamma) \to \mathrm{Vect}(\gamma)$, one for every smooth curve $\gamma : I \to M$, and these operators are uniquely characterized by the axioms of Theorem 3.7.8. This family of linear operators is the *Levi-Civita connection*.

3.1 Second Fundamental Form

Let $M \subset \mathbb{R}^n$ be a smooth m-manifold. Then each tangent space of M is an m-dimensional real vector space and hence is isomorphic to $\mathbb{R}^m$. Thus any two tangent spaces $T_p M$ and $T_q M$ are of course isomorphic to each other. While there is no canonical isomorphism from $T_p M$ to $T_q M$ we shall see that every smooth curve γ in M connecting p to q induces an isomorphism between the tangent spaces via parallel transport of tangent vectors along γ.

Throughout we use the standard inner product on $\mathbb{R}^n$ given by

$$\langle v, w \rangle = v_1 w_1 + v_2 w_2 + \cdots + v_n w_n$$

© The Editor(s) (if applicable) and The Author(s), under exclusive license to
Springer-Verlag GmbH, DE, part of Springer Nature 2022
J.W. Robbin, D.A. Salamon, *Introduction to Differential Geometry*,
Springer Studium Mathematik (Master), https://doi.org/10.1007/978-3-662-64340-2_3

for $v = (v_1, \ldots, v_n) \in \mathbb{R}^n$ and $w = (w_1, \ldots, w_n) \in \mathbb{R}^n$. The associated Euclidean norm will be denoted by

$$|v| = \sqrt{\langle v, v \rangle} = \sqrt{v_1^2 + v_2^2 + \cdots + v_n^2}$$

for $v = (v_1, \ldots, v_n) \in \mathbb{R}^n$. When $M \subset \mathbb{R}^n$ is a smooth m-dimensional submanifold, a first observation is that each tangent space of M inherits an inner product from the ambient space $\mathbb{R}^n$. The resulting *field of inner products* is called the first fundamental form.

Definition 3.1.1   Let $M \subset \mathbb{R}^n$ be a smooth m-dimensional submanifold. The **first fundamental form on M** is the field which assigns to each $p \in M$ the bilinear map

$$g_p : T_p M \times T_p M \to \mathbb{R}$$

defined by

$$g_p(v, w) = \langle v, w \rangle \tag{3.1.1}$$

for $v, w \in T_p M$.

A second observation is that the inner product on the ambient space also determines an orthogonal projection of $\mathbb{R}^n$ onto the tangent space $T_p M$ for each $p \in M$. This projection can be represented by the matrix $\Pi(p) \in \mathbb{R}^{n \times n}$ which is uniquely determined by the conditions

$$\Pi(p) = \Pi(p)^2 = \Pi(p)^\mathsf{T}, \tag{3.1.2}$$

and

$$\Pi(p)v = v \qquad \Longleftrightarrow \qquad v \in T_p M \tag{3.1.3}$$

for $p \in M$ and $v \in \mathbb{R}^n$ (see Exercise 2.6.9).

Lemma 3.1.2   *The map $\Pi : M \to \mathbb{R}^{n \times n}$ defined by (3.1.2) and (3.1.3) is smooth.*

Proof   This follows directly from Theorem 2.6.10 and Corollary 2.6.12. More explicitly, if $U \subset M$ is an open set and $\phi : U \to \Omega$ is a coordinate chart onto an open subset $\Omega \subset \mathbb{R}^m$ with the inverse $\psi := \phi^{-1} : \Omega \to U$, then

$$\Pi(p) = d\psi(\phi(p)) \Big(d\psi(\phi(p))^\mathsf{T} d\psi(\phi(p)) \Big)^{-1} d\psi(\phi(p))^\mathsf{T}$$

for $p \in U$ and this proves Lemma 3.1.2.   □

Fig. 3.1 A unit normal vector
field

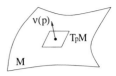

Example 3.1.3 (Gauß map) Let $M \subset \mathbb{R}^{m+1}$ be a submanifold of codimension
one. Then $TM^{\perp}$ is a vector bundle of rank one (Corollary 2.6.13), and so each
fiber $T_p M^{\perp}$ is spanned by a unit vector $v(p) \in \mathbb{R}^m$, determined by $T_p M$ up to
a sign. By Theorem 2.6.10 each $p_0 \in M$ has an open neighborhood $U \subset M$ on
which there exists a smooth map

$$v : U \to \mathbb{R}^{m+1}$$

satisfying

$$v(p) \perp T_p M, \qquad |v(p)| = 1 \tag{3.1.4}$$

for all $p \in U$ (see Fig. 3.1). Such a map v is called a **Gauß map**. The function
$\Pi : M \to \mathbb{R}^{n \times n}$ is in this case given by

$$\Pi(p) = \mathbb{1} - v(p)v(p)^{\mathsf{T}} \tag{3.1.5}$$

for $p \in U$.

Example 3.1.4 Let $M = S^2 \subset \mathbb{R}^3$. Then $v(p) = p$ and so

$$\Pi(p) = \mathbb{1} - pp^{\mathsf{T}} = \begin{pmatrix} 1 - x^2 & -xy & -xz \\ -yx & 1 - y^2 & -yz \\ -zx & -zy & 1 - z^2 \end{pmatrix}$$

for $p = (x, y, z) \in S^2$.

Example 3.1.5 (Möbius strip) Consider the submanifold

$$M := \left\{ (x, y, z) \in \mathbb{R}^3 \,\middle|\, \begin{array}{l} x = (1 + r\cos(\theta/2))\cos(\theta), \\ y = (1 + r\cos(\theta/2))\sin(\theta), \\ z = r\sin(\theta/2), \ r, \theta \in \mathbb{R}, \ |r| < \varepsilon \end{array} \right\}$$

for $\varepsilon > 0$ sufficiently small. Show that there does not exist a global smooth function
$v : M \to \mathbb{R}^3$ satisfying (3.1.4).

Example 3.1.6 Let $U \subset \mathbb{R}^n$ be an open set and $f : U \to \mathbb{R}^{n-m}$ be a smooth
function such that $0 \in \mathbb{R}^{n-m}$ is a regular value of f and $U \cap M = f^{-1}(0)$. Then
$T_p M = \ker df(p)$ and

$$\Pi(p) = \mathbb{1} - df(p)^{\mathsf{T}} \left(df(p)df(p)^{\mathsf{T}} \right)^{-1} df(p)$$

for every $p \in U \cap M$.

Example 3.1.7 Let $\Omega \subset \mathbb{R}^m$ be an open set and $\psi : \Omega \to M$ be a smooth embedding. Then $T_{\psi(x)}M = \mathrm{im}\, d\psi(x)$ and

$$\Pi(\psi(x)) = d\psi(x)\left(d\psi(x)^\mathsf{T} d\psi(x)\right)^{-1} d\psi(x)^\mathsf{T}$$

for every $x \in \Omega$.

Next we differentiate the map $\Pi : M \to \mathbb{R}^{n \times n}$ in Lemma 3.1.2. The derivative at $p \in M$ takes the form of a linear map

$$d\Pi(p) : T_p M \to \mathbb{R}^{n \times n}$$

which, as usual, is defined by

$$d\Pi(p)v := \left.\frac{d}{dt}\right|_{t=0} \Pi(\gamma(t)) \in \mathbb{R}^{n \times n}$$

for $v \in T_p M$, where $\gamma : \mathbb{R} \to M$ is chosen such that $\gamma(0) = p$ and $\dot\gamma(0) = v$ (see Definition 2.2.13). We emphasize that the expression $d\Pi(p)v$ is a matrix and can therefore be multiplied by a vector in $\mathbb{R}^n$.

Lemma 3.1.8 *For all $p \in M$ and $v, w \in T_p M$ we have*

$$\left(d\Pi(p)v\right)w = \left(d\Pi(p)w\right)v \in T_p M^\perp.$$

Proof Choose a smooth path $\gamma : \mathbb{R} \to M$ and a vector field $X : \mathbb{R} \to \mathbb{R}^n$ along γ such that

$$\gamma(0) = p, \qquad \dot\gamma(0) = v, \qquad X(0) = w.$$

For example, we can choose $X(t) := \Pi(\gamma(t))w$. Then

$$X(t) = \Pi(\gamma(t))X(t)$$

for every $t \in \mathbb{R}$. Differentiate this equation to obtain

$$\dot X(t) = \Pi(\gamma(t))\dot X(t) + \left(d\Pi(\gamma(t))\dot\gamma(t)\right)X(t). \tag{3.1.6}$$

Hence

$$\left(d\Pi(\gamma(t))\dot\gamma(t)\right)X(t) = \left(\mathbb{1} - \Pi(\gamma(t))\right)\dot X(t) \in T_{\gamma(t)}M^\perp \tag{3.1.7}$$

for every $t \in \mathbb{R}$ and, with $t = 0$, we obtain $(d\Pi(p)v)w \in T_p M^\perp$.

Now choose a smooth map

$$\mathbb{R}^2 \to M : (s, t) \mapsto \gamma(s, t)$$

satisfying

$$\gamma(0, 0) = p, \qquad \frac{\partial \gamma}{\partial s}(0, 0) = v, \qquad \frac{\partial \gamma}{\partial t}(0, 0) = w,$$

(for example by doing this in local coordinates) and denote

$$X(s, t) := \frac{\partial \gamma}{\partial s}(s, t) \in T_{\gamma(s,t)} M, \qquad Y(s, t) := \frac{\partial \gamma}{\partial t}(s, t) \in T_{\gamma(s,t)} M.$$

Then

$$\frac{\partial Y}{\partial s} = \frac{\partial^2 \gamma}{\partial s \partial t} = \frac{\partial X}{\partial t}$$

and hence, using (3.1.7), we obtain

$$
\begin{aligned}
\left(d\Pi(\gamma)\frac{\partial \gamma}{\partial t}\right)\frac{\partial \gamma}{\partial s} &= \left(d\Pi(\gamma)\frac{\partial \gamma}{\partial t}\right)X \\
&= (\mathbb{1} - \Pi(\gamma))\frac{\partial X}{\partial t} \\
&= (\mathbb{1} - \Pi(\gamma))\frac{\partial Y}{\partial s} \\
&= \left(d\Pi(\gamma)\frac{\partial \gamma}{\partial s}\right)Y \\
&= \left(d\Pi(\gamma)\frac{\partial \gamma}{\partial s}\right)\frac{\partial \gamma}{\partial t}.
\end{aligned}
$$

With $s = t = 0$ we obtain

$$\left(d\Pi(p)w\right)v = \left(d\Pi(p)v\right)w \in T_p M^\perp$$

and this proves Lemma 3.1.8.   □

Definition 3.1.9   The collection of symmetric bilinear maps

$$h_p : T_p M \times T_p M \to T_p M^\perp,$$

defined by

$$h_p(v, w) := (d\Pi(p)v)w = (d\Pi(p)w)v \tag{3.1.8}$$

for $p \in M$ and $v, w \in T_p M$ is called the **second fundamental form** on M.

Example 3.1.10 Let $M \subset \mathbb{R}^{m+1}$ be an m-manifold and $\nu : M \to S^m$ be a Gauß map so that $T_p M = \nu(p)^\perp$ for every $p \in M$ (see Example 3.1.3). Then $\Pi(p) = \mathbb{1} - \nu(p)\nu(p)^\mathsf{T}$ and hence

$$h_p(v, w) = -\nu(p)\langle d\nu(p)v, w\rangle$$

for $p \in M$ and $v, w \in T_p M$.

Exercise 3.1.11 Choose a splitting $\mathbb{R}^n = \mathbb{R}^m \times \mathbb{R}^{n-m}$ and write the elements of $\mathbb{R}^n$ as tuples $(x, y) = (x_1, \ldots, x_m, y_1, \ldots, y_{n-m})$ Let $M \subset \mathbb{R}^n$ be a smooth m-dimensional submanifold such that $p = 0 \in M$ and

$$T_0 M = \mathbb{R}^m \times \{0\}, \qquad T_0 M^\perp = \{0\} \times \mathbb{R}^{n-m}.$$

By the implicit function theorem, there are open neighborhoods $\Omega \subset \mathbb{R}^m$ and $V \subset \mathbb{R}^{n-m}$ of zero and a smooth map $f : \Omega \to V$ such that

$$M \cap (\Omega \times V) = \text{graph}(f) = \{(x, f(x)) \,|\, x \in \Omega\}.$$

Thus $f(0) = 0$ and $df(0) = 0$. Prove that the second fundamental form $h_p : T_p M \times T_p M \to T_p M^\perp$ is given by the second derivatives of f, i.e.

$$h_p(v, w) = \left(0, \sum_{i,j=1}^{m} \frac{\partial^2 f}{\partial x_i \partial x_j}(0) v_i w_j\right)$$

for $v, w \in T_p M = \mathbb{R}^m \times \{0\}$.

Exercise 3.1.12 Let $M \subset \mathbb{R}^n$ be an m-manifold. Fix a point $p \in M$ and a unit tangent vector $v \in T_p M$ so that $|v| = 1$ and define

$$L := \{p + tv + w \,|\, t \in \mathbb{R}, \, w \perp T_p M\}.$$

Let $\gamma : (-\varepsilon, \varepsilon) \to M \cap L$ be a smooth curve such that $\gamma(0) = p$, $\dot{\gamma}(0) = v$, and $|\dot{\gamma}(t)| = 1$ for all t. Prove that

$$\ddot{\gamma}(0) = h_p(v, v).$$

Draw a picture of M and L in the case $n = 3$ and $m = 2$.

3.2 Covariant Derivative

Definition 3.2.1 Let $I \subset \mathbb{R}$ be an open interval and let $\gamma : I \to M$ be a smooth curve. A **vector field along** γ is a smooth map $X : I \to \mathbb{R}^n$ such that $X(t) \in T_{\gamma(t)} M$ for every $t \in I$ (see Fig. 3.2). The set of smooth vector fields along γ is a real vector space and will be denoted by

$$\text{Vect}(\gamma) := \{ X : I \to \mathbb{R}^n \mid X \text{ is smooth and } X(t) \in T_{\gamma(t)} M \ \forall \ t \in I \}.$$

The first derivative $\dot{X}(t)$ of a vector field along γ at $t \in I$ will, in general, not be tangent to M. We may decompose it as a sum of a tangent vector and a normal vector in the form

$$\dot{X}(t) = \Pi(\gamma(t))\dot{X}(t) + \big(\mathbb{1} - \Pi(\gamma(t))\big)\dot{X}(t),$$

where $\Pi : M \to \mathbb{R}^{n \times n}$ is defined by (3.1.2) and (3.1.3). The tangential component of this decomposition plays an important geometric role. It is called the covariant derivative of X at t.

Definition 3.2.2 (Covariant derivative) Let $I \subset \mathbb{R}$ be an open interval, let $\gamma : I \to M$ be a smooth curve, and let $X \in \text{Vect}(\gamma)$. The **covariant derivative of** X is the vector field $\nabla X \in \text{Vect}(\gamma)$, defined by

$$\nabla X(t) := \Pi(\gamma(t))\dot{X}(t) \in T_{\gamma(t)} M \tag{3.2.1}$$

for $t \in I$.

Lemma 3.2.3 (Gauß–Weingarten formula) *The derivative of a vector field X along a curve γ is given by*

$$\dot{X}(t) = \nabla X(t) + h_{\gamma(t)}(\dot{\gamma}(t), X(t)). \tag{3.2.2}$$

Here the first summand is tangent to M and the second summand is orthogonal to the tangent space of M at $\gamma(t)$.

Proof This is equation (3.1.6) in the proof of Lemma 3.1.8. □

It follows directly from the definition that the covariant derivative along a curve $\gamma : I \to M$ is a linear operator $\nabla : \text{Vect}(\gamma) \to \text{Vect}(\gamma)$. The following lemma summarizes the basic properties of this operator.

Fig. 3.2 A vector field along a curve

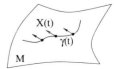

Lemma 3.2.4 (Covariant derivative)   *The covariant derivative satisfies the following axioms for any two open intervals $I, J \subset \mathbb{R}$.*

(i)   *Let $\gamma : I \to M$ be a smooth curve, let $\lambda : I \to \mathbb{R}$ be a smooth function, and let $X \in \text{Vect}(\gamma)$. Then*

$$\nabla(\lambda X) = \dot{\lambda} X + \lambda \nabla X. \qquad (3.2.3)$$

(ii)   *Let $\gamma : I \to M$ be a smooth curve, let $\sigma : J \to I$ be a smooth function and let $X \in \text{Vect}(\gamma)$. Then*

$$\nabla(X \circ \sigma) = \dot{\sigma}(\nabla X \circ \sigma). \qquad (3.2.4)$$

(iii)   *Let $\gamma : I \to M$ be a smooth curve and let $X, Y \in \text{Vect}(\gamma)$. Then*

$$\frac{d}{dt}\langle X, Y \rangle = \langle \nabla X, Y \rangle + \langle X, \nabla Y \rangle. \qquad (3.2.5)$$

(iv)   *Let $\gamma : I \times J \to M$ be a smooth map, denote by ∇_s the covariant derivative along the curve $s \mapsto \gamma(s, t)$ (with t fixed), and denote by ∇_t the covariant derivative along the curve $t \mapsto \gamma(s, t)$ (with s fixed). Then*

$$\nabla_s \partial_t \gamma = \nabla_t \partial_s \gamma. \qquad (3.2.6)$$

Proof   Part (i) follows from the Leibniz rule $\frac{d}{dt}(\lambda X) = \dot{\lambda} X + \lambda \dot{X}$ and (ii) follows from the chain rule $\frac{d}{dt}(X \circ \sigma) = \dot{\sigma}(\dot{X} \circ \sigma)$. To prove part (iii), use the orthogonal projections $\Pi(\gamma(t)) : \mathbb{R}^n \to T_{\gamma(t)} M$ to obtain

$$
\begin{aligned}
\frac{d}{dt}\langle X, Y \rangle &= \langle \dot{X}, Y \rangle + \langle X, \dot{Y} \rangle \\
&= \langle \dot{X}, \Pi(\gamma)Y \rangle + \langle \Pi(\gamma)X, \dot{Y} \rangle \\
&= \langle \Pi(\gamma)\dot{X}, Y \rangle + \langle X, \Pi(\gamma)\dot{Y} \rangle \\
&= \langle \nabla X, Y \rangle + \langle X, \nabla Y \rangle.
\end{aligned}
$$

Part (iv) holds because the second derivatives commute. This completes the proof of Lemma 3.2.4.   □

Part (i) in Lemma 3.2.4 asserts that the operator ∇ is what is called a *connection*, part (iii) asserts that it is compatible with the first fundamental form, and part (iv) asserts that it is *torsion-free*. Theorem 3.7.8 below asserts that these conditions (together with an extended chain rule) determine the covariant derivative uniquely.

3.3 Parallel Transport

Definition 3.3.1 (Parallel vector field) Let $I \subset \mathbb{R}$ be an interval and let $\gamma : I \to M$ be a smooth curve. A vector field X along γ is called **parallel** iff

$$\nabla X(t) = 0$$

for all $t \in I$.

Example 3.3.2 Assume $m = n$ so that $M \subset \mathbb{R}^m$ is an open set. Then a vector field along a smooth curve $\gamma : I \to M$ is a smooth map $X : I \to \mathbb{R}^m$. Its covariant derivative is equal to the ordinary derivative $\nabla X(t) = \dot{X}(t)$ and hence X is is parallel if and only if it is constant.

Remark 3.3.3 For every $X \in \mathrm{Vect}(\gamma)$ and every $t \in I$ we have

$$\nabla X(t) = 0 \qquad \Longleftrightarrow \qquad \dot{X}(t) \perp T_{\gamma(t)} M.$$

In particular, $\dot{\gamma}$ is a vector field along γ and $\nabla \dot{\gamma}(t) = \Pi(\gamma(t))\ddot{\gamma}(t)$. Hence $\dot{\gamma}$ is a parallel vector field along γ if and only if $\ddot{\gamma}(t) \perp T_{\gamma(t)} M$ for all $t \in I$. We will return to this observation in Chap. 4.

In general, a vector field X along a smooth curve $\gamma : I \to M$ is parallel if and only if $\dot{X}(t)$ is orthogonal to $T_{\gamma(t)} M$ for every t and, by the Gauß–Weingarten formula (3.2.2), we have

$$\nabla X = 0 \qquad \Longleftrightarrow \qquad \dot{X} = h_\gamma(\dot{\gamma}, X).$$

The next theorem shows that any given tangent vector $v_0 \in T_{\gamma(t_0)} M$ extends uniquely to a parallel vector field along γ.

Theorem 3.3.4 (Existence and uniqueness) Let $I \subset \mathbb{R}$ be an interval and $\gamma : I \to M$ be a smooth curve. Let $t_0 \in I$ and $v_0 \in T_{\gamma(t_0)} M$ be given. Then there is a unique parallel vector field $X \in \mathrm{Vect}(\gamma)$ such that $X(t_0) = v_0$.

Proof Choose a basis $e_1, \dots, e_m$ of the tangent space $T_{\gamma(t_0)} M$ and let

$$X_1, \dots, X_m \in \mathrm{Vect}(\gamma)$$

be vector fields along γ such that

$$X_i(t_0) = e_i, \qquad i = 1, \dots, m.$$

(For example choose $X_i(t) := \Pi(\gamma(t))e_i$.) Then the vectors $X_i(t_0)$ are linearly independent. Since linear independence is an open condition there is a constant $\varepsilon > 0$ such that the vectors $X_1(t), \dots, X_m(t) \in T_{\gamma(t)} M$ are linearly independent for every $t \in I_0 := (t_0 - \varepsilon, t_0 + \varepsilon) \cap I$. Since $T_{\gamma(t)} M$ is an m-dimensional real vector space this implies that the vectors $X_i(t)$ form a basis of $T_{\gamma(t)}M$ for every $t \in I_0$. We express the vector $\nabla X_i(t) \in T_{\gamma(t)}M$ in this basis and denote the coefficients by $a_i^k(t)$ so that

$$\nabla X_i(t) = \sum_{k=1}^{m} a_i^k(t) X_k(t).$$

The resulting functions $a_i^k : I_0 \to \mathbb{R}$ are smooth. Likewise, if $X : I \to \mathbb{R}^n$ is any vector field along γ, then there are smooth functions $\xi^i : I_0 \to \mathbb{R}$ such that

$$X(t) = \sum_{i=1}^{m} \xi^i(t) X_i(t) \qquad \text{for all } t \in I_0.$$

The derivative of X is given by

$$\dot{X}(t) = \sum_{i=1}^{m} \left(\dot{\xi}^i(t) X_i(t) + \xi^i(t) \dot{X}_i(t) \right)$$

and the covariant derivative by

$$\nabla X(t) = \sum_{i=1}^{m} \left(\dot{\xi}^i(t) X_i(t) + \xi^i(t) \nabla X_i(t) \right)$$

$$= \sum_{i=1}^{m} \dot{\xi}^i(t) X_i(t) + \sum_{i=1}^{m} \xi^i(t) \sum_{k=1}^{m} a_i^k(t) X_k(t)$$

$$= \sum_{k=1}^{m} \left(\dot{\xi}^k(t) + \sum_{i=1}^{m} a_i^k(t) \xi^i(t) \right) X_k(t)$$

for $t \in I_0$. Hence $\nabla X(t) = 0$ if and only if

$$\dot{\xi}(t) + A(t)\xi(t) = 0, \qquad A(t) := \begin{pmatrix} a_1^1(t) & \cdots & a_m^1(t) \\ \vdots & & \vdots \\ a_1^m(t) & \cdots & a_m^m(t) \end{pmatrix}.$$

Thus we have translated the equation $\nabla X = 0$ over the interval I_0 into a time dependent linear ordinary differential equation. By a theorem in Analysis II (see [64, Lemma 4.4.3]), this equation has a unique solution for any initial condition at any point in I_0. Thus we have proved that every $t_0 \in I$ is contained in an interval $I_0 \subset I$, open in the relative topology of I, such that, for every $t_1 \in I_0$ and every $v_1 \in T_{\gamma(t_1)}M$, there exists a unique parallel vector field $X : I_0 \to \mathbb{R}^n$ along $\gamma|_{I_0}$ satisfying $X(t_1) = v_1$. We formulate this condition on the interval I_0 as a logical formula:

$$\forall\, t_1 \in I_0 \ \ \forall\, v_1 \in T_{\gamma(t_1)}M \ \ \exists!\ X \in \text{Vect}(\gamma|_{I_0})$$

$$\text{such that } \nabla X = 0 \text{ and } X(t_1) = v_1. \tag{3.3.1}$$

If two I-open intervals $I_0, I_1 \subset I$ satisfy this condition and have nonempty inter-section, then their union $I_0 \cup I_1$ also satisfies (3.3.1). (Prove this!) Now define

$$J := \bigcup \{ I_0 \subset I \mid I_0 \text{ is an } I\text{-open interval}, \ I_0 \text{ satisfies (3.3.1)}, \ t_0 \in I_0 \}.$$

This interval J satisfies (3.3.1). Moreover, it is nonempty and, by definition, it is open in the relative topology of I. We prove that it is also closed in the relative topology of I. Thus let $(t_i)_{i \in \mathbb{N}}$ be a sequence in J converging to a point $t^* \in I$. By what we have proved above, there exists a constant $\varepsilon > 0$ such that the in-terval $I^* := (t^* - \varepsilon, t^* + \varepsilon) \cap I$ satisfies (3.3.1). Since the sequence $(t_i)_{i \in \mathbb{N}}$ con-verges to t^*, there exists an $i \in \mathbb{N}$ such that $t_i \in I^*$. Since $t_i \in J$ there exists an interval $I_0 \subset I$, open in the relative topology of I, that contains t_0 and t_i and satisfies (3.3.1). Hence the interval $I_0 \cup I^*$ is open in the relative topology of I, contains t_0 and t^*, and satisfies (3.3.1). This shows that $t^* \in J$. Thus we have proved that the interval J is nonempty, and open and closed in the relative topology of I. Hence $J = I$ and this proves Theorem 3.3.4. $\qquad \square$

Definition 3.3.5 (Parallel transport) Let $I \subset \mathbb{R}$ be an interval and let $\gamma : I \to M$ be a smooth curve. For $t_0, t \in I$ we define the map

$$\Phi_\gamma(t, t_0) : T_{\gamma(t_0)} M \to T_{\gamma(t)} M$$

by $\Phi_\gamma(t, t_0) v_0 := X(t)$ where $X \in \mathrm{Vect}(\gamma)$ is the unique parallel vector field along γ satisfying $X(t_0) = v_0$. The collection of maps $\Phi_\gamma(t, t_0)$ for $t, t_0 \in I$ is called **parallel transport along** γ.

Recall the notation

$$\gamma^* TM = \{ (s, v) \mid s \in I, \ v \in T_{\gamma(s)} M \}$$

for the pullback tangent bundle. This set is a smooth submanifold of $I \times \mathbb{R}^n$. (See Theorem 2.6.10 and Corollary 2.6.13.) The next theorem summarizes the proper-ties of parallel transport. In particular, the last assertion shows that the covariant derivative can be recovered from the parallel transport maps.

Theorem 3.3.6 (Parallel transport) *Let $\gamma : I \to M$ be a smooth curve on an interval $I \subset \mathbb{R}$.*

(i)   *The map $\Phi_\gamma(t, s) : T_{\gamma(s)} M \to T_{\gamma(t)} M$ is linear for all $s, t \in I$.*
(ii) *For all $r, s, t \in I$ we have*

$$\Phi_\gamma(t, s) \circ \Phi_\gamma(s, r) = \Phi_\gamma(t, r), \qquad \Phi_\gamma(t, t) = \mathrm{id}.$$

(iii) *For all $s, t \in I$ and all $v, w \in T_{\gamma(s)} M$ we have*

$$\langle \Phi_\gamma(t, s) v, \Phi_\gamma(t, s) w \rangle = \langle v, w \rangle.$$

Thus $\Phi_\gamma(t, s) : T_{\gamma(s)} M \to T_{\gamma(t)} M$ is an orthogonal transformation.

(iv) *If $J \subset \mathbb{R}$ is an interval and $\sigma : J \to I$ is a smooth map, then*

$$\Phi_{\gamma \circ \sigma}(t, s) = \Phi_\gamma(\sigma(t), \sigma(s)).$$

for all $s, t \in J$.
(v) *The map*

$$I \times \gamma^* TM \to \gamma^* TM : (t, (s, v)) \mapsto (t, \Phi_\gamma(t, s)v)$$

is smooth.
(vi) *For all $X \in \mathrm{Vect}(\gamma)$ and $t, t_0 \in I$ we have*

$$\frac{d}{dt}\Phi_\gamma(t_0, t)X(t) = \Phi_\gamma(t_0, t)\nabla X(t).$$

Proof Assertion (i) holds because the sum of two parallel vector fields along γ is again parallel and the product of a parallel vector field with a constant real number is again parallel. Assertion (ii) follows directly from the uniqueness statement in Theorem 3.3.4.

We prove (iii). Fix a number $s \in I$ and two tangent vectors

$$v, w \in T_{\gamma(s)}M.$$

Define the vector fields $X, Y \in \mathrm{Vect}(\gamma)$ along γ by

$$X(t) := \Phi_\gamma(t, s)v, \qquad Y(t) := \Phi_\gamma(t, s)w.$$

These vector fields are parallel. Thus, by equation (3.2.5) in Lemma 3.2.4, we have

$$\frac{d}{dt}\langle X, Y \rangle = \langle \nabla X, Y \rangle + \langle X, \nabla Y \rangle = 0.$$

Hence the function $I \to \mathbb{R} : t \mapsto \langle X(t), Y(t) \rangle$ is constant and this proves (iii).

We prove (iv). Fix an element $s \in J$ and a tangent vector $v \in T_{\gamma(\sigma(s))}M$. Define the vector field X along γ by

$$X(t) := \Phi_\gamma(t, \sigma(s))v$$

for $t \in I$. Thus X is the unique parallel vector field along γ that satisfies

$$X(\sigma(s)) = v.$$

Denote

$$\widetilde{\gamma} := \gamma \circ \sigma : J \to M, \qquad \widetilde{X} := X \circ \sigma : I \to \mathbb{R}^n.$$

Then $\widetilde{X}$ is a vector field along $\widetilde{\gamma}$ and, by the chain rule, we have

$$\frac{d}{dt}\widetilde{X}(t) = \frac{d}{dt}X(\sigma(t)) = \dot{\sigma}(t)\dot{X}(\sigma(t)).$$

Projecting orthogonally onto the tangent space $T_{\gamma(\sigma(t))}M$ we obtain

$$\nabla\widetilde{X}(t) = \dot{\sigma}(t)\nabla X(\sigma(t)) = 0$$

for every $t \in J$. Hence $\widetilde{X}$ is the unique parallel vector field along $\widetilde{\gamma}$ that satisfies $\widetilde{X}(s) = v$. Thus

$$\Phi_{\widetilde{\gamma}}(t,s)v = \widetilde{X}(t) = X(\sigma(t)) = \Phi_{\gamma}(\sigma(t),\sigma(s))v.$$

This proves (iv).

We prove (v). Fix a point $t_0 \in I$, choose an orthonormal basis $e_1,\dots,e_m$ of $T_{\gamma(t_0)}M$, and define $X_i(t) := \Phi_{\gamma}(t,t_0)e_i$ for $t \in I$ and $i = 1,\dots,m$. Thus $X_i \in \mathrm{Vect}(\gamma)$ is the unique parallel vector field along γ such that $X_i(t_0) = e_i$. Then by (iii) we have

$$\langle X_i(t), X_j(t)\rangle = \delta_{ij}$$

for all $i, j \in \{1,\dots,m\}$ and all $t \in I$. Hence the vectors $X_1(t),\dots,X_m(t)$ form an orthonormal basis of $T_{\gamma(t)}M$ for every $t \in I$. This implies that, for each $s \in I$ and each tangent vector $v \in T_{\gamma(s)}M$, we have

$$v = \sum_{i=1}^{m}\langle X_i(s), v\rangle X_i(s).$$

Since each vector field X_i is parallel it satisfies $X_i(t) = \Phi_{\gamma}(t,s)X_i(s)$. Hence

$$\Phi_{\gamma}(t,s)v = \sum_{i=1}^{m}\langle X_i(s), v\rangle X_i(t) \qquad (3.3.2)$$

for all $s, t \in I$ and $v \in T_{\gamma(s)}M$. This proves (v).

We prove (vi). Let $X_1,\dots,X_m \in \mathrm{Vect}(\gamma)$ be as in the proof of (v). Thus every vector field X along γ can be written in the form

$$X(t) = \sum_{i=1}^{m}\xi^i(t)X_i(t), \qquad \xi^i(t) := \langle X_i(t), X(t)\rangle.$$

Since the vector fields X_i are parallel we have

$$\nabla X(t) = \sum_{i=1}^{m}\dot{\xi}^i(t)X_i(t)$$

for all $t \in I$. Hence

$$\Phi_\gamma(t_0, t) X(t) = \sum_{i=1}^{m} \xi^i(t) X_i(t_0), \qquad \Phi_\gamma(t_0, t) \nabla X(t) = \sum_{i=1}^{m} \dot{\xi}^i(t) X_i(t_0).$$

Evidently, the derivative of the first sum with respect to t is equal to the second sum. This proves (vi) and Theorem 3.3.6. □

Remark 3.3.7 For $s, t \in I$ we can think of the linear map

$$\Phi_\gamma(t, s) \Pi(\gamma(s)) : \mathbb{R}^n \to T_{\gamma(t)} M \subset \mathbb{R}^n$$

as a real $n \times n$ matrix. The formula (3.3.2) in the proof of (v) shows that this matrix can be expressed in the form

$$\Phi_\gamma(t, s) \Pi(\gamma(s)) = \sum_{i=1}^{m} X_i(t) X_i(s)^\top \in \mathbb{R}^{n \times n}.$$

The right hand side defines a smooth matrix valued function on $I \times I$ and this is equivalent to the assertion in (v).

Remark 3.3.8 It follows from assertions (ii) and (iii) in Theorem 3.3.6 that

$$\Phi_\gamma(t, s)^{-1} = \Phi_\gamma(s, t) = \Phi_\gamma(t, s)^*$$

for all $s, t \in I$. Here the linear map $\Phi_\gamma(t, s)^* : T_{\gamma(t)} M \to T_{\gamma(s)} M$ is understood as the adjoint operator of $\Phi_\gamma(t, s) : T_{\gamma(s)} M \to T_{\gamma(t)} M$ with respect to the inner products on the two subspaces of $\mathbb{R}^n$ inherited from the Euclidean inner product on the ambient space.

The two theorems in this section carry over verbatim to any smooth vector bundle $E \subset M \times \mathbb{R}^n$ over a manifold. As in the case of the tangent bundle one can define the covariant derivative of a section of E along γ as the orthogonal projection of the ordinary derivative in the ambient space $\mathbb{R}^n$ onto the fiber $E_{\gamma(t)}$. Instead of *parallel vector fields* one then speaks about *horizontal sections* and one proves as in Theorem 3.3.4 that there is a unique horizontal section along γ through any point in any of the fibers $E_{\gamma(t_0)}$. This gives parallel transport maps from $E_{\gamma(s)}$ to $E_{\gamma(t)}$ for any pair $s, t \in I$ and Theorem 3.3.6 carries over verbatim to all vector bundles $E \subset M \times \mathbb{R}^n$. We spell this out in more detail in the case where $E = TM^\perp \subset M \times \mathbb{R}^n$ is the normal bundle of M.

Let $\gamma : I \to M$ be a smooth curve. A **normal vector field along** γ is a smooth map $Y : I \to \mathbb{R}^n$ such that $Y(t) \perp T_{\gamma(t)} M$ for every $t \in I$. The set of normal vector fields along γ will be denoted by

$$\mathrm{Vect}^\perp(\gamma) := \{Y : I \to \mathbb{R}^n \mid Y \text{ is smooth and } Y(t) \perp T_{\gamma(t)} M \text{ for all } t \in I\}.$$

This is again a real vector space. The **covariant derivative** of a normal vector field $Y \in \text{Vect}^\perp(\gamma)$ at $t \in I$ is defined as the orthogonal projection of the ordinary derivative onto the orthogonal complement of $T_{\gamma(t)}M$ and will be denoted by

$$\nabla^\perp Y(t) := \left(\mathbb{1} - \Pi(\gamma(t))\right)\dot{Y}(t). \tag{3.3.3}$$

Thus the covariant derivative defines a linear operator

$$\nabla^\perp : \text{Vect}^\perp(\gamma) \to \text{Vect}^\perp(\gamma).$$

There is a version of the Gauß–Weingarten formula for the covariant derivative of a normal vector field. This is the content of the next lemma.

Lemma 3.3.9 *Let $M \subset \mathbb{R}^n$ be a smooth m-manifold. For $p \in M$ and $u \in T_pM$ define the linear map $h_p(u) : T_pM \to T_pM^\perp$ by*

$$h_p(u)v := h_p(u, v) = \left(d\Pi(p)u\right)v \tag{3.3.4}$$

for $v \in T_pM$. Then the following holds.

(i) *The adjoint operator $h_p(u)^* : T_pM^\perp \to T_pM$ is given by*

$$h_p(u)^*w = \left(d\Pi(p)u\right)w, \qquad w \in T_pM^\perp. \tag{3.3.5}$$

(ii) *If $I \subset \mathbb{R}$ is an interval, $\gamma : I \to M$ is a smooth curve, and $Y \in \text{Vect}^\perp(\gamma)$, then the derivative of Y satisfies the **Gauß–Weingarten formula***

$$\dot{Y}(t) = \nabla^\perp Y(t) - h_{\gamma(t)}(\dot{\gamma}(t))^* Y(t). \tag{3.3.6}$$

Proof Since $\Pi(p) \in \mathbb{R}^{n \times n}$ is a symmetric matrix for every $p \in M$ so is the matrix $d\Pi(p)u$ for every $p \in M$ and every $u \in T_pM$. Hence

$$\begin{aligned}
\langle v, h_p(u)^*w \rangle &= \langle h_p(u)v, w \rangle \\
&= \langle \left(d\Pi(p)u\right)v, w \rangle \\
&= \langle v, \left(d\Pi(p)u\right)w \rangle
\end{aligned}$$

for every $v \in T_pM$ and every $w \in T_pM^\perp$. This proves (i).

To prove (ii) we observe that, for $Y \in \text{Vect}^\perp(\gamma)$ and $t \in I$, we have

$$\Pi(\gamma(t))Y(t) = 0.$$

Differentiating this identity we obtain

$$\Pi(\gamma(t))\dot{Y}(t) + \left(d\Pi(\gamma(t))\dot{\gamma}(t)\right)Y(t) = 0$$

and hence

$$\dot{Y}(t) = \dot{Y}(t) - \Pi(\gamma(t))\dot{Y}(t) - \big(d\Pi(\gamma(t))\dot{\gamma}(t)\big)Y(t)$$
$$= \nabla^\perp Y(t) - h_{\gamma(t)}(\dot{\gamma}(t))^* Y(t)$$

for $t \in I$. Here the last equation follows from (i) and the definition of $\nabla^\perp$. This proves Lemma 3.3.9. □

Theorem 3.3.4 and its proof carry over to the normal bundle $TM^\perp$. Thus, if $\gamma : I \to M$ is a smooth curve, then for all $s \in I$ and $w \in T_{\gamma(s)}M^\perp$ there is a unique normal vector field $Y \in \text{Vect}^\perp(\gamma)$ such that

$$\nabla^\perp Y \equiv 0, \qquad Y(s) = w.$$

This gives rise to parallel transport maps

$$\Phi_\gamma^\perp(t,s) : T_{\gamma(s)}M^\perp \to T_{\gamma(t)}M^\perp$$

defined by

$$\Phi_\gamma^\perp(t,s)w := Y(t)$$

for $s, t \in I$ and $w \in T_{\gamma(s)}M^\perp$, where Y is the unique normal vector field along γ satisfying $\nabla^\perp Y \equiv 0$ and $Y(s) = w$. These parallel transport maps satisfy exactly the same conditions that have been spelled out in Theorem 3.3.6 for the tangent bundle and the proof carries over verbatim to the present setting.

3.4   The Frame Bundle

Each tangent space of an m-manifold M is isomorphic to the Euclidean space $\mathbb{R}^m$, however, in general there is no canonical isomorphism. The space of all pairs consisting of a point p in the manifold M and an isomorphism from $\mathbb{R}^m$ to the tangent space of M at p is itself a smooth manifold, called the frame bundle of M.

3.4.1   Frames of a Vector Space

Let V be an m-dimensional real vector space. A **frame of V** is a basis $e_1, \ldots, e_m$ of V. It determines a vector space isomorphism $e : \mathbb{R}^m \to V$ via

$$e\xi := \sum_{i=1}^m \xi^i e_i, \qquad \xi = (\xi^1, \ldots, \xi^m) \in \mathbb{R}^m.$$

Conversely, each isomorphism $e : \mathbb{R}^m \to V$ determines a basis $e_1, \dots, e_m$ of V via $e_i = e(0, \dots, 0, 1, 0 \dots, 0)$, where the coordinate 1 appears in the ith place. The set of vector space isomorphisms from $\mathbb{R}^m$ to V will be denoted by

$$\mathcal{L}_{\mathrm{iso}}(\mathbb{R}^m, V) := \{e : \mathbb{R}^m \to V \mid e \text{ is a vector space isomorphism}\}.$$

The general linear group $\mathrm{GL}(m) = \mathrm{GL}(m, \mathbb{R})$ (of nonsingular real $m \times m$-matrices) acts on this space by composition on the right via

$$\mathrm{GL}(m) \times \mathcal{L}_{\mathrm{iso}}(\mathbb{R}^m, V) \to \mathcal{L}_{\mathrm{iso}}(\mathbb{R}^m, V) : (a, e) \mapsto a^* e := e \circ a.$$

This is a **contravariant group action** in that

$$a^* b^* e = (ba)^* e, \qquad 1^* e = e$$

for $a, b \in \mathrm{GL}(m)$ and $e \in \mathcal{L}_{\mathrm{iso}}(\mathbb{R}^m, V)$. Moreover, the action is **free**, i.e. for all $a \in \mathrm{GL}(m)$ and $e \in \mathcal{L}_{\mathrm{iso}}(\mathbb{R}^m, V)$, we have

$$a^* e = e \qquad \Longleftrightarrow \qquad a = 1.$$

It is **transitive** in that for all $e, e' \in \mathcal{L}_{\mathrm{iso}}(\mathbb{R}^m, V)$ there is a group element $a \in \mathrm{GL}(m)$ such that $e' = a^* e$. Thus we can identify the space $\mathcal{L}_{\mathrm{iso}}(\mathbb{R}^m, V)$ with the group $\mathrm{GL}(m)$ via the bijection

$$\mathrm{GL}(m) \to \mathcal{L}_{\mathrm{iso}}(\mathbb{R}^m, V) : a \mapsto a^* e_0$$

induced by a fixed element $e_0 \in \mathcal{L}_{\mathrm{iso}}(\mathbb{R}^m, V)$. This identification is not canonical; it depends on the choice of e_0. The space $\mathcal{L}_{\mathrm{iso}}(\mathbb{R}^m, V)$ admits a bijection to a group but is not itself a group.

3.4.2 The Frame Bundle

Definition 3.4.1 (Frame bundle) Let $M \subset \mathbb{R}^n$ be a smooth m-manifold. The **frame bundle** of M is the set

$$\mathcal{F}(M) := \{(p, e) \mid p \in M, \ e \in \mathcal{F}(M)_p\}, \tag{3.4.1}$$

where $\mathcal{F}(M)_p$ is the space of frames of the tangent space at p, i.e.

$$\mathcal{F}(M)_p := \mathcal{L}_{\mathrm{iso}}(\mathbb{R}^m, T_p M).$$

Define a right action of $\mathrm{GL}(m)$ on $\mathcal{F}(M)$ by

$$a^*(p, e) := (p, a^* e) = (p, e \circ a) \tag{3.4.2}$$

for $a \in \mathrm{GL}(m)$ and $(p, e) \in \mathcal{F}(M)$.

One can think of a frame $e \in \mathcal{L}_{\mathrm{iso}}(\mathbb{R}^m, T_p M)$ as a linear map from $\mathbb{R}^m$ to $\mathbb{R}^n$ whose image is $T_p M$ and hence as an $n \times m$-matrix of rank m. The basis of $T_p M$ associated to this frame is given by the columns of the matrix $e \in \mathbb{R}^{n \times m}$. Thus the frame bundle $\mathcal{F}(M)$ of an embedded manifold $M \subset \mathbb{R}^n$ is a subset of the Euclidean space $\mathbb{R}^n \times \mathbb{R}^{n \times m}$.

Lemma 3.4.2   *The frame bundle*

$$\mathcal{F}(M) \subset \mathbb{R}^n \times \mathbb{R}^{n \times m}$$

is a smooth manifold of dimension $m + m^2$, the group action

$$\mathrm{GL}(m) \times \mathcal{F}(M) \to \mathcal{F}(M) : (a, (p, e)) \mapsto a^*(p, e)$$

is smooth, and the projection

$$\pi : \mathcal{F}(M) \to M$$

defined by $\pi(p, e) := p$ for $(p, e) \in \mathcal{F}(M)$ is a surjective submersion. The orbits of the $\mathrm{GL}(m)$-action on $\mathcal{F}(M)$ are the fibers of this projection, i.e.

$$\mathrm{GL}(m)^*(p, e) = \pi^{-1}(p) \cong \mathcal{F}(M)_p$$

for $(p, e) \in \mathcal{F}(M)$, and the group $\mathrm{GL}(m)$ acts freely and transitively on each of these fibers.

Proof   Let $U \subset M$ be an M-open set. A **moving frame** over U is a sequence of m smooth vector fields $E_1, \ldots, E_m \in \mathrm{Vect}(U)$ on U such that the vectors $E_1(p), \ldots, E_m(p)$ form a basis of $T_p M$ for each $p \in U$. Any such moving frame gives a bijection

$$U \times \mathrm{GL}(m) \to \mathcal{F}(U) : (p, a) \mapsto a^*(p, E(p)) = (p, E(p) \circ a),$$

where

$$E(p) := (E_1(p), \ldots, E_m(p)) \in \mathcal{F}(M)_p$$

for $p \in U$. This bijection (when composed with a parametrization of U) gives a parametrization of the open set $\mathcal{F}(U)$ in $\mathcal{F}(M)$. The assertions of the lemma then follow from the fact that the diagram

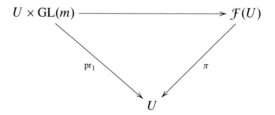

commutes. More precisely, suppose that there exists a coordinate chart

$$\phi : U \to \Omega$$

with values in an open set $\Omega \subset \mathbb{R}^m$, and denote its inverse by

$$\psi := \phi^{-1} : \Omega \to U.$$

Then the open set

$$\mathcal{F}(U) = \pi^{-1}(U) = \{(p, e) \in \mathcal{F}(M) \mid p \in U\} = (U \times \mathbb{R}^{n \times m}) \cap \mathcal{F}(M)$$

is parametrized by the map

$$\Omega \times \mathrm{GL}(m) \to \mathcal{F}(U) : (x, a) \mapsto \big(\psi(x), d\psi(x) \circ a\big).$$

This map is amooth and so is its inverse

$$\mathcal{F}(U) \to \Omega \times \mathrm{GL}(m) : (p, e) \mapsto \big(\phi(p), d\phi(p) \circ e\big).$$

These are the desired coordinate chart on $\mathcal{F}(M)$. Thus $\mathcal{F}(M)$ is a smooth manifold of dimension $m + m^2$. Moreover, in these coordinates the projection $\pi : \mathcal{F}(U) \to U$ is the map $\Omega \times \mathrm{GL}(m) \to \Omega : (x, a) \mapsto x$ and so π is a submersion. The remaining assertions follow directly from the definitions and this proves Lemma 3.4.2. $\qquad\square$

The frame bundle $\mathcal{F}(M)$ is a **principal bundle** over M with **structure group** $\mathrm{GL}(m)$. More generally, a principal bundle over a manifold B with structure group G is a smooth manifold P equipped with a surjective submersion $\pi : P \to B$ and a smooth contravariant action

$$\mathrm{G} \times P \to P : (g, p) \mapsto pg$$

by a Lie group G such that $\pi(pg) = \pi(p)$ for all $p \in P$ and $g \in \mathrm{G}$ and such that the group G acts freely and transitively on the fiber $P_b := \pi^{-1}(b)$ for each $b \in B$. In this book we shall mostly be concerned with the frame bundle of a manifold M and the orthonormal frame bundle.

Definition 3.4.3 The **orthonormal frame bundle of M** is the set

$$\mathcal{O}(M) := \big\{(p, e) \in \mathbb{R}^n \times \mathbb{R}^{n \times m} \mid p \in M, \; \mathrm{im}\, e = T_p M, \; e^\mathsf{T} e = 1_m \big\}.$$

If we denote by $e_i := e(0, \ldots, 0, 1, 0, \ldots, 0)$ (with 1 as the ith argument) the basis of $T_p M$ induced by the isomorphism $e : \mathbb{R}^m \to T_p M$, then we have

$$e^\mathsf{T} e = 1 \quad \Longleftrightarrow \quad \langle e_i, e_j \rangle = \delta_{ij} \quad \Longleftrightarrow \quad \begin{matrix} e_1, \ldots, e_m \text{ is an} \\ \text{orthonormal basis.} \end{matrix}$$

Thus $\mathcal{O}(M)$ is the bundle of orthonormal frames of the tangent spaces $T_p M$ or the bundle of orthogonal isomorphisms $e : \mathbb{R}^m \to T_p M$. It is a principal bundle over M with structure group $\mathrm{O}(m)$.

Exercise 3.4.4 Prove that $\mathcal{O}(M)$ is a submanifold of $\mathcal{F}(M)$ and that the obvious projection $\pi : \mathcal{O}(M) \to M$ is a submersion. Prove that the action of $\mathrm{GL}(m)$ on $\mathcal{F}(M)$ restricts to an action of the orthogonal group $\mathrm{O}(m)$ on $\mathcal{O}(M)$ whose orbits are the fibers

$$\mathcal{O}(M)_p := \left\{ e \in \mathbb{R}^{n \times m} \,\middle|\, (p, e) \in \mathcal{O}(M) \right\}$$
$$= \left\{ e \in \mathcal{L}_{\mathrm{iso}}(\mathbb{R}^m, T_p M) \,\middle|\, e^{\mathsf{T}} e = \mathbb{1} \right\}.$$

Hint: If $\phi : U \to \Omega$ is a coordinate chart on M with inverse $\psi : \Omega \to U$, then

$$e_x := d\psi(x)(d\psi(x)^{\mathsf{T}} d\psi(x))^{-1/2} : \mathbb{R}^m \to T_{\psi(x)} M$$

is an orthonormal frame of the tangent space $T_{\psi(x)} M$ for every $x \in \Omega$.

3.4.3 Horizontal Lifts

We have seen in Lemma 3.4.2 that the frame bundle $\mathcal{F}(M)$ is a smooth submanifold of $\mathbb{R}^n \times \mathbb{R}^{n \times m}$. Next we examine the tangent space of $\mathcal{F}(M)$ at a point $(p, e) \in \mathcal{F}(M)$. By Definition 2.2.1, this tangent space is given by

$$T_{(p,e)} \mathcal{F}(M) = \left\{ (\dot{\gamma}(0), \dot{e}(0)) \,\middle|\, \begin{array}{l} \mathbb{R} \to \mathcal{F}(M) : t \mapsto (\gamma(t), e(t)) \\ \text{is a smooth curve satisfying} \\ \gamma(0) = p \text{ and } e(0) = e \end{array} \right\}.$$

The next lemma gives an explicit formula for this tangent space in terms of the second fundamental form $h_p : T_p M \times T_p M \to T_p M^{\perp}$ in Definition 3.1.9. Compare this formula with Lemma 4.3.1 in the next chapter.

Lemma 3.4.5 *Let $M \subset \mathbb{R}^n$ be a smooth m-dimensional submanifold. Then the tangent space of $\mathcal{F}(M)$ at (p, e) is given by*

$$T_{(p,e)} \mathcal{F}(M) = \left\{ (\widehat{p}, \widehat{e}) \,\middle|\, \begin{array}{l} \widehat{p} \in T_p M, \ \widehat{e} \in \mathbb{R}^{n \times m}, \text{ and} \\ (\mathbb{1} - \Pi(p))\widehat{e} = h_p(\widehat{p})e \end{array} \right\}. \tag{3.4.3}$$

Proof We prove the inclusion "$\subset$" in (3.4.3). Let $(\widehat{p}, \widehat{e}) \in T_{(p,e)} \mathcal{F}(M)$ and choose a smooth curve $\mathbb{R} \to \mathcal{F}(M) : t \mapsto (\gamma(t), e(t))$ such that

$$\gamma(0) = p, \qquad e(0) = e, \qquad \dot{\gamma}(0) = \widehat{p}, \qquad \dot{e}(0) = \widehat{e}.$$

Fix a vector $\xi \in \mathbb{R}^m$ and define the vector field $X \in \mathrm{Vect}(\gamma)$ by $X(t) := e(t)\xi$ for $t \in \mathbb{R}$. Then the Gauß–Weingarten formula (3.2.2) asserts that

$$\dot{e}(t)\xi = \dot{X}(t)$$
$$= \nabla X(t) + h_{\gamma(t)}(\dot{\gamma}(t), X(t))$$
$$= \Pi(\gamma(t))\dot{e}(t)\xi + h_{\gamma(t)}(\dot{\gamma}(t), e(t)\xi)$$

for all $t \in \mathbb{R}$. Take $t = 0$ to obtain

$$(1 - \Pi(p))\widehat{e}\xi = h_p(\widehat{p}, e\xi) = h_p(\widehat{p})e\xi$$

for all $\xi \in \mathbb{R}^m$. This proves the inclusion "$\subset$" in (3.4.3). Equality holds because both sides of the equation are $(m + m^2)$-dimensional linear subspaces of $\mathbb{R}^n \times \mathbb{R}^{n \times m}$. This proves Lemma 3.4.5. $\qquad\square$

It is convenient to consider two kinds of curves in $\mathcal{F}(M)$, namely vertical curves with constant projections to M and horizontal lifts of curves in M. We denote by $\mathcal{L}(\mathbb{R}^m, T_p M)$ the space of linear maps from $\mathbb{R}^m$ to $T_p M$.

Definition 3.4.6 (Horizontal lift) Let $\gamma : \mathbb{R} \to M$ be a smooth curve. A smooth curve $\beta : \mathbb{R} \to \mathcal{F}(M)$ is called a **lift of** γ iff

$$\pi \circ \beta = \gamma.$$

Any such lift has the form $\beta(t) = (\gamma(t), e(t))$ with $e(t) \in \mathcal{L}_{\mathrm{iso}}(\mathbb{R}^m, T_{\gamma(t)} M)$. The associated curve of frames $e(t)$ of the tangent spaces $T_{\gamma(t)} M$ is called a **moving frame along** γ. A curve

$$\beta(t) = (\gamma(t), e(t)) \in \mathcal{F}(M)$$

is called **horizontal** or a **horizontal lift of** γ iff the vector field

$$X(t) := e(t)\xi$$

along γ is parallel for every $\xi \in \mathbb{R}^m$. Thus a horizontal lift of γ has the form

$$\beta(t) = (\gamma(t), \Phi_\gamma(t, 0)e) \tag{3.4.4}$$

for some $e \in \mathcal{L}_{\mathrm{iso}}(\mathbb{R}^m, T_{\gamma(0)} M)$.

Lemma 3.4.7

(i) *The tangent space of $\mathcal{F}(M)$ at $(p, e) \in \mathcal{F}(M)$ is the direct sum*

$$T_{(p,e)} \mathcal{F}(M) = H_{(p,e)} \oplus V_{(p,e)}$$

*of the **horizontal space***

$$H_{(p,e)} := \{(v, h_p(v)e) \mid v \in T_p M\} \tag{3.4.5}$$

*and the **vertical space***

$$V_{(p,e)} := \{0\} \times \mathcal{L}(\mathbb{R}^m, T_p M). \tag{3.4.6}$$

(ii) *The vertical space $V_{(p,e)}$ at $(p,e) \in \mathcal{F}(M)$ is the kernel of the linear map*

$$d\pi(p,e) : T_{(p,e)}\mathcal{F}(M) \to T_p M.$$

(iii) *A curve $\beta : \mathbb{R} \to \mathcal{F}(M)$ is horizontal if and only if it is tangent to the horizontal spaces, i.e. $\dot{\beta}(t) \in H_{\beta(t)}$ for every $t \in \mathbb{R}$.*
(iv) *If $\beta : \mathbb{R} \to \mathcal{F}(M)$ is a horizontal curve, so is $a^*\beta$ for every $a \in \mathrm{GL}(m)$.*

Proof   The proof has four steps.

Step 1 *Let $(p,e) \in \mathcal{F}(M)$. Then $V_{(p,e)} = \ker d\pi(p,e) \subset T_{(p,e)}\mathcal{F}(M)$.*

Since π is a submersion, the fiber $\pi^{-1}(p)$ is a submanifold of $\mathcal{F}(M)$ by Theorem 2.2.19 and $T_{(p,e)}\pi^{-1}(p) = \ker d\pi(p,e)$. Now let $(\widehat{p},\widehat{e}) \in \ker d\pi(p,e)$. Then there exists a vertical curve $\beta : \mathbb{R} \to \mathcal{F}(M)$ with $\pi \circ \beta \equiv p$ such that

$$\beta(0) = (p,e), \qquad \dot{\beta}(0) = (\widehat{p},\widehat{e}).$$

Any such curve has the form $\beta(t) := (p, e(t))$ where $e(t) \in \mathcal{L}_{\mathrm{iso}}(\mathbb{R}^m, T_p M)$. Hence $\widehat{p} = 0$ and $\widehat{e} = \dot{e}(0) \in \mathcal{L}(\mathbb{R}^m, T_p M)$. This shows that

$$\ker d\pi(p,e) \subset V_{(p,e)}. \tag{3.4.7}$$

Conversely, for every $\widehat{e} \in \mathcal{L}(\mathbb{R}^m, T_p M)$, the curve

$$\mathbb{R} \to \mathcal{L}(\mathbb{R}^m, T_p M) : t \mapsto e(t) := e + t\widehat{e}$$

takes values in the open set $\mathcal{L}_{\mathrm{iso}}(\mathbb{R}^m, T_p M)$ for t sufficiently small and hence $\beta(t) := (p, e(t))$ is a vertical curve with $\dot{\beta}(0) = (0, \widehat{e})$. Thus

$$V_{(p,e)} \subset \ker d\pi(p,e) \subset T_{(p,e)}\mathcal{F}(M). \tag{3.4.8}$$

Combining (3.4.7) and (3.4.8) we obtain Step 1 and part (ii).

Step 2 *Let $(p,e) \in \mathcal{F}(M)$. Then $H_{(p,e)} \subset T_{(p,e)}\mathcal{F}(M)$. Moreover, every horizontal curve $\beta : \mathbb{R} \to \mathcal{F}(M)$ satisfies $\dot{\beta}(t) \in H_{\beta(t)}$ for all $t \in \mathbb{R}$.*

Fix a tangent vector $v \in T_p M$, let $\gamma : \mathbb{R} \to M$ be a smooth curve satisfying $\gamma(0) = p$ and $\dot{\gamma}(0) = v$, and let $\beta : \mathbb{R} \to \mathcal{F}(M)$ be the horizontal lift of γ with $\beta(0) = (p,e)$. Then

$$\beta(t) = (\gamma(t), e(t)), \qquad e(t) := \Phi_\gamma(t,0)e.$$

Fix a vector $\xi \in \mathbb{R}^m$ and consider the vector field

$$X(t) := e(t)\xi = \Phi_\gamma(t,0)e\xi$$

along γ. This vector field is parallel and hence, by the Gauß–Weingarten formula, it satisfies

$$\dot{e}(0)\xi = \dot{X}(0) = h_{\gamma(0)}(\dot{\gamma}(0), X(0)) = h_p(v)e\xi.$$

Here we have used (3.3.4). Thus

$$(v, h_p(v)e) = (\dot{\gamma}(0), \dot{e}(0)) = \dot{\beta}(0) \in T_{\beta(0)}\mathcal{F}(M) = T_{(p,e)}\mathcal{F}(M)$$

and so $H_{(p,e)} \subset T_{(p,e)}\mathcal{F}(M)$. Moreover, $\dot{\beta}(0) = (v, h_p(v)e) \in H_{(p,e)} = H_{\beta(0)}$ and this proves Step 2.

Step 3 *We prove part (i).*

We have $V_{(p,e)} \subset T_{(p,e)}\mathcal{F}(M)$ by Step 1 and $H_{(p,e)} \subset T_{(p,e)}\mathcal{F}(M)$ by Step 2. Moreover $H_{(p,e)} \cap V_{(p,e)} = \{0\}$ and so $T_{(p,e)}\mathcal{F}(M) = H_{(p,e)} \oplus V_{(p,e)}$ for dimensional reasons. This proves Step 3.

Step 4 *We prove parts (iii) and (iv).*

By Step 2 every horizontal curve $\beta : \mathbb{R} \to \mathcal{F}(M)$ satisfies $\dot{\beta}(t) \in H_{\beta(t)}$. Conversely, let $\mathbb{R} \to \mathcal{F}(M) : t \mapsto \beta(t) = (\gamma(t), e(t))$ be a smooth curve satisfying $\dot{\beta}(t) \in H_{\beta(t)}$ for all t. Then $\dot{e}(t) = h_{\gamma(t)}(\dot{\gamma}(t))e(t)$ for all t. By the Gauß–Weingarten formula (3.2.2) this implies that the vector field $X(t) = e(t)\xi$ along γ is parallel for every $\xi \in \mathbb{R}^m$, so β is horizontal. This proves part (iii). Part (iv) follows from (iii) and the fact that the horizontal tangent bundle $H \subset T\mathcal{F}(M)$ is invariant under the induced action of the group $\mathrm{GL}(m)$ on $T\mathcal{F}(M)$. This proves Lemma 3.4.7. $\square$

The reason for the terminology introduced in Definition 3.4.6 is that one draws the extremely crude picture of the frame bundle displayed in Fig. 3.3. One thinks of $\mathcal{F}(M)$ as "lying over" M. One would then represent the equation $\gamma = \pi \circ \beta$ by the following commutative diagram:

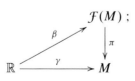

hence the word "lift". The vertical space is tangent to the vertical line in Fig. 3.3 while the horizontal space is transverse to the vertical space. This crude imagery can be extremely helpful.

Exercise 3.4.8 The group $\mathrm{GL}(m)$ acts on $\mathcal{F}(M)$ by diffeomorphisms. Thus for each $a \in \mathrm{GL}(m)$ the map

$$\mathcal{F}(M) \to \mathcal{F}(M) : (p, e) \mapsto a^*(p, e) = (p, e \circ a)$$

Fig. 3.3 The frame bundle

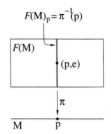

$F(M)_p = \pi^{-1}(p)$

$F(M)$

(p,e)

π

M   p

is a diffeomorphism of $\mathcal{F}(M)$. The derivative of this diffeomorphism is a diffeomorphism of the tangent bundle $T\mathcal{F}(M)$ and this is called the induced action of
$GL(m)$ on $T\mathcal{F}(M)$. Prove that the horizontal and vertical subbundles are invariant
under the induced action of $GL(m)$ on $T\mathcal{F}(M)$.

Exercise 3.4.9   Prove that $H_{(p,e)} \subset T_{(p,e)}\mathcal{O}(M)$ and that

$$T_{(p,e)}\mathcal{O}(M) = H_{(p,e)} \oplus V'_{(p,e)}, \qquad V'_{(p,e)} := V_{(p,e)} \cap T_{(p,e)}\mathcal{O}(M)$$

for every $(p,e) \in \mathcal{O}(M)$.

The following definition introduces an important class of vector fields on the
frame bundle that will play a central role in Sect. 3.5. They will be used to prove
the Development Theorem 3.5.21 in Sect. 3.5.4 below.

Definition 3.4.10 (Basic vector field)   Every vector $\xi \in \mathbb{R}^m$ determines a vector
field $B_\xi \in \text{Vect}(\mathcal{F}(M))$ defined by

$$B_\xi(p,e) := (e\xi, h_p(e\xi)e) \tag{3.4.9}$$

for $(p,e) \in \mathcal{F}(M)$. This vector field is horizontal, i.e.

$$B_\xi(p,e) \in H_{(p,e)},$$

and projects to $e\xi$, i.e.

$$d\pi(p,e)B_\xi(p,e) = e\xi$$

for all $(p,e) \in \mathcal{F}(M)$. These two conditions determine the vector field B_ξ uniquely.
It is called the **basic vector field** corresponding to ξ.

Exercise 3.4.11
(i)   Prove that every basic vector field $B_\xi \in \text{Vect}(\mathcal{F}(M))$ is tangent to the orthonor-
      mal frame bundle $\mathcal{O}(M)$.
(ii) Let $\mathbb{R} \to \mathcal{F}(M) : t \mapsto (\gamma(t), e(t))$ be an integral curve of the vector field B_ξ
      and $a \in GL(m)$. Prove that $\mathbb{R} \to \mathcal{F}(M) : t \mapsto a^*\beta(t) = (\gamma(t), a^*e(t))$ is an
      integral curve of $B_{a^{-1}\xi}$.

(iii) Prove that the vector field $B_\xi \in \text{Vect}(\mathcal{F}(M))$ is complete for all $\xi \in \mathbb{R}^m$ if and only if the restricted vector field $B_\xi|_{\mathcal{O}(M)} \in \text{Vect}(\mathcal{O}(M))$ on the orthonormal frame bundle is complete for all $\xi \in \mathbb{R}^m$.

Definition 3.4.12 (Complete manifold) A smoth m-manifold $M \subset \mathbb{R}^n$ is called **complete** iff, for every smooth curve $\xi : \mathbb{R} \to \mathbb{R}^m$ and every element $(p_0, e_0) \in \mathcal{F}(M)$, there exists a smooth curve $\beta : \mathbb{R} \to \mathcal{F}(M)$ such that $\beta(0) = (p_0, e_0)$ and $\dot{\beta}(t) = B_{\xi(t)}(\beta(t))$ for all $t \in \mathbb{R}$.

3.5 Motions and Developments

Our aim in this sections is to define motion without sliding, twisting, or wobbling. This is the motion that results when a heavy object is rolled, with a minimum of friction, along the floor. It is also the motion of the large snowball a child creates as it rolls it into the bottom part of a snowman.

We shall eventually justify mathematically the physical intuition that either of the curves of contact in such ideal rolling may be specified arbitrarily; the other is then determined uniquely. Thus for example the heavy object may be rolled along an arbitrary curve on the floor; if that curve is marked in wet ink, another curve will be traced in the object. Conversely, if a curve is marked in wet ink on the object, the object may be rolled so as to trace a curve on the floor. However, if both curves are prescribed, it will be necessary to slide the object as it is being rolled, if one wants to keep the curves in contact.

We assume throughout this section that M and M' are two m-dimensional submanifolds of $\mathbb{R}^n$. Objects on M' will be denoted by the same letter as the corresponding objects on M with primes affixed. Thus for example $\Pi'(p') \in \mathbb{R}^{n \times n}$ denotes the orthogonal projection of $\mathbb{R}^n$ onto the tangent space $T_{p'} M'$, ∇' denotes the covariant derivative of a vector field along a curve in M', and $\Phi'_{\gamma'}$ denotes parallel transport along a curve in M'.

3.5.1 Motion

Definition 3.5.1 A **motion** of M along M' (on an interval $I \subset \mathbb{R}$) is a triple (Ψ, γ, γ') of smooth maps

$$\Psi : I \to O(n), \qquad \gamma : I \to M, \qquad \gamma' : I \to M'$$

such that

$$\Psi(t) T_{\gamma(t)} M = T_{\gamma'(t)} M' \qquad \forall\, t \in I.$$

Note that a motion also matches normal vectors, i.e.

$$\Psi(t) T_{\gamma(t)} M^\perp = T_{\gamma'(t)} M'^\perp \qquad \forall\, t \in I.$$

Remark 3.5.2 Associated to a motion (Ψ, γ, γ') of M along M' is a family of (affine) isometries $\psi_t : \mathbb{R}^n \to \mathbb{R}^n$ defined by

$$\psi_t(p) := \gamma'(t) + \Psi(t)(p - \gamma(t)) \tag{3.5.1}$$

for $t \in I$ and $p \in \mathbb{R}^n$. These isometries satisfy

$$\psi_t(\gamma(t)) = \gamma'(t), \qquad d\psi_t(\gamma(t))T_{\gamma(t)}M = T_{\gamma'(t)}M' \qquad \forall\, t \in I.$$

Remark 3.5.3 There are three operations on motions.

Reparametrization If (Ψ, γ, γ') is a motion of M along M' on an interval $I \subset \mathbb{R}$ and $\sigma : J \to I$ is a smooth map between intervals, then the triple

$$(\Psi \circ \sigma, \gamma \circ \sigma, \gamma' \circ \sigma)$$

is a motion of M along M' on the interval J.

Inversion If (Ψ, γ, γ') is a motion of M along M', then

$$(\Psi^{-1}, \gamma', \gamma)$$

is a motion of M' along M.

Composition If (Ψ, γ, γ') is a motion of M along M' on an interval I and $(\Psi', \gamma', \gamma'')$ is a motion of M' along M'' on the same interval, then

$$(\Psi'\Psi, \gamma, \gamma'')$$

is a motion of M along M''.

We now give the three simplest examples of "bad" motions; i.e. motions which do not satisfy the concepts we are about to define. In all three of these examples, p is a point of M and M' is the affine tangent space to M at p:

$$M' := p + T_pM = \{p + v \mid v \in T_pM\}.$$

Example 3.5.4 (Pure sliding) Take a nonzero tangent vector $v \in T_pM$ and let

$$\gamma(t) := p, \qquad \gamma'(t) = p + tv, \qquad \Psi(t) := \mathbb{1}.$$

Then $\dot{\gamma}(t) = 0$, $\dot{\gamma}'(t) = v \neq 0$, and so

$$\Psi(t)\dot{\gamma}(t) \neq \dot{\gamma}'(t).$$

(See Fig. 3.4.)

Fig. 3.4 Pure sliding

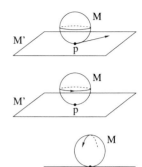

Fig. 3.5 Pure twisting

Fig. 3.6 Pure wobbling

Example 3.5.5 (Pure twisting) Let γ and γ' be the constant curves

$$\gamma(t) = \gamma'(t) = p$$

and take $\Psi(t)$ to be the identity on $T_pM^\perp$ and any curve of rotations on the tangent space T_pM. As a concrete example with $m = 2$ and $n = 3$ one can take M to be the sphere of radius one centered at the point $(0, 1, 0)$ and p to be the origin:

$$M := \{(x, y, z) \in \mathbb{R}^3 \mid x^2 + (y - 1)^2 + z^2 = 1\}, \qquad p := (0, 0, 0).$$

Then M' is the (x, z)-plane and $A(t)$ is any curve of rotations in the (x, z)-plane, i.e. about the y-axis $T_pM^\perp$. (See Fig. 3.5.)

Example 3.5.6 (Pure wobbling) This is the same as pure twisting except that $\Psi(t)$ is the identity on T_pM and any curve of rotations on $T_pM^\perp$. As a concrete example with $m = 1$ and $n = 3$ one can take M to be the circle of radius one in the (x, y)-plane centered at the point $(0, 1, 0)$ and p to be the origin:

$$M := \{(x, y, 0) \in \mathbb{R}^3 \mid x^2 + (y - 1)^2 = 1\}, \qquad p := (0, 0, 0).$$

Then M' is the x-axis and $\Psi(t)$ is any curve of rotations in the (y, z)-plane, i.e. about the axis M'. (See Fig. 3.6.)

3.5.2 Sliding

When a train slides on the track (e.g. in the process of stopping suddenly), there is a terrific screech. Since we usually do not hear a screech, this means that the wheel moves along without sliding. In other words the velocity of the point of contact in the train wheel M equals the velocity of the point of contact in the track M'. But the track is not moving; hence the point of contact in the wheel is not moving. One may explain the paradox this way: the train is moving forward and the wheel is rotating around the axle. The velocity of a point on the wheel is the sum of these two velocities. When the point is on the bottom of the wheel, the two velocities cancel.

Definition 3.5.7   A motion (Ψ, γ, γ') is said to be **without sliding** iff it satisfies $\Psi(t)\dot{\gamma}(t) = \dot{\gamma}'(t)$ for every t.

Here is the geometric picture of the no sliding condition. As explained in Remark 3.5.2 we can view a motion as a smooth family of isometries

$$\psi_t(p) := \gamma'(t) + \Psi(t)(p - \gamma(t))$$

acting on the manifold M with $\gamma(t) \in M$ being the point of contact with M'. Differentiating the curve $t \mapsto \psi_t(p)$ which describes the motion of the point $p \in M$ in the space $\mathbb{R}^n$ we obtain

$$\frac{d}{dt}\psi_t(p) = \dot{\gamma}'(t) - \Psi(t)\dot{\gamma}(t) + \dot{\Psi}(t)(p - \gamma(t)).$$

Taking $p = \gamma(t_0)$ we find

$$\frac{d}{dt}\bigg|_{t=t_0} \psi_t(\gamma(t_0)) = \dot{\gamma}'(t_0) - \Psi(t_0)\dot{\gamma}(t_0).$$

This expression vanishes under the no sliding condition. In general the curve $t \mapsto \psi_t(\gamma(t_0))$ will be non-constant, but (when the motion is without sliding) its velocity will vanish at the instant $t = t_0$; i.e. at the instant when it becomes the point of contact. In other words *the motion is without sliding if and only if the point of contact is motionless.*

We remark that, if the motion is without sliding, we have:

$$|\dot{\gamma}'(t)| = |\Psi(t)\dot{\gamma}(t)| = |\dot{\gamma}(t)|$$

so that the curves γ and γ' have the same arclength:

$$\int_{t_0}^{t_1} |\dot{\gamma}'(t)|\, dt = \int_{t_0}^{t_1} |\dot{\gamma}(t)|\, dt$$

on any interval $[t_0, t_1] \subset I$. Hence any motion with $\dot{\gamma} = 0$ and $\dot{\gamma}' \neq 0$ is not without sliding (such as the example of pure sliding above).

Exercise 3.5.8   Give an example of a motion where $|\dot{\gamma}(t)| = |\dot{\gamma}'(t)|$ for every t but which is not without sliding.

Example 3.5.9   We describe mathematically the motion of the train wheel. Let the center of the wheel move right parallel to the x-axis at height one and the wheel have radius one and make one revolution in 2π units of time. Then the track M' is the x-axis and we take

$$M := \{(x, y) \in \mathbb{R}^2 \mid x^2 + (y - 1)^2 = 1\}.$$

Choose

$$\gamma(t) := (\cos(t - \pi/2), 1 + \sin(t - \pi/2))$$
$$= (\sin(t), 1 - \cos(t)),$$
$$\gamma'(t) := (t, 0),$$

and define $\Psi(t) \in GL(2)$ by

$$\Psi(t) := \begin{pmatrix} \cos(t) & \sin(t) \\ -\sin(t) & \cos(t) \end{pmatrix}.$$

The reader can easily verify that this is a motion without sliding. A fixed point p_0 on M, say $p_0 = (0, 0)$, sweeps out a cycloid with parametric equations

$$x = t - \sin(t), \qquad y = 1 - \cos(t).$$

(Check that $(\dot{x}, \dot{y}) = (0, 0)$ when $y = 0$; i.e. for $t = 2n\pi$.)

Remark 3.5.10   These same formulas give a motion of a sphere M rolling without sliding along a straight line in a plane M'. Namely in coordinates (x, y, z) the sphere is given by the equation

$$x^2 + (y - 1)^2 + z^2 = 1,$$

the plane is $y = 0$ and the line is the x-axis. The z-coordinate of a point is unaffected by the motion. Note that the curve γ' traces out a straight line in the plane M' and the curve γ traces out a great circle on the sphere M.

Exercise 3.5.11   The operations of reparametrization, inversion, and composition respect motion without sliding; i.e. if (Ψ, γ, γ') and $(\Psi', \gamma', \gamma'')$ are motions without sliding on an interval I and $\sigma : J \to I$ is a smooth map between intervals, then the motions $(\Psi \circ \sigma, \gamma \circ \sigma, \gamma' \circ \sigma)$, $(\Psi^{-1}, \gamma', \gamma)$, and $(\Psi'\Psi, \gamma, \gamma'')$ are also without sliding.

3.5.3   Twisting and Wobbling

A motion (Ψ, γ, γ') on an intervall $I \subset \mathbb{R}$ transforms vector fields along γ into vector fields along γ' by the formula

$$X'(t) = (\Psi X)(t) := \Psi(t)X(t) \in T_{\gamma'(t)}M'$$

for $t \in I$ and $X \in \text{Vect}(\gamma)$; so $X' \in \text{Vect}(\gamma')$.

Lemma 3.5.12 *Let (Ψ, γ, γ') be a motion of M along M' on an interval $I \subset \mathbb{R}$. Then the following are equivalent.*

(i)   *The instantaneous velocity of each tangent vector is normal, i.e. for $t \in I$*

$$\dot{\Psi}(t) T_{\gamma(t)} M \subset T_{\gamma'(t)} M'^{\perp}.$$

(ii) *Ψ intertwines covariant differentiation, i.e. for $X \in \mathrm{Vect}(\gamma)$*

$$\nabla'(\Psi X) = \Psi \nabla X.$$

(iii) *Ψ transforms parallel vector fields along γ into parallel vector fields along γ', i.e. for $X \in \mathrm{Vect}(\gamma)$*

$$\nabla X = 0 \qquad \Longrightarrow \qquad \nabla'(\Psi X) = 0.$$

(iv) *Ψ intertwines parallel transport, i.e. for $s, t \in I$ and $v \in T_{\gamma(s)} M$*

$$\Psi(t) \Phi_{\gamma}(t, s) v = \Phi'_{\gamma'}(t, s) \Psi(s) v.$$

*A motion that satisfies these conditions is called **without twisting**.*

Proof   We prove that (i) is equivalent to (ii). A motion satisfies the equation

$$\Psi(t) \Pi(\gamma(t)) = \Pi'(\gamma'(t)) \Psi(t)$$

for every $t \in I$. This restates the condition that $\Psi(t)$ maps tangent vectors of M to tangent vectors of M' and normal vectors of M to normal vectors of M'. Differentiating the equation $X'(t) = \Psi(t) X(t)$ we obtain

$$\dot{X}'(t) = \Psi(t) \dot{X}(t) + \dot{\Psi}(t) X(t).$$

Applying $\Pi'(\gamma'(t))$ this gives

$$\nabla' X' = \Psi \nabla X + \Pi'(\gamma') \dot{\Psi} X.$$

Hence (ii) holds if and only if $\Pi'(\gamma'(t)) \dot{\Psi}(t) = 0$ for every $t \in I$. Thus we have proved that (i) is equivalent to (ii). That (ii) implies (iii) is obvious.

We prove that (iii) implies (iv). Let $t_0 \in I$ and $v_0 \in T_{\gamma(t_0)} M$. Define the vector field $X \in \mathrm{Vect}(\gamma)$ by $X(t) := \Phi_{\gamma}(t, t_0) v_0$ for $t \in I$ and let $X' := \Psi X$. Then $\nabla X = 0$, hence $\nabla' X' = 0$ by (iii), and hence

$$X'(t) = \Phi'_{\gamma'}(t, t_0) X'(t_0) = \Phi'_{\gamma'}(t, t_0) \Psi(t_0) v_0$$

for all $t \in I$. Since $X'(t) = \Psi(t) X(t) = \Psi(t) \Phi_{\gamma}(t, t_0) v_0$, this implies (iv).

We prove that (iv) implies (ii). Let $X \in \mathrm{Vect}(\gamma)$ and $X' := \Psi X$. By (iv) we have

$$\Phi'_{\gamma'}(t_0, t) X'(t) = \Psi(t_0) \Phi_{\gamma}(t_0, t) X(t).$$

Differentiating this equation with respect to t at $t = t_0$ and using Theorem 3.3.6, we obtain $\nabla' X'(t_0) = \Psi(t_0) \nabla X(t_0)$. This proves the lemma. $\square$

Lemma 3.5.13 *Let (Ψ, γ, γ') be a motion of M along M' on an interval $I \subset \mathbb{R}$. Then the following are equivalent.*

(i) *The instantaneous velocity of each normal vector is tangent, i.e. for $t \in I$*

$$\dot{\Psi}(t) T_{\gamma(t)} M^{\perp} \subset T_{\gamma'(t)} M'.$$

(ii) *Ψ intertwines normal covariant differentiation, i.e. for $Y \in \text{Vect}^{\perp}(\gamma)$*

$$\nabla'^{\perp}(\Psi Y) = \Psi \nabla^{\perp} Y.$$

(iii) *Ψ transforms parallel normal vector fields along γ into parallel normal vector fields along γ', i.e. for $Y \in \text{Vect}^{\perp}(\gamma)$*

$$\nabla^{\perp} Y = 0 \qquad \Longrightarrow \qquad \nabla'^{\perp}(\Psi Y) = 0.$$

(iv) *Ψ intertwines parallel transport of normal vector fields, i.e. for $s, t \in I$ and $w \in T_{\gamma(s)} M^{\perp}$*

$$\Psi(t) \Phi_{\gamma}^{\perp}(t, s) w = \Phi'^{\perp}_{\gamma'}(t, s) \Psi(s) w.$$

*A motion that satisfies these conditions is called **without wobbling**.*

The proof that the four conditions in Lemma 3.5.13 are equivalent is word for word analogous to the proof of Lemma 3.5.12 and will be omitted.

In summary a *motion is without twisting iff tangent vectors at the point of contact are rotating towards the normal space and it is without wobbling iff normal vectors at the point of contact are rotating towards the tangent space.* In case $m = 2$ and $n = 3$ motion without twisting means that the instantaneous axis of rotation is parallel to the tangent plane.

Remark 3.5.14 The operations of reparametrization, inversion, and composition respect motion without twisting, respectively without wobbling; i.e. if (Ψ, γ, γ') and $(\Psi', \gamma', \gamma'')$ are motions without twisting, respectively without wobbling, on an interval I and $\sigma : J \to I$ is a smooth map between intervals, then the motions $(\Psi \circ \sigma, \gamma \circ \sigma, \gamma' \circ \sigma)$, $(\Psi^{-1}, \gamma', \gamma)$, and $(\Psi'\Psi, \gamma, \gamma'')$ are also without twisting, respectively without wobbling.

Remark 3.5.15 Let $I \subset \mathbb{R}$ be an interval and $t_0 \in I$. Given curves $\gamma : I \to M$ and $\gamma' : I \to M'$ and an orthogonal matrix $\Psi_0 \in O(n)$ such that

$$\Psi_0 T_{\gamma(t_0)} M = T_{\gamma'(t_0)} M'$$

there is a unique motion (Ψ, γ, γ') of M along M' (with the given γ and γ') **without twisting or wobbling** satisfying the initial condition:

$$\Psi(t_0) = \Psi_0.$$

Indeed, the path of matrices $\Psi : I \to O(n)$ is uniquely determined by the conditions (iv) in Lemma 3.5.12 and Lemma 3.5.13. It is given by the explicit formula

$$
\begin{aligned}
\Psi(t)v &= \Phi'_{\gamma'}(t, t_0)\Psi_0\Phi_\gamma(t_0, t)\Pi(\gamma(t))v \\
&\quad + \Phi'^{\perp}_{\gamma'}(t, t_0)\Psi_0\Phi^{\perp}_\gamma(t_0, t)(v - \Pi(\gamma(t))v)
\end{aligned}
\tag{3.5.2}
$$

for $t \in I$ and $v \in \mathbb{R}^n$. We prove below a somewhat harder result where the motion is without twisting, wobbling, or sliding. It is in this situation that γ and γ' determine one another (up to an initial condition).

Remark 3.5.16 We can now give another interpretation of parallel transport. Given $\gamma : \mathbb{R} \to M$ and $v_0 \in T_{\gamma(t_0)}M$ take M' to be an affine subspace of the same dimension as M. Let (Ψ, γ, γ') be a motion of M along M' without twisting (and, if you like, without sliding or wobbling). Let $X' \in \text{Vect}(\gamma')$ be the constant vector field along γ' (so that $\nabla' X' = 0$) with value

$$
X'(t) = \Psi_0 v_0, \qquad \Psi_0 := \Psi(t_0).
$$

Let $X \in \text{Vect}(\gamma)$ be the corresponding vector field along γ so that

$$
\Psi(t)X(t) = \Psi_0 v_0
$$

Then $X(t) = \Phi_\gamma(t, t_0)v_0$. To put it another way, imagine that M is a ball. To define parallel transport along a given curve γ roll the ball (without sliding) along a plane M' keeping the curve γ in contact with the plane M'. Let γ' be the curve traced out in M'. If a constant vector field in the plane M' is drawn in wet ink along the curve γ', it will mark off a (covariant) parallel vector field along γ in M.

Exercise 3.5.17 Describe parallel transport along a great circle in a sphere.

3.5.4   Development

A development is an intrinsic version of motion without sliding or twisting.

Definition 3.5.18 A **development of M along M' (on an interval I)** is a triple (Φ, γ, γ') where $\gamma : I \to M$ and $\gamma' : I \to M'$ are smooth paths and Φ is a family of orthogonal isomorphisms

$$
\Phi(t) : T_{\gamma(t)}M \to T_{\gamma'(t)}M'
$$

parametrized by $t \in I$, such that

$$
\Phi(t)\dot\gamma(t) = \dot\gamma'(t)
\tag{3.5.3}
$$

for all $t \in I$ and Φ intertwines parallel transport, i.e.

$$\Phi(t)\Phi_\gamma(t, s) = \Phi'_{\gamma'}(t, s)\Phi(s) \tag{3.5.4}$$

for all $s, t \in I$. In particular, the family Φ of isomorphisms is smooth, i.e. if X is a smooth vector field along γ, then the formula $X'(t) := \Phi(t)X(t)$ defines a smooth vector field along γ'.

Lemma 3.5.19 *Let $I \subset \mathbb{R}$ be an interval, $\gamma : I \to M$ and $\gamma' : I \to M'$ be smooth curves, and $\Phi(t) : T_{\gamma(t)}M \to T_{\gamma'(t)}M'$ be a family of orthogonal isomorphisms parametrized by $t \in I$. Then the following are equivalent.*

(i) *(Φ, γ, γ') is a development.*
(ii) *Φ satisfies (3.5.3) and*

$$\nabla'(\Phi X) = \Phi \nabla X \tag{3.5.5}$$

for all $X \in \text{Vect}(\gamma)$.
(iii) *There exists a motion (Ψ, γ, γ') without sliding and twisting such that*

$$\Phi(t) = \Psi(t)|_{T_{\gamma(t)}M} \qquad \text{for all } t \in I. \tag{3.5.6}$$

(iv) *There exists a motion (Ψ, γ, γ') of M along M' without sliding, twisting, and wobbling that satisfies (3.5.6).*

Proof That (3.5.4) is equivalent to (3.5.5) was proved in Lemma 3.5.12. This (i) is equivalent to (ii). That (iv) implies (iii) and (iii) implies (i) is obvious. To prove that (i) implies (iv) choose any $t_0 \in I$ and any orthogonal matrix $\Psi_0 \in O(n)$ such that $\Psi_0|_{T_{\gamma(t_0)}M} = \Phi(t_0)$ and define $\Psi(t) : \mathbb{R}^n \to \mathbb{R}^n$ by (3.5.2). This proves Lemma 3.5.19. $\qquad\square$

Remark 3.5.20 The operations of reparametrization, inversion, and composition yield developments when applied to developments; i.e. if (Φ, γ, γ') is a development of M along M' on an interval I, $(\Phi', \gamma', \gamma'')$ is a development of M' along M'' on the same interval I, and $\sigma : J \to I$ is a smooth map of intervals, then the triples

$$(\Phi \circ \sigma, \gamma \circ \sigma, \gamma' \circ \sigma), \qquad (\Phi^{-1}, \gamma', \gamma)), \qquad (\Phi'\Phi, \gamma, \gamma'')$$

are all developments.

Theorem 3.5.21 (Development Theorem) *Let $p_0 \in M$ and $t_0 \in \mathbb{R}$, let $\gamma' : \mathbb{R} \to M'$ be a smooth curve, and let*

$$\Phi_0 : T_{p_0}M \to T_{\gamma'(t_0)}M'$$

be an orthogonal isomorphism. Then the following holds.

(i) *There exists a development* $(\Phi, \gamma, \gamma'|_I)$ *on some open interval* $I \subset \mathbb{R}$ *containing* t_0 *that satisfies the initial condition*

$$\gamma(t_0) = p_0, \qquad \Phi(t_0) = \Phi_0. \tag{3.5.7}$$

(ii) *Any two developments* $(\Phi_1, \gamma_1, \gamma'|_{I_1})$ *and* $(\Phi_2, \gamma_2, \gamma'|_{I_2})$ *as in (i) on two intervals* I_1 *and* I_2 *agree on the intersection* $I_1 \cap I_2$, *i.e.*

$$\gamma_1(t) = \gamma_2(t), \qquad \Phi_1(t) = \Phi_2(t)$$

for every $t \in I_1 \cap I_2$.
(iii) *If* M *is complete, then (i) holds with* $I = \mathbb{R}$.

Proof Let $\gamma : \mathbb{R} \to M$ be any smooth curve such that

$$\gamma(t_0) = p_0$$

and, for $t \in \mathbb{R}$, define the linear map

$$\Phi(t) : T_{\gamma(t)}M \to T_{\gamma'(t)}M'$$

by

$$\Phi(t) := \Phi'_{\gamma'}(t, t_0)\Phi_0\Phi_\gamma(t_0, t). \tag{3.5.8}$$

This is an orthogonal transformation for every t and it intertwines parallel transport. However, in general $\Phi(t)\dot{\gamma}(t)$ will not be equal to $\dot{\gamma}'(t)$.

To construct a development that satisfies (3.5.3), we choose an orthonormal frame $e_0 : \mathbb{R}^m \to T_{p_0}M$ and, for $t \in \mathbb{R}$, define $e(t) : \mathbb{R}^m \to T_{\gamma(t)}M$ by

$$e(t) := \Phi_\gamma(t, t_0)e_0. \tag{3.5.9}$$

We can think of $e(t)$ as a real $n \times m$-matrix and the map

$$\mathbb{R} \to \mathbb{R}^{n \times m} : t \mapsto e(t)$$

is smooth. In fact, the map $t \mapsto (\gamma(t), e(t))$ is a smooth path in the frame bundle $\mathcal{F}(M)$. Define the smooth map $\xi : \mathbb{R} \to \mathbb{R}^m$ by

$$\dot{\gamma}'(t) = \Phi'_{\gamma'}(t, t_0)\Phi_0 e_0 \xi(t). \tag{3.5.10}$$

We prove the following.

Claim: *The triple* (Φ, γ, γ') *is a development on an interval* $I \subset \mathbb{R}$ *if and only if the path* $t \mapsto (\gamma(t), e(t))$ *satisfies the differential equation*

$$(\dot{\gamma}(t), \dot{e}(t)) = B_{\xi(t)}(\gamma(t), e(t)) \tag{3.5.11}$$

for every $t \in I$, where $B_{\xi(t)} \in \mathrm{Vect}(\mathcal{F}(M))$ denotes the basic vector field associated to $\xi(t) \in \mathbb{R}^m$ (see equation (3.4.9)).

The triple (Φ, γ, γ') is a development on I if and only if

$$\Phi(t)\dot{\gamma}(t) = \dot{\gamma}'(t)$$

for every $t \in I$. By (3.5.8) and (3.5.10) this is equivalent to the condition

$$\Phi'_{\gamma'}(t, t_0)\Phi_0\Phi_\gamma(t_0, t)\dot{\gamma}(t) = \dot{\gamma}'(t) = \Phi'_{\gamma'}(t, t_0)\Phi_0 e_0 \xi(t),$$

hence to

$$\Phi_\gamma(t_0, t)\dot{\gamma}(t) = e_0 \xi(t),$$

and hence to

$$\dot{\gamma}(t) = \Phi_\gamma(t, t_0)e_0 \xi(t) = e(t)\xi(t) \qquad (3.5.12)$$

for every $t \in I$. By (3.5.9) and the Gauß–Weingarten formula, we have

$$\dot{e}(t) = h_{\gamma(t)}(\dot{\gamma}(t))e(t)$$

for every $t \in \mathbb{R}$. Hence it follows from (3.4.9) that (3.5.12) is equivalent to (3.5.11). This proves the claim.

Parts (i) and (ii) follow directly from the claim. Part (iii) follows from the claim and Definition 3.4.12. This proves Theorem 3.5.21. $\square$

Remark 3.5.22 As any two developments $(\Phi_1, \gamma_1, \gamma'|_{I_1})$ and $(\Phi_2, \gamma_2, \gamma'|_{I_2})$ on two intervals I_1 and I_2 that satisfy the initial condition (3.5.7) agree on $I_1 \cap I_2$ there is a development defined on $I_1 \cup I_2$. Hence there is a unique *maximally defined development* $(\Phi, \gamma, \gamma'|_I)$, defined on a maximal interval $I = I(t_0, p_0, \Phi_0)$, associated to the initial data t_0, p_0, Φ_0.

Denote the space of initial data by

$$\mathcal{P} := \left\{ (t, p, \Phi) \,\middle|\, \begin{array}{l} t \in \mathbb{R}, \ p \in M, \ \Phi : T_pM \to T_{\gamma'(t)}M' \\ \text{is an orthogonal transformation} \end{array} \right\}, \qquad (3.5.13)$$

define the set $\mathcal{D} \subset \mathbb{R} \times \mathcal{P}$ by

$$\mathcal{D} := \{(t, t_0, p_0, \Phi_0) \mid (t_0, p_0, \Phi_0) \in \mathcal{P}, \ t \in I(t_0, p_0, \Phi_0)\} \qquad (3.5.14)$$

and let

$$\mathcal{D} \to \mathcal{P} : (t, t_0, p_0, \Phi_0) \mapsto (t, \gamma(t), \Phi(t)), \qquad (3.5.15)$$

be the map which assigns to each $(t_0, p_0, \Phi_0) \in \mathcal{P}$ and each $t \in I(t_0, p_0, \Phi_0)$ the value at time t of the unique development $(\Phi, \gamma, \gamma'|_I)$ associated to the inital condition (t_0, p_0, Φ_0) on the maximal time interval $I = I(t_0, p_0, \Phi_0)$. Then the space $\mathcal{P}$ has a natural structure of a smooth manifold (in the intrinsic setting), and it follows from Theorem 2.4.9 and the proof of Theorem 3.5.21 that $\mathcal{D}$ is an open subset of $\mathbb{R} \times \mathcal{P}$ and the map (3.5.15) is smooth.

The smooth structure on $\mathcal{P}$ can be understood as follows. The space

$$\mathcal{O}(\gamma') = \{(t, e') \mid (\gamma'(t), e') \in \mathcal{O}(M')\}$$

is the pullback of the orthonormal frame bundle $\mathcal{O}(M') \to M'$ under the curve $\gamma' : \mathbb{R} \to M'$ or, equivalently, is the orthonormal frame bundle of the pullback tangent bundle $(\gamma')^* TM$. Thus $\mathcal{O}(\gamma')$ is a smooth submanifold of $\mathbb{R} \times \mathbb{R}^{n \times m}$. The group $O(m)$ acts diagonally on $\mathcal{O}(\gamma') \times \mathcal{O}(M)$ and the action is free. Hence the quotient $(\mathcal{O}(\gamma') \times \mathcal{O}(M))/O(m)$ is a smooth manifold by Theorem 2.9.14, and it can be naturally identified with $\mathcal{P}$ via the bijection $[(t, e'), (p, e)] \mapsto (t, p, e' \circ e^{-1})$.

Remark 3.5.23 The statement of Theorem 3.5.21 is essentially symmetric in M and M' as the operation of inversion carries developments to developments. Hence given

$$\gamma : \mathbb{R} \to M, \qquad p_0' \in M', \qquad t_0 \in \mathbb{R}, \qquad \Phi_0 : T_{\gamma(t_0)} M \to T_{p_0'} M',$$

we may speak of the development (Φ, γ, γ') corresponding to γ with initial conditions $\gamma'(t_0) = p_0'$ and $\Phi(t_0) = \Phi_0$.

Corollary 3.5.24 (Motions) *Let $p_0 \in M$ and $t_0 \in \mathbb{R}$, let $\gamma' : \mathbb{R} \to M'$ be a smooth curve, and let $\Psi_0 \in O(n)$ be a matrix such that*

$$\Psi_0 T_{p_0} M = T_{\gamma'(t_0)} M'.$$

Then the following holds.

(i) *There exists a motion $(\Psi, \gamma, \gamma'|_I)$ without sliding, twisting and wobbling on some open interval $I \subset \mathbb{R}$ containing t_0 that satisfies the initial condition $\gamma(t_0) = p_0$ and $\Psi(t_0) = \Psi_0$.*

(ii) *Any two motions as in (i) on two intervals I_1 and I_2 agree on the intersection $I_1 \cap I_2$.*

(iii) *If M is complete, then (i) holds with $I = \mathbb{R}$.*

Proof Theorem 3.5.21 and Remark 3.5.15. $\qquad \qquad \qquad \qquad \qquad \qquad \square$

Corollary 3.5.25 (Completeness) *The following are equivalent.*

(i) *M is complete, i.e. for every smooth curve $\xi : \mathbb{R} \to \mathbb{R}^m$ and every element $(p_0, e_0) \in \mathcal{F}(M)$, there exists a smooth curve $\beta : \mathbb{R} \to \mathcal{F}(M)$ such that $\dot{\beta}(t) = B_{\xi(t)}(\beta(t))$ for all $t \in \mathbb{R}$ and $\beta(0) = (p_0, e_0)$ (Definition 3.4.12).*

(ii) *For every smooth curve $\xi : \mathbb{R} \to \mathbb{R}^m$ and every element $(p_0, e_0) \in \mathcal{O}(M)$, there is a smooth curve $\alpha : \mathbb{R} \to \mathcal{O}(M)$ such that $\dot{\alpha}(t) = B_{\xi(t)}(\alpha(t))$ for every $t \in \mathbb{R}$ and $\alpha(0) = (p_0, e_0)$.*

(iii) *For every smooth curve $\gamma' : \mathbb{R} \to \mathbb{R}^m$, every $p_0 \in M$, and every orthogonal isomorphism $\Phi_0 : T_{p_0} M \to \mathbb{R}^m$ there exists a development (Φ, γ, γ') of M along $M' = \mathbb{R}^m$ on all of $\mathbb{R}$ that satisfies $\gamma(0) = p_0$ and $\Phi(0) = \Phi_0$.*

Proof We have already noted that the basic vector fields are all tangent to the orthonormal frame bundle $\mathcal{O}(M) \subset \mathcal{F}(M)$. Now note that if a smooth curve $I \to \mathcal{F}(M) : t \mapsto \beta(t) = (\gamma(t), e(t))$ on an interval $I \subset \mathbb{R}$ satisfies the differential equation

$$\dot{\beta}(t) = B_{\xi(t)}(\beta(t))$$

for all t, then so does the curve

$$I \to \mathcal{F}(M) : t \mapsto a^*\beta(t) = (\gamma(t), e(t) \circ a)$$

for every $a \in \mathrm{GL}(m, \mathbb{R})$. Since any frame $e_0 : \mathbb{R}^m \to T_{p_0} M$ can be carried to any other (in particular an orthonormal one) by a suitable matrix $a \in \mathrm{GL}(m, \mathbb{R})$, this shows that (i) is equivalent to (ii).

That (i) implies (iii) was proved in Theorem 3.5.21.

We prove that (iii) implies (ii). Fix a smooth map $\xi : \mathbb{R} \to \mathbb{R}^m$ and an element $(p_0, e_0) \in \mathcal{O}(M)$. Define

$$\Phi_0 := e_0^{-1} : T_{p_0} M \to \mathbb{R}^m$$

and

$$\gamma'(t) := \int_0^t \xi(s)\, ds \in \mathbb{R}^m \qquad \text{for } t \in \mathbb{R}.$$

By (ii) there exists a development (Φ, γ, γ') of M along $\mathbb{R}^m$ on all of $\mathbb{R}$ that satisfies the initial conditions

$$\gamma(0) = p_0, \qquad \Phi(0) = \Phi_0.$$

Then

$$\Phi(t) = \Phi_0 \Phi_\gamma(0, t) : T_{\gamma(t)} M \to \mathbb{R}^m, \qquad \Phi(t)\dot{\gamma}(t) = \dot{\gamma}'(t) = \xi(t)$$

for all $t \in \mathbb{R}$ by Definition 3.5.18. Define

$$e(t) := \Phi_\gamma(t, 0)e_0 = \Phi(t)^{-1} : \mathbb{R}^m \to T_{\gamma(t)} M$$

for $t \in \mathbb{R}$. Then $(\gamma, e) : \mathbb{R} \to \mathcal{F}(M)$ is a smooth curve that satisfies the initial condition $(\gamma(0), e(0)) = (p_0, e_0)$ and the differential equation

$$\dot{\gamma}(t) = \Phi(t)^{-1}\xi(t) = e(t)\xi(t),$$
$$\dot{e}(t) = h_{\gamma(t)}(\dot{\gamma}(t))e(t) = h_{\gamma(t)}(e(t)\xi(t))e(t)$$

by the Gauß–Weingarten formula. Thus

$$(\dot{\gamma}(t), \dot{e}(t)) = B_{\xi(t)}(\gamma(t), e(t))$$

for all $t \in \mathbb{R}$. This proves Corollary 3.5.25.   □

It is of course easy to give an example of a manifold which is not complete; e.g. if (Φ, γ, γ') is any development of M along M', then $M \setminus \{\gamma(t_1)\}$ is not complete as the given development is only defined for $t \neq t_1$. In Sect. 4.6 we give equivalent characterizations of completeness. In particular, we will see that any compact submanifold of $\mathbb{R}^n$ is complete.

Exercise 3.5.26   An **affine subspace** of $\mathbb{R}^n$ is a subset of the form

$$E = p + \mathbb{E} = \{p + v \mid v \in \mathbb{E}\}$$

where $\mathbb{E} \subset \mathbb{R}^n$ is a linear subspace and $p \in \mathbb{R}^n$. Prove that every affine subspace of $\mathbb{R}^n$ is a complete submanifold.

3.6   Christoffel Symbols

The goal of this subsection is to examine the covariant derivative in local coordinates on an embedded manifold $M \subset \mathbb{R}^n$ of dimension m. Let

$$\phi : U \to \Omega$$

be a coordinate chart, defined on an M-open subset $U \subset M$ with values in an open set $\Omega \subset \mathbb{R}^m$, and denote its inverse by

$$\psi := \phi^{-1} : \Omega \to U \subset M.$$

At this point it is convenient to use superscripts for the coordinates of a vector $x \in \Omega$. Thus we write

$$x = (x^1, \ldots, x^m) \in \Omega.$$

If $p = \psi(x) \in U$ is the corresponding element of M, then the tangent space of M at p is the image of the linear map $d\psi(x) : \mathbb{R}^m \to \mathbb{R}^n$ (Theorem 2.2.3) and thus two tangent vectors $v, w \in T_pM$ can be written in the form

$$v = d\psi(x)\xi = \sum_{i=1}^{m} \xi^i \frac{\partial \psi}{\partial x^i}(x),$$

$$w = d\psi(x)\eta = \sum_{j=1}^{m} \eta^j \frac{\partial \psi}{\partial x^j}(x)$$

$$(3.6.1)$$

Fig. 3.7 A vector field along
a curve in local coordinates

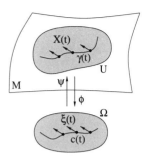

for $\xi = (\xi^1, \ldots, \xi^m) \in \mathbb{R}^m$ and $\eta = (\eta^1, \ldots, \eta^m) \in \mathbb{R}^m$. Recall that the restriction of the inner product in the ambient space $\mathbb{R}^n$ to the tangent space is the first funda-mental form $g_p : T_p M \times T_p M \to \mathbb{R}$ (Definition 3.1.1). Thus

$$g_p(v, w) = \langle v, w \rangle = \sum_{i,j=1}^{m} \xi^i g_{ij}(x) \eta^j, \tag{3.6.2}$$

where the functions $g_{ij} : \Omega \to \mathbb{R}$ are defined by

$$g_{ij}(x) := \left\langle \frac{\partial \psi}{\partial x^i}(x), \frac{\partial \psi}{\partial x^j}(x) \right\rangle \qquad \text{for } x \in \Omega. \tag{3.6.3}$$

In other words, the first fundamental form is in local coordinates represented by the matrix valued function $g = (g_{ij})_{i,j=1}^{m} : \Omega \to \mathbb{R}^{m \times m}$.

   Now let $c = (c^1, \ldots, c^m) : I \to \Omega$ be a smooth curve in Ω, defined on an inter-val $I \subset \mathbb{R}$, and consider the curve

$$\gamma = \psi \circ c : I \to M$$

(see Fig. 3.7). Our goal is to describe the operator $X \mapsto \nabla X$ on the space of vector fields along γ in local coordinates. Let $X : I \to \mathbb{R}^n$ be a vector field along γ. Then

$$X(t) \in T_{\gamma(t)} M = T_{\psi(c(t))} M = \operatorname{im}\left(d\psi(c(t)) : \mathbb{R}^m \to \mathbb{R}^n\right)$$

for every $t \in I$ and hence there exists a unique smooth function

$$\xi = (\xi^1, \ldots, \xi^m) : I \to \mathbb{R}^m$$

such that

$$X(t) = d\psi(c(t))\xi(t) = \sum_{i=1}^{m} \xi^i(t) \frac{\partial \psi}{\partial x^i}(c(t)). \tag{3.6.4}$$

Differentiate this identity to obtain

$$\dot{X}(t) = \sum_{i=1}^{m} \dot{\xi}^i(t) \frac{\partial \psi}{\partial x^i}(c(t)) + \sum_{i,j=1}^{m} \xi^i(t) \dot{c}^j(t) \frac{\partial^2 \psi}{\partial x^i \partial x^j}(c(t)). \tag{3.6.5}$$

We examine the projection $\nabla X(t) = \Pi(\gamma(t))\dot{X}(t)$ of this vector onto the tangent space of M at $\gamma(t)$. The first summand on the right in (3.6.5) is already tangent to M. For the second summand we simply observe that the vector $\Pi(\psi(x))\partial^2\psi/\partial x^i \partial x^j$ lies in tangent space $T_{\psi(x)}M$ and can therefore be expressed as a linear combination of the basis vectors $\partial\psi/\partial x^1, \ldots, \partial\psi/\partial x^m$. The coefficients will be denoted by $\Gamma_{ij}^k(x)$. Thus there exist smooth functions $\Gamma_{ij}^k : \Omega \to \mathbb{R}$ for $i, j, k = 1, \ldots, m$ such that

$$\Pi(\psi(x)) \frac{\partial^2 \psi}{\partial x^i \partial x^j}(x) = \sum_{k=1}^{m} \Gamma_{ij}^k(x) \frac{\partial \psi}{\partial x^k}(x) \tag{3.6.6}$$

for all $x \in \Omega$ and all $i, j \in \{1, \ldots, m\}$. The coefficients $\Gamma_{ij}^k : \Omega \to \mathbb{R}$ are called the **Christoffel symbols** associated to the coordinate chart $\phi : U \to \Omega$. To sum up we have proved the following.

Lemma 3.6.1 *Let $c : I \to \Omega$ be a smooth curve and define*

$$\gamma := \psi \circ c : I \to M.$$

If $\xi : I \to \mathbb{R}^m$ is a smooth map and $X \in \text{Vect}(\gamma)$ is given by (3.6.4), then its covariant derivative at time $t \in I$ is given by

$$\nabla X(t) = \sum_{k=1}^{m} \left(\dot{\xi}^k(t) + \sum_{i,j=1}^{m} \Gamma_{ij}^k(c(t))\xi^i(t)\dot{c}^j(t) \right) \frac{\partial \psi}{\partial x^k}(c(t)), \tag{3.6.7}$$

where the Γ_{ij}^k are the Christoffel symbols defined by (3.6.6).

Our next goal is to understand how the Christoffel symbols are determined by the metric in local coordinates. Recall from equation (3.6.2) that the inner products on the tangent spaces inherited from the standard Euclidean inner product on the ambient space $\mathbb{R}^n$ are in local coordinates represented by the matrix valued function

$$g = (g_{ij})_{i,j=1}^{m} : \Omega \to \mathbb{R}^{m \times m}$$

given by

$$g_{ij} := \left\langle \frac{\partial \psi}{\partial x^i}, \frac{\partial \psi}{\partial x^j} \right\rangle_{\mathbb{R}^n}. \tag{3.6.8}$$

We shall see that the Christoffel symbols are completely determined by the functions $g_{ij} : \Omega \to \mathbb{R}$. Here are first some elementary observations.

Remark 3.6.2   The matrix $g(x) \in \mathbb{R}^{m \times m}$ is symmetric and positive definite for every $x \in \Omega$. This follows from the fact that the matrix $d\psi(x) \in \mathbb{R}^{n \times m}$ has rank m and the matrix $g(x)$ is given by

$$g(x) = d\psi(x)^{\mathsf{T}} d\psi(x)$$

Thus $\xi^{\mathsf{T}} g(x) \xi = |d\psi(x)\xi|^2 > 0$ for all $\xi \in \mathbb{R}^m \setminus \{0\}$.

Remark 3.6.3   For $x \in \Omega$ we have $\det(g(x)) > 0$ by Remark 3.6.2 and so the matrix $g(x)$ is invertible. Denote the entries of the inverse matrix $g(x)^{-1} \in \mathbb{R}^{m \times m}$ by $g^{k\ell}(x)$. They are determined by the condition

$$\sum_{j=1}^{m} g_{ij}(x) g^{jk}(x) = \delta_i^k = \begin{cases} 1, & \text{if } i = k, \\ 0, & \text{if } i \neq k. \end{cases}$$

Since $g(x)$ is symmetric and positive definite, so is its inverse matrix $g(x)^{-1}$. In particular, we have $g^{k\ell}(x) = g^{\ell k}(x)$ for all $x \in \Omega$ and all $k, \ell \in \{1, \ldots, m\}$.

Remark 3.6.4   Suppose that $X, Y \in \text{Vect}(\gamma)$ are vector fields along our curve $\gamma = \psi \circ c : I \to M$ and $\xi, \eta : I \to \mathbb{R}^m$ are defined by

$$X(t) = \sum_{i=1}^{m} \xi^i(t) \frac{\partial \psi}{\partial x^i}(c(t)), \qquad Y(t) = \sum_{j=1}^{m} \eta^j(t) \frac{\partial \psi}{\partial x^j}(c(t)).$$

Then the inner product of $X(t)$ and $Y(t)$ is given by

$$\langle X(t), Y(t) \rangle = \sum_{i,j=1}^{m} \xi^i(t) g_{ij}(c(t)) \eta^j(t).$$

Lemma 3.6.5 (Christoffel symbols)   *Let $\Omega \subset \mathbb{R}^m$ be an open set and let $g_{ij} : \Omega \to \mathbb{R}$ for $i, j = 1, \ldots, m$ be smooth functions such that each matrix $(g_{ij}(x))_{i,j=1}^{m}$ is symmetric and positive definite. Let $\Gamma_{ij}^k : \Omega \to \mathbb{R}$ be smooth functions for $i, j, k = 1, \ldots, m$. Then the Γ_{ij}^k satisfy the conditions*

$$\Gamma_{ij}^k = \Gamma_{ji}^k, \qquad \frac{\partial g_{ij}}{\partial x^\ell} = \sum_{k=1}^{m} \left(g_{ik} \Gamma_{j\ell}^k + g_{jk} \Gamma_{i\ell}^k \right) \qquad (3.6.9)$$

for $i, j, k, \ell = 1, \ldots, m$ if and only if they are given by

$$\Gamma_{ij}^k = \sum_{\ell=1}^{m} g^{k\ell} \frac{1}{2} \left(\frac{\partial g_{\ell i}}{\partial x^j} + \frac{\partial g_{\ell j}}{\partial x^i} - \frac{\partial g_{ij}}{\partial x^\ell} \right). \qquad (3.6.10)$$

If the Γ_{ij}^k are defined by (3.6.6) and the g_{ij} by (3.6.8), then the Γ_{ij}^k satisfy (3.6.9) and hence are given by (3.6.10).

Proof Suppose that the Γ_{ij}^k are given by (3.6.6) and the g_{ij} by (3.6.8). Let

$$c : I \to \Omega, \qquad \xi, \eta : I \to \mathbb{R}^m$$

be smooth functions and suppose that the vector fields X, Y along the curve

$$\gamma := \psi \circ c : I \to M$$

are given by

$$X(t) := \sum_{i=1}^m \xi^i(t) \frac{\partial \psi}{\partial x^i}(c(t)), \qquad Y(t) := \sum_{j=1}^m \eta^j(t) \frac{\partial \psi}{\partial x^j}(c(t)).$$

Dropping the argument t in each term, we obtain from Remark 3.6.4 and Lemma 3.6.1 that

$$\langle X, Y \rangle = \sum_{i,j} g_{ij}(c) \xi^i \eta^j,$$

$$\langle X, \nabla Y \rangle = \sum_{i,k} g_{ik}(c) \xi^i \left(\dot{\eta}^k + \sum_{j,\ell} \Gamma_{j\ell}^k(c) \eta^j \dot{c}^\ell \right),$$

$$\langle \nabla X, Y \rangle = \sum_{j,k} g_{kj}(c) \left(\dot{\xi}^k + \sum_{i,\ell} \Gamma_{i\ell}^k(c) \xi^i \dot{c}^\ell \right) \eta^j.$$

Hence it follows from equation (3.2.5) in Lemma 3.2.4 that

$$0 = \frac{d}{dt} \langle X, Y \rangle - \langle X, \nabla Y \rangle - \langle \nabla X, Y \rangle$$

$$= \sum_{i,j} \left(g_{ij}(c) \dot{\xi}^i \eta^j + g_{ij}(c) \xi^i \dot{\eta}^j + \sum_\ell \frac{\partial g_{ij}}{\partial x^\ell}(c) \xi^i \eta^j \dot{c}^\ell \right)$$

$$- \sum_{i,k} g_{ik}(c) \xi^i \dot{\eta}^k - \sum_{i,j,k,\ell} g_{ik}(c) \Gamma_{j\ell}^k(c) \xi^i \eta^j \dot{c}^\ell$$

$$- \sum_{j,k} g_{kj}(c) \dot{\xi}^k \eta^j - \sum_{i,j,k,\ell} g_{kj}(c) \Gamma_{i\ell}^k(c) \xi^i \eta^j \dot{c}^\ell$$

$$= \sum_{i,j,\ell} \left(\frac{\partial g_{ij}}{\partial x^\ell}(c) - \sum_k g_{ik}(c) \Gamma_{j\ell}^k(c) - \sum_k g_{jk}(c) \Gamma_{i\ell}^k(c) \right) \xi^i \eta^j \dot{c}^\ell.$$

This holds for all smooth maps $c : I \to \Omega$ and $\xi, \eta : I \to \mathbb{R}^m$, so the Γ_{ij}^k satisfy the second equation in (3.6.9). That they are symmetric in i and j is obvious.

To prove that (3.6.9) is equivalent to (3.6.10), define

$$\Gamma_{\ell ij} := \sum_{k=1}^{m} g_{\ell k} \Gamma_{ij}^{k}. \tag{3.6.11}$$

Then (3.6.9) is equivalent to

$$\Gamma_{\ell ij} = \Gamma_{\ell ji}, \qquad \frac{\partial g_{ij}}{\partial x^\ell} = \Gamma_{ij\ell} + \Gamma_{ji\ell}. \tag{3.6.12}$$

and (3.6.10) is equivalent to

$$\Gamma_{\ell ij} = \frac{1}{2}\left(\frac{\partial g_{\ell i}}{\partial x^j} + \frac{\partial g_{\ell j}}{\partial x^i} - \frac{\partial g_{ij}}{\partial x^\ell} \right). \tag{3.6.13}$$

If the $\Gamma_{\ell ij}$ are given by (3.6.13), then they satisfy

$$\Gamma_{\ell ij} = \Gamma_{\ell ji}$$

and

$$2\Gamma_{ij\ell} + 2\Gamma_{ji\ell} = \frac{\partial g_{ij}}{\partial x^\ell} + \frac{\partial g_{i\ell}}{\partial x^j} - \frac{\partial g_{j\ell}}{\partial x^i} + \frac{\partial g_{ji}}{\partial x^\ell} + \frac{\partial g_{j\ell}}{\partial x^i} - \frac{\partial g_{i\ell}}{\partial x^j}$$
$$= 2\frac{\partial g_{ij}}{\partial x^\ell}$$

for all i, j, ℓ. Conversely, if the $\Gamma_{\ell ij}$ satisfy (3.6.12), then

$$\frac{\partial g_{ij}}{\partial x^\ell} = \Gamma_{ij\ell} + \Gamma_{ji\ell},$$
$$\frac{\partial g_{\ell i}}{\partial x^j} = \Gamma_{\ell ij} + \Gamma_{i\ell j} = \Gamma_{\ell ij} + \Gamma_{ij\ell},$$
$$\frac{\partial g_{\ell j}}{\partial x^i} = \Gamma_{\ell ji} + \Gamma_{j\ell i} = \Gamma_{\ell ij} + \Gamma_{ji\ell}.$$

Take the sum of the last two minus the first of these equations to obtain

$$\frac{\partial g_{\ell i}}{\partial x^j} + \frac{\partial g_{\ell j}}{\partial x^i} - \frac{\partial g_{ij}}{\partial x^\ell} = 2\Gamma_{\ell ij}.$$

Thus (3.6.12) is equivalent to (3.6.13) and so (3.6.9) is equivalent to (3.6.10). This proves Lemma 3.6.5.   □

3.7   Riemannian Metrics*

We wish to carry over the fundamental notions of differential geometry to the intrinsic setting. First we need an inner product on the tangent spaces to replace the first fundamental form in Definition 3.1.1. This is the content of Definition 3.7.1 and Lemma 3.7.4 below. Second we must introduce the covariant derivative of a vector field along a curve. With this understood all the definitions, theorems, and proofs in this chapter carry over in an almost word by word fashion to the intrinsic setting.

3.7.1   Existence of Riemannian Metrics

We will always consider norms that are induced by inner products. But in general there is no ambient space that can induce an inner product on each tangent space. This leads to the following definition.

Definition 3.7.1 Let M be a smooth m-manifold. A **Riemannian metric** on M is a collection of inner products

$$T_p M \times T_p M \to \mathbb{R} : (v, w) \mapsto g_p(v, w),$$

one for every $p \in M$, such that the map

$$M \to \mathbb{R} : p \mapsto g_p(X(p), Y(p))$$

is smooth for every pair of vector fields $X, Y \in \mathrm{Vect}(M)$. We will also denote the inner product by $\langle v, w \rangle_p$ and drop the subscript p if the base point is understood from the context. A smooth manifold equipped with a Riemannian metric is called a **Riemannian manifold**.

Example 3.7.2 If $M \subset \mathbb{R}^n$ is a smooth submanifold, then a Riemannian metric on M is given by restricting the standard inner product on $\mathbb{R}^n$ to the tangent spaces $T_p M \subset \mathbb{R}^n$. This is the first fundamental form of an embedded manifold (see Definition 3.1.1).

More generally, assume that M is a Riemannian m-manifold in the intrinsic sense of Definition 3.7.1 with an atlas $\mathcal{A} = \{(\phi_\alpha, U_\alpha)\}_{\alpha \in A}$. Then the Riemannian metric g determines a collection of smooth functions

$$g_\alpha = (g_{\alpha, ij})_{i,j=1}^m : \phi_\alpha(U_\alpha) \to \mathbb{R}^{m \times m}, \tag{3.7.1}$$

one for each $\alpha \in A$, defined by

$$\xi^\mathsf{T} g_\alpha(x)\eta := g_p(v, w), \quad \phi_\alpha(p) = x, \quad d\phi_\alpha(p)v = \xi, \quad d\phi_\alpha(p)w = \eta, \tag{3.7.2}$$

for $x \in \phi_\alpha(U_\alpha)$ and $\xi, \eta \in \mathbb{R}^m$.

Each matrix $g_\alpha(x)$ is symmetrix and positive definite. Note that the tangent vectors v and w in (3.7.2) can also be written in the form

$$v = [\alpha, \xi]_p, \qquad w = [\alpha, \eta]_p.$$

Choosing standard basis vectors

$$\xi = e_i, \qquad \eta = e_j$$

in $\mathbb{R}^m$ we obtain

$$[\alpha, e_i]_p = d\phi_\alpha(p)^{-1} e_i =: \frac{\partial}{\partial x^i}(p)$$

and hence

$$g_{\alpha,ij}(x) = \left\langle \frac{\partial}{\partial x^i}(\phi_\alpha^{-1}(x)), \frac{\partial}{\partial x^j}(\phi_\alpha^{-1}(x)) \right\rangle. \tag{3.7.3}$$

For different coordinate charts the maps g_α and g_β are related through the transition map

$$\phi_{\beta\alpha} := \phi_\beta \circ \phi_\alpha^{-1} : \phi_\alpha(U_\alpha \cap U_\beta) \to \phi_\beta(U_\alpha \cap U_\beta)$$

via

$$g_\alpha(x) = d\phi_{\beta\alpha}(x)^\mathsf{T} g_\beta(\phi_{\beta\alpha}(x)) d\phi_{\beta\alpha}(x) \tag{3.7.4}$$

for $x \in \phi_\alpha(U_\alpha \cap U_\beta)$. Equation (3.7.4) can also be written in the shorthand notation

$$g_\alpha = \phi_{\beta\alpha}^* g_\beta$$

for $\alpha, \beta \in A$.

Exercise 3.7.3 Every collection of smooth maps

$$g_\alpha : \phi_\alpha(U_\alpha) \to \mathbb{R}^{m \times m}$$

with values in the set of positive definite symmetric matrices that satisfies (3.7.4) for all $\alpha, \beta \in A$ determines a global Riemannian metric via (3.7.2).

In this intrinsic setting there is no canonical metric on M (such as the metric induced by $\mathbb{R}^n$ on an embedded manifold). In fact, it is not completely obvious that a manifold admits a Riemannian metric and this is the content of the next lemma.

Lemma 3.7.4 *Every paracompact Hausdorff manifold admits a Riemannian metric.*

Proof   Let m be the dimension of M and let $\mathcal{A} = \{(\phi_\alpha, U_\alpha)\}_{\alpha \in A}$ be an atlas on M. By Theorem 2.9.9 there is a partition of unity $\{\theta_\alpha\}_{\alpha \in A}$ subordinate to the open cover $\{U_\alpha\}_{\alpha \in A}$. Now there are two equivalent ways to construct a Riemannian metric on M.

The first method is to carry over the standard inner product on $\mathbb{R}^m$ to the tangent spaces $T_p M$ for $p \in U_\alpha$ via the coordinate chart ϕ_α, multiply the resulting Riemannian metric on U_α by the compactly supported function θ_α, extend it by zero to all of M, and then take the sum over all α. This leads to the following formula. The inner product of two tangent vectors $v, w \in T_p M$ is defined by

$$\langle v, w \rangle_p := \sum_{p \in U_\alpha} \theta_\alpha(p) \langle d\phi_\alpha(p)v, d\phi_\alpha(p)w \rangle, \tag{3.7.5}$$

where the sum runs over all $\alpha \in A$ with $p \in U_\alpha$ and the inner product is the standard inner product on $\mathbb{R}^m$. Since $\mathrm{supp}(\theta_\alpha) \subset U_\alpha$ for each α and the sum is locally finite we find that the function

$$M \to \mathbb{R} : p \mapsto \langle X(p), Y(p) \rangle_p$$

is smooth for every pair of vector fields $X, Y \in \mathrm{Vect}(M)$. Moreover, the right hand side of (3.7.5) is symmetric in v and w and is positive for $v = w \neq 0$ because each summand is nonnegative and each summand with $\theta_\alpha(p) > 0$ is positive. Thus equation (3.7.5) defines a Riemannian metric on M.

The second method is to define the functions

$$g_\alpha : \phi_\alpha(U_\alpha) \to \mathbb{R}^{m \times m}$$

by

$$g_\alpha(x) := \sum_{\gamma \in A} \theta_\gamma(\phi_\alpha^{-1}(x)) d\phi_{\gamma\alpha}(x)^{\mathsf{T}} d\phi_{\gamma\alpha}(x) \tag{3.7.6}$$

for $x \in \phi_\alpha(U_\alpha)$ where each summand is defined on $\phi_\alpha(U_\alpha \cap U_\gamma)$ and is understood to be zero for $x \notin \phi_\alpha(U_\alpha \cap U_\gamma)$. We leave it to the reader to verify that these functions are smooth and satisfy the condition (3.7.4) for all $\alpha, \beta \in A$. Moreover, the formulas (3.7.5) and (3.7.6) determine the same Riemannian metric on M. (Prove this!) This proves Lemma 3.7.4. $\square$

3.7.2   Two Examples

Example 3.7.5 (Fubini–Study metric)   The complex projective space carries a natural Riemannian metric, defined as follows. Identify $\mathbb{C}P^n$ with the quotient of the unit sphere $S^{2n+1} \subset \mathbb{C}^{n+1}$ by the diagonal action of the circle S^1, i.e. $\mathbb{C}P^n = S^{2n+1}/S^1$. Then the tangent space of $\mathbb{C}P^n$ at the equivalence class

$$[z] = [z_0 : \cdots : z_n] \in \mathbb{C}P^n$$

of a point $z = (z_0, \ldots, z_n) \in S^{2n+1}$ can be identified with the orthogonal complement of $\mathbb{C}z$ in $\mathbb{C}^{n+1}$. Now choose the inner product on $T_{[z]}\mathbb{C}\mathrm{P}^n$ to be the one inherited from the standard inner product on $\mathbb{C}^{n+1}$ via this identification. The resulting metric on $\mathbb{C}\mathrm{P}^n$ is called the **Fubini–Study metric**. **Exercise:** Prove that the action of $\mathrm{U}(n+1)$ on $\mathbb{C}^{n+1}$ induces a transitive action of the quotient group

$$\mathrm{PSU}(n+1) := \mathrm{U}(n+1)/S^1$$

by isometries. If $z \in S^{2n+1}$, prove that the unitary matrix

$$g := 2zz^* - \mathbb{1}$$

descends to an isometry ϕ on $\mathbb{C}\mathrm{P}^n$ with fixed point $p := [z]$ and $d\phi(p) = -\mathrm{id}$. Show that, in the case $n = 1$, the pullback of the Fubini–Study metric on $\mathbb{C}\mathrm{P}^1$ under the stereographic projection

$$S^2 \setminus \{(0,0,1)\} \to \mathbb{C}\mathrm{P}^1 \setminus \{[0:1]\} : (x_1, x_2, x_3) \mapsto \left[1 : \frac{x_1 + \mathrm{i}x_2}{1 - x_3}\right]$$

is one quarter of the standard metric on S^2.

Example 3.7.6 Think of the complex Grassmannian $G_k(\mathbb{C}^n)$ of k-planes in $\mathbb{C}^n$ as a quotient of the space

$$\mathcal{F}_k(\mathbb{C}^n) := \{D \in \mathbb{C}^{n \times k} \mid D^*D = \mathbb{1}\}$$

of unitary k-frames in $\mathbb{C}^n$ by the right action of the unitary group $\mathrm{U}(k)$. The space $\mathcal{F}_k(\mathbb{C}^n)$ inherits a Riemannian metric from the ambient Euclidean space $\mathbb{C}^{n \times k}$. Show that the tangent space of $G_k(\mathbb{C}^n)$ at a point $\Lambda = \mathrm{im}D$, with $D \in \mathcal{F}_k(\mathbb{C}^n)$ can be identified with the space

$$T_\Lambda G_k(\mathbb{C}^n) = \left\{\widehat{D} \in \mathbb{C}^{n \times k} \mid D^*\widehat{D} = 0\right\}.$$

Define the inner product on this tangent space to be the restriction of the standard inner product on $\mathbb{C}^{n \times k}$ to this subspace. **Exercise:** Prove that the unitary group $\mathrm{U}(n)$ acts on $G_k(\mathbb{C}^n)$ by isometries.

3.7.3 The Levi-Civita Connection

A subtle point in this discussion is how to extend the notion of *covariant derivative* to general Riemannian manifolds. In this case the idea of projecting the derivative in the ambient space orthogonally onto the tangent space has no obvious analogue. Instead we shall see how the covariant derivatives of vector fields along curves can be characterized by several axioms and these can be used to define the covariant derivative in the intrinsic setting. An alternative, but somewhat less satisfactory, approach is to carry over the formula for the covariant derivative in local coordinates to the intrinsic setting and show that the result is independent of the choice of the coordinate chart. Of course, these approaches are equivalent and lead to the same result. We formulate them as a series of exercises. The details are straightforward.

Assume throughout that M is a Riemannian m-manifold with an atlas

$$\mathcal{A} = \{(\phi_\alpha, U_\alpha)\}_{\alpha \in A}$$

and suppose that the Riemannian metric is in local coordinates given by

$$g_\alpha = (g_{\alpha,ij})_{i,j=1}^m : \phi_\alpha(U_\alpha) \to \mathbb{R}^{m \times m}$$

for $\alpha \in A$. These functions satisfy (3.7.4) for all $\alpha, \beta \in A$.

Definition 3.7.7 Let $f : N \to M$ be a smooth map between manifolds. A **vector field along** f is a collection of tangent vectors

$$X(q) \in T_{f(q)}M,$$

one for each $q \in N$, such that the map

$$N \to TM : q \mapsto (f(q), X(q))$$

is smooth. The space of vector fields along f will be denoted by $\mathrm{Vect}(f)$.

As before we will not distinguish in notation between the collection of tangent vectors $X(q) \in T_{f(q)}M$ and the associated map $N \to TM$ and denote them both by X. The following theorem introduces the Levi-Civita connection as a collection of linear operators $\nabla : \mathrm{Vect}(\gamma) \to \mathrm{Vect}(\gamma)$, one for each smooth curve $\gamma : I \to M$.

Theorem 3.7.8 (Levi-Civita connection) *There exists a unique collection of linear operators*

$$\nabla : \mathrm{Vect}(\gamma) \to \mathrm{Vect}(\gamma)$$

*(called the **covariant derivative**), one for every smooth curve $\gamma : I \to M$ on an open interval $I \subset \mathbb{R}$, satisfying the following axioms.*

(Leibniz Rule) *For every smooth curve $\gamma : I \to M$, every smooth function $\lambda : I \to \mathbb{R}$, and every vector field $X \in \mathrm{Vect}(\gamma)$, we have*

$$\nabla(\lambda X) = \dot{\lambda} X + \lambda \nabla X. \tag{3.7.7}$$

(Chain Rule) *Let $\Omega \subset \mathbb{R}^n$ be an open set, let $c : I \to \Omega$ be a smooth curve, let $\gamma : \Omega \to M$ be a smooth map, and let X be a smooth vector field along γ.*

Denote by $\nabla_i X$ the covariant derivative of X along the curve $x^i \mapsto \gamma(x)$ (with the other coordinates fixed). Then $\nabla_i X$ is a smooth vector field along γ and the covariant derivative of the vector field $X \circ c \in \text{Vect}(\gamma \circ c)$ is

$$\nabla(X \circ c) = \sum_{j=1}^{n} \dot{c}^j(t) \nabla_j X(c(t)). \tag{3.7.8}$$

(Riemannian) *For any two vector fields $X, Y \in \text{Vect}(\gamma)$ we have*

$$\frac{d}{dt} \langle X, Y \rangle = \langle \nabla X, Y \rangle + \langle X, \nabla Y \rangle. \tag{3.7.9}$$

(Torsion-free) *Let $I, J \subset \mathbb{R}$ be open intervals and $\gamma : I \times J \to M$ be a smooth map. Denote by ∇_s the covariant derivative along the curve $s \mapsto \gamma(s,t)$ (with t fixed) and by ∇_t the covariant derivative along the curve $t \mapsto \gamma(s,t)$ (with s fixed). Then*

$$\nabla_s \partial_t \gamma = \nabla_t \partial_s \gamma. \tag{3.7.10}$$

Proof The proof is based on a reformulation of the axioms in local coordinates. The (Leibniz Rule) and (Chain Rule) axioms assert that the covariant derivative is in local coordinates given by Christoffel symbols Γ_{ij}^k as in equation (3.6.7) in Lemma 3.6.1. The (Riemannian) and (Torsion-free) axioms assert that the Christoffel symbols satisfy the equations in (3.6.9) and hence, by Lemma 3.6.5, are given by (3.6.10). (See also Exercise 3.7.10.) This proves Theorem 3.7.8. $\square$

Exercise 3.7.9 The **Christoffel symbols** of the Riemannian metric are the functions $\Gamma_{\alpha,ij}^k : \phi_\alpha(U_\alpha) \to \mathbb{R}$. defined by

$$\Gamma_{\alpha,ij}^k := \sum_{\ell=1}^{m} g_\alpha^{k\ell} \frac{1}{2} \left(\frac{\partial g_{\alpha,\ell i}}{\partial x^j} + \frac{\partial g_{\alpha,\ell j}}{\partial x^i} - \frac{\partial g_{\alpha,ij}}{\partial x^\ell} \right) \tag{3.7.11}$$

(see Lemma 3.6.5). Prove that they are related by the equation

$$\sum_k \frac{\partial \phi_{\beta\alpha}^{k'}}{\partial x^k} \Gamma_{\alpha,ij}^k = \frac{\partial^2 \phi_{\beta\alpha}^{k'}}{\partial x^i \partial x^j} + \sum_{i',j'} \left(\Gamma_{\beta,i'j'}^{k'} \circ \phi_{\beta\alpha} \right) \frac{\partial \phi_{\beta\alpha}^{i'}}{\partial x^i} \frac{\partial \phi_{\beta\alpha}^{j'}}{\partial x^j}.$$

for all $\alpha, \beta \in A$.

Exercise 3.7.10 Denote $\psi_\alpha := \phi_\alpha^{-1} : \phi_\alpha(U_\alpha) \to M$. Prove that the covariant derivative of a vector field

$$X(t) = \sum_{i=1}^{m} \xi_\alpha^i(t) \frac{\partial \psi_\alpha}{\partial x^i} (c_\alpha(t))$$

along $\gamma = \psi_\alpha \circ c_\alpha : I \to M$ is given by

$$\nabla X(t) = \sum_{k=1}^{m} \left(\dot{\xi}_\alpha^k(t) + \sum_{i,j=1}^{m} \Gamma_{\alpha,ij}^k(c(t)) \xi_\alpha^i(t) \dot{c}_\alpha^j(t) \right) \frac{\partial \psi_\alpha}{\partial x^k}(c_\alpha(t)). \qquad (3.7.12)$$

Prove that ∇X is independent of the choice of the coordinate chart.

Exercise 3.7.11 Let $\Omega \subset \mathbb{R}^2$ be open and $\lambda : \Omega \to (0, \infty)$ be a smooth function. Let $g : \Omega \to \mathbb{R}^{2\times2}$ be given by

$$g(x) = \begin{pmatrix} \lambda(x) & 0 \\ 0 & \lambda(x) \end{pmatrix}.$$

Compute the Christoffel symbols Γ_{ij}^k via (3.6.10).

Exercise 3.7.12 Let $\phi : S^2 \setminus \{(0,0,1)\} \to \mathbb{C}$ be the stereographic projection, given by

$$\phi(p) := \left(\frac{p_1}{1 - p_3}, \frac{p_2}{1 - p_3} \right)$$

Prove that the metric $g : \mathbb{R}^2 \to \mathbb{R}^{2\times2}$ has the form $g(x) = \lambda(x)\mathbb{1}$ where

$$\lambda(x) := \frac{4}{(1 + |x|^2)^2} \qquad \text{for } x = (x^1, x^2) \in \mathbb{R}^2.$$

3.7.4   Basic Vector Fields in the Intrinsic Setting

Let M be a Riemannian m-manifold with an atlas $\mathcal{A} = \{(\phi_\alpha, U_\alpha)\}_{\alpha \in A}$. Then the **frame bundle** (3.4.1) admits the structure of a smooth manifold with the open cover $\widetilde{U}_\alpha := \pi^{-1}(U_\alpha)$ and coordinate charts

$$\widetilde{\phi}_\alpha : \widetilde{U}_\alpha \to \phi_\alpha(U_\alpha) \times \mathrm{GL}(m)$$

given by

$$\widetilde{\phi}_\alpha(p, e) := (\phi_\alpha(p), d\phi_\alpha(p)e).$$

The derivatives of the horizontal curves in Definition 3.4.6 form a **horizontal subbundle** $H \subset T\mathcal{F}(M)$ of the tangent bundle of the Frame bundle whose fibers $H_{(p,e)}$ over an element $(p, e) \in \mathcal{F}(M)$ can in local coordinates be described as follows. Let

$$x := \phi_\alpha(p), \qquad a := d\phi_\alpha(p)e \in \mathrm{GL}(m), \qquad (3.7.13)$$

and let $(\widehat{x}, \widehat{a}) \in \mathbb{R}^m \times \mathbb{R}^{m \times m}$. This pair has the form

$$(\widehat{x}, \widehat{a}) = d\widetilde{\phi}_\alpha(p, e)(\widehat{p}, \widehat{e}), \qquad (\widehat{p}, \widehat{e}) \in H_{(p,e)}, \tag{3.7.14}$$

if and only if

$$\widehat{a}_\ell^k = - \sum_{i,j=1}^m \Gamma_{\alpha,ij}^k(x)\widehat{x}^i a_\ell^j \tag{3.7.15}$$

for $k, \ell = 1, \dots, m$, where the functions $\Gamma_{\alpha,ij}^k : \phi_\alpha(U_\alpha) \to \mathbb{R}$ are the Christoffel symbols defined by (3.7.11). Thus a tangent vector $(\widehat{p}, \widehat{e}) \in T_{(p,e)}\mathcal{F}(M)$ is **horizontal** if and only if its coordinates $(\widehat{x}, \widehat{a})$ in (3.7.14) satisfy (3.7.15). Hence, for every vector $\xi \in \mathbb{R}^m$, there exists a unique horizontal vector field $B_\xi \in \mathrm{Vect}(\mathcal{F}(M))$ (the **basic vector field** associated to ξ) such that

$$d\pi(p, e)B_\xi(p, e) = e\xi$$

for all $(p, e) \in \mathcal{F}(M)$. This vector field assigns to a pair $(p, e) \in \mathcal{F}(M)$ with the coordinates $(x, a) \in \mathbb{R}^m \times \mathrm{GL}(m)$ as in (3.7.13) the horizontal tangent vector $(\widehat{p}, \widehat{e}) \in H_{(p,e)} \subset T_{(p,e)}\mathcal{F}(M)$ whose coordinates $(\widehat{x}, \widehat{a}) \in \mathbb{R}^m \times \mathbb{R}^{m \times m}$ satisfy (3.7.15) and $\widehat{x} = a\xi$.

Exercise 3.7.13 Verify the equivalence of (3.7.14) and (3.7.15). Prove that the notion of a horizontal tangent vector of $\mathcal{F}(M)$ is independent of the choice of the coordinate chart. **Hint:** Use Exercise 3.7.9.

Exercise 3.7.14 Examine the orthonormal frame bundle $\mathcal{O}(M)$ in the intrinsic setting.

Exercise 3.7.15 Carry over the proofs of Theorem 3.3.4, Theorem 3.3.6, and Theorem 3.5.21 to the intrinsic setting.

Geodesics

4

This chapter introduces geodesics in Riemannian manifolds. It begins in Sect. 4.1 by introducing geodesics as extremals of the energy and length functionals and characterizing them as solutions of a second order differential equation. In Sect. 4.2 we show that minimizing the length with fixed endpoints gives rise to an intrinsic distance function $d : M \times M \to \mathbb{R}$ which induces the topology M inherits from the ambient space $\mathbb{R}^n$. Section 4.3 introduces the exponential map, Sect. 4.4 shows that geodesics minimize the length on short time intervals, Sect. 4.5 establishes the existence of geodesically convex neighborhoods, and Sect. 4.6 shows that the geodesic flow is complete if and only if (M, d) is a complete metric space, and that in the complete case any two points are joined by a minimal geodesic. Section 4.7 discusses geodesics in the intrinsic setting.

4.1 Length and Energy

This section explains the length and energy functionals on the space of paths with fixed endpoints and introduces geodesics as their extremal points.

4.1.1 The Length and Energy Functionals

The concept of a geodesic in a manifold generalizes that of a straight line in Euclidean space. A straight line has parametrizations of form $t \mapsto p + \sigma(t)v$ where $\sigma : \mathbb{R} \to \mathbb{R}$ is a diffeomorphism and $p, v \in \mathbb{R}^n$. Different choices of σ yield different parametrizations of the same line. Certain parametrizations are preferred, for example those parametrizations which are "proportional to the arclength", i.e. where $\sigma(t) = at + b$ for constants $a, b \in \mathbb{R}$, so that the tangent vector $\dot{\sigma}(t)v$ has constant length. The same distinctions can be made for geodesics. Some authors use the term geodesic to include all parametrizations of a geodesic while others restrict the term to cover only geodesics parametrized proportional to the arclength. We

© The Editor(s) (if applicable) and The Author(s), under exclusive license to
Springer-Verlag GmbH, DE, part of Springer Nature 2022
J.W. Robbin, D.A. Salamon, *Introduction to Differential Geometry*,
Springer Studium Mathematik (Master), https://doi.org/10.1007/978-3-662-64340-2_4

follow the latter course, referring to the more general concept as a "reparametrized geodesic". Thus a reparametrized geodesic need not be a geodesic.

We assume throughout that $M \subset \mathbb{R}^n$ is a smooth m-manifold.

Definition 4.1.1 (Length and energy) Let $I = [a, b] \subset \mathbb{R}$ be a compact interval with $a < b$ and let $\gamma : I \to M$ be a smooth curve in M. The **length** $L(\gamma)$ and the **energy** $E(\gamma)$ are defined by

$$L(\gamma) := \int_a^b |\dot{\gamma}(t)| \, dt, \tag{4.1.1}$$

$$E(\gamma) := \frac{1}{2} \int_a^b |\dot{\gamma}(t)|^2 \, dt. \tag{4.1.2}$$

A **variation of** γ is a family of smooth curves $\gamma_s : I \to M$, where s ranges over the reals, such that the map $\mathbb{R} \times I \to M : (s, t) \mapsto \gamma_s(t)$ is smooth and $\gamma_0 = \gamma$. The variation $\{\gamma_s\}_{s \in \mathbb{R}}$ is said to have **fixed endpoints** iff $\gamma_s(a) = \gamma(a)$ and $\gamma_s(b) = \gamma(b)$ for all $s \in \mathbb{R}$.

Remark 4.1.2 The length of a continuous function $\gamma : [a, b] \to \mathbb{R}^n$ can be defined as the supremum of the numbers $\sum_{i=1}^N |\gamma(t_i) - \gamma(t_{i-1})|$ over all partitions $a = t_0 < t_1 < \cdots < t_N = b$ of the interval $[a, b]$. By a theorem in first year analysis [64] this supremum is finite whenever γ is continuously differentiable and is given by (4.1.1).

We shall sometimes suppress the notation for the endpoints $a, b \in I$. When $\gamma(a) = p$ and $\gamma(b) = q$ we say that γ **is a curve from** p to q. One can always compose γ with an affine reparametrization $t' = a + (b - a)t$ to obtain a new curve $\gamma'(t) := \gamma(t')$ on the unit interval $0 \le t \le 1$. This new curve satisfies $L(\gamma') = L(\gamma)$ and $E(\gamma') = (b - a)E(\gamma)$. More generally, the length $L(\gamma)$, but not the energy $E(\gamma)$, is invariant under reparametrization.

Remark 4.1.3 (Reparametrization) Let $I = [a, b]$ and $I' = [a', b']$ be compact intervals. If $\gamma : I \to \mathbb{R}^n$ is a smooth curve and $\sigma : I' \to I$ is a smooth function such that $\sigma(a') = a$, $\sigma(b') = b$, and $\dot{\sigma}(t') \ge 0$ for all $t' \in I'$, then the curves γ and $\gamma \circ \sigma$ have the same length. Namely,

$$L(\gamma \circ \sigma) = \int_{a'}^{b'} \left| \frac{d}{dt'} \gamma(\sigma(t')) \right| dt' = \int_{a'}^{b'} \left| \dot{\gamma}(\sigma(t')) \right| \dot{\sigma}(t') \, dt' = L(\gamma). \tag{4.1.3}$$

Here second equality follows from the chain rule and the third equality follows from the change of variables formula for the Riemann integral.

Theorem 4.1.4 (Characterization of geodesics) *Let $I = [a, b] \subset \mathbb{R}$ be a compact interval and let $\gamma : I \to M$ be a smooth curve. Then the following are equivalent.*

(i)   *γ is an **extremal of the energy functional**, i.e. every variation $\{\gamma_s\}_{s \in \mathbb{R}}$ of γ with fixed endpoints satisfies*

$$\tfrac{d}{ds}\big|_{s=0} E(\gamma_s) = 0.$$

(ii) *γ is **parametrized proportional to the arclength**, i.e. the velocity $|\dot\gamma(t)| \equiv c \geq 0$ is constant, and either γ is constant, i.e. $\gamma(t) = p = q$ for all $t \in I$, or $c > 0$ and γ is an **extremal of the length functional**, i.e. every variation $\{\gamma_s\}_{s \in \mathbb{R}}$ of γ with fixed endpoints satisfies*

$$\tfrac{d}{ds}\big|_{s=0} L(\gamma_s) = 0.$$

(iii) *The velocity vector of γ is parallel, i.e. $\nabla \dot\gamma(t) = 0$ for all $t \in I$.*
(iv) *The acceleration of γ is normal to M, i.e. $\ddot\gamma(t) \perp T_{\gamma(t)} M$ for all $t \in I$.*
(v)   *If (Φ, γ, γ') is a development of M along $M' = \mathbb{R}^m$, then $\gamma' : I \to \mathbb{R}^m$ is a straight line parametrized proportional to the arclength, i.e. $\ddot\gamma' \equiv 0$.*

Proof   See Sect. 4.1.3.   □

Definition 4.1.5 (Geodesic)   A smooth curve $\gamma : I \to M$ on an interval I is called a **geodesic** iff its restriction to each compact subinterval satisfies the equivalent conditions of Theorem 4.1.4. So γ is a geodesic if and only if

$$\nabla \dot\gamma(t) = 0 \qquad \text{for all } t \in I. \tag{4.1.4}$$

By the Gauß–Weingarten formula (3.2.2) with $X = \dot\gamma$ this is equivalent to

$$\ddot\gamma(t) = h_{\gamma(t)}(\dot\gamma(t), \dot\gamma(t)) \qquad \text{for all } t \in I. \tag{4.1.5}$$

Remark 4.1.6

(i)   The conditions (i) and (ii) in Theorem 4.1.4 are meaningless when I is not compact because then the curve has at most one endpoint and the length and energy integrals may be infinite. However, the conditions (iii), (iv), and (v) in Theorem 4.1.4 are equivalent for smooth curves on any interval, compact or not.

(ii) The function $s \mapsto E(\gamma_s)$ associated to a smooth variation is always smooth and so condition (i) in Theorem 4.1.4 is meaningful. However, more care has to be taken in part (ii) because the function $s \mapsto L(\gamma_s)$ need not be differentiable. It is differentiable at $s = 0$ whenever $\dot\gamma(t) \neq 0$ for all $t \in I$.

4.1.2 The Space of Paths

Fix two points $p, q \in M$ and a compact interval $I = [a, b]$ and denote by

$$\Omega_{p,q} := \Omega_{p,q}(I) := \{\gamma : I \to M \mid \gamma \text{ is smooth and } \gamma(a) = p, \, \gamma(b) = q\}$$

the space of smooth curves in M from p to q, defined on the interval I. Then the length and energy are functionals $L, E : \Omega_{p,q} \to \mathbb{R}$ and their extremal points can be understood as *critical points* as we now explain.

We may think of the space $\Omega_{p,q}$ as a kind of *"infinite-dimensional manifold"*. This is to be understood in a heuristic sense and we use these terms here to emphasize an analogy. Of course, the space $\Omega_{p,q}$ is not a manifold in the strict sense of the word. To begin with it is not embedded in any finite-dimensional Euclidean space. However, it has many features in common with manifolds. The first is that we can speak of *smooth curves in* $\Omega_{p,q}$. Of course $\Omega_{p,q}$ is itself a space of curves in M. Thus a smooth curve in $\Omega_{p,q}$ would then be a curve of curves, namly a map $\mathbb{R} \to \Omega_{p,q} : s \mapsto \gamma_s$ that assigns to each real number s a smooth curve $\gamma_s : I \to M$ satisfying $\gamma_s(a) = p$ and $\gamma_s(b) = q$. We shall call such a curve of curves **smooth** iff the associated map $\mathbb{R} \times I \to M : (s, t) \mapsto \gamma_s(t)$ is smooth. Thus smooth curves in $\Omega_{p,q}$ are the variations of γ with fixed endpoints introduced in Definition 4.1.1.

Having defined what we mean by a smooth curve in $\Omega_{p,q}$ we can also differentiate such a curve with respect to s. Here we can simply recall that, since $M \subset \mathbb{R}^n$, we have a smooth map $\mathbb{R} \times I \to \mathbb{R}^n$ and the derivative of the curve $s \mapsto \gamma_s$ in $\Omega_{p,q}$ can simply be understood as the partial derivative of the map $(s, t) \mapsto \gamma_s(t)$ with respect to s. Thus, in analogy with embedded manifolds, we define the **tangent space** of the space of curves $\Omega_{p,q}$ at γ as the set of all derivatives of smooth curves $\mathbb{R} \to \Omega_{p,q} : s \mapsto \gamma_s$ passing through γ, i.e.

$$T_\gamma \Omega_{p,q} := \left\{ \frac{\partial}{\partial s}\Big|_{s=0} \gamma_s \;\middle|\; \mathbb{R} \to \Omega_{p,q} : s \mapsto \gamma_s \text{ is smooth and } \gamma_0 = \gamma \right\}.$$

Let us denote such a partial derivative by $X(t) := \frac{\partial}{\partial s}\big|_{s=0} \gamma_s(t) \in T_{\gamma(t)}M$. Thus we obtain a smooth vector field along γ. Since $\gamma_s(a) = p$ and $\gamma_s(b) = q$ for all s, this vector field must vanish at $t = a, b$. This suggests the formula

$$T_\gamma \Omega_{p,q} = \{X \in \mathrm{Vect}(\gamma) \mid X(a) = 0, \, X(b) = 0\}. \tag{4.1.6}$$

That every tangent vector of the path space $\Omega_{p,q}$ at γ is a vector field along γ vanishing at the endpoints follows from the above discussion. The converse inclusion is the content of the next lemma.

Lemma 4.1.7 *Let $p, q \in M$, $\gamma \in \Omega_{p,q}$, and $X \in \mathrm{Vect}(\gamma)$ with $X(a) = 0$ and $X(b) = 0$. Then there exists a smooth map $\mathbb{R} \to \Omega_{p,q} : s \mapsto \gamma_s$ such that*

$$\gamma_0(t) = \gamma(t), \qquad \frac{\partial}{\partial s}\Big|_{s=0} \gamma_s(t) = X(t) \qquad \text{for all } t \in I. \tag{4.1.7}$$

Proof The proof has two steps.

Step 1 *There exists smooth map $M \times I \to \mathbb{R}^n : (r, t) \mapsto Y_t(r)$ with compact support such that $Y_t(r) \in T_r M$ for all $t \in I$ and $r \in M$ and $Y_a(r) = Y_b(r) = 0$ for all $r \in M$.*

Define $Z_t(r) := \Pi(r) X(t)$ for $t \in I$ and $r \in M$. Choose an open set $U \subset \mathbb{R}^n$ such that $\gamma(I) \subset U$ and $\overline{U} \cap M$ is compact (e.g. take $U := \bigcup_{a \le t \le b} B_\varepsilon(\gamma(t))$ for $\varepsilon > 0$ sufficiently small). Now let $\beta : \mathbb{R}^n \to [0, 1]$ be a smooth cutoff function with support in the unit ball such that $\beta(0) = 1$ and define the vector fields Y_t by $Y_t(r) := \beta(\varepsilon^{-1}(r - \gamma(t))) Z_t(r)$ for $t \in I$ and $r \in M$.

Step 2 *We prove the lemma.*

The vector field $Y_t : M \to TM$ in Step 1 is complete for each t. Thus there exists a unique smooth map $\mathbb{R} \times I \to M : (s, t) \mapsto \gamma_s(t)$ such that, for each $t \in I$, the curve $\mathbb{R} \to M : s \mapsto \gamma_s(t)$ is the unique solution of the differential equation $\frac{\partial}{\partial s} \gamma_s(t) = Y_t(\gamma_s(t))$ with $\gamma_0(t) = \gamma(t)$. These maps γ_s satisfies (4.1.7) by Step 1. $\square$

We can now define the **derivative of the energy functional** E at γ in the direction of a tangent vector $X \in T_\gamma \Omega_{p,q}$ by

$$dE(\gamma)X := \frac{d}{ds}\bigg|_{s=0} E(\gamma_s), \qquad (4.1.8)$$

where $s \mapsto \gamma_s$ is as in Lemma 4.1.7. Similarly, the **derivative of the length functional** L at γ in the direction of $X \in T_\gamma \Omega_{p,q}$ is defined by

$$dL(\gamma)X := \frac{d}{ds}\bigg|_{s=0} L(\gamma_s). \qquad (4.1.9)$$

To define (4.1.8) and (4.1.9) the functions $s \mapsto E(\gamma_s)$ and $s \mapsto L(\gamma_s)$ must be differentiable at $s = 0$. This is true for E but it only holds for L when $\dot{\gamma}(t) \ne 0$ for all $t \in I$. Second, we must show that the right hand sides of (4.1.8) and (4.1.9) depend only on X and not on the choice of $\{\gamma_s\}_{s \in \mathbb{R}}$. Third, we must verify that $dE(\gamma) : T_\gamma \Omega_{p,q} \to \mathbb{R}$ and $dL(\gamma) : \Omega_{p,q} \to \mathbb{R}$ are linear maps. This is an exercise in first year analysis (see also the proof of Theorem 4.1.4). A curve $\gamma \in \Omega_{p,q}$ is then an extremal point of E (respectively L when $\dot{\gamma}(t) \ne 0$ for all t) if and only if $dE(\gamma) = 0$ (respectively $dL(\gamma) = 0$). Such a curve is also called a **critical point** of E (respectively L).

4.1.3 Characterization of Geodesics

Proof of Theorem 4.1.4 The equivalence of (iii) and (iv) follows directly from the equations $\nabla \dot{\gamma}(t) = \Pi(\gamma(t)) \ddot{\gamma}(t)$ and $\ker(\Pi(\gamma(t))) = T_{\gamma(t)} M^\perp$.

We prove that (i) is equivalent to (iii) and (iv). Let $X \in T_\gamma \Omega_{p,q}$ and choose a smooth curve of curves $\mathbb{R} \to \Omega_{p,q} : s \mapsto \gamma_s$ satisfying (4.1.7). Then the function $(s, t) \mapsto |\dot{\gamma}_s(t)|^2$ is smooth and hence

$$
\begin{aligned}
dE(\gamma)X &= \frac{d}{ds}\bigg|_{s=0} E(\gamma_s) \\
&= \frac{d}{ds}\bigg|_{s=0} \frac{1}{2} \int_a^b |\dot{\gamma}_s(t)|^2 \, dt \\
&= \frac{1}{2} \int_a^b \frac{\partial}{\partial s}\bigg|_{s=0} |\dot{\gamma}_s(t)|^2 \, dt \\
&= \int_a^b \left\langle \dot{\gamma}(t), \frac{\partial}{\partial s}\bigg|_{s=0} \dot{\gamma}_s(t) \right\rangle dt \\
&= \int_a^b \left\langle \dot{\gamma}(t), \dot{X}(t) \right\rangle dt \\
&= - \int_a^b \left\langle \ddot{\gamma}(t), X(t) \right\rangle dt.
\end{aligned}
\tag{4.1.10}
$$

That (iii) implies (i) follows directly from this identity. To prove that (i) implies (iv) we argue indirectly and assume that there exists a point $t_0 \in [0, 1]$ such that $\ddot{\gamma}(t_0)$ is not orthogonal to $T_{\gamma(t_0)}M$. Then there exists a vector $v_0 \in T_{\gamma(t_0)}M$ such that $\langle \ddot{\gamma}(t_0), v_0 \rangle > 0$. We may assume without loss of generality that $a < t_0 < b$. Then there exists a constant $\varepsilon > 0$ such that $a < t_0 - \varepsilon < t_0 + \varepsilon < b$ and

$$
t_0 - \varepsilon < t < t_0 + \varepsilon \qquad \Longrightarrow \qquad \langle \ddot{\gamma}(t), \Pi(\gamma(t))v_0 \rangle > 0.
$$

Now choose a smooth cutoff function $\beta : I \to [0, 1]$ such that $\beta(t) = 0$ for all $t \in I$ with $|t - t_0| \geq \varepsilon$ and $\beta(t_0) = 1$. Define $X \in T_\gamma \Omega_{p,q}$ by

$$
X(t) := \beta(t)\Pi(\gamma(t))v_0 \qquad \text{for } t \in I.
$$

Then $\langle \ddot{\gamma}(t), X(t) \rangle \geq 0$ for all t and $\langle \ddot{\gamma}(t_0), X(t_0) \rangle > 0$. Hence

$$
dE(\gamma)X = - \int_a^b \langle \ddot{\gamma}(t), X(t) \rangle \, dt < 0
$$

and so γ does not satisfy (i). Thus (i) is equivalent to (iii) and (iv).

We prove that (i) is equivalent to (ii). Assume first that γ satisfies (i). Then γ also satisfies (iv) and hence $\ddot{\gamma}(t) \perp T_{\gamma(t)}M$ for all $t \in I$. This implies

$$0 = \langle \ddot{\gamma}(t), \dot{\gamma}(t) \rangle = \frac{1}{2}\frac{d}{dt}|\dot{\gamma}(t)|^2.$$

Hence the function $I \to \mathbb{R} : t \mapsto |\dot{\gamma}(t)|^2$ is constant. Choose $c \geq 0$ such that $|\dot{\gamma}(t)| \equiv c$. If $c = 0$, then $\gamma(t)$ is constant and so $\gamma(t) \equiv p = q$. If $c > 0$, then

$$dL(\gamma)X = \frac{d}{ds}\bigg|_{s=0} \int_a^b |\dot{\gamma}_s(t)|\, dt$$

$$= \int_a^b \frac{\partial}{\partial s}\bigg|_{s=0} |\dot{\gamma}_s(t)|\, dt$$

$$= \int_a^b |\dot{\gamma}(t)|^{-1} \left\langle \dot{\gamma}(t), \frac{\partial}{\partial s}\bigg|_{s=0} \dot{\gamma}_s(t) \right\rangle dt$$

$$= \frac{1}{c} \int_a^b \left\langle \dot{\gamma}(t), \dot{X}(t) \right\rangle dt$$

$$= \frac{1}{c} dE(\gamma)X.$$

Thus, in the case $c > 0$, γ is an extremal point of E if and only if it is an extremal point of L. Hence (i) is equivalent to (ii).

We prove that (iii) is equivalent to (v). Let (Φ, γ, γ') be a development of M along $M' = \mathbb{R}^m$. Then

$$\dot{\gamma}'(t) = \Phi(t)\dot{\gamma}(t), \qquad \frac{d}{dt}\Phi(t)X(t) = \Phi(t)\nabla X(t)$$

for all $X \in \mathrm{Vect}(\gamma)$ and all $t \in I$. Take $X = \dot{\gamma}$ to obtain $\ddot{\gamma}'(t) = \Phi(t)\nabla\dot{\gamma}(t)$ for all $t \in I$. Thus $\nabla\dot{\gamma} \equiv 0$ if and only if $\ddot{\gamma}' \equiv 0$. This proves Theorem 4.1.4. $\qquad\square$

Remark 4.1.3 shows that reparametrization by a nondecreasing surjective map $\sigma : I' \to I$ gives rise to map

$$\Omega_{p,q}(I) \to \Omega_{p,q}(I') : \gamma \mapsto \gamma \circ \sigma$$

which preserves the length functional, i.e.

$$L(\gamma \circ \sigma) = L(\gamma)$$

for all $\gamma \in \Omega_{p,q}(I)$. Thus the chain rule in infinite dimensions should assert that if $\gamma \circ \sigma$ is an extremal (i.e. critical) point of L, then γ is an extremal point of L. Moreover, if σ is a diffeomorphism, the map $\gamma \mapsto \gamma \circ \sigma$ is bijective and should give rise to a bijective correspondence between the extremal points of L on $\Omega_{p,q}(I)$ and those on $\Omega_{p,q}(I')$. Finally, if the tangent vector field $\dot{\gamma}$ vanishes nowhere, then γ can be parametrized by the arclength. This is spelled out in more detail in the next exercise.

Exercise 4.1.8 Let $\gamma : I = [a, b] \to M$ be a smooth curve such that

$$\dot{\gamma}(t) \neq 0$$

for all $t \in I$ and define

$$T := L(\gamma) = \int_a^b |\dot{\gamma}(t)| \, dt.$$

(i)   Prove that there exists a unique diffeomorphism $\sigma : [0, T] \to I$ such that

$$\sigma(t') = t \qquad \Longleftrightarrow \qquad t' = \int_a^t |\dot{\gamma}(s)| \, ds$$

for all $t' \in [0, T]$ and all $t \in [a, b]$. Prove that $\gamma' := \gamma \circ \sigma : [0, T] \to M$ is **parametrized by the arclength**, i.e. $|\dot{\gamma}'(t')| = 1$ for all $t' \in [0, T]$.

(ii) Prove that

$$dL(\gamma)X = -\int_a^b \langle \dot{V}(t), X(t) \rangle \, dt, \qquad V(t) := |\dot{\gamma}(t)|^{-1}\dot{\gamma}(t). \qquad (4.1.11)$$

Hint: See the relevant formula in the proof of Theorem 4.1.4.

(iii) Prove that γ is an extremal point of L if and only if the curve γ' in part (i) is a geodesic.

(iv) Prove that γ is an extremal point of L if and only if there exists a geodesic $\gamma' : I' \to M$ and a diffeomorphism $\sigma : I' \to I$ such that $\gamma' = \gamma \circ \sigma$.

Next we generalize this exercise to cover the case where $\dot{\gamma}$ is allowed to vanish. Recall from Remark 4.1.6 that the function $s \mapsto L(\gamma_s)$ need not be differentiable. As an example consider the case where $\gamma = \gamma_0$ is constant (see also Exercise 4.4.12 below).

Exercise 4.1.9 Let $\gamma : I \to M$ be a smooth curve and let $X \in T_\gamma \Omega_{p,q}(I)$. Choose a smooth curve of curves $\mathbb{R} \to \Omega_{p,q}(I) : s \mapsto \gamma_s$ that satisfies (4.1.7). Prove that the one-sided derivatives of the function $s \mapsto L(\gamma_s)$ exist at $s = 0$ and satisfy the inequalities

$$- \int_I |\dot{X}(t)| \, dt \leq \frac{d}{ds} L(\gamma_s) \bigg|_{s=0} \leq \int_I |\dot{X}(t)| \, dt.$$

Exercise 4.1.10 Let (Φ, γ, γ') be a development of M along M'. Show that γ is a geodesic in M if and only if γ' is a geodesic in M'.

4.2 Distance

Assume that $M \subset \mathbb{R}^n$ is a connected smooth m-dimensional submanifold. Two points $p, q \in M$ are of distance $|p - q|$ apart in the ambient Euclidean space $\mathbb{R}^n$. In this section we define a distance function which is more intimately tied to M by minimizing the length functional over the space of curves in M with fixed endpoints. Thus it may happen that two points in M have a very short distance in $\mathbb{R}^n$ but can only be joined by very long curves in M (see Fig. 4.1). This leads to the *intrinsic distance in M*. Throughout we denote by $I = [0, 1]$ the unit interval and, for $p, q \in M$, by

$$\Omega_{p,q} := \{\gamma : [0, 1] \to M \mid \gamma \text{ is smooth and } \gamma(0) = p, \; \gamma(1) = q\} \qquad (4.2.1)$$

the space of smooth paths on the unit interval joining p to q. Since M is connected the set $\Omega_{p,q}$ is nonempty for all $p, q \in M$. (Prove this!)

Definition 4.2.1 The **intrinsic distance** between two points $p, q \in M$ is the real number $d(p, q) \geq 0$ defined by

$$d(p, q) := \inf_{\gamma \in \Omega_{p,q}} L(\gamma). \qquad (4.2.2)$$

The inequality $d(p, q) \geq 0$ holds because each curve has nonnegative length and the inequality $d(p, q) < \infty$ holds because $\Omega_{p,q} \neq \emptyset$.

Fig. 4.1 Curves in M

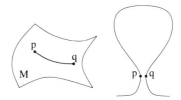

Remark 4.2.2 Every smooth curve $\gamma : [0, 1] \to \mathbb{R}^n$ with endpoints $\gamma(0) = p$ and $\gamma(1) = q$ satisfies the inequality

$$L(\gamma) = \int_0^1 |\dot{\gamma}(t)| \, dt \geq \left| \int_0^1 \dot{\gamma}(t) \, dt \right| = |p - q|.$$

Thus $d(p, q) \geq |p - q|$. For $\gamma(t) := p + t(q - p)$ we have equality and hence the straight lines minimize the length among all curves from p to q.

Lemma 4.2.3 *The function $d : M \times M \to [0, \infty)$ defines a metric on M:*

(i)   *If $p, q \in M$ satisfy $d(p, q) = 0$, then $p = q$.*
(ii) *For all $p, q \in M$ we have $d(p, q) = d(q, p)$.*
(iii) *For all $p, q, r \in M$ we have $d(p, r) \leq d(p, q) + d(q, r)$.*

Proof By Remark 4.2.2 we have $d(p, q) \geq |p - q|$ for all $p, q \in M$ and this proves part (i). Part (ii) follows from the fact that the curve $\widetilde{\gamma}(t) := \gamma(1 - t)$ has the same length as γ and belongs to $\Omega_{q,p}$ whenever $\gamma \in \Omega_{p,q}$. To prove part (iii) fix a constant $\varepsilon > 0$ and choose curves $\gamma_0 \in \Omega_{p,q}$ and $\gamma_1 \in \Omega_{q,r}$ such that $L(\gamma_0) < d(p, q) + \varepsilon$ and $L(\gamma_1) < d(q, r) + \varepsilon$. By Remark 4.1.3 we may assume without loss of generality that $\gamma_0(1 - t) = \gamma_1(t) = q$ for $t > 0$ sufficiently small. Under this assumption the curve

$$\gamma(t) := \begin{cases} \gamma_0(2t), & \text{for } 0 \leq t \leq 1/2, \\ \gamma_1(2t - 1), & \text{for } 1/2 < t \leq 1 \end{cases}$$

is smooth. Moreover, $\gamma(0) = p$ and $\gamma(1) = r$ and so $\gamma \in \Omega_{p,r}$. Thus

$$d(p, r) \leq L(\gamma) = L(\gamma_0) + L(\gamma_1) < d(p, q) + d(q, r) + 2\varepsilon.$$

Hence $d(p, r) < d(p, q) + d(q, r) + 2\varepsilon$ for every $\varepsilon > 0$. This proves part (iii) and Lemma 4.2.3.   □

Remark 4.2.4 It is natural to ask if the infimum in (4.2.2) is always attained. This is easily seen not to be the case in general. For example, let M result from the Euclidean space $\mathbb{R}^m$ by removing a point p_0. Then the distance $d(p, q) = |p - q|$ is equal to the length of the line segment from p to q and any other curve from p to q is longer. Hence if p_0 is in the interior of this line segment, the infimum is not attained. We shall prove below that the infimum is attained whenever M is complete.

Example 4.2.5 Let

$$M := S^2 = \left\{ p \in \mathbb{R}^3 \,\middle|\, |p| = 1 \right\}$$

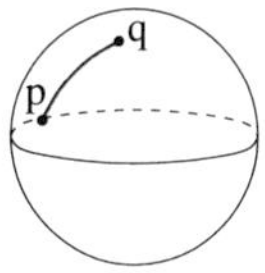

Fig. 4.2 A geodesic on the
2-sphere

be the unit sphere in $\mathbb{R}^3$ and fix two points $p, q \in S^2$. Then $d(p, q)$ is the length of
the shortest curve on the 2-sphere connecting p and q. Such a curve is a segment
on a great circle through p and q (see Fig. 4.2) and its length is

$$d(p, q) = \cos^{-1}(\langle p, q \rangle), \qquad (4.2.3)$$

where $\langle p, q \rangle$ denotes the standard inner product, and we have

$$0 \le d(p, q) \le \pi.$$

(See Example 4.3.11 below for details.) By Lemma 4.2.3 this defines a metric on S^2.
Exercise: Prove directly that (4.2.3) is a distance function on S^2.

We now have two topologies on our manifold $M \subset \mathbb{R}^n$, namely the topology
determined by the metric d in Lemma 4.2.3 and the relative topology inherited
from $\mathbb{R}^n$. The latter is also determined by a distance function, namely the *extrinsic
distance function* defined as the restriction of the Euclidean distance function on $\mathbb{R}^n$
to the subset M. We denote it by

$$d_0 : M \times M \to [0, \infty), \qquad d_0(p, q) := |p - q|. \qquad (4.2.4)$$

A natural question is if these two metrics d and d_0 induce the same topology on M.
In other words is a subset $U \subset M$ open with respect to d_0 if and only if it is open
with respect to d? Or, equivalently, does a sequence $p_\nu \in M$ converge to $p_0 \in M$
with respect to d if and only if it converges to p_0 with respect to d_0? Lemma 4.2.7
answers this question in the affirmative.

Exercise 4.2.6 Prove that every translation of $\mathbb{R}^n$ and every orthogonal transformation preserves the lengths of curves.

Lemma 4.2.7 *For every $p_0 \in M$ we have*

$$\lim_{p, q \to p_0} \frac{d(p, q)}{|p - q|} = 1.$$

Lemma 4.2.8 *Let $p_0 \in M$ and let $\phi_0 : U_0 \to \Omega_0$ be a coordinate chart onto an
open subset of $\mathbb{R}^m$ such that its derivative $d\phi_0(p_0) : T_{p_0}M \to \mathbb{R}^m$ is an orthogonal
transformation. Then*

$$\lim_{p, q \to p_0} \frac{d(p, q)}{|\phi_0(p) - \phi_0(q)|} = 1.$$

The proofs will be given below. The lemmas imply that the topology M inherits as a subset of $\mathbb{R}^m$, the topology on M determined by the metric d, and the topology on M induced by the local coordinate systems on M are all the same.

Corollary 4.2.9 *For every subset $U \subset M$ the following are equivalent.*

(i)   *U is open with respect to the metric d in (4.2.2).*
(ii)   *U is open with respect to the metric d_0 in (4.2.4).*
(iii) *For every coordinate chart $\phi_0 : U_0 \to \Omega_0$ of M onto an open subset $\Omega_0 \subset \mathbb{R}^m$ the set $\phi_0(U_0 \cap U)$ is an open subset of $\mathbb{R}^m$.*

Proof By Remark 4.2.2 we have

$$|p - q| \le d(p, q) \tag{4.2.5}$$

for all $p, q \in M$. Thus the identity $\mathrm{id}_M : (M, d) \to (M, d_0)$ is Lipschitz continuous with Lipschitz constant one and so every d_0-open subset of M is d-open. Conversely, let $U \subset M$ be a d-open subset of M and let $p_0 \in U$ and $\varepsilon > 0$. Then, by Lemma 4.2.7, there exists a constant $\delta > 0$ such that all $p, q \in M$ with $|p - p_0| < \delta$ and $|q - p_0| < \delta$ satisfy

$$d(p, q) \le (1 + \varepsilon)|p - q|.$$

Since U is d-open, there exists a constant $\rho > 0$ such that

$$B_\rho(p_0, d) \subset U.$$

With

$$\rho_0 := \min\left\{\delta, \frac{\rho}{1 + \varepsilon}\right\}$$

this implies $B_{\rho_0}(p_0, d_0) \subset U$. Namely, if $p \in M$ satisfies

$$|p - p_0| < \rho_0 \le \delta,$$

then

$$d(p, p_0) \le (1 + \varepsilon)|p - p_0| < (1 + \varepsilon)\rho_0 \le \rho$$

and so $p \in U$. Thus U is d_0-open and this proves that (i) is equivalent to (ii).

That (ii) implies (iii) follows from the fact that each coordinate chart ϕ_0 is a homeomorphism. To prove that (iii) implies (i), we argue indirectly and assume that U is not d-open. Then there exists a sequence $p_\nu \in M \setminus U$ that converges to an element $p_0 \in U$. Let $\phi_0 : U_0 \to \Omega_0$ be a coordinate chart with $p_0 \in U_0$. Then $\lim_{\nu \to \infty} |\phi_0(p_\nu) - \phi_0(p_0)| = 0$ by Lemma 4.2.8. Thus $\phi_0(U_0 \cap U)$ is not open and so U does not satisfy (iii). This proves Corollary 4.2.9. $\square$

Fig. 4.3 Locally, M is the graph of f

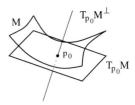

Proof of Lemma 4.2.7 By Remark 4.2.2 the estimate $|p - q| \le d(p, q)$ holds for all $p, q \in M$. The lemma asserts that, for all $p_0 \in M$ and all $\varepsilon > 0$, there exists a d_0-open neighborhood $U_0 \subset M$ of p_0 such that all $p, q \in U_0$ satisfy

$$|p - q| \le d(p, q) \le (1 + \varepsilon)|p - q|. \tag{4.2.6}$$

Let $p_0 \in M$ and $\varepsilon > 0$, and define $x : \mathbb{R}^n \to T_{p_0} M$ and $y : \mathbb{R}^n \to T_{p_0} M^\perp$ by

$$x(p) := \Pi(p_0)(p - p_0), \qquad y(p) := (\mathbb{1} - \Pi(p_0))(p - p_0),$$

where $\Pi(p_0) : \mathbb{R}^n \to T_{p_0} M$ denotes the orthogonal projection as usual. Then the derivative of the map $x|_M : M \to T_{p_0} M$ at $p = p_0$ is the identity on $T_{p_0} M$. Hence the Inverse Function Theorem 2.2.17 asserts that the map $x|_M : M \to T_{p_0} M$ is locally invertible near p_0. Extending this inverse to a smooth map from $T_{p_0} M$ to $\mathbb{R}^n$ and composing it with the map $y : M \to T_{p_0} M^\perp$, we obtain a smooth map

$$f : T_{p_0} M \to T_{p_0} M^\perp$$

and an open neighborhood $W \subset \mathbb{R}^n$ of p_0 such that

$$p \in M \qquad \Longleftrightarrow \qquad y(p) = f(x(p))$$

for all $p \in W$ (see Fig. 4.3). Moreover, by definition the map f satisfies

$$f(0) = 0 \in T_{p_0} M^\perp, \qquad df(0) = 0 : T_{p_0} M \to T_{p_0} M^\perp.$$

Hence there exists a constant $\delta > 0$ such that, for every $x \in T_{p_0} M$, we have

$$|x| < \delta \qquad \Longrightarrow \qquad x + f(x) \in W \text{ and } \|df(x)\| = \sup_{0 \ne \widehat{x} \in T_{p_0} M} \frac{|df(x)\widehat{x}|}{|\widehat{x}|} < \varepsilon.$$

Define

$$U_0 := \{p \in M \cap W \mid |x(p)| < \delta\}.$$

Given $p, q \in U_0$ let $\gamma : [0, 1] \to M$ be the curve whose projection to the x-axis is the straight line joining $x(p)$ to $x(q)$, i.e.

$$x(\gamma(t)) = x(p) + t(x(q) - x(p)) =: x(t),$$
$$y(\gamma(t)) = f(x(\gamma(t))) = f(x(t)) =: y(t).$$

Then $\gamma(t) \in U_0$ for all $t \in [0, 1]$ and

$$
\begin{aligned}
L(\gamma) &= \int_0^t |\dot{x}(t) + \dot{y}(t)| \, dt \\
&= \int_0^t |\dot{x}(t) + df(x(t))\dot{x}(t)| \, dt \\
&\leq \int_0^t \left(1 + \|df(x(t))\|\right)|\dot{x}(t)| \, dt \\
&\leq (1 + \varepsilon) \int_0^t |\dot{x}(t)| \, dt \\
&= (1 + \varepsilon)|x(p) - x(q)| \\
&= (1 + \varepsilon)|\Pi(p_0)(p - q)| \\
&\leq (1 + \varepsilon)|p - q|.
\end{aligned}
$$

Hence $d(p, q) \leq L(\gamma) \leq (1 + \varepsilon)|p - q|$ and this proves Lemma 4.2.7.   □

Proof of Lemma 4.2.8 By assumption we have

$$
|d\phi_0(p_0)v| = |v|
$$

for all $v \in T_{p_0}M$. Fix a constant $\varepsilon > 0$. Then, by continuity of the derivative, there exists a d_0-open neighborhood $M_0 \subset M$ of p_0 such that for all $p \in M_0$ and all $v \in T_p M$ we have

$$
(1 - \varepsilon)|d\phi_0(p)v| \leq |v| \leq (1 + \varepsilon)|d\phi_0(p)v|.
$$

Thus for every curve $\gamma : [0, 1] \to M_0$ we have

$$
(1 - \varepsilon)L(\phi_0 \circ \gamma)) \leq L(\gamma) \leq (1 + \varepsilon)L(\phi_0 \circ \gamma).
$$

One is tempted to take the infimum over all curves $\gamma : [0, 1] \to M_0$ joining two pints $p, q \in M_0$ to obtain the inequality

$$
(1 - \varepsilon)|\phi_0(p) - \phi_0(q)| \leq d(p, q) \leq (1 + \varepsilon)|\phi_0(p) - \phi_0(q)|. \tag{4.2.7}
$$

However, we must justify these inequalities by showing that the infimum over all curves in M_0 agrees with the infimum over all curves in M joining the points p and q.

It suffices to show that the inequalities hold on a smaller neighborhood $M_1 \subset M_0$ of p_0. Choose such a smaller neighborhood M_1 such that the open set $\phi_0(M_1)$ is a convex subset of Ω_0. Then the right inequality in (4.2.7) follows by taking the curve $\gamma : [0, 1] \to M_1$ from $\gamma(0) = p$ to $\gamma(1) = q$ such that $\phi_0 \circ \gamma : [0, 1] \to \phi_0(M_1)$ is a straight line. To prove the left inequality in (4.2.7) we use the fact that M_0 is d-open by Lemma 4.2.7. Hence, after shrinking M_1 if necessary, there exists a constant $r > 0$ such that

$$p_0 \in M_1 \subset B_r(p_0, d) \subset B_{3r}(p_0, d) \subset M_0.$$

Then, for $p, q \in M_1$ we have $d(p, q) \leq 2r$ while $L(\gamma) \geq 4r$ for any curve γ from p to q which leaves M_0. Hence the distance $d(p, q)$ of $p, q \in M_1$ is the infimum of the lengths $L(\gamma)$ over all curves $\gamma : [0, 1] \to M_0$ that join $\gamma(0) = p$ to $\gamma(1) = q$. This proves the left inequality in (4.2.7) and Lemma 4.2.8. $\qquad\square$

A next question one might ask is the following. Can we choose a coordinate chart $\phi : U \to \Omega$ on M with values in an open set $\Omega \subset \mathbb{R}^m$ so that the length of each smooth curve $\gamma : [0, 1] \to U$ is equal to the length of the curve $c := \phi \circ \gamma : [0, 1] \to \Omega$? We examine this question by considering the inverse map $\psi := \phi^{-1} : \Omega \to U$. Denote the components of x and $\psi(x)$ by

$$x = (x^1, \ldots, x^m) \in \Omega, \qquad \psi(x) = (\psi^1(x), \ldots, \psi^n(x)) \in U.$$

Given a smooth curve $[0, 1] \to \Omega : t \mapsto c(t) = (c^1(t), \ldots, c^m(t))$ we can write the length of the composition $\gamma = \psi \circ c : [0, 1] \to M$ in the form

$$
\begin{aligned}
L(\psi \circ c) &= \int_0^1 \left| \frac{d}{dt} \psi(c(t)) \right| dt \\
&= \int_0^1 \sqrt{\sum_{v=1}^n \left(\frac{d}{dt} \psi^v(c(t)) \right)^2} \, dt \\
&= \int_0^1 \sqrt{\sum_{v=1}^n \left(\sum_{i=1}^m \frac{\partial \psi^v}{\partial x^i}(c(t)) \dot{c}^i(t) \right)^2} \, dt \\
&= \int_0^1 \sqrt{\sum_{v=1}^n \sum_{i,j=1}^m \frac{\partial \psi^v}{\partial x^i}(c(t)) \frac{\partial \psi^v}{\partial x^j}(c(t)) \dot{c}^i(t) \dot{c}^j(t)} \, dt \\
&= \int_0^1 \sqrt{\sum_{i,j=1}^m \dot{c}^i(t) g_{ij}(c(t)) \dot{c}^j(t)} \, dt.
\end{aligned}
$$

Here the functions $g_{ij} : \Omega \to \mathbb{R}$ are defined by

$$g_{ij}(x) := \sum_{v=1}^n \frac{\partial \psi^v}{\partial x^i}(x) \frac{\partial \psi^v}{\partial x^j}(x) = \left\langle \frac{\partial \psi}{\partial x^i}(x), \frac{\partial \psi}{\partial x^j}(x) \right\rangle. \qquad (4.2.8)$$

Fig. 4.4 A spherical triangle

Thus we have a smooth function $g = (g_{ij}) : \Omega \to \mathbb{R}^{m \times m}$ with values in the positive definite matrices given by $g(x) = d\psi(x)^{\mathsf{T}} d\psi(x)$ such that

$$L(\psi \circ c) = \int_0^1 \sqrt{\dot{c}(t)^{\mathsf{T}} g(c(t)) \dot{c}(t)} \, dt \qquad (4.2.9)$$

for every smooth curve $c : [0, 1] \to \Omega$. Thus the condition $L(\psi \circ c) = L(c)$ for every such curve is equivalent to

$$g_{ij}(x) = \delta_{ij}$$

for all $x \in \Omega$ or, equivalently,

$$d\psi(x)^{\mathsf{T}} d\psi(x) = \mathbb{1}. \qquad (4.2.10)$$

This means that ψ preserves angles and areas. The next example shows that for $M = S^2$ it is impossible to find such coordinates.

Example 4.2.10 Consider the manifold $M = S^2$. If there is a diffeomorphism $\psi : \Omega \to U$ from an open set $\Omega \subset \mathbb{R}^2$ onto an open set $U \subset S^2$ that satisfies (4.2.10), it has to map straight lines onto arcs of great circles and it preserves the area. However, the area A of a spherical triangle bounded by three arcs on great circles satisfies the angle sum formula

$$\alpha + \beta + \gamma = \pi + A.$$

(See Fig. 4.4.) Hence there can be no such map ψ.

4.3 The Exponential Map

Geodesics give rise to a flow on the tangent bundle, the *geodesic flow*. It is generated by a vector field on the tangent bundle, called the *geodesic spray*. The time-1-map of the geodesic flow gives rise to the *exponential map*.

4.3.1 Geodesic Spray

The tangent bundle TM is a smooth $2m$-dimensional manifold in $\mathbb{R}^n \times \mathbb{R}^n$ by Corollary 2.6.12. The next lemma characterizes the tangent bundle of the tangent bundle. Compare this with Lemma 3.4.5.

Lemma 4.3.1 *The tangent space of TM at $(p, v) \in TM$ is given by*

$$T_{(p,v)}TM = \left\{ (\widehat{p}, \widehat{v}) \in \mathbb{R}^n \times \mathbb{R}^n \, \middle| \, \begin{array}{l} \widehat{p} \in T_pM \text{ and} \\ (\mathbb{1} - \Pi(p))\widehat{v} = h_p(\widehat{p}, v) \end{array} \right\}. \qquad (4.3.1)$$

Proof We prove the inclusion "$\subset$" in (4.3.1). Let $(\widehat{p}, \widehat{v}) \in T_{(p,v)}TM$ and choose a smooth curve $\mathbb{R} \to TM : t \mapsto (\gamma(t), X(t))$ such that

$$\gamma(0) = p, \qquad X(0) = v, \qquad \dot{\gamma}(0) = \widehat{p}, \qquad \dot{X}(0) = \widehat{v}.$$

Then $\dot{X} = \nabla X + h_\gamma(\dot{\gamma}, X)$ by the Gauß–Weingarten formula (3.2.2) and hence $(\mathbb{1} - \Pi(\gamma(t)))\dot{X}(t) = h_{\gamma(t)}(\dot{\gamma}(t), X(t))$ for all $t \in \mathbb{R}$. Take $t = 0$ to obtain $(\mathbb{1} - \Pi(p))\widehat{v} = h_p(\widehat{p}, v)$. This proves the inclusion "$\subset$" in (4.3.1). Equality holds because both sides of the equation are $2m$-dimensional linear subspaces of $\mathbb{R}^n \times \mathbb{R}^n$. $\qquad \square$

By Lemma 4.3.1 a smooth map $S = (S_1, S_2) : TM \to \mathbb{R}^n \times \mathbb{R}^n$ is a vector field on TM if and only if

$$S_1(p, v) \in T_pM, \qquad (\mathbb{1} - \Pi(p))S_2(p, v) = h_p(S_1(p, v), v)$$

for all $(p, v) \in TM$. A special case is where $S_1(p, v) = v$. Such vector fields correspond to second order differential equations on M.

Definition 4.3.2 (Spray) A vector field $S \in \text{Vect}(TM)$ is called a **spray** iff it has the form $S(p, v) = (v, S_2(p, v))$ where $S_2 : TM \to \mathbb{R}^n$ is a smooth map satisfying

$$(\mathbb{1} - \Pi(p))S_2(p, v) = h_p(v, v), \qquad S_2(p, \lambda v) = \lambda^2 S_2(p, v) \qquad (4.3.2)$$

for all $(p, v) \in TM$ and $\lambda \in \mathbb{R}$. The vector field $S \in \text{Vect}(TM)$ defined by

$$S(p, v) := (v, h_p(v, v)) \quad \in \quad T_{(p,v)}TM \qquad (4.3.3)$$

for $p \in M$ and $v \in T_pM$ is called the **geodesic spray**.

4.3.2 The Exponential Map

Lemma 4.3.3 *Let $\gamma : I \to M$ be a smooth curve on an open interval $I \subset \mathbb{R}$. Then γ is a geodesic if and only if the curve $I \to TM : t \mapsto (\gamma(t), \dot{\gamma}(t))$ is an integral curve of the geodesic spray S in (4.3.3).*

Proof A smooth curve $I \to TM : t \mapsto (\gamma(t), X(t))$ is an integral curve of S if and only if $\dot{\gamma}(t) = X(t)$ and $\dot{X}(t) = h_{\gamma(t)}(X(t), X(t))$ for all $t \in I$. By equation (4.1.5), this holds if and only if γ is a geodesic and $\dot{\gamma} = X$. $\qquad \square$

Combining Lemma 4.3.3 with Theorem 2.4.7 we obtain the following existence and uniqueness result for geodesics.

Lemma 4.3.4 *Let $M \subset \mathbb{R}^n$ be an m-dimensional submanifold.*

(i) *For every $p \in M$ and every $v \in T_p M$ there is an $\varepsilon > 0$ and a smooth curve $\gamma : (-\varepsilon, \varepsilon) \to M$ such that*

$$\nabla \dot{\gamma} \equiv 0, \qquad \gamma(0) = p, \qquad \dot{\gamma}(0) = v. \tag{4.3.4}$$

(ii) *If $\gamma_1 : I_1 \to M$ and $\gamma_2 : I_2 \to M$ are geodesics and $t_0 \in I_1 \cap I_2$ with*

$$\gamma_1(t_0) = \gamma_2(t_0), \qquad \dot{\gamma}_1(t_0) = \dot{\gamma}_2(t_0),$$

then $\gamma_1(t) = \gamma_2(t)$ for all $t \in I_1 \cap I_2$.

Proof Lemma 4.3.3 and Theorem 2.4.7. $\square$

Definition 4.3.5 (Exponential map) For $p \in M$ and $v \in T_p M$ the interval

$$I_{p,v} := \bigcup \left\{ I \subset \mathbb{R} \,\middle|\, \begin{array}{l} I \text{ is an open interval containing } 0 \text{ and there is a} \\ \text{geodesic } \gamma : I \to M \text{ satisfying } \gamma(0) = p, \; \dot{\gamma}(0) = v \end{array} \right\}.$$

is called the **maximal existence interval** for the geodesic through p in the direction v. For $p \in M$ define the set $V_p \subset T_p M$ by

$$V_p := \{ v \in T_p M \mid 1 \in I_{p,v} \}. \tag{4.3.5}$$

The **exponential map** at p is the map $\exp_p : V_p \to M$ that assigns to every tangent vector $v \in V_p$ the point $\exp_p(v) := \gamma(1)$, where $\gamma : I_{p,v} \to M$ is the unique geodesic satisfying $\gamma(0) = p$ and $\dot{\gamma}(0) = v$.

Lemma 4.3.6

(i) *The set*

$$V := \{ (p, v) \mid p \in M, \; v \in V_p \} \subset TM$$

is open and the map $V \to M : (p, v) \mapsto \exp_p(v)$ is smooth.

(ii) *If $p \in M$ and $v \in V_p$, then*

$$I_{p,v} = \{ t \in \mathbb{R} \mid tv \in V_p \}$$

and the geodesic $\gamma : I_{p,v} \to M$ with $\gamma(0) = p$ and $\dot{\gamma}(0) = v$ is given by

$$\gamma(t) = \exp_p(tv), \qquad t \in I_{p,v}.$$

Fig. 4.5 The exponential map

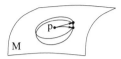

Proof   Part (i) follows directly from Lemma 4.3.3 and Theorem 2.4.9. To prove part (ii), fix an element $p \in M$ and a tangent vector $v \in V_p$, and let $\gamma : I_{p,v} \to M$ be the unique geodesic with $\gamma(0) = p$ and $\dot{\gamma}(0) = v$. Fix a nonzero real number λ and define the map $\gamma_\lambda : \lambda^{-1} I_{p,v} \to M$ by

$$\gamma_\lambda(t) := \gamma(\lambda t) \qquad \text{for } t \in \lambda^{-1} I_{p,v}.$$

Then $\dot{\gamma}_\lambda(t) = \lambda \dot{\gamma}(\lambda t)$ and $\ddot{\gamma}_\lambda(t) = \lambda^2 \ddot{\gamma}(\lambda t)$ and hence

$$\nabla \dot{\gamma}_\lambda(t) = \Pi(\gamma_\lambda(t)) \ddot{\gamma}_\lambda(t) = \lambda^2 \Pi(\gamma(\lambda t)) \ddot{\gamma}(\lambda t) = \lambda^2 \nabla \dot{\gamma}(\lambda t) = 0$$

for every $t \in \lambda^{-1} I_{p,v}$. This shows that γ_λ is a geodesic with

$$\gamma_\lambda(0) = p, \qquad \dot{\gamma}_\lambda(0) = \lambda v.$$

In particular, we have $\lambda^{-1} I_{p,v} \subset I_{p,\lambda v}$. Interchanging the roles of v and λv we obtain $\lambda^{-1} I_{p,v} = I_{p,\lambda v}$. Thus

$$\lambda \in I_{p,v} \qquad \Longleftrightarrow \qquad 1 \in I_{p,\lambda v} \qquad \Longleftrightarrow \qquad \lambda v \in V_p$$

and

$$\gamma(\lambda) = \gamma_\lambda(1) = \exp_p(\lambda v)$$

for $\lambda \in I_{p,v}$. This proves Lemma 4.3.6. $\square$

Since $\exp_p(0) = p$ by definition, the derivative of the exponential map at $v = 0$ is a linear map from $T_p M$ to itself. This derivative is the identity map as illustrated in Fig. 4.5 and proved in the following corollary.

Corollary 4.3.7   *The map* $\exp_p : V_p \to M$ *is smooth and its derivative at the origin is* $d \exp_p(0) = \mathrm{id} : T_p M \to T_p M$.

Proof   The set V_p is an open subset of the linear subspace $T_p M \subset \mathbb{R}^n$, with respect to the relative topology, and hence is a manifold. The tangent space of V_p at each point is $T_p M$. By Lemma 4.3.6 the exponential map $\exp_p : V_p \to M$ is smooth and its derivative at the origin is given by

$$d \exp_p(0) v = \left. \frac{d}{dt} \right|_{t=0} \exp_p(t v) = \dot{\gamma}(0) = v,$$

where $\gamma : I_{p,v} \to M$ is once again the unique geodesic through p in the direction v. This proves Corollary 4.3.7. $\square$

Corollary 4.3.8 *Let $p \in M$ and, for $r > 0$, denote*

$$B_r(p) := \{v \in T_p M \mid |v| < r\}.$$

If $r > 0$ is sufficiently small, then $B_r(p) \subset V_p$, the set

$$U_r(p) := \exp_p(B_r(p))$$

is an open subset of M, and the restriction of the exponential map to $B_r(p)$ is a diffeomorphism from $B_r(p)$ to $U_r(p)$.

Proof This follows directly from Corollary 4.3.7 and Theorem 2.2.17.   □

Definition 4.3.9 (Injectivity radius)   Let $M \subset \mathbb{R}^n$ be a smooth m-manifold. The **injectivity radius** of M at p is the supremum of all real numbers $r > 0$ such that $B_r(p) \subset V_p$ and the restriction of the exponential map $\exp_p$ to $B_r(p)$ is a diffeomorphism onto its image

$$U_r(p) := \exp_p(B_r(p)).$$

It will be denoted by

$$\mathrm{inj}(p) := \mathrm{inj}(p; M) := \sup\left\{ r > 0 \,\middle|\, \begin{array}{l} B_r(p) \subset V_p \text{ and} \\ \exp_p : B_r(p) \to U_r(p) \\ \text{is a diffeomorphism} \end{array} \right\}.$$

The **injectivity radius of** M is the infimum of the injectivity radii of M at p over all $p \in M$. It will be denoted by

$$\mathrm{inj}(M) := \inf_{p \in M} \mathrm{inj}(p; M).$$

4.3.3   Examples and Exercises

Example 4.3.10   The exponential map on $\mathbb{R}^m$ is given by

$$\exp_p(v) = p + v \qquad \text{for } p, v \in \mathbb{R}^m.$$

For every $p \in \mathbb{R}^m$ this map is a diffeomorphism from $T_p \mathbb{R}^m = \mathbb{R}^m$ to $\mathbb{R}^m$ and hence the injectivity radius of $\mathbb{R}^m$ is infinity.

Example 4.3.11 The exponential map on S^m is given by

$$\exp_p(v) = \cos(|v|)p + \frac{\sin(|v|)}{|v|}v$$

for every $p \in S^m$ and every nonzero tangent vector $v \in T_pS^m = p^\perp$. The restriction of this map to the open ball of radius r in T_pM is a diffeomorphism onto its image if and only if $r \leq \pi$. Hence the injectivity radius of S^m at every point is π.

 Exercise: Given $p \in S^m$ and $0 \neq v \in T_pS^m = p^\perp$, prove that the geodesic $\gamma : \mathbb{R} \to S^m$ with $\gamma(0) = p$ and $\dot{\gamma}(0) = v$ is given by

$$\gamma(t) = \cos(t|v|)p + \frac{\sin(t|v|)}{|v|}v$$

for $t \in \mathbb{R}$. Show that in the case $0 \leq |v| \leq \pi$ there is no shorter curve in S^m connecting p and $q := \gamma(1)$ and deduce that the intrinsic distance on S^m is given by

$$d(p,q) = \cos^{-1}(\langle p, q \rangle)$$

for $p, q \in S^m$ (see Example 4.2.5 for $m = 2$).

Example 4.3.12 Consider the orthogonal group $O(n) \subset \mathbb{R}^{n \times n}$ with the standard inner product $\langle v, w \rangle := \text{trace}(v^T w)$ on $\mathbb{R}^{n \times n}$. The orthogonal projection $\Pi(g) : \mathbb{R}^{n \times n} \to T_gO(n)$ is given by

$$\Pi(g)v := \frac{1}{2}(v - gv^Tg)$$

and the second fundamental form by

$$h_g(v, v) = -gv^Tv.$$

Hence a curve $\gamma : \mathbb{R} \to O(n)$ is a geodesic if and only if $\gamma^T\ddot{\gamma} + \dot{\gamma}^T\dot{\gamma} = 0$ or, equivalently, $\gamma^T\dot{\gamma}$ is constant. This shows that geodesics in $O(n)$ have the form $\gamma(t) = g \exp(t\xi)$ for $g \in O(n)$ and $\xi \in \mathfrak{o}(n)$. It follows that

$$\exp_g(v) = g \exp(g^{-1}v) = \exp(vg^{-1})g$$

for $g \in O(n)$ and $v \in T_gO(n)$. In particular, for $g = 1$ the exponential map $\exp_1 : \mathfrak{o}(n) \to O(n)$ agrees with the exponential matrix.

Exercise 4.3.13 What is the injectivity radius of the 2-torus $\mathbb{T}^2 = S^1 \times S^1$, the punctured 2-plane $\mathbb{R}^2 \setminus \{(0, 0)\}$, and the orthogonal group $O(n)$?

4.3.4 Geodesics in Local Coordinates

Lemma 4.3.14 *Let $M \subset \mathbb{R}^n$ be an m-dimensional manifold and choose a coordinate chart $\phi : U \to \Omega$ with inverse*

$$\psi := \phi^{-1} : \Omega \to U.$$

Let $\Gamma_{ij}^k : \Omega \to \mathbb{R}$ be the Christoffel symbols defined by (3.6.6) and let $c : I \to \Omega$ be a smooth curve. Then the curve

$$\gamma := \psi \circ c : I \to M$$

is a geodesic if and only if c satisfies the 2nd order differential equation

$$\ddot{c}^k + \sum_{i,j=1}^{m} \Gamma_{ij}^k(c)\dot{c}^i \dot{c}^j = 0 \tag{4.3.6}$$

for $k = 1, \ldots, m$.

Proof This follows immediately from the definition of geodesics and equation (3.6.7) in Lemma 3.6.1 with $X = \dot{\gamma}$ and $\xi = \dot{c}$.   □

We remark that Lemma 4.3.14 gives rise to another proof of Lemma 4.3.4 that is based on the existence and uniqueness of solutions of second order differential equations in local coordinates.

Exercise 4.3.15 Let $\Omega \subset \mathbb{R}^m$ be an open set and $g = (g_{ij}) : \Omega \to \mathbb{R}^{m \times m}$ be a smooth map with values in the space of positive definite symmetric matrices. Consider the energy functional

$$E(c) := \int_0^1 L(c(t), \dot{c}(t))\, dt$$

on the space of paths $c : [0, 1] \to \Omega$, where $L : \Omega \times \mathbb{R}^m \to \mathbb{R}$ is defined by

$$L(x, \xi) := \frac{1}{2} \sum_{i,j=1}^{m} \xi^i g_{ij}(x)\xi^j. \tag{4.3.7}$$

The **Euler–Lagrange equations** of this variational problem have the form

$$\frac{d}{dt} \frac{\partial L}{\partial \xi^k}(c(t), \dot{c}(t)) = \frac{\partial L}{\partial x^k}(c(t), \dot{c}(t)), \qquad k = 1, \ldots, m. \tag{4.3.8}$$

Prove that the Euler–Lagrange equations (4.3.8) are equivalent to the geodesic equations (4.3.6), where the $\Gamma_{ij}^k : \Omega \to \mathbb{R}$ are given by (3.6.10).

4.4 Minimal Geodesics

Any straight line segment in Euclidean space is the shortest curve joining its end-points. The analogous assertion for geodesics in a manifold M is false; consider for example an arc which is more than half of a great circle on a sphere. In this section we consider curves which realize the shortest distance between their endpoints.

4.4.1 Characterization of Minimal Geodesics

Lemma 4.4.1 *Let $I = [a, b]$ be a compact interval, let $\gamma : I \to M$ be a smooth curve, and define $p := \gamma(a)$ and $q := \gamma(b)$. Then the following are equivalent.*

(i) *γ is parametrized proportional to the arclength, i.e. $|\dot{\gamma}(t)| = c$ is constant, and γ minimizes the length, i.e. $L(\gamma) \leq L(\gamma')$ for every smooth curve γ' in M joining p and q.*

(ii) *γ minimizes the energy, i.e. $E(\gamma) \leq E(\gamma')$ for every smooth curve γ' in M joining p and q.*

Definition 4.4.2 (Minimal geodesic) A smooth curve $\gamma : I \to M$ on a compact interval $I \subset \mathbb{R}$ is called a **minimal geodesic** iff it satisfies the equivalent conditions of Lemma 4.4.1.

Remark 4.4.3

(i) Condition (i) says that (the velocity $|\dot{\gamma}|$ is constant and) $L(\gamma) = d(p, q)$, i.e. that γ is a shortest curve from p to q. It is not precluded that there be more than one such γ; consider for example the case where M is a sphere and p and q are antipodal.

(ii) Condition (ii) implies that

$$\frac{d}{ds}\Big|_{s=0} E(\gamma_s) = 0$$

for every smooth variation $\mathbb{R} \times I \to M : s \mapsto \gamma_s(t)$ of γ with fixed endpoints. Hence a minimal geodesic is a geodesic.

(iii) Finally, we remark that $L(\gamma)$ (but not $E(\gamma)$) is independent of the parametrization of γ. Hence, if γ is a minimal geodesic, then $L(\gamma) \leq L(\gamma')$ for every γ' (from p to q) whereas $E(\gamma) \leq E(\gamma')$ for those γ' defined on (an interval the same length as) I.

Proof of Lemma 4.4.1 We prove that (i) implies (ii). Let c be the (constant) value of $|\dot{\gamma}(t)|$. Then

$$L(\gamma) = (b - a)c, \qquad E(\gamma) = \frac{(b - a)c^2}{2}.$$

Then, for every smooth curve $\gamma' : I \to M$ with $\gamma'(a) = p$ and $\gamma'(b) = q$, we have

$$
\begin{aligned}
4E(\gamma)^2 &= c^2 L(\gamma)^2 \\
&\le c^2 L(\gamma')^2 \\
&= c^2 \left(\int_a^b |\dot{\gamma}'(t)| \, dt \right)^2 \\
&\le c^2 (b-a) \int_a^b |\dot{\gamma}'(t)|^2 \, dt \\
&= 2(b-a)c^2 E(\gamma') \\
&= 4E(\gamma)E(\gamma').
\end{aligned}
$$

Here the fourth step follows from the Cauchy–Schwarz inequality. Now divide by $4E(\gamma)$ to obtain $E(\gamma) \le E(\gamma')$.

We prove that (ii) implies (i). We have already shown in Remark 4.4.3 that (ii) implies that γ is a geodesic. It is easy to dispose of the case where M is one-dimensional. In that case any γ minimizing $E(\gamma)$ or $L(\gamma)$ must be monotonic onto a subarc; otherwise it could be altered so as to make the integral smaller. Hence suppose M is of dimension at least two. Suppose, by contradiction, that $L(\gamma') < L(\gamma)$ for some curve γ' from p to q. Since the dimension of M is bigger than one, we may approximate γ' by a curve whose tangent vector nowhere vanishes, i.e. we may assume without loss of generality that $\dot{\gamma}'(t) \ne 0$ for all t. Then we can reparametrize γ' proportional to arclength without changing its length, and by a further transformation we can make its domain equal to I. Thus we may assume without loss of generality that $\gamma' : I \to M$ is a smooth curve with $\gamma'(a) = p$ and $\gamma'(b) = q$ such that $|\dot{\gamma}'(t)| = c'$ and

$$(b-a)c' = L(\gamma') < L(\gamma) = (b-a)c.$$

This implies $c' < c$ and hence

$$E(\gamma') = \frac{(b-a)c'^2}{2} < \frac{(b-a)c^2}{2} = E(\gamma).$$

This contradicts (ii) and proves Lemma 4.4.1.   □

4.4.2 Local Existence of Minimal Geodesics

The next theorem asserts the existence of minimal geodesics joining two points that are sufficiently close to each other. It also shows that the set $U_r(p) = \exp_p(B_r(p))$ that was introduced in Definition 4.3.9 is actually the open ball $U_r(p) = \{q \in M \mid d(p,q) < r\}$ whenever $r \le \mathrm{inj}(p; M)$.

Fig. 4.6 The Gauß Lemma

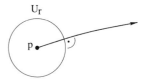

U_r

p

Theorem 4.4.4 (Existence of minimal geodesics) *Let $M \subset \mathbb{R}^n$ be a smooth m-manifold, fix a point $p \in M$, and let $r > 0$ be smaller than the injectivity radius of M at p. Let $v \in T_p M$ such that $|v| < r$. Then*

$$d(p, q) = |v|, \qquad q := \exp_p(v),$$

and a curve $\gamma \in \Omega_{p,q}$ has minimal length $L(\gamma) = |v|$ if and only if there is a smooth map $\beta : [0, 1] \to [0, 1]$ satisfying

$$\beta(0) = 0, \qquad \beta(1) = 1, \qquad \dot{\beta} \geq 0$$

such that $\gamma(t) = \exp_p(\beta(t)v)$ for $0 \leq t \leq 1$.

The proof is based on the following lemma.

Lemma 4.4.5 (Gauß Lemma) *Let M, p, r be as in Theorem 4.4.4, let $I \subset \mathbb{R}$ be an open interval, and let $w : I \to V_p$ be a smooth curve whose norm*

$$|w(t)| =: r$$

is constant. Define

$$\alpha(s, t) := \exp_p(sw(t))$$

for $(s, t) \in \mathbb{R} \times I$ with $sw(t) \in V_p$. Then

$$\left\langle \frac{\partial \alpha}{\partial s}, \frac{\partial \alpha}{\partial t} \right\rangle \equiv 0.$$

Thus the geodesics through p are orthogonal to the boundaries of the embedded balls $U_r(p)$ in Corollary 4.3.8 (see Fig. 4.6).

Proof of Lemma 4.4.5 For every $t \in I$ we have

$$\alpha(0, t) = \exp_p(0) = p$$

and so the assertion holds for $s = 0$, i.e.

$$\left\langle \frac{\partial \alpha}{\partial s}(0, t), \frac{\partial \alpha}{\partial t}(0, t) \right\rangle = 0.$$

Moreover, each curve $s \mapsto \alpha(s, t)$ is a geodesic, i.e.

$$\nabla_s \frac{\partial \alpha}{\partial s} = \Pi(\alpha) \frac{\partial^2 \alpha}{\partial s^2} \equiv 0.$$

By Theorem 4.1.4, the function

$$s \mapsto \left| \frac{\partial \alpha}{\partial s}(s, t) \right|$$

is constant for every t, so that

$$\left| \frac{\partial \alpha}{\partial s}(s, t) \right| = \left| \frac{\partial \alpha}{\partial s}(0, t) \right| = |w(t)| = r \qquad \text{for } (s, t) \in \mathbb{R} \times I.$$

This implies

$$\frac{\partial}{\partial s} \left\langle \frac{\partial \alpha}{\partial s}, \frac{\partial \alpha}{\partial t} \right\rangle = \left\langle \nabla_s \frac{\partial \alpha}{\partial s}, \frac{\partial \alpha}{\partial t} \right\rangle + \left\langle \frac{\partial \alpha}{\partial s}, \nabla_s \frac{\partial \alpha}{\partial t} \right\rangle$$

$$= \left\langle \frac{\partial \alpha}{\partial s}, \Pi(\alpha) \frac{\partial^2 \alpha}{\partial s \partial t} \right\rangle$$

$$= \left\langle \Pi(\alpha) \frac{\partial \alpha}{\partial s}, \frac{\partial^2 \alpha}{\partial s \partial t} \right\rangle$$

$$= \left\langle \frac{\partial \alpha}{\partial s}, \frac{\partial^2 \alpha}{\partial s \partial t} \right\rangle$$

$$= \frac{1}{2} \frac{\partial}{\partial t} \left| \frac{\partial \alpha}{\partial s} \right|^2$$

$$= 0.$$

Since the function $\langle \frac{\partial \alpha}{\partial s}, \frac{\partial \alpha}{\partial t} \rangle$ vanishes for $s = 0$ we obtain

$$\left\langle \frac{\partial \alpha}{\partial s}(s, t), \frac{\partial \alpha}{\partial t}(s, t) \right\rangle = 0$$

for all s and t. This proves Lemma 4.4.5.   □

Proof of Theorem 4.4.4 Let $r > 0$ be as in Corollary 4.3.8 and let $v \in T_p M$ such that $0 < |v| =: \varepsilon < r$. Denote $q := \exp_p(v)$ and let $\gamma \in \Omega_{p,q}$. Assume first that

$$\gamma(t) \in \exp_p\left(\overline{B}_\varepsilon(p)\right) = \overline{U}_\varepsilon \qquad \forall\, t \in [0, 1].$$

Then there is a unique smooth function $[0, 1] \to T_p M : t \mapsto v(t)$ such that $|v(t)| \leq \varepsilon$ and $\gamma(t) = \exp_p(v(t))$ for every t. The set

$$I := \{t \in [0, 1] \,|\, \gamma(t) \neq p\} = \{t \in [0, 1] \,|\, v(t) \neq 0\} \subset (0, 1]$$

is open in the relative topology of $(0, 1]$. Thus I is a union of open intervals in $(0, 1)$ and one half open interval containing 1. Define $\beta : [0, 1] \to [0, 1]$ and $w : I \to T_p M$ by

$$\beta(t) := \frac{|v(t)|}{\varepsilon}, \qquad w(t) := \varepsilon \frac{v(t)}{|v(t)|}.$$

Then β is continuous, both β and w are smooth on I,

$$\beta(0) = 0, \qquad \beta(1) = 1, \qquad w(1) = v,$$

and

$$|w(t)| = \varepsilon, \qquad \gamma(t) = \exp_p(\beta(t)w(t))$$

for all $t \in I$. We prove that $L(\gamma) \geq \varepsilon$. To see this let $\alpha : [0, 1] \times I \to M$ be the map of Lemma 4.4.5, i.e.

$$\alpha(s, t) := \exp_p(sw(t)).$$

Then $\gamma(t) = \alpha(\beta(t), t)$ and hence

$$\dot{\gamma}(t) = \dot{\beta}(t) \frac{\partial \alpha}{\partial s}(\beta(t), t) + \frac{\partial \alpha}{\partial t}(\beta(t), t)$$

for every $t \in I$. Hence it follows from Lemma 4.4.5 that

$$|\dot{\gamma}(t)|^2 = \dot{\beta}(t)^2 \left| \frac{\partial \alpha}{\partial s}(\beta(t), t) \right|^2 + \left| \frac{\partial \alpha}{\partial t}(\beta(t), t) \right|^2 \geq \dot{\beta}(t)^2 \varepsilon^2$$

for every $t \in I$. Hence

$$L(\gamma) = \int_0^1 |\dot{\gamma}(t)| \, dt = \int_I |\dot{\gamma}(t)| \, dt \geq \varepsilon \int_I \left| \dot{\beta}(t) \right| \, dt \geq \varepsilon \int_I \dot{\beta}(t) \, dt = \varepsilon.$$

Here the last equality follows by applying the fundamental theorem of calculus to each interval in I and using the fact that $\beta(0) = 0$ and $\beta(1) = 1$. If $L(\gamma) = \varepsilon$, we must have

$$\frac{\partial \alpha}{\partial t}(\beta(t), t) = 0, \qquad \dot{\beta}(t) \geq 0 \qquad\qquad \text{for all } t \in I.$$

Thus I is a single half open interval containing 1 and on this interval the condition $\frac{\partial \alpha}{\partial t}(\beta(t), t) = 0$ implies $\dot{w}(t) = 0$. Since $w(1) = v$ we have $w(t) = v$ for every $t \in I$. Hence $\gamma(t) = \exp_p(\beta(t)v)$ for every $t \in [0, 1]$. It follows that β is smooth on the closed interval $[0, 1]$ (and not just on I). Thus we have proved that every $\gamma \in \Omega_{p,q}$ with values in $\overline{U}_\varepsilon$ has length $L(\gamma) \geq \varepsilon$ with equality if and only if γ is a reparametrized geodesic. But if $\gamma \in \Omega_{p,q}$ does not take values only in $\overline{U}_\varepsilon$, there must be a $T \in (0, 1)$ such that $\gamma([0, T]) \subset \overline{U}_\varepsilon$ and $\gamma(T) \in \partial U_\varepsilon$. Then $L(\gamma|_{[0,T]}) \geq \varepsilon$, by what we have just proved, and $L(\gamma|_{[T,1]}) > 0$ because the restriction of γ to $[T, 1]$ cannot be constant; so in this case we have $L(\gamma) > \varepsilon$. This proves Theorem 4.4.4. □

The next corollary gives a partial answer to our problem of finding length minimizing curves. It asserts that geodesics minimize the length *locally*.

Corollary 4.4.6   *Let $M \subset \mathbb{R}^n$ be a smooth m-manifold, let $I \subset \mathbb{R}$ be an open interval, and let $\gamma : I \to M$ be a geodesic. Fix a point $t_0 \in I$. Then there exists a constant $\varepsilon > 0$ such that*

$$t_0 - \varepsilon < s < t < t_0 + \varepsilon \qquad \Longrightarrow \qquad L(\gamma|_{[s,t]}) = d(\gamma(s), \gamma(t)).$$

Proof   Since γ is a geodesic its derivative has constant norm $|\dot{\gamma}(t)| \equiv c$ (see Theorem 4.1.4). Choose $\delta > 0$ so small that the interval $[t_0 - \delta, t_0 + \delta]$ is contained in I. Then there is a constant $r > 0$ such that $r \leq \text{inj}(\gamma(t))$ whenever $|t - t_0| \leq \delta$. Choose $\varepsilon > 0$ such that

$$\varepsilon < \delta, \qquad 2\varepsilon c < r.$$

If $t_0 - \varepsilon < s < t < t_0 + \varepsilon$, then

$$\gamma(t) = \exp_{\gamma(s)}((t - s)\dot{\gamma}(s))$$

and

$$|(t - s)\dot{\gamma}(s)| = |t - s|c < 2\varepsilon c < r \leq \text{inj}(\gamma(s)).$$

Hence it follows from Theorem 4.4.4 that

$$L(\gamma|_{[s,t]}) = |t - s|c = d(\gamma(s), \gamma(t)).$$

This proves Corollary 4.4.6.   □

4.4.3   Examples and Exercises

Exercise 4.4.7   How large can the constant ε in Corollary 4.4.6 be chosen in the case $M = S^2$? Compare this with the injectivity radius.

Remark 4.4.8 We conclude from Theorem 4.4.4 that

$$S_r(p) := \{q \in M \mid d(p,q) = r\} = \exp_p(\{v \in T_p M \mid |v| = r\}) \qquad (4.4.1)$$

for $0 < r < \mathrm{inj}(p; M)$. The Gauß Lemma 4.4.5 shows that the geodesic rays $[0, 1] \to M : s \mapsto \exp_p(sv)$ emanating from p are the orthogonal trajectories to the concentric spheres $S_r(p)$.

Exercise 4.4.9 Let

$$M \subset \mathbb{R}^3$$

be of dimension two and suppose that M is invariant under the (orthogonal) reflection about some plane $E \subset \mathbb{R}^3$. Show that E intersects M in a geodesic. (**Hint:** Otherwise there would be points $p, q \in M$ very close to one another joined by two distinct minimal geodesics.) Conclude for example that the coordinate planes intersect the ellipsoid $(x/a)^2 + (y/b)^2 + (z/c)^2 = 1$ in geodesics.

Exercise 4.4.10 Choose geodesic normal coordinates near $p \in M$ via

$$q = \exp_p\left(\sum_{i=1}^m x^i(q)e_i\right),$$

where $e_1, \ldots, e_m$ is an orthonormal basis of $T_p M$ (see Corollary 4.5.4 below). Then we have $x^i(p) = 0$ and

$$B_r(p) = \{q \in M \mid d(p,q) < r\} = \left\{q \in M \,\middle|\, \sum_{i=1}^m |x^i(q)|^2 < r^2\right\} \qquad (4.4.2)$$

for $0 < r < \mathrm{inj}(p; M)$. Hence Theorem 4.5.3 below asserts that $B_r(p)$ is convex for $r > 0$ sufficiently small.

(i) Show that it can happen that a geodesic in $B_r(p)$ is not minimal. **Hint:** Take M to be the hemisphere $\{(x, y, z) \in \mathbb{R}^3 \mid x^2 + y^2 + z^2 = 1, z > 0\}$ together with the disc $\{(x, y, z) \in \mathbb{R}^3 \mid x^2 + y^2 \leq 1, z = 0\}$, but smooth the corners along the circle $x^2 + y^2 = 1$, $z = 0$. Take $p = (0, 0, 1)$ and $r = \pi/2$.

(ii) Show that, if $r > 0$ is sufficiently small, then the unique geodesic γ in $B_r(p)$ joining two points $q, q' \in B_r(p)$ is minimal and that in fact any curve γ' from q to q' which is not a reparametrization of γ is strictly longer, i.e. $L(\gamma') > L(\gamma) = d(q, q')$.

Exercise 4.4.11 Let $\gamma : I = [a, b] \to M$ be a smooth curve with endpoints $\gamma(a) = p$ and $\gamma(b) = q$ and nowhere vanishing derivative, i.e. $\dot{\gamma}(t) \neq 0$ for all $t \in I$. Prove that the following are equivalent.

(i)   The curve γ is an **extremal of the length functional**, i.e. every smooth map $\mathbb{R} \times I \to M : (s, t) \mapsto \gamma_s(t)$ with $\gamma_s(a) = p$ and $\gamma_s(b) = q$ for all s satisfies

$$\frac{d}{ds} L(\gamma_s) \bigg|_{s=0} = 0.$$

(ii) The curve γ is a reparametrized geodesic, i.e. there exists a smooth map $\sigma : [a, b] \to [0, 1]$ with $\sigma(a) = 0$, $\sigma(b) = 1$, $\dot{\sigma}(t) \geq 0$ for all $t \in I$, and a vector $v \in T_p M$ such that

$$q = \exp_p(v), \qquad \gamma(t) = \exp_p(\sigma(t)v)$$

for all $t \in I$. (We remark that the hypothesis $\dot{\gamma}(t) \neq 0$ implies that σ is actually a diffeomorphism, i.e. $\dot{\sigma}(t) > 0$ for all $t \in I$.)

(iii) The curve γ minimizes the length functional locally, i.e. there exists an $\varepsilon > 0$ such that $L(\gamma|_{[s,t]}) = d(\gamma(s), \gamma(t))$ for every closed subinterval $[s, t] \subset I$ of length $t - s < \varepsilon$.

It is often convenient to consider curves γ where $\dot{\gamma}(t)$ is allowed to vanish for some values of t; then γ cannot (in general) be parametrized by arclength. Such a curve $\gamma : I \to M$ can be smooth (as a map) and yet its image may have corners (where $\dot{\gamma}$ necessarily vanishes). Note that a curve with corners can never minimize the distance, even locally.

Exercise 4.4.12 Show that conditions (ii) and (iii) in Exercise 4.4.11 are equivalent, even without the assumption that $\dot{\gamma}$ is nowhere vanishing. Deduce that, if $\gamma : I \to M$ is a shortest curve joining p to q, i.e. $L(\gamma) = d(p, q)$, then γ is a reparametrized geodesic.

Show by example that one can have a variation $\{\gamma_s\}_{s \in \mathbb{R}}$ of a reparametrized geodesic $\gamma_0 = \gamma$ for which the map $s \mapsto L(\gamma_s)$ is not even differentiable at $s = 0$. (**Hint:** Take γ to be constant. See also Exercise 4.1.9.)

Show, however, that conditions (i), (ii) and (iii) in Exercise 4.4.11 remain equivalent if the hypothesis that $\dot{\gamma}$ is nowhere vanishing is weakened to the hypothesis that $\dot{\gamma}(t) \neq 0$ for all but finitely many $t \in I$. Conclude that a broken geodesic is a reparametrized geodesic if and only if it minimizes arclength locally. (A **broken geodesic** is a continuous map $\gamma : I = [a, b] \to M$ for which there exist $a = t_0 < t_1 < \cdots < t_n = b$ such that $\gamma|_{[t_{i-1}, t_i]}$ is a geodesic for $i = 1, \ldots, n$. It is thus a geodesic if and only if $\dot{\gamma}$ is continuous at the break points, i.e. $\dot{\gamma}(t_i^-) = \dot{\gamma}(t_i^+)$ for $i = 1, \ldots, n - 1$.)

4.5 Convex Neighborhoods

A subset of an affine space is called convex iff it contains the line segment joining any two of its points. The definition carries over to a submanifold M of Euclidean space (or indeed more generally to any manifold M equipped with a spray) once we reword the definition so as to confront the difficulty that a geodesic joining two points might not exist nor, if it does, need it be unique.

Definition 4.5.1 (Geodesically convex set) Let $M \subset \mathbb{R}^n$ be a smooth m-dimensional manifold. A subset $U \subset M$ is called **geodesically convex** iff, for all $p_0, p_1 \in U$, there exists a unique geodesic $\gamma : [0, 1] \to U$ such that $\gamma(0) = p_0$ and $\gamma(1) = p_1$.

It is not precluded in Definition 4.5.1 that there be other geodesics from p to q which leave and then re-enter U, and these may even be shorter than the geodesic in U.

Exercise 4.5.2

(a) Find a geodesically convex set U in a manifold M and points $p_0, p_1 \in U$ such that the unique geodesic $\gamma : [0, 1] \to U$ with $\gamma(0) = p_0$ and $\gamma(1) = p_1$ has length $L(\gamma) > d(p_0, p_1)$. **Hint:** An interval of length bigger than π in S^1.
(b) Find a set U in a manifold M such that any two points in U can be joined by a minimal geodesic in U, but U is not geodesically convex. **Hint:** A closed hemisphere in S^2.

Theorem 4.5.3 (Convex Neighborhood Theorem) *Let $M \subset \mathbb{R}^n$ be a smooth m-dimensional submanifold and fix a point $p_0 \in M$. Let $\phi : U \to \Omega$ be any coordinate chart on an open neighborhood $U \subset M$ of p_0 with values in an open set $\Omega \subset \mathbb{R}^m$. Then the set*

$$U_r := \{p \in U \mid |\phi(p) - \phi(p_0)| < r\} \tag{4.5.1}$$

is geodesically convex for $r > 0$ sufficiently small.

Before giving the proof of Theorem 4.5.3 we derive a useful corollary.

Corollary 4.5.4 *Let $M \subset \mathbb{R}^n$ be a smooth m-manifold and let $p_0 \in M$. Then, for $r > 0$ sufficiently small, the open ball*

$$U_r(p_0) := \{p \in M \mid d(p_0, p) < r\} \tag{4.5.2}$$

is geodesically convex.

Proof Choose an orthonormal basis $e_1, \ldots, e_m$ of $T_{p_0} M$ and define

$$\Omega := \{x \in \mathbb{R}^m \mid |x| < \mathrm{inj}(p_0; M)\},$$
$$U := \{p \in M \mid d(p_0, p) < \mathrm{inj}(p_0; M)\}. \tag{4.5.3}$$

Define the map $\psi : \Omega \to U$ by

$$\psi(x) := \exp_{p_0}\left(\sum_{i=1}^m x^i e_i\right) \tag{4.5.4}$$

for $x = (x^1, \ldots, x^m) \in \Omega$. Then ψ is a diffeomorphism and $d(p_0, \psi(x)) = |x|$ for all $x \in \Omega$ by Theorem 4.4.4. Hence its inverse

$$\phi := \psi^{-1} : U \to \Omega \tag{4.5.5}$$

satisfies $\phi(p_0) = 0$ and $|\phi(p)| = d(p_0, p)$ for all $p \in U$. Thus

$$U_r(p_0) = \{p \in U \mid |\phi(p) - \phi(p_0)| < r\} \qquad \text{for } 0 < r < \mathrm{inj}(p_0; M)$$

and so Corollary 4.5.4 follows from Theorem 4.5.3. $\qquad\qquad\qquad\qquad\qquad\square$

Definition 4.5.5 (Geodesically normal coordinates) The coordinate chart $\phi : U \to \Omega$ in (4.5.4) and (4.5.5) sends geodesics through p_0 to straight lines through the origin. Its components $x^1, \ldots, x^m : U \to \mathbb{R}$ are called **geodesically normal coordinates** at p_0.

Proof of Theorem 4.5.3 Assume without loss of generality that $\phi(p_0) = 0$. Let $\Gamma_{ij}^k : \Omega \to \mathbb{R}$ be the Christoffel symbols of the coordinate chart and, for $x \in \Omega$, define the quadratic function $Q_x : \mathbb{R}^m \to \mathbb{R}$ by

$$Q_x(\xi) := \sum_{k=1}^m (\xi^k)^2 - \sum_{i,j,k=1}^m x^k \Gamma_{ij}^k(x) \xi^i \xi^j.$$

Shrinking U, if necessary, we may assume that

$$\max_{i,j=1,\ldots,m} \left| \sum_{k=1}^m x^k \Gamma_{ij}^k(x) \right| \leq \frac{1}{2m} \qquad \text{for all } x \in \Omega.$$

Then, for all $x \in \Omega$ and all $\xi \in \mathbb{R}^m$ we have

$$Q_x(\xi) \geq |\xi|^2 - \frac{1}{2m}\left(\sum_{i=1}^m |\xi^i|\right)^2 \geq \frac{1}{2}|\xi|^2 \geq 0.$$

Hence Q_x is positive definite for every $x \in \Omega$.

Now let $\gamma : [0, 1] \to U$ be a geodesic and define

$$c(t) := \phi(\gamma(t))$$

for $0 \leq t \leq 1$. Then, by Lemma 4.3.14, c satisfies the differential equation

$$\ddot{c}^k + \sum_{i,j} \Gamma_{ij}^k(c) \dot{c}^i \dot{c}^j = 0.$$

Hence

$$\frac{d^2}{dt^2} \frac{|c|^2}{2} = \frac{d}{dt} \langle \dot{c}, c \rangle = |\dot{c}|^2 + \langle \ddot{c}, c \rangle = Q_c(\dot{c}) \geq \frac{|\dot{c}|^2}{2} \geq 0$$

and so the function $t \mapsto |\phi(\gamma(t))|^2$ is convex. Thus, if $\gamma(0), \gamma(1) \in U_r$ for some $r > 0$, it follows that $\gamma(t) \in U_r$ for all $t \in [0, 1]$.

Consider the exponential map

$$V = \{(p, v) \in TM \mid v \in V_p\} \to M : (p, v) \mapsto \exp_p(v)$$

in Lemma 4.3.6. Its domain V is open and the exponential map is smooth. Since it sends the pair $(p_0, 0) \in V$ to $\exp_{p_0}(0) = p_0 \in U$, it follows from continuity that there exist constants $\varepsilon > 0$ and $r > 0$ such that

$$p \in U_r, \quad v \in T_p M, \quad |v| < \varepsilon \qquad \Longrightarrow \qquad v \in V_p, \quad \exp_p(v) \in U. \qquad (4.5.6)$$

Moreover, we have

$$d \exp_{p_0}(0) = \mathrm{id} : T_{p_0} M \to T_{p_0} M$$

by Corollary 4.3.7. Hence the Implicit Function Theorem 2.6.15 asserts that the constants $\varepsilon > 0$ and $r > 0$ can be chosen such that (4.5.6) holds and there exists a smooth map $h : U_r \times U_r \to \mathbb{R}^n$ that satisfies the conditions

$$h(p, q) \in T_p M, \qquad |h(p, q)| < \varepsilon \qquad (4.5.7)$$

for all $p, q \in U_r$ and

$$\exp_p(v) = q \qquad \Longleftrightarrow \qquad v = h(p, q) \qquad (4.5.8)$$

for all $p, q \in U_r$ and all $v \in T_p M$ with $|v| < \varepsilon$. In particular, we have

$$h(p_0, p_0) = 0$$

and $\exp_p(h(p, q)) = q$ for all $p, q \in U_r$.

Fix two constants $\varepsilon > 0$ and $r > 0$ and a smooth map $h : U_r \times U_r \to \mathbb{R}^n$ such that (4.5.6), (4.5.7), (4.5.8) are satisfied. We show that any two points $p, q \in U_r$ are joined by a geodesic in U_r. Let $p, q \in U_r$ and define

$$\gamma(t) := \exp_p(th(p,q)) \qquad \text{for } 0 \le t \le 1.$$

This curve $\gamma : [0, 1] \to M$ is well defined by (4.5.6) and (4.5.7), it is a geodesic satisfying $\gamma(0) = p \in U_r$ by Lemma 4.3.6, it satisfies $\gamma(1) = q \in U_r$ by (4.5.8), it takes values in U by (4.5.6) and (4.5.7), and so $\gamma([0, 1]) \subset U_r$ because the function $[0, 1] \to \mathbb{R} : t \mapsto |\phi(\gamma(t))|^2$ is convex.

We show that there exists at most one geodesic in U_r joining p and q. Let $p, q \in U_r$ and let $\gamma : [0, 1] \to U_r$ be any geodesic such that $\gamma(0) = p$ and $\gamma(1) = q$. Define $v := \dot{\gamma}(0) \in T_p M$. Then $\gamma(t) = \exp_p(tv)$ for $0 \le t \le 1$ by Lemma 4.3.6. We claim that $|v| < \varepsilon$. Suppose, by contradiction, that

$$|v| \ge \varepsilon.$$

Then

$$T := \frac{\varepsilon}{|v|} \le 1$$

and, for $0 < t < T$, we have $|tv| < \varepsilon$ and $\exp_p(tv) = \gamma(t) \in U_r$ and so

$$h(p, \gamma(t)) = tv.$$

by (4.5.8). Thus

$$|h(p, \gamma(t))| = t|v| \qquad \text{for } 0 < t < T.$$

Take the limit $t \nearrow T$ to obtain

$$|h(p, \gamma(T))| = T|v| = \varepsilon$$

in contradiction to (4.5.7). This contradiction shows that $|v| < \varepsilon$. Since

$$\exp_p(v) = \gamma(1) = q \in U_r$$

it follows from (4.5.8) that $v = h(p, q)$. This proves Theorem 4.5.3. $\square$

Remark 4.5.6   Theorem 4.5.3 and its proof carry over to general sprays (see Definition 4.3.2).

Exercise 4.5.7   Consider the set $U_r(p) = \{q \in M \mid d(p,q) < r\}$ for $p \in M$ and $r > 0$. Corollary 4.5.4 asserts that this set is geodesically convex for r sufficiently small. How large can you choose r in the cases

$$M = S^2, \qquad M = \mathbb{T}^2 = S^1 \times S^1, \qquad M = \mathbb{R}^2, \qquad M = \mathbb{R}^2 \setminus \{0\}.$$

Compare this with the injectivity radius. If the set $U_r(p)$ in these examples is geodesically convex, does it follow that every geodesic in $U_r(p)$ is minimizing?

4.6   Completeness and Hopf–Rinow

For a Riemannian manifold there are different notions of completeness. First, in Sect. 3.4 completeness was defined in terms of the completeness of time dependent basic vector fields on the frame bundle (Definition 3.4.10). Second, there is a distance function

$$d : M \times M \to [0, \infty)$$

defined by equation (4.2.2) so that we can speak of completeness of the metric space (M, d) in the sense that every Cauchy sequence converges. Third, there is the question of whether geodesics through any point in any direction exist for all time; if so we call a Riemannian manifold geodesically complete. The remarkable fact is that these three rather different notions of completeness are actually equivalent and that, in the complete case, any two points in M can be joined by a shortest geodesic. This is the content of the Hopf–Rinow theorem. We will spell out the details of the proof for embedded manifolds and leave it to the reader (as a straight forward exercise) to extend the proof to the intrinsic setting.

4.6.1   Geodesic Completeness

Definition 4.6.1 (Geodesically complete manifold)   Let $M \subset \mathbb{R}^n$ be an m-dimensional manifold. Given a point $p \in M$ we say that M **is geodesically complete at** p iff, for every tangent vector $v \in T_p M$, there exists a geodesic $\gamma : \mathbb{R} \to M$ (on the entire real axis) satisfying $\gamma(0) = p$ and $\dot{\gamma}(0) = v$ (or equivalently $V_p = T_p M$ where $V_p \subset T_p M$ is defined by (4.3.5)). The manifold M is called **geodesically complete** iff it is geodesically complete at every point $p \in M$.

Definition 4.6.2   Let (M, d) be a metric space. A subset $A \subset M$ is called **bounded** iff

$$\sup_{p \in A} d(p, p_0) < \infty$$

for some (and hence every) point $p_0 \in M$.

Example 4.6.3   A manifold $M \subset \mathbb{R}^n$ can be contained in a bounded subset of $\mathbb{R}^n$ and still not be bounded with respect to the metric (4.2.2). An example is the 1-manifold $M = \{(x, y) \in \mathbb{R}^2 \,|\, 0 < x < 1, \ y = \sin(1/x)\}$.

Exercise 4.6.4   Let (M, d) be a metric space. Prove that every compact subset $K \subset M$ is closed and bounded. Find an example of a metric space that contains a closed and bounded subset that is not compact.

Theorem 4.6.5 (Completeness) *Let $M \subset \mathbb{R}^n$ be a connected m-dimensional manifold and let $d : M \times M \to [0, \infty)$ be the distance function defined by (4.1.1), (4.2.1), and (4.2.2). Then the following are equivalent.*

(i)   *M is geodesically complete.*
(ii) *There exists a point $p \in M$ such that M is geodesically complete at p.*
(iii) *Every closed and bounded subset of M is compact.*
(iv) *(M, d) is a complete metric space.*
(v)   *M is complete, i.e. for every smooth curve $\xi : \mathbb{R} \to \mathbb{R}^m$ and every element $(p_0, e_0) \in \mathcal{F}(M)$ there exists a smooth curve $\beta : \mathbb{R} \to \mathcal{F}(M)$ satisfying*

$$\dot{\beta}(t) = B_{\xi(t)}(\beta(t)), \qquad \beta(0) = (p_0, e_0). \tag{4.6.1}$$

(vi) *The basic vector field $B_\xi \in \mathrm{Vect}(\mathcal{F}(M))$ is complete for every $\xi \in \mathbb{R}^m$.*
(vii) *For every smooth curve $\gamma' : \mathbb{R} \to \mathbb{R}^m$, every $p_0 \in M$, and every orthogonal isomorphism $\Phi_0 : T_{p_0}M \to \mathbb{R}^m$ there exists a development (Φ, γ, γ') of M along $\mathbb{R}^m$ on all of $\mathbb{R}$ that satisfies $\gamma(0) = p_0$ and $\Phi(0) = \Phi_0$.*

Proof The proof relies on Theorem 4.6.6 below.   □

4.6.2 Global Existence of Minimal Geodesics

Theorem 4.6.6 (Hopf–Rinow) *Let $M \subset \mathbb{R}^n$ be a connected m-manifold and let $p \in M$. Assume M is geodesically complete at p. Then, for every $q \in M$, there exists a geodesic $\gamma : [0, 1] \to M$ such that*

$$\gamma(0) = p, \qquad \gamma(1) = q, \qquad L(\gamma) = d(p, q).$$

Before giving the proof of the Hopf–Rinow Theorem we show that it implies Theorem 4.6.5.

Theorem 4.6.6 implies Theorem 4.6.5 That (i) implies (ii) follows directly from the definitions.

We prove that (ii) implies (iii). Thus assume that M is geodesically complete at the point $p_0 \in M$ and let $K \subset M$ be a closed and bounded subset. Then $r := \sup_{q \in K} d(p_0, q) < \infty$. Hence Theorem 4.6.6 asserts that, for every $q \in K$, there exists a vector $v \in T_{p_0}M$ such that $|v| = d(p_0, q) \leq r$ and $\exp_{p_0}(v) = q$. Thus

$$K \subset \exp_{p_0}(\overline{B}_r(p_0)), \qquad \overline{B}_r(p_0) = \{v \in T_{p_0}M \mid |v| \leq r\}.$$

Then $B := \{v \in T_{p_0}M \mid |v| \leq r, \ \exp_{p_0}(v) \in K\}$ is a closed and bounded subset of the Euclidean space $T_{p_0}M$. Hence B is compact and $K = \exp_{p_0}(B)$. Since the exponential map $\exp_{p_0} : T_{p_0}M \to M$ is continuous it follows that K is compact. This shows that (ii) implies (iii).

We prove that (iii) implies (iv). Thus assume that every closed and bounded subset of M is compact and choose a Cauchy sequence $p_i \in M$. Choose $i_0 \in \mathbb{N}$ such that $d(p_i, p_j) \le 1$ for all $i, j \in \mathbb{N}$ with $i, j \ge i_0$. Define

$$c := \max_{1 \le i \le i_0} d(p_1, p_i) + 1.$$

Then $d(p_1, p_i) \le d(p_1, p_{i_0}) + d(p_{i_0}, p_i) \le d(p_1, p_{i_0}) + 1 \le c$ for all $i \ge i_0$ and so $d(p_1, p_i) \le c$ for all $i \in \mathbb{N}$. Hence the set $\{p_i \mid i \in \mathbb{N}\}$ is bounded and so is its closure. By (iii) this implies that the sequence p_i has a convergent subsequence. Since p_i is a Cauchy sequence, this implies that p_i converges. Thus we have proved that (iii) implies (iv).

We prove that (iv) implies (v). Fix a smooth curve $\xi : \mathbb{R} \to \mathbb{R}^m$ and an element $(p_0, e_0) \in \mathcal{F}(M)$. Assume, by contradiction, that there exists a real number $T > 0$ such that there exists a solution $\beta : [0, T) \to \mathcal{F}(M)$ of equation (4.6.1) that cannot be extended to the interval $[0, T + \varepsilon)$ for any $\varepsilon > 0$. Write $\beta(t) =: (\gamma(t), e(t))$ so that γ and e satisfy the equations

$$\dot{\gamma}(t) = e(t)\xi(t), \quad \dot{e}(t) = h_{\gamma(t)}(\dot{\gamma}(t))e(t), \quad \gamma(0) = p_0, \quad e(0) = e_0.$$

This implies $e(t)\eta \in T_{\gamma(t)}M$ and $\dot{e}(t)\eta \in T_{\gamma(t)}^{\perp}M$ for all $\eta \in \mathbb{R}^m$ and therefore

$$\frac{d}{dt}\langle \eta, e(t)^{\mathsf{T}}e(t)\zeta \rangle = \frac{d}{dt}\langle e(t)\eta, e(t)\zeta \rangle = \langle \dot{e}(t)\eta, e(t)\zeta \rangle + \langle e(t)\eta, \dot{e}(t)\zeta \rangle = 0$$

for all $\eta, \zeta \in \mathbb{R}^m$ and all $t \in [0, T)$. Thus the function $t \mapsto e(t)^{\mathsf{T}}e(t)$ is constant, hence

$$e(t)^{\mathsf{T}}e(t) = e_0^{\mathsf{T}}e_0, \quad \|e(t)\| = \sup_{0 \ne \eta \in \mathbb{R}^m} \frac{|e(t)\eta|}{|\eta|} = \|e_0\| \qquad (4.6.2)$$

for $0 \le t < T$, hence

$$|\dot{\gamma}(t)| = |e(t)\xi(t)| \le \|e_0\|\|\xi(t)\| \le \|e_0\| \sup_{0 \le s \le T} |\xi(s)| =: c_T$$

and so $d(\gamma(s), \gamma(t)) \le L(\gamma|_{[s,t]}) \le (t - s)c_T$ for $0 \le s < t < T$. Since (M, d) is a complete metric space, this shows that the limit $p_1 := \lim_{t \nearrow T} \gamma(t) \in M$ exists. Thus the set $K := \gamma([0, T)) \cup \{p_1\} \subset M$ is compact and so is the set

$$\widetilde{K} := \{(p, e) \in \mathcal{F}(M) \mid p \in K, \ e^{\mathsf{T}}e = e_0^{\mathsf{T}}e_0\} \subset \mathcal{F}(M).$$

By equation (4.6.2) the curve $[0, T) \to \mathbb{R} \times \mathcal{F}(M) : t \mapsto (t, \gamma(t), e(t))$ takes values in the compact set $[0, T] \times \widetilde{K}$ and is the integral curve of a vector field on the manifold $\mathbb{R} \times \mathcal{F}(M)$. Hence Corollary 2.4.15 asserts that $[0, T)$ cannot be the maximal existence interval of this integral curve, a contradiction. This shows that (iv) implies (v).

That (v) implies (vi) follows by taking $\xi(t) \equiv \xi$ in (v).

We prove that (vi) implies (i). Fix an element $p_0 \in M$ and a tangent vector $v_0 \in T_{p_0} M$. Let $e_0 \in \mathcal{L}_{iso}(\mathbb{R}^m, T_{p_0} M)$ be any isomorphism and choose $\xi \in \mathbb{R}^m$ such that $e_0 \xi = v_0$. By (vi) the vector field B_ξ has a unique integral curve $\mathbb{R} \to \mathcal{F}(M)$: $t \mapsto \beta(t) = (\gamma(t), e(t))$ with

$$\beta(0) = (p_0, e_0).$$

Thus

$$\dot{\gamma}(t) = e(t)\xi, \qquad \dot{e}(t) = h_{\gamma(t)}(e(t)\xi)e(t),$$

and hence

$$\ddot{\gamma}(t) = \dot{e}(t)\xi = h_{\gamma(t)}(e(t)\xi)e(t)\xi = h_{\gamma(t)}(\dot{\gamma}(t), \dot{\gamma}(t)).$$

By the Gauß–Weingarten formula, this implies $\nabla \dot{\gamma}(t) = 0$ for every t and hence $\gamma : \mathbb{R} \to M$ is a geodesic with $\gamma(0) = p_0$ and $\dot{\gamma}(0) = e_0 \xi = v_0$. Thus M is geodesically complete and this shows that (vi) implies (i).

The equivalence of (v) and (vii) was established in Corollary 3.5.25 and this shows that Theorem 4.6.6 implies Theorem 4.6.5. $\square$

4.6.3   Proof of the Hopf–Rinow Theorem

The proof of Theorem 4.6.6 relies on the next two lemmas.

Lemma 4.6.7 *Let $M \subset \mathbb{R}^n$ be a connected m-manifold and $p \in M$. Suppose $\varepsilon > 0$ is smaller than the injectivity radius of M at p and denote*

$$\Sigma_1(p) := \{v \in T_p M \mid |v| = 1\}, \qquad S_\varepsilon(p) := \{p' \in M \mid d(p, p') = \varepsilon\}.$$

Then the map $\Sigma_1(p) \to S_\varepsilon(p) : v \mapsto \exp_p(\varepsilon v)$ is a diffeomorphism and, for all $q \in M$, we have

$$d(p, q) > \varepsilon \qquad \Longrightarrow \qquad d(S_\varepsilon(p), q) = d(p, q) - \varepsilon.$$

Proof By Theorem 4.4.4, we have

$$d(p, \exp_p(v)) = |v| \qquad \text{for all } v \in T_p M \text{ with } |v| \le \varepsilon$$

and

$$d(p, p') > \varepsilon \qquad \text{for all } p' \in M \setminus \{\exp_p(v) \mid v \in T_p M, |v| \le \varepsilon\}.$$

This shows that $S_\varepsilon(p) = \exp_p(\varepsilon\Sigma_1(p))$ and, since ε is smaller than the injectivity radius, the map

$$\Sigma_1(p) \to S_\varepsilon(p) : v \mapsto \exp_p(\varepsilon v)$$

is a diffeomorphism.

To prove the second assertion, let $q \in M$ such that

$$r := d(p, q) > \varepsilon.$$

Fix a constant $\delta > 0$ and choose a smooth curve $\gamma : [0, 1] \to M$ such that

$$\gamma(0) = p, \qquad \gamma(1) = q, \qquad L(\gamma) \le r + \delta.$$

Choose $t_0 > 0$ such that $\gamma(t_0)$ is the last point of the curve on $S_\varepsilon(p)$, i.e.

$$\gamma(t_0) \in S_\varepsilon(p), \qquad \gamma(t) \notin S_\varepsilon(p) \text{ for } t_0 < t \le 1.$$

Then

$$
\begin{aligned}
d(\gamma(t_0), q) &\le L(\gamma|_{[t_0,1]}) \\
&= L(\gamma) - L(\gamma|_{[0,t_0]}) \\
&\le L(\gamma) - \varepsilon \\
&\le r + \delta - \varepsilon.
\end{aligned}
$$

This shows that $d(S_\varepsilon(p), q) \le r + \delta - \varepsilon$ for every $\delta > 0$ and therefore

$$d(S_\varepsilon(p), q) \le r - \varepsilon.$$

Moreover,

$$d(p', q) \ge d(p, q) - d(p, p') = r - \varepsilon$$

for all $p' \in S_\varepsilon(p)$. Thus

$$d(S_\varepsilon(p), q) = r - \varepsilon$$

and this proves Lemma 4.6.7. □

Lemma 4.6.8 (Curve Shortening Lemma) *Let $M \subset \mathbb{R}^n$ be an m-manifold, let $p \in M$, and let ε be a real number such that*

$$0 < \varepsilon < \mathrm{inj}(p; M).$$

Then, for all $v, w \in T_p M$, we have

$$|v| = |w| = \varepsilon, \quad d(\exp_p(v), \exp_p(w)) = 2\varepsilon \qquad \Longrightarrow \qquad v + w = 0.$$

Fig. 4.7 Two unit tangent
vectors

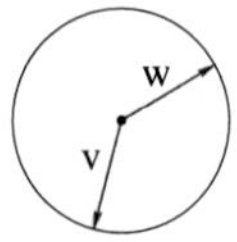

Proof We will prove that, for all $v, w \in T_p M$, we have

$$\lim_{\delta \to 0} \frac{d(\exp_p(\delta v), \exp_p(\delta w))}{\delta} = |v - w|. \tag{4.6.3}$$

Assume this holds and suppose, by contradiction, that there exist two tangent vectors $v, w \in T_p M$ such that

$$|v| = |w| = 1, \qquad d(\exp_p(\varepsilon v), \exp_p(\varepsilon w)) = 2\varepsilon, \qquad v + w \neq 0.$$

Then

$$|v - w| < 2$$

(see Fig. 4.7). Thus by (4.6.3) there exists a constant $0 < \delta < \varepsilon$ such that

$$d(\exp_p(\delta v), \exp_p(\delta w)) < 2\delta.$$

Then

$$d(\exp_p(\varepsilon v), \exp_p(\varepsilon w))$$
$$\leq d(\exp_p(\varepsilon v), \exp_p(\delta v)) + d(\exp_p(\delta v), \exp_p(\delta w)) + d(\exp_p(\delta w), \exp_p(\varepsilon w))$$
$$< \varepsilon - \delta + 2\delta + \varepsilon - \delta = 2\varepsilon$$

and this contradicts our assumption.

It remains to prove (4.6.3). For this we observe that

$$\lim_{\delta \to 0} \frac{d(\exp_p(\delta v), \exp_p(\delta w))}{\delta}$$
$$= \lim_{\delta \to 0} \frac{d(\exp_p(\delta v), \exp_p(\delta w))}{|\exp_p(\delta v) - \exp_p(\delta w)|} \frac{|\exp_p(\delta v) - \exp_p(\delta w)|}{\delta}$$
$$= \lim_{\delta \to 0} \frac{|\exp_p(\delta v) - \exp_p(\delta w)|}{\delta}$$
$$= \lim_{\delta \to 0} \left| \frac{\exp_p(\delta v) - p}{\delta} - \frac{\exp_p(\delta w) - p}{\delta} \right|$$
$$= |v - w|.$$

Here the second equality follows from Lemma 4.2.7.   □

Proof of Theorem 4.6.6 By assumption $M \subset \mathbb{R}^n$ is a connected submanifold, and $p \in M$ is given such that the exponential map $\exp_p : T_pM \to M$ is defined on the entire tangent space at p. Fix a point $q \in M \setminus \{p\}$ so that

$$0 < r := d(p, q) < \infty.$$

Choose a constant $\varepsilon > 0$ smaller than the injectivity radius of M at p and smaller than r. Then, by Lemma 4.6.7, we have

$$d(S_\varepsilon(p), q) = r - \varepsilon.$$

Hence there exists a tangent vector $v \in T_pM$ such that

$$d(\exp_p(\varepsilon v), q) = r - \varepsilon, \qquad |v| = 1.$$

Define the curve $\gamma : [0, r] \to M$ by

$$\gamma(t) := \exp_p(tv) \qquad \text{for } 0 \le t \le r.$$

By Lemma 4.3.6, this is a geodesic and it satisfies $\gamma(0) = p$. We must prove that $\gamma(r) = q$ and $L(\gamma) = d(p, q)$. Instead we will prove the following stronger statement.

Claim *For every $t \in [0, r]$ we have*

$$d(\gamma(t), q) = r - t.$$

In particular, $\gamma(r) = q$ and $L(\gamma) = r = d(p, q)$.

Consider the subset

$$I := \{t \in [0, r] \mid d(\gamma(t), q) = r - t\} \subset [0, r].$$

This set is nonempty, because $\varepsilon \in I$, it is obviously closed, and

$$t \in I \qquad \Longrightarrow \qquad [0, t] \subset I. \tag{4.6.4}$$

Namely, if $t \in I$ and $0 \le s \le t$, then

$$d(\gamma(s), q) \le d(\gamma(s), \gamma(t)) + d(\gamma(t), q) \le t - s + r - t = r - s$$

and

$$d(\gamma(s), q) \ge d(p, q) - d(p, \gamma(s)) \ge r - s.$$

Hence $d(\gamma(s), q) = r - s$ and hence $s \in I$. This proves (4.6.4).

Fig. 4.8 The proof of the
Hopf–Rinow theorem

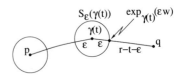

We prove that I is open (in the relative topology of $[0, r]$). Let $t \in I$ be given with $t < r$. Choose a constant $\varepsilon > 0$ smaller than the injectivity radius of M at $\gamma(t)$ and smaller than $r - t$. Then, by Lemma 4.6.7 with p replaced by $\gamma(t)$, we have

$$d(S_\varepsilon(\gamma(t)), q) = r - t - \varepsilon.$$

Next we choose $w \in T_{\gamma(t)} M$ such that

$$|w| = 1, \qquad d(\exp_{\gamma(t)}(\varepsilon w), q) = r - t - \varepsilon.$$

Then

$$
\begin{aligned}
d(\gamma(t - \varepsilon), \exp_{\gamma(t)}(\varepsilon w)) &\geq d(\gamma(t - \varepsilon), q) - d(\exp_{\gamma(t)}(\varepsilon w), q) \\
&= (r - t + \varepsilon) - (r - t - \varepsilon) \\
&= 2\varepsilon.
\end{aligned}
$$

The converse inequality is obvious, because both points have distance ε to $\gamma(t)$ (see Fig. 4.8).

Thus we have proved that

$$d(\gamma(t - \varepsilon), \exp_{\gamma(t)}(\varepsilon w)) = 2\varepsilon.$$

Since

$$\gamma(t - \varepsilon) = \exp_{\gamma(t)}(-\varepsilon \dot{\gamma}(t)),$$

it follows from Lemma 4.6.8 that

$$w = \dot{\gamma}(t).$$

Hence $\exp_{\gamma(t)}(sw) = \gamma(t + s)$ and this implies that

$$d(\gamma(t + \varepsilon), q) = r - t - \varepsilon.$$

Thus $t + \varepsilon \in I$ and, by (4.6.4), we have $[0, t + \varepsilon] \in I$. Thus we have proved that I is open. In other words, I is a nonempty subset of $[0, r]$ which is both open and closed, and hence $I = [0, r]$. This proves the claim and Theorem 4.6.6. $\square$

4.7 Geodesics in the Intrinsic Setting*

This section examines the distance function on a Riemannian manifold, shows how the results of this chapter extend to the intrinsic setting, and discusses several examples.

4.7.1 Intrinsic Distance

Let M be a connected smooth manifold (Sect. 2.8) equipped with a Riemannian metric (Sect. 3.7). Then we can define the length of a curve $\gamma : [0, 1] \to M$ by the formula (4.1.1) and it is invariant under reparametrization as in Remark 4.1.3. The **distance function** $d : M \times M \to \mathbb{R}$ is then given by the same formula (4.2.2). We prove that it still defines a metric on M and that this metric induces the same topology as the smooth structure.

Lemma 4.7.1 *Let M be a connected smooth Riemannian manifold and define the function $d : M \times M \to [0, \infty)$ by (4.1.1), (4.2.1), and (4.2.2). Then d is a metric and induces the same topology as the smooth structure.*

Proof The proof has three steps.

Step 1 *Fix a point $p_0 \in M$ and let $\phi : U \to \Omega$ be a coordinate chart of M onto an open subset $\Omega \subset \mathbb{R}^m$ such that $p_0 \in U$. Then there exists an open neighborhood $V \subset U$ of p_0 and constants $\delta, r > 0$ such that*

$$\delta|\phi(p) - \phi(p_0)| \le d(p, p_0) \le \delta^{-1}|\phi(p) - \phi(p_0)| \tag{4.7.1}$$

for every $p \in V$ and $d(p, p_0) \ge \delta r$ for every $p \in M \setminus V$.

Denote the inverse of the coordinate chart ϕ by $\psi := \phi^{-1} : \Omega \to M$ and define the map $g = (g_{ij})_{i,j=1}^m : \Omega \to \mathbb{R}^{m \times m}$ by

$$g_{ij}(x) := \left\langle \frac{\partial \psi}{\partial x^i}(x), \frac{\partial \psi}{\partial x^j}(x) \right\rangle_{\psi(x)}$$

for $x \in \Omega$. Then a smooth curve $\gamma : [0, 1] \to U$ has the length

$$L(\gamma) = \int_0^1 \sqrt{\dot{c}(t)^\mathsf{T} g(c(t))\dot{c}(t)} \, dt, \qquad c(t) := \phi(\gamma(t)). \tag{4.7.2}$$

Let $x_0 := \phi(p_0) \in \Omega$ and choose $r > 0$ such that $\overline{B}_r(x_0) \subset \Omega$. Then there is a constant $\delta \in (0, 1]$ such that

$$\delta|\xi| \le \sqrt{\xi^\mathsf{T} g(x)\xi} \le \delta^{-1}|\xi| \tag{4.7.3}$$

for all $x \in B_r(x_0)$ and $\xi, \eta \in \mathbb{R}^m$. Define $V := \phi^{-1}(B_r(x_0)) \subset U$.

Now let $p \in V$ and denote $x := \phi(p) \in B_r(x_0)$. Then, for every smooth curve $\gamma : [0, 1] \to V$ with $\gamma(0) = p_0$ and $\gamma(1) = p$, the curve

$$c := \phi \circ \gamma$$

takes values in $B_r(x_0)$ and satisfies $c(0) = x_0$ and $c(1) = x$. Hence, by (4.7.2) and (4.7.3), we have

$$L(\gamma) \geq \delta \int_0^1 |\dot{c}(t)|\, dt \geq \delta \left| \int_0^1 \dot{c}(t)\, dt \right| = \delta |x - x_0|.$$

If $\gamma : [0, 1] \to M$ is a smooth curve with endpoints $\gamma(0) = p_0$ and $\gamma(1) = p$ whose image is not entirely contained in V, then there exists a $T \in (0, 1]$ such that $\gamma(t) \in V$ for $0 \leq t < T$ and $\gamma(T) \in \partial V$, so $c(t) = \phi(\gamma(t)) \in B_r(x_0)$ for $0 \leq t < T$ and $|c(T) - x_0| = r$. Hence, by the above argument, we have

$$L(\gamma) \geq \delta r.$$

This shows that $d(p_0, p) \geq \delta r$ for $p \in M \setminus V$ and $d(p_0, p) \geq \delta |\phi(p) - \phi(p_0)|$ for $p \in V$. If $p \in V$, $x := \phi(p)$, and $c(t) := x_0 + t(x - x_0)$, then $\gamma := \psi \circ c$ is a smooth curve in V with $\gamma(0) = p_0$ and $\gamma(1) = p$ and, by (4.7.2) and (4.7.3),

$$L(\gamma) \leq \delta^{-1} \int_0^1 |\dot{c}(t)|\, dt = \delta^{-1} |x - x_0|.$$

This proves Step 1.

Step 2 *d is a distance function.*

Step 1 shows that $d(p, p_0) > 0$ for every $p \in M \setminus \{p_0\}$ and hence d satisfies condition (i) in Lemma 4.2.3. The proofs of (ii) and (iii) remain unchanged in the intrinsic setting and this proves Step 2.

Step 3 *The topology on M induced by d agrees with the topology induced by the smooth structure.*

Assume first that $W \subset M$ is open with respect to the manifold topology and let $p_0 \in W$. Let $\phi : U \to \Omega$ be a coordinate chart of M onto an open subset $\Omega \subset \mathbb{R}^m$ such that $p_0 \in U$, and choose $V \subset U$ and δ, r as in Step 1. Then $\phi(V \cap W)$ is an open subset of Ω containing the point $\phi(p_0)$. Hence there exists a constant $0 < \varepsilon \leq \delta r$ such that $B_{\delta^{-1}\varepsilon}(\phi(p_0)) \subset \phi(V \cap W)$. Thus by Step 1 we have $d(p, p_0) \geq \delta r \geq \varepsilon$ for all $p \in M \setminus V$. Hence, if $p \in M$ satisfies $d(p, p_0) < \varepsilon$, then $p \in V$, so $|\phi(p) - \phi(p_0)| < \delta^{-1} d(p, p_0) < \delta^{-1}\varepsilon$ by (4.7.1), and therefore

$\phi(p) \in \phi(V \cap W)$. Thus $B_\varepsilon(p_0; d) \subset W$ and this shows that W is open with respect to d.

Conversely, assume that $W \subset M$ be open with respect to d and choose a coordinate chart $\phi : U \to \Omega$ onto an open set $\Omega \subset \mathbb{R}^m$. We must prove that $\phi(U \cap W)$ is an open subset of Ω. To see this, choose $x_0 \in \phi(U \cap W)$ and let $p_0 := \phi^{-1}(x_0) \in U \cap W$. Now choose $V \subset U$ and δ, r as in Step 1. Choose $\varepsilon > 0$ such that $B_{\delta^{-1}\varepsilon}(p_0; d) \subset W$ and $B_\varepsilon(x_0) \subset \phi(V)$. Let $x \in \mathbb{R}^n$ such that $|x - x_0| < \varepsilon$. Then $x \in \phi(V)$ and therefore $p := \phi^{-1}(x) \in V$. This implies $d(p, p_0) < \delta^{-1}|\phi(p) - \phi(p_0)| = \delta^{-1}|x - x_0| < \delta^{-1}\varepsilon$, thus $p \in W \cap U$, and so $x = \phi(p) \in \phi(W \cap U)$. Thus $\phi(W \cap U)$ is open, and so W is open in the manifold topology of M. This proves Step 3 and Lemma 4.7.1. $\square$

4.7.2 Geodesics and the Levi-Civita Connection

With the covariant derivative understood (Theorem 3.7.8), we can define geodesics on M as smooth curves $\gamma : I \to M$ that satisfy the equation $\nabla\dot{\gamma} = 0$, as in Definition 4.1.5. Then all the above results about geodesics, as well as their proofs, carry over almost verbatim to the intrinsic setting. In particular, geodesics are in local coordinates described by equation (4.3.6) (Lemma 4.3.14) and they are the critical points of the energy functional

$$E(\gamma) := \frac{1}{2} \int_0^1 |\dot{\gamma}(t)|^2 \, dt$$

on the space $\Omega_{p,q}$ of all paths $\gamma : [0, 1] \to M$ with fixed endpoints $\gamma(0) = p$ and $\gamma(1) = q$. Here we use the fact that Lemma 4.1.7 extends to the intrinsic setting via the Embedding Theorem 2.9.12. So for every vector field $X \in \mathrm{Vect}(\gamma)$ along γ with $X(0) = 0$ and $X(1) = 0$ there exists a curve of curves $\mathbb{R} \to \Omega_{p,q} : s \mapsto \gamma_s$ with $\gamma_0 = \gamma$ and $\partial_s\gamma_s|_{s=0} = X$. Then, by the properties of the Levi-Civita connection, we have

$$dE(\gamma)X = \frac{1}{2} \int_0^1 \partial_s|\partial_t\gamma_s(t)|^2 \, dt$$

$$= \int_0^1 \langle \dot{\gamma}(t), \nabla_t X(t) \rangle \, dt$$

$$= -\int_0^1 \langle \nabla_t\dot{\gamma}(t), X(t) \rangle \, dt.$$

The right hand side vanishes for all X if and only if $\nabla\dot{\gamma} \equiv 0$ (Theorem 4.1.4). With this understood, we find that, for all $p \in M$ and $v \in T_pM$, there exists a unique geodesic $\gamma : I_{p,v} \to M$ on a maximal open interval $I_{p,v} \subset \mathbb{R}$ containing zero that satisfies $\gamma(0) = p$ and $\dot{\gamma}(0) = v$ (Lemma 4.3.4).

This gives rise to a smooth exponential map

$$\exp_p : V_p = \{v \in T_pM \mid 1 \in I_{p,v}\} \to M$$

as in Sect. 4.3 which satisfies

$$d \exp_p(0) = \mathrm{id} : T_pM \to T_pM$$

as in Corollary 4.3.7. This leads directly to the injectivity radius, the Gauß Lemma 4.4.5, the local length minimizing property of geodesics in Theorem 4.4.4, and the Convex Neighborhood Theorem 4.5.3. Also the proof of the equivalence of metric and geodesic completeness in Theorem 4.6.5 and of the Hopf–Rinow Theorem 4.6.6 carry over verbatim to the intrinsic setting of general Riemannian manifolds. The only place where some care must be taken is in the proof of the Curve Shortening Lemma 4.6.8 as is spelled out in Exercise 4.7.2 below.

4.7.3 Examples and Exercises

Exercise 4.7.2 Choose a coordinate chart $\phi : U \to \Omega$ with $\phi(p_0) = 0$ such that the metric in local coordinates satisfies

$$g_{ij}(0) = \delta_{ij}.$$

Refine the estimate (4.7.1) in the proof of Lemma 4.7.1 and show that

$$\lim_{p,q \to p_0} \frac{d(p,q)}{|\phi(p) - \phi(q)|} = 1.$$

This is the intrinsic analogue of Lemma 4.2.8. Use this to prove that equation (4.6.3) continues to hold for all Riemannian manifolds, i.e.

$$\lim_{\delta \to 0} \frac{d(\exp_p(\delta v), \exp_p(\delta w))}{\delta} = |v - w|$$

for $p \in M$ and $v, w \in T_pM$. With this understood, the proof of the **Curve Shortening Lemma** 4.6.8 carries over verbatim to the intrinsic setting.

Exercise 4.7.3 The real projective space $\mathbb{R}P^n$ inherits a Riemannian metric from S^n as it is a quotient of S^n by an isometric involution. Prove that each geodesic in S^n with its standard metric descends to a geodesic in $\mathbb{R}P^n$.

Exercise 4.7.4 Let $f : S^3 \to S^2$ be the **Hopf fibration** defined by

$$f(z, w) = \left(|z|^2 - |w|^2, 2\operatorname{Re}\bar{z}w, 2\operatorname{Im}\bar{z}w\right)$$

Prove that the image of a great circle in S^3 is a nonconstant geodesic in S^2 if and only if it is orthogonal to the fibers of f, which are also great circles. Here we identify S^3 with the unit sphere in $\mathbb{C}^2$. (See also Exercise 2.5.22.)

Exercise 4.7.5 Prove that a nonconstant geodesic $\gamma : \mathbb{R} \to S^{2n+1}$ descends to a nonconstant geodesic in $\mathbb{C}P^n$ with the Fubini–Study metric (see Example 3.7.5) if and only if $\dot{\gamma}(t) \perp \mathbb{C}\gamma(t)$ for every $t \in \mathbb{R}$.

Exercise 4.7.6 Consider the manifold

$$\mathcal{F}_k(\mathbb{R}^n) := \left\{D \in \mathbb{R}^{n \times k} \,\middle|\, D^\mathsf{T}D = \mathbb{1}\right\}$$

of orthonormal k-frames in $\mathbb{R}^n$, equipped with the Riemannian metric inherited from the standard inner product $\langle X, Y \rangle := \operatorname{trace}(X^\mathsf{T}Y)$ on the space of real $n \times k$-matrices.

(a) Prove that

$$T_D\mathcal{F}_k(\mathbb{R}^n) = \left\{X \in \mathbb{R}^{n \times k} \,\middle|\, D^\mathsf{T}X + X^\mathsf{T}D = 0\right\},$$
$$T_D\mathcal{F}_k(\mathbb{R}^n)^\perp = \left\{DA \,\middle|\, A = A^\mathsf{T} \in \mathbb{R}^{k \times k}\right\}.$$

and that the orthogonal projection $\Pi(D) : \mathbb{R}^{n \times k} \to T_D\mathcal{F}_k(\mathbb{R}^n)$ is given by

$$\Pi(D)X = X - \frac{1}{2}D\left(D^\mathsf{T}X + X^\mathsf{T}D\right).$$

(b) Prove that the second fundamental form of $\mathcal{F}_k(\mathbb{R}^n)$ is given by

$$h_D(X)Y = -\frac{1}{2}D\left(X^\mathsf{T}Y + Y^\mathsf{T}X\right)$$

for $D \in \mathcal{F}_k(\mathbb{R}^n)$ and $X, Y \in T_D\mathcal{F}_k(\mathbb{R}^n)$.

(c) Prove that a smooth map $\mathbb{R} \to \mathcal{F}_k(\mathbb{R}^n) : t \mapsto D(t)$ is a geodesic if and only if it satisfies the differential equation

$$\ddot{D} = -D\dot{D}^\mathsf{T}\dot{D}. \tag{4.7.4}$$

Prove that the function $D^\mathsf{T}\dot{D}$ is constant for every geodesic in $\mathcal{F}_k(\mathbb{R}^n)$. Compare this with Example 4.3.12.

Exercise 4.7.7 Let $G_k(\mathbb{R}^n) = \mathcal{F}_k(\mathbb{R}^n)/O(k)$ be the real Grassmannian of k-dimensional subspaces in $\mathbb{R}^n$, equipped with a Riemannian metric as in Example 3.7.6. Prove that a geodesics $\mathbb{R} \to \mathcal{F}_k(\mathbb{R}^n) : t \mapsto D(t)$ descends to a nonconstant geodesic in $G_k(\mathbb{R}^n)$ if and only if $D^{\mathsf{T}}\dot{D} \equiv 0$ and $\dot{D} \not\equiv 0$. Deduce that the exponential map on $G_k(\mathbb{R}^n)$ is given by

$$
\exp_\Lambda(\widehat{\Lambda}) = \mathrm{im}\left(D \cos\left(\left(\widehat{D}^{\mathsf{T}}\widehat{D}\right)^{1/2} \right) + \widehat{D}\left(\widehat{D}^{\mathsf{T}}\widehat{D}\right)^{-1/2} \sin\left(\left(\widehat{D}^{\mathsf{T}}\widehat{D}\right)^{1/2} \right) \right)
$$

for $\Lambda \in \mathcal{F}_k(\mathbb{R}^n)$ and $\widehat{\Lambda} \in T_\Lambda \mathcal{F}_k(\mathbb{R}^n) \setminus \{0\}$. Here we identify the tangent space $T_\Lambda \mathcal{F}_k(\mathbb{R}^n)$ with the space of linear maps from Λ to $\Lambda^\perp$, and choose the matrices $D \in \mathcal{F}_k(\mathbb{R}^n)$ and $\widehat{D} \in \mathbb{R}^{n \times k}$ such that

$$
\Lambda = \mathrm{im}\, D, \qquad D^{\mathsf{T}}\widehat{D} = 0, \qquad \widehat{\Lambda} \circ D = \widehat{D} : \mathbb{R}^k \to \Lambda^\perp = \ker D^{\mathsf{T}}.
$$

Prove that the group $O(n)$ acts on $G_k(\mathbb{R}^n)$ by isometries. Which subgroup acts trivially?

Exercise 4.7.8 Carry over Exercises 4.7.6 and 4.7.7 to the complex Grassmannian $G_k(\mathbb{C}^n)$. Prove that the group $U(n)$ acts on $G_k(\mathbb{C}^n)$ by isometries. Which subgroup acts trivially?

Curvature

5

This chapter begins by introducing the notion of an isometry (Sect. 5.1). It shows that isometries of embedded manifolds preserve the lengths of curves and can be characterized as diffeomorphisms whose derivatives preserve the inner products. The chapter then moves on to the Riemann curvature tensor and establishes its symmetry properties (Sect. 5.2). That section also includes a discussion of the covariant derivative of a global vector field. The next section is devoted to the generalized Gauß Theorema Egregium which asserts that isometries preserve geodesics, the covariant derivative, and the Riemann curvature tensor (Sect. 5.3). The final section examines the Riemann curvature tensor in local coordinates and shows how the definitions and results of the present chapter carry over to the intrinsic setting of Riemannian manifolds (Sect. 5.4).

5.1 Isometries

Let M and M' be connected submanifolds of $\mathbb{R}^n$. An isometry is an isomorphism of the intrinsic geometries of M and M'. Recall the definition of the intrinsic distance function

$$d : M \times M \to [0, \infty)$$

in Sect. 4.2 by

$$d(p,q) := \inf_{\gamma \in \Omega_{p,q}} L(\gamma), \qquad L(\gamma) = \int_0^1 |\dot{\gamma}(t)| \, dt$$

for $p, q \in M$. Let d' denote the intrinisic distance function on M'.

© The Editor(s) (if applicable) and The Author(s), under exclusive license to
Springer-Verlag GmbH, DE, part of Springer Nature 2022
J.W. Robbin, D.A. Salamon, *Introduction to Differential Geometry*,
Springer Studium Mathematik (Master), https://doi.org/10.1007/978-3-662-64340-2_5

Theorem 5.1.1 (Isometries) *Let $\phi : M \to M'$ be a bijective map. Then the following are equivalent.*

(i)   *ϕ intertwines the distance functions on M and M', i.e.*

$$d'(\phi(p), \phi(q)) = d(p, q)$$

for all $p, q \in M$.
(ii) *ϕ is a diffeomorphism and*

$$d\phi(p) : T_p M \to T_{\phi(p)} M'$$

is an orthogonal isomorphism for every $p \in M$.
(iii) *ϕ is a diffeomorphism and*

$$L(\phi \circ \gamma) = L(\gamma)$$

for every smooth curve $\gamma : [a, b] \to M$.

*The bijection ϕ is called an **isometry** iff it satisfies these equivalent conditions. In the case $M = M'$ the isometries $\phi : M \to M$ form a group denoted by $\mathcal{I}(M)$ and called the **isometry group** of M.*

The proof is based on the following lemma.

Lemma 5.1.2 *For every $p \in M$ there exists a constant $\varepsilon > 0$ such that, for all $v, w \in T_p M$ with $0 < |w| < |v| < \varepsilon$, we have*

$$d(\exp_p(w), \exp_p(v)) = |v| - |w| \qquad \Longrightarrow \qquad w = \frac{|w|}{|v|} v. \qquad (5.1.1)$$

Remark 5.1.3 It follows from the triangle inequality and Theorem 4.4.4 that

$$d(\exp_p(v), \exp_p(w)) \geq d(\exp_p(v), p) - d(\exp_p(w), p)$$
$$= |v| - |w|$$

whenever $0 < |w| < |v| < \mathrm{inj}(p)$. Lemma 5.1.2 asserts that equality can only hold when w is a positive multiple of v or, to put it differently, that the distance between $\exp_p(v)$ and $\exp_p(w)$ must be strictly bigger that $|v| - |w|$ whenever w is not a positive multiple of v.

Proof of Lemma 5.1.2 As in Corollary 4.3.8 we denote

$$B_\varepsilon(p) := \{v \in T_p M \mid |v| < \varepsilon\},$$
$$U_\varepsilon(p) := \{q \in M \mid d(p,q) < \varepsilon\}.$$

By Theorem 4.4.4 and the definition of the injectivity radius, the exponential map at p is a diffeomorphism $\exp_p : B_\varepsilon(p) \to U_\varepsilon(p)$ for $\varepsilon < \mathrm{inj}(p)$. Choose $0 < r < \mathrm{inj}(p)$. Then the closure of $U_r(p)$ is a compact subset of M. Hence there is a constant $\varepsilon > 0$ such that $\varepsilon < r$ and $\varepsilon < \mathrm{inj}(p')$ for every $p' \in \overline{U_r(p)}$. Since $\varepsilon < r$ we have

$$\varepsilon < \mathrm{inj}(p') \qquad \forall\, p' \in U_\varepsilon(p). \tag{5.1.2}$$

Thus $\exp_{p'} : B_\varepsilon(p') \to U_\varepsilon(p')$ is a diffeomorphism for every $p' \in U_\varepsilon(p)$. Define $p_1 := \exp_p(w)$ and $p_2 := \exp_p(v)$. Then, by assumption, we have $d(p_1, p_2) = |v| - |w| < \varepsilon$. Since $p_1 \in U_\varepsilon(p)$ it follows from our choice of ε that $\varepsilon < \mathrm{inj}(p_1)$. Hence there is a unique tangent vector $v_1 \in T_{p_1} M$ such that

$$|v_1| = d(p_1, p_2) = |v| - |w|, \qquad \exp_{p_1}(v_1) = p_2.$$

Following first the shortest geodesic from p to p_1 and then the shortest geodesic from p_1 to p_2 we obtain (after suitable reparametrization) a smooth $\gamma : [0, 2] \to M$ such that

$$\gamma(0) = p, \qquad \gamma(1) = p_1, \qquad \gamma(2) = p_2,$$

and

$$L(\gamma|_{[0,1]}) = d(p, p_1) = |w|, \qquad L(\gamma|_{[1,2]}) = d(p_1, p_2) = |v| - |w|.$$

Thus $L(\gamma) = |v| = d(p, p_2)$. Hence, by Theorem 4.4.4, there is a smooth function $\beta : [0, 2] \to [0, 1]$ satisfying

$$\beta(0) = 0, \qquad \beta(2) = 1, \qquad \dot{\beta}(t) \geq 0, \qquad \gamma(t) = \exp_p(\beta(t)v)$$

for every $t \in [0, 2]$. This implies

$$\exp_p(w) = p_1 = \gamma(1) = \exp_p(\beta(1)v), \qquad 0 \leq \beta(1) \leq 1.$$

Since w and $\beta(1)v$ are both elements of $B_\varepsilon(p)$ and $\exp_p$ is injective on $B_\varepsilon(p)$, this implies $w = \beta(1)v$. Since $\beta(1) \geq 0$ we have $\beta(1) = |w|/|v|$. This proves (5.1.1) and Lemma 5.1.2. $\square$

Proof of Theorem 5.1.1   That (ii) implies (iii) follows from the definition of the length of a curve. Namely

$$L(\phi \circ \gamma) = \int_a^b \left| \frac{d}{dt} \phi(\gamma(t)) \right| dt$$

$$= \int_a^b |d\phi(\gamma(t))\dot{\gamma}(t)| \, dt$$

$$= \int_a^b |\dot{\gamma}(t)| \, dt$$

$$= L(\gamma).$$

In the third equation we have used (ii). That (iii) implies (i) follows immediately from the definition of the intrinsic distance functions d and d'.

We prove that (i) implies (ii). Fix a point $p \in M$ and choose $\varepsilon > 0$ so small that $\varepsilon < \min\{\text{inj}(p; M), \text{inj}(\phi(p); M')\}$ and that the assertion of Lemma 5.1.2 holds for the point $p' := \phi(p) \in M'$. Then there is a unique homeomorphism

$$\Phi_p : B_\varepsilon(p) \to B_\varepsilon(\phi(p))$$

such that the following diagram commutes.

$$
\begin{array}{ccccccc}
T_p M & \supset & B_\varepsilon(p) & \xrightarrow{\ \Phi_p\ } & B_\varepsilon(\phi(p)) & \subset & T_{\phi(p)} M' \ . \\
 & & {\scriptstyle \exp_p} \downarrow & & \downarrow {\scriptstyle \exp'_{\phi(p)}} & & \\
M & \supset & U_\varepsilon(p) & \xrightarrow{\ \phi\ } & U_\varepsilon(\phi(p)) & \subset & M'
\end{array}
$$

Here the vertical maps are diffeomorphisms and $\phi : U_\varepsilon(p) \to U_\varepsilon(\phi(p))$ is a homeomorphism by (i). Hence $\Phi_p : B_\varepsilon(p) \to B_\varepsilon(\phi(p))$ is a homeomorphism.

Claim 1 *The map Φ_p satisfies the equations*

$$\exp'_{\phi(p)}(\Phi_p(v)) = \phi(\exp_p(v)), \qquad (5.1.3)$$
$$|\Phi_p(v)| = |v|, \qquad (5.1.4)$$
$$\Phi_p(tv) = t\Phi_p(v) \qquad (5.1.5)$$

for every $v \in B_\varepsilon(p)$ and every $t \in [0, 1]$.

Equation (5.1.3) holds by definition. To prove (5.1.4) we observe that, by Theorem 4.4.4, we have

$$\begin{aligned}
\left|\Phi_p(v)\right| &= d'(\phi(p), \exp'_{\phi(p)}(\Phi_p(v))) \\
&= d'(\phi(p), \phi(\exp_p(v))) \\
&= d(p, \exp_p(v)) \\
&= |v|.
\end{aligned}$$

Here the second equation follows from (5.1.3) and the third equation from (i). Equation (5.1.5) holds for $t = 0$ because $\Phi_p(0) = 0$ and for $t = 1$ it is a tautology. Hence assume $0 < t < 1$. Then

$$\begin{aligned}
d'(\exp'_{\phi(p)}(\Phi_p(tv)), \exp'_{\phi(p)}(\Phi_p(v))) &= d'(\phi(\exp_p(tv)), \phi(\exp_p(v))) \\
&= d(\exp_p(tv), \exp_p(v)) \\
&= |v| - |tv| \\
&= \left|\Phi_p(v)\right| - \left|\Phi_p(tv)\right|.
\end{aligned}$$

Here the first equation follows from (5.1.3), the second equation from (i), the third equation from Theorem 4.4.4 and the fact that $|v| < \operatorname{inj}(p)$, and the last equation follows from (5.1.4). Since $0 < \left|\Phi_p(tv)\right| < \left|\Phi_p(v)\right| < \varepsilon$ we can apply Lemma 5.1.2 and obtain

$$\Phi_p(tv) = \frac{\left|\Phi_p(tv)\right|}{\left|\Phi_p(v)\right|} \Phi_p(v) = t\Phi_p(v).$$

This proves Claim 1.

By Claim 1, Φ_p extends to a bijective map $\Phi_p : T_p M \to T_{\phi(p)} M'$ via

$$\Phi_p(v) := \frac{1}{\delta}\Phi_p(\delta v),$$

where $\delta > 0$ is chosen so small that $\delta|v| < \varepsilon$. The right hand side of this equation is independent of the choice of δ. Hence the extension is well defined. It is bijective because the original map Φ_p is a bijection from $B_\varepsilon(p)$ to $B_\varepsilon(\phi(p))$. The reader may verify that the extended map satisfies the conditions (5.1.4) and (5.1.5) for all $v \in T_p M$ and all $t \geq 0$.

Claim 2 *The extended map* $\Phi_p : T_pM \to T_{\phi(p)}M'$ *is linear and preserves the inner product.*

It follows from the equation (4.6.3) in the proof of Lemma 4.6.8 that

$$
\begin{aligned}
|v - w| &= \lim_{t \to 0} \frac{d(\exp_p(tv), \exp_p(tw))}{t} \\
&= \lim_{t \to 0} \frac{d'(\phi(\exp_p(tv)), \phi(\exp_p(tw)))}{t} \\
&= \lim_{t \to 0} \frac{d'(\exp'_{\phi(p)}(\Phi_p(tv)), \exp'_{\phi(p)}(\Phi_p(tw)))}{t} \\
&= \lim_{t \to 0} \frac{d'(\exp'_{\phi(p)}(t\Phi_p(v)), \exp'_{\phi(p)}(t\Phi_p(w)))}{t} \\
&= \left| \Phi_p(v) - \Phi_p(w) \right|.
\end{aligned}
$$

Here the second equation follows from (i), the third from (5.1.3), the fourth from (5.1.4), and the last equation follows again from (4.6.3). By polarization we obtain

$$
\begin{aligned}
2\langle v, w \rangle &= |v|^2 + |w|^2 - |v - w|^2 \\
&= \left| \Phi_p(v) \right|^2 + \left| \Phi_p(w) \right|^2 - \left| \Phi_p(v) - \Phi_p(w) \right|^2 \\
&= 2\langle \Phi_p(v), \Phi_p(w) \rangle.
\end{aligned}
$$

Thus Φ_p preserves the inner product. Hence, for all $v_1, v_2, w \in T_pM$, we have

$$
\begin{aligned}
\langle \Phi_p(v_1 + v_2), \Phi_p(w) \rangle &= \langle v_1 + v_2, w \rangle \\
&= \langle v_1, w \rangle + \langle v_2, w \rangle \\
&= \langle \Phi_p(v_1), \Phi_p(w) \rangle + \langle \Phi_p(v_2), \Phi_p(w) \rangle \\
&= \langle \Phi_p(v_1) + \Phi_p(v_2), \Phi_p(w) \rangle.
\end{aligned}
$$

Since Φ_p is surjective, this implies

$$
\Phi_p(v_1 + v_2) = \Phi_p(v_1) + \Phi_p(v_2)
$$

for all $v_1, v_2 \in T_pM$. With $v_1 = v$ and $v_2 = -v$ we obtain

$$
\Phi_p(-v) = -\Phi_p(v)
$$

for every $v \in T_pM$ and by (5.1.5) this gives

$$
\Phi_p(tv) = t\Phi_p(v)
$$

for all $v \in T_pM$ and $t \in \mathbb{R}$. This proves Claim 2.

Claim 3 ϕ is smooth and $d\phi(p) = \Phi_p$.

By (5.1.3) we have

$$\phi = \exp'_{\phi(p)} \circ \Phi_p \circ \exp_p^{-1} : U_\varepsilon(p) \to U_\varepsilon(\phi(p)).$$

Since Φ_p is linear, this shows that the restriction of ϕ to the open set $U_\varepsilon(p)$ is smooth. Moreover, for every $v \in T_p M$ we have

$$d\phi(p)v = \frac{d}{dt}\bigg|_{t=0} \phi(\exp_p(tv)) = \frac{d}{dt}\bigg|_{t=0} \exp'_{\phi(p)}(t\Phi_p(v)) = \Phi_p(v).$$

Here we have used equations (5.1.3) and (5.1.5) as well as Lemma 4.3.6. This proves Claim 3 and Theorem 5.1.1. □

Exercise 5.1.4 Prove that every isometry $\psi : \mathbb{R}^n \to \mathbb{R}^n$ is an affine map

$$\psi(p) = Ap + b$$

where $A \in O(n)$ and $b \in \mathbb{R}^n$. Thus ψ is a composition of translation and rotation. **Hint:** Let $e_1, \ldots, e_n$ be the standard basis of $\mathbb{R}^n$. Prove that any two vectors $v, w \in \mathbb{R}^n$ that satisfy

$$|v| = |w|$$

and

$$|v - e_i| = |w - e_i| \qquad \text{for } i = 1, \ldots, n$$

must be equal.

Remark 5.1.5 If $\psi : \mathbb{R}^n \to \mathbb{R}^n$ is an isometry of the ambient Euclidean space with $\psi(M) = M'$, then certainly $\phi := \psi|_M$ is an isometry from M onto M'. On the other hand, if M is a plane manifold

$$M = \{(0, y, z) \in \mathbb{R}^3 \mid 0 < y < \pi/2\}$$

and M' is the cylindrical manifold

$$M' = \{(x, y, z) \in \mathbb{R}^3 \mid x^2 + y^2 = 1, \, x > 0, \, y > 0\},$$

Then the map $\phi : M \to M'$ defined by

$$\phi(0, y, z) := (\cos(y), \sin(y), z)$$

is an isometry which is *not* of the form $\phi = \psi|_M$. Indeed, an isometry of the form $\phi = \psi|_M$ necessarily preserves the second fundamental form (as well as the first) in the sense that

$$d\psi(p)h_p(v, w) = h'_{\psi(p)}(d\psi(p)v, d\psi(p)w)$$

for $v, w \in T_pM$ but in the example h vanishes identically while h' does not.

We may thus distinguish two fundamental question:

I. Given M and M' when are they extrinsically isomorphic, i.e. when is there an ambient isometry $\psi : \mathbb{R}^n \to \mathbb{R}^n$ with $\psi(M) = M'$?
II. Given M and M' when are they intrinsically isomorphic, i.e. when is there an isometry $\phi : M \to M'$ from M onto M'?

As we have noted, both the first and second fundamental forms are preserved by extrinsic isomorphisms while only the first fundamental form need be preserved by an intrinsic isomorphism (i.e. an isometry).

A question which occurred to Gauß (who worked for a while as a cartographer) is this: Can one draw a perfectly accurate map of a portion of the earth? (i.e. a map for which the distance between points on the map is proportional to the distance between the corresponding points on the surface of the earth). We can now pose this question as follows: Is there an isometry from an open subset of a sphere to an open subset of a plane? Gauß answered this question negatively by associating an invariant, the Gaußian curvature $K : M \to \mathbb{R}$, to a surface $M \subset \mathbb{R}^3$. According to his *Theorema Egregium*

$$K' \circ \phi = K$$

for an isometry $\phi : M \to M'$. The sphere has positive curvature; the plane has zero curvature; hence the perfectly accurate map does not exist. Our aim is to explain these ideas.

Local Isometries
We shall need a concept slightly more general than that of "isometry".

Definition 5.1.6 (Local isometry) A smooth map $\phi : M \to M'$ is called a **local isometry** iff its derivative

$$d\phi(p) : T_pM \to T_{\phi(p)}M'$$

is an orthogonal linear isomorphism for every $p \in M$.

Remark 5.1.7 Let $M \subset \mathbb{R}^n$ and $M' \subset \mathbb{R}^{n'}$ be manifolds and $\phi : M \to M'$ be a map. The following are equivalent.

(i) ϕ is a local isometry.
(ii) For every $p \in M$ there are open neighborhoods $U \subset M$ and $U' \subset M'$ such that the restriction of ϕ to U is an isometry from U onto U'.

That (ii) implies (i) follows immediately from Theorem 5.1.1. On the other hand (i) implies that $d\phi(p)$ is invertible so that (ii) follows from the inverse function theorem.

Example 5.1.8 The map

$$\mathbb{R} \to S^1 : \theta \mapsto e^{i\theta}$$

is a local isometry but not an isometry.

Exercise 5.1.9 Let $M \subset \mathbb{R}^n$ be a compact connected 1-manifold. Prove that M is diffeomorphic to the circle S^1. Define the length of a compact connected Riemannian 1-manifold. Prove that two compact connected 1-manifolds $M, M' \subset \mathbb{R}^n$ are isometric if and only if they have the same length. **Hint:** Let $\gamma : \mathbb{R} \to M$ be a geodesic with $|\dot{\gamma}(t)| \equiv 1$. Show that γ is not injective; otherwise construct an open cover of M without finite subcover. If $t_0 < t_1$ with $\gamma(t_0) = \gamma(t_1)$, show that $\dot{\gamma}(t_0) = \dot{\gamma}(t_1)$; otherwise show that $\gamma(t_0 + t) = \gamma(t_1 - t)$ for all t and find a contradiction.

The next result asserts that two local isometries that have the same value and the same derivative at a single point must agree everywhere, provided that the domain is connected.

Lemma 5.1.10 *Let $M \subset \mathbb{R}^n$ and $M' \subset \mathbb{R}^{n'}$ be smooth m-manifolds and assume that M is connected. Let $\phi : M \to M'$ and $\psi : M \to M'$ be local isometries and let $p_0 \in M$ such that*

$$\phi(p_0) = \psi(p_0) =: p_0', \qquad d\phi(p_0) = d\psi(p_0) : T_{p_0}M \to T_{p_0'}M'.$$

Then $\phi(p) = \psi(p)$ for every $p \in M$.

Proof Define the set

$$M_0 := \{p \in M \mid \phi(p) = \psi(p), d\phi(p) = d\psi(p)\}.$$

This set is obviously closed. We prove that M_0 is open. Let $p \in M_0$ and choose $U \subset M$ and $U' \subset M'$ as in Remark 5.1.7 (ii). Denote

$$\Phi_p := d\phi(p) = d\psi(p) : T_pM \to T_{p'}M', \qquad p' := \phi(p) = \psi(p)$$

Then it follows from equation (5.1.3) in the proof of Theorem 5.1.1 that there exists a constant $\varepsilon > 0$ such that $U_\varepsilon(p) \subset U$ and $U_\varepsilon(p') \subset U'$ and

$$q \in U_\varepsilon(p) \qquad \Longrightarrow \qquad \phi(q) = \exp'_{p'} \circ \Phi_p \circ \exp_p^{-1}(q) = \psi(q).$$

Hence $U_\varepsilon(p) \subset M_0$. Thus M_0 is open, closed, and nonempty. Since M is connected it follows that $M_0 = M$ and this proves Lemma 5.1.10.   □

Exercise 5.1.11

(i)   *If a sequence of local isometries $\phi_i : M \to M'$ converges uniformly to a local isometry $\phi : M \to M'$, then it converges in the C^∞ topology.* **Hint:** Let $p \in M$. Then every sufficiently small tangent vector $v \in T_p M$ satisfies the equation $d\phi(p)v = (\exp'_{\phi(p)})^{-1}(\phi(\exp_p(v)))$. Use this to prove that $d\phi_i(p)$ converges to $d\phi(p)$. Deduce that ϕ_i converges to ϕ uniformly with all derivatives in a neighborhood of p.

(ii) *The C^∞ topology on the space of local isometries from M to M' agrees with the C^0 topology.*

5.2   The Riemann Curvature Tensor

This section defines the Riemann curvature tensor and proves the Gauß–Codazzi formula (Sect. 5.2.1), introduces the covariant derivative of a global vector field (Sect. 5.2.2), expresses the curvature tensor in terms of a global formula (Sect. 5.2.3), establishes its symmetry properties (Sect. 5.2.4), and examines the curvature for a class of Riemannian metrics on Lie groups (Sect. 5.2.5).

5.2.1   Definition and Gauß–Codazzi

Let $M \subset \mathbb{R}^n$ be a smooth manifold and $\gamma : \mathbb{R}^2 \to M$ be a smooth map. Denote by (s, t) the coordinates on $\mathbb{R}^2$. Let $Z \in \mathrm{Vect}(\gamma)$ be a smooth vector field along γ, i.e. $Z : \mathbb{R}^2 \to \mathbb{R}^n$ is a smooth map such that $Z(s, t) \in T_{\gamma(s,t)} M$ for all s and t. The **covariant partial derivatives** of Z with respect to the variables s and t are defined by

$$\nabla_s Z := \Pi(\gamma)\frac{\partial Z}{\partial s}, \qquad \nabla_t Z := \Pi(\gamma)\frac{\partial Z}{\partial t}.$$

In particular $\partial_s \gamma = \partial\gamma/\partial s$ and $\partial_t \gamma = \partial\gamma/\partial t$ are vector fields along γ and we have $\nabla_s \partial_t \gamma - \nabla_t \partial_s \gamma = 0$ as both terms on the left are equal to $\Pi(\gamma)\partial_s \partial_t \gamma$. Thus ordinary partial differentiation and covariant partial differentiation commute. The analogous formula (which results on replacing ∂ by ∇ and γ by Z) is in general false. Instead we have the following.

Definition 5.2.1   The **Riemann curvature tensor** assigns to each $p \in M$ the bi-linear map $R_p : T_p M \times T_p M \to \mathcal{L}(T_p M, T_p M)$ characterized by the equation

$$R_p(u, v)w = \left(\nabla_s \nabla_t Z - \nabla_t \nabla_s Z \right)(0, 0) \tag{5.2.1}$$

for $u, v, w \in T_p M$ where $\gamma : \mathbb{R}^2 \to M$ is a smooth map and $Z \in \mathrm{Vect}(\gamma)$ is a smooth vector field along γ such that

$$\gamma(0, 0) = p, \qquad \partial_s \gamma(0, 0) = u, \qquad \partial_t \gamma(0, 0) = v, \qquad Z(0, 0) = w. \tag{5.2.2}$$

We must prove that R is well defined, i.e. that the right hand side of equation (5.2.1) is independent of the choice of γ and Z. This follows from the Gauß–Codazzi formula which we prove next. Recall that the second fundamental form can be viewed as a linear map $h_p : T_p M \to \mathcal{L}(T_p M, T_p M^\perp)$ and that, for $u \in T_p M$, the linear map $h_p(u) \in \mathcal{L}(T_p M, T_p M^\perp)$ and its dual $h_p(u)^* \in \mathcal{L}(T_p M^\perp, T_p M)$ are given by

$$h_p(u)v = \left(d\Pi(p)u \right)v, \qquad h_p(u)^*w = \left(d\Pi(p)u \right)w$$

for $v \in T_p M$ and $w \in T_p M^\perp$.

Theorem 5.2.2   *The Riemann curvature tensor is well defined and given by the Gauß–Codazzi formula*

$$R_p(u, v) = h_p(u)^* h_p(v) - h_p(v)^* h_p(u) \tag{5.2.3}$$

for $u, v \in T_p M$.

Proof   Let $u, v, w \in T_p M$ and choose a smooth map $\gamma : \mathbb{R}^2 \to M$ and a smooth vector field Z along γ such that (5.2.2) holds. Then, by the Gauß–Weingarten formula (3.2.2), we have

$$\begin{aligned}
\nabla_t Z &= \partial_t Z - h_\gamma(\partial_t \gamma)Z \\
&= \partial_t Z - \left(d\Pi(\gamma)\partial_t \gamma \right)Z \\
&= \partial_t Z - \left(\partial_t (\Pi \circ \gamma) \right)Z.
\end{aligned}$$

Hence

$$\begin{aligned}
\partial_s \nabla_t Z &= \partial_s \partial_t Z - \partial_s \left(\left(\partial_t (\Pi \circ \gamma) \right)Z \right) \\
&= \partial_s \partial_t Z - \left(\partial_s \partial_t (\Pi \circ \gamma) \right)Z - \left(\partial_t (\Pi \circ \gamma) \right)\partial_s Z \\
&= \partial_s \partial_t Z - \left(\partial_s \partial_t (\Pi \circ \gamma) \right)Z - \left(d\Pi(\gamma)\partial_t \gamma \right)\left(\nabla_s Z + h_\gamma(\partial_s \gamma)Z \right) \\
&= \partial_s \partial_t Z - \left(\partial_s \partial_t (\Pi \circ \gamma) \right)Z - h_\gamma(\partial_t \gamma)\nabla_s Z - h_\gamma(\partial_t \gamma)^* h_\gamma(\partial_s \gamma)Z.
\end{aligned}$$

Interchanging s and t and taking the difference we obtain

$$\partial_s \nabla_t Z - \partial_t \nabla_s Z = h_\gamma(\partial_s \gamma)^* h_\gamma(\partial_t \gamma)Z - h_\gamma(\partial_t \gamma)^* h_\gamma(\partial_s \gamma)Z$$
$$+ h_\gamma(\partial_s \gamma)\nabla_t Z - h_\gamma(\partial_t \gamma)\nabla_s Z.$$

Here the first two terms on the right are tangent to M and the last two terms on the right are orthogonal to $T_\gamma M$. Hence

$$\nabla_s \nabla_t Z - \nabla_t \nabla_s Z = \Pi(\gamma)\big(\partial_s \nabla_t Z - \partial_t \nabla_s Z\big)$$
$$= h_\gamma(\partial_s \gamma)^* h_\gamma(\partial_t \gamma)Z - h_\gamma(\partial_t \gamma)^* h_\gamma(\partial_s \gamma)Z.$$

Evaluating the right hand side at $s = t = 0$ we find that

$$\big(\nabla_s \nabla_t Z - \nabla_t \nabla_s Z\big)(0,0) = h_p(u)^* h_p(v)w - h_p(v)^* h_p(u)w.$$

This proves the Gauß–Codazzi equation and shows that the left hand side is independent of the choice of γ and Z. This proves Theorem 5.2.2. $\qquad\square$

5.2.2   Covariant Derivative of a Global Vector Field

So far we have only defined the covariant derivatives of vector fields along curves. The same method can be applied to global vector fields. This leads to the following definition.

Definition 5.2.3 (Covariant derivative)   Let $M \subset \mathbb{R}^n$ be an m-dimensional submanifold and X be a vector field on M. Fix a point $p \in M$ and a tangent vector $v \in T_p M$. The **covariant derivative of X at p in the direction v** is the tangent vector

$$\nabla_v X(p) := \Pi(p)dX(p)v \in T_p M,$$

where $\Pi(p) \in \mathbb{R}^{n \times n}$ denotes the orthogonal projection onto $T_p M$.

Remark 5.2.4   Let $\gamma : I \to M$ be a smooth curve on an interval $I \subset \mathbb{R}$ and let $X \in \mathrm{Vect}(M)$ be a smooth vector field on M. Then $X \circ \gamma$ is a smooth vector field along γ and the covariant derivative of $X \circ \gamma$ is related to the covariant derivative of X by the formula

$$\nabla(X \circ \gamma)(t) = \nabla_{\dot\gamma(t)} X(\gamma(t)). \tag{5.2.4}$$

Remark 5.2.5 (Gauß–Weingarten formula)   Differentiating the equation $X = \Pi X$ (understood as a function from M to $\mathbb{R}^n$) and using the notation $\partial_v X(p) := dX(p)v$ for the derivative of X at p in the direction v we obtain the **Gauß–Weingarten formula** for global vector fields:

$$\partial_v X(p) = \nabla_v X(p) + h_p(v)X(p). \tag{5.2.5}$$

Remark 5.2.6 (Levi-Civita connection) Differentiating a vector field Y on M covariantly in the direction of another vector field X we obtain a vector field $\nabla_X Y \in$ Vect(M) defined by

$$(\nabla_X Y)(p) := \nabla_{X(p)} Y(p)$$

for $p \in M$. This gives rise to a family of linear operators

$$\nabla_X : \text{Vect}(M) \to \text{Vect}(M),$$

one for each vector field $X \in$ Vect(M), and the assignment

$$\text{Vect}(M) \to \mathcal{L}(\text{Vect}(M), \text{Vect}(M)) : X \mapsto \nabla_X \qquad (5.2.6)$$

is itself a linear operator. This linear operator is called the **Levi-Civita connection** on the tangent bundle TM.

The Levi-Civita connection (5.2.6) satisfies the conditions

$$\nabla_{fX}(Y) = f \nabla_X Y, \qquad (5.2.7)$$
$$\nabla_X (f Y) = f \nabla_X Y + (\mathcal{L}_X f) Y, \qquad (5.2.8)$$
$$\mathcal{L}_X \langle Y, Z \rangle = \langle \nabla_X Y, Z \rangle + \langle Y, \nabla_X Z \rangle, \qquad (5.2.9)$$
$$\nabla_Y X - \nabla_X Y = [X, Y] \qquad (5.2.10)$$

for all $X, Y, Z \in$ Vect(M) and all $f \in \mathcal{F}(M)$, where $\mathcal{L}_X f = df \circ X$ and $[X, Y] \in$ Vect(M) denotes the Lie bracket of the vector fields X and Y. The conditions (5.2.7) and (5.2.8) assert that the linear operator (5.2.6) is a **connection** on the tangent bundle TM, condition (5.2.9) asserts that the connection (5.2.6) is **Riemannian** (i.e. it is compatible with the first fundamental form), and condition (5.2.10) asserts that it is **torsion-free**.

The next lemma shows that the Levi-Civita connection (5.2.6) is uniquely determined by (5.2.9) and (5.2.10), and hence is the unique torsion-free Riemannian connection on the tangent bundle TM.

Lemma 5.2.7 (Uniqueness Lemma) *There is a unique linear operator*

$$\text{Vect}(M) \to \mathcal{L}(\text{Vect}(M), \text{Vect}(M)) : X \mapsto \nabla_X$$

satisfying equations (5.2.9) and (5.2.10) for all $X, Y, Z \in$ Vect(M).

Proof Existence follows from the properties of the Levi-Civita connection. We prove uniqueness. Let $X \mapsto D_X$ be any linear operator from Vect(M) to $\mathcal{L}(\text{Vect}(M), \text{Vect}(M))$ that satisfies (5.2.9) and (5.2.10). Then we have

$$\mathcal{L}_X \langle Y, Z \rangle = \langle D_X Y, Z \rangle + \langle Y, D_X Z \rangle,$$
$$\mathcal{L}_Y \langle X, Z \rangle = \langle D_Y X, Z \rangle + \langle X, D_Y Z \rangle,$$
$$-\mathcal{L}_Z \langle X, Y \rangle = -\langle D_Z X, Y \rangle - \langle X, D_Z Y \rangle.$$

Adding these three equations we find

$$
\begin{aligned}
&\mathcal{L}_X\langle Y, Z\rangle + \mathcal{L}_Y\langle Z, X\rangle - \mathcal{L}_Z\langle X, Y\rangle \\
&= 2\langle D_X Y, Z\rangle + \langle D_Y X - D_X Y, Z\rangle \\
&\quad + \langle X, D_Y Z - D_Z Y\rangle + \langle Y, D_X Z - D_Z X\rangle \\
&= 2\langle D_X Y, Z\rangle + \langle [X, Y], Z\rangle + \langle X, [Z, Y]\rangle + \langle Y, [Z, X]\rangle.
\end{aligned}
$$

The same equation holds for the Levi-Civita connection and hence

$$
\langle D_X Y, Z\rangle = \langle \nabla_X Y, Z\rangle.
$$

This implies $D_X Y = \nabla_X Y$ for all $X, Y \in \mathrm{Vect}(M)$.   □

Exercise 5.2.8 In the proof of Lemma 5.2.7 we did not actually use the assumption that the operator $D_X : \mathrm{Vect}(M) \to \mathrm{Vect}(M)$ is linear nor that the operator $X \mapsto D_X$ is linear. Prove directly that if a map

$$
D_X : \mathcal{L}(M) \to \mathcal{L}(M)
$$

satisfies (5.2.9) for all $Y, Z \in \mathrm{Vect}(M)$, then D_X is linear. Prove that every map $\mathrm{Vect}(M) \to \mathcal{L}(\mathrm{Vect}(M), \mathrm{Vect}(M)) : X \mapsto D_X$ that satisfies (5.2.10) is linear.

Exercise 5.2.9 Let ϕ^t be the flow of a complete vector field X on M and let ψ^t be the flow of a complete vector field Y on M.

(i)   Prove that the formula $\widetilde{X}(p, v) := (X(p), dX(p)v)$ defines a vector field on the tangent bundle TM. **Hint:** Lemma 4.3.1 and equation (5.2.5).
(ii) Prove that the flow of $\widetilde{X}$ is given by $\widetilde{\phi}^t(p, v) := (\phi^t(p), d\phi^t(p)v)$.
(iii) Prove that the vector fields $t^{-1}((\psi^t)^* X - X)$ converge to $[X, Y]$ in the C^1 topology as t tends to zero. **Hint:** Establish C^0 convergence in Lemma 2.4.18 and then use this result for the vector fields $\widetilde{X}$ and $\widetilde{Y}$.

Remark 5.2.10 (The Levi-Civita connection in local coordinates) Let $\phi : U \to \Omega$ be a coordinate chart on an open set $U \subset M$ with values in an open set $\Omega \subset \mathbb{R}^m$. In such a coordinate chart a vector field $X \in \mathrm{Vect}(M)$ is represented by a smooth map $\xi = (\xi^1, \ldots, \xi^m) : \Omega \to \mathbb{R}^m$ defined by

$$
\xi(\phi(p)) = d\phi(p)X(p)
$$

for $p \in U$. If $Y \in \mathrm{Vect}(M)$ is represented by η, then $\nabla_X Y$ is represented by the map

$$
(\nabla_\xi \eta)^k := \sum_{i=1}^m \frac{\partial \eta^k}{\partial x^i} \xi^i + \sum_{i,j=1}^m \Gamma_{ij}^k \xi^i \eta^j. \tag{5.2.11}
$$

Here the $\Gamma_{ij}^k : \Omega \to \mathbb{R}$ are the Christoffel symbols defined by

$$
\Gamma_{ij}^k := \sum_{\ell=1}^m g^{k\ell} \frac{1}{2}\left(\frac{\partial g_{\ell i}}{\partial x^j} + \frac{\partial g_{\ell j}}{\partial x^i} - \frac{\partial g_{ij}}{\partial x^\ell} \right), \tag{5.2.12}
$$

where g_{ij} is the metric tensor and g^{ij} is the inverse matrix so that

$$\sum_j g_{ij} g^{jk} = \delta_i^k$$

(see Lemma 3.6.5). This formula can be used to prove the existence statement in Lemma 5.2.7 and hence define the Levi-Civita connection in the intrinsic setting.

5.2.3 A Global Formula

Lemma 5.2.11 *For $X, Y, Z \in \mathrm{Vect}(M)$ we have*

$$R(X, Y)Z = \nabla_X \nabla_Y Z - \nabla_Y \nabla_X Z + \nabla_{[X,Y]} Z. \qquad (5.2.13)$$

Proof Fix a point $p \in M$. Then the right hand side of equation (5.2.13) at p remains unchanged if we multiply each of the vector fields X, Y, Z by a smooth function $f : M \to [0, 1]$ that is equal to one near p. Choosing f with compact support we may therefore assume that the vector fields X and Y are complete. Let ϕ^s denote the flow of X and ψ^t the flow of Y. Define the map $\gamma : \mathbb{R}^2 \to M$ by

$$\gamma(s, t) := \phi^s \circ \psi^t(p), \qquad s, t \in \mathbb{R}.$$

Then

$$\partial_s \gamma = X(\gamma), \qquad \partial_t \gamma = (\phi_*^s Y)(\gamma).$$

Hence, by Remark 5.2.4 we have

$$\nabla_s(Z \circ \gamma) = (\nabla_X Z)(\gamma), \qquad \nabla_t(Z \circ \gamma) = \left(\nabla_{\phi_*^s Y} Z\right)(\gamma).$$

Using Remark 5.2.4 again we obtain

$$\nabla_s \nabla_t(Z \circ \gamma) = \nabla_{\partial_s \gamma}\left(\nabla_{\phi_*^s Y} Z\right)(\gamma) + \left(\nabla_{\partial_s \phi_*^s Y} Z\right)(\gamma),$$
$$\nabla_t \nabla_s(Z \circ \gamma) = \left(\nabla_{\phi_*^s Y} \nabla_X Z\right)(\gamma).$$

Since

$$\left.\frac{\partial}{\partial s}\right|_{s=0} \phi_*^s Y = [X, Y]$$

and $\partial_s \gamma = X(\gamma)$, it follows that

$$\nabla_s \nabla_t(Z \circ \gamma)(0, 0) = \left(\nabla_X \nabla_Y Z + \nabla_{[X,Y]} Z\right)(p),$$
$$\nabla_t \nabla_s(Z \circ \gamma)(0, 0) = \left(\nabla_Y \nabla_X Z\right)(p).$$

Hence

$$R_p(X(p), Y(p))Z(p) = \big(\nabla_s \nabla_t (Z \circ \gamma) - \nabla_t \nabla_s (Z \circ \gamma)\big)(0, 0)$$
$$= \big(\nabla_X \nabla_Y Z - \nabla_Y \nabla_X Z + \nabla_{[X,Y]} Z\big)(p).$$

This proves Lemma 5.2.11.   □

Remark 5.2.12 Equation (5.2.13) can be written succinctly as

$$[\nabla_X, \nabla_Y] + \nabla_{[X,Y]} = R(X, Y). \tag{5.2.14}$$

This can be contrasted with the equation

$$[\mathcal{L}_X, \mathcal{L}_Y] + \mathcal{L}_{[X,Y]} = 0 \tag{5.2.15}$$

for the operator $\mathcal{L}_X$ on the space of real valued functions on M.

Remark 5.2.13 Equation (5.2.13) can be used to define the Riemann curvature tensor. To do this one must again prove that the right hand side of equation (5.2.13) at p depends only on the values $X(p), Y(p), Z(p)$ of the vector fields X, Y, Z at the point p. For this it suffices to prove that the map

$$\text{Vect}(M) \times \text{Vect}(M) \times \text{Vect}(M) \to \text{Vect}(M) : (X, Y, Z) \mapsto R(X, Y)Z$$

is linear over the Ring $\mathcal{F}(M)$ of smooth real valued functions on M, i.e.

$$R(fX, Y)Z = R(X, fY)Z = R(X, Y)fZ = fR(X, Y)Z \tag{5.2.16}$$

for $X, Y, Z \in \text{Vect}(M)$ and $f \in \mathcal{F}(M)$. The formula (5.2.16) follows from the equations (5.2.7), (5.2.8), (5.2.15), and $[X, fY] = f[X, Y] - (\mathcal{L}_X f)Y$. It follows from (5.2.16) that the right hand side of (5.2.13) at p depends only on the vectors $X(p), Y(p), Z(p)$. The proof requires two steps. One first shows that if X vanishes near p, then the right hand side of (5.2.13) vanishes at p (and similarly for Y and Z). Just multiply X by a smooth function equal to zero at p and equal to one on the support of X; then $fX = X$ and hence the vector field $R(X, Y)Z = R(fX, Y)Z = fR(X, Y)Z$ vanishes at p. Second, we choose a local frame $E_1, \dots, E_m \in \text{Vect}(M)$, i.e. vector fields that form a basis of $T_p M$ for each p in some open set $U \subset M$. Then we may write

$$X = \sum_{i=1}^{m} \xi^i E_i, \qquad Y = \sum_{j=1}^{m} \eta^j E_j, \qquad Z = \sum_{k=1}^{m} \zeta^k E_k$$

in U. Using the first step and the $\mathcal{F}(M)$-multilinearity we obtain

$$R(X, Y)Z = \sum_{i,j,k=1}^{m} \xi^i \eta^j \zeta^k R(E_i, E_j)E_k$$

in U. If $X'(p) = X(p)$, then $\xi^i(p) = \xi'^i(p)$ so if $X(p) = X'(p), Y(p) = Y'(p), Z(p) = Z'(p)$, then $(R(X, Y)Z)(p) = (R(X', Y')Z')(p)$ as required.

5.2.4 Symmetries

Theorem 5.2.14 *The Riemann curvature tensor satisfies*

$$R(Y, X) = -R(X, Y) = R(X, Y)^*, \tag{5.2.17}$$
$$R(X, Y)Z + R(Y, Z)X + R(Z, X)Y = 0, \tag{5.2.18}$$
$$\langle R(X, Y)Z, W \rangle = \langle R(Z, W)X, Y \rangle \tag{5.2.19}$$

for $X, Y, Z, W \in \mathrm{Vect}(M)$. *Equation* (5.2.18) *is the **first Bianchi identity**.*

Proof The first equation in (5.2.17) is obvious from the definition and the second follows from the Gauß–Codazzi formula (5.2.3). Alternatively, choose a smooth map $\gamma : \mathbb{R}^2 \to M$ and two vector fields Z, W along γ. Then

$$\begin{aligned}
0 &= \partial_s \partial_t \langle Z, W \rangle - \partial_t \partial_s \langle Z, W \rangle \\
&= \partial_s \langle \nabla_t Z, W \rangle + \partial_s \langle Z, \nabla_t W \rangle - \partial_t \langle \nabla_s Z, W \rangle - \partial_t \langle Z, \nabla_s W \rangle \\
&= \langle \nabla_s \nabla_t Z, W \rangle + \langle Z, \nabla_s \nabla_t W \rangle - \langle \nabla_t \nabla_s Z, W \rangle - \langle Z, \nabla_t \nabla_s W \rangle \\
&= \langle R(\partial_s \gamma, \partial_t \gamma) Z, W \rangle + \langle Z, R(\partial_s \gamma, \partial_t \gamma) W \rangle.
\end{aligned}$$

This proof has the advantage that it carries over to the intrinsic setting. We prove the first Bianchi identity using (5.2.10) and (5.2.13):

$$\begin{aligned}
R(X,&Y)Z + R(Y, Z)X + R(Z, X)Y \\
&= \nabla_X \nabla_Y Z - \nabla_Y \nabla_X Z + \nabla_{[X,Y]} Z + \nabla_Y \nabla_Z X - \nabla_Z \nabla_Y X + \nabla_{[Y,Z]} X \\
&\quad + \nabla_Z \nabla_X Y - \nabla_X \nabla_Z Y + \nabla_{[Z,X]} Y \\
&= \nabla_{[Y,Z]} X - \nabla_X [Y, Z] + \nabla_{[Z,X]} Y - \nabla_Y [Z, X] + \nabla_{[X,Y]} Z - \nabla_Z [X, Y] \\
&= [X, [Y, Z]] + [Y, [Z, X]] + [Z, [X, Y]].
\end{aligned}$$

The last term vanishes by the Jacobi identity. We prove (5.2.19) by combining the first Bianchi identity with (5.2.17):

$$\begin{aligned}
\langle R(X, &Y)Z, W \rangle - \langle R(Z, W)X, Y \rangle \\
&= -\langle R(Y, Z)X, W \rangle - \langle R(Z, X)Y, W \rangle - \langle R(Z, W)X, Y \rangle \\
&= \langle R(Y, Z)W, X \rangle + \langle R(Z, X)W, Y \rangle + \langle R(W, Z)X, Y \rangle \\
&= \langle R(Y, Z)W, X \rangle - \langle R(X, W)Z, Y \rangle \\
&= \langle R(Y, Z)W, X \rangle - \langle R(W, X)Y, Z \rangle.
\end{aligned}$$

Note that the first line is related to the last by a cyclic permutation. Repeating this argument we find

$$\langle R(Y, Z)W, X \rangle - \langle R(W, X)Y, Z \rangle = \langle R(Z, W)X, Y \rangle - \langle R(X, Y)Z, W \rangle.$$

Combining the last two identities we obtain (5.2.19). This proves Theorem 5.2.14.

$\square$

Remark 5.2.15 We may think of a vector field X on M as a section of the tangent bundle. This is reflected in the alternative notation

$$\Omega^0(M, TM) := \text{Vect}(M).$$

A 1-**form on** M **with values in the tangent bundle** is a collection of linear maps $A(p) : T_pM \to T_pM$, one for every $p \in M$, which is smooth in the sense that for every smooth vector field X on M the assignment $p \mapsto A(p)X(p)$ defines again a smooth vector field on M. We denote by

$$\Omega^1(M, TM)$$

the space of smooth 1-forms on M with values in TM. The covariant derivative of a vector field Y is such a 1-form with values in the tangent bundle which assigns to every $p \in M$ the linear map $T_pM \to T_pM : v \mapsto \nabla_v Y(p)$. Thus we can think of the covariant derivative as a linear operator

$$\nabla : \Omega^0(M, TM) \to \Omega^1(M, TM).$$

The equation (5.2.7) asserts that the operators $X \mapsto \nabla_X$ indeed determine a linear operator from $\Omega^0(M, TM)$ to $\Omega^1(M, TM)$. Equation (5.2.8) asserts that this linear operator ∇ is a **connection** on the tangent bundle of M. Equation (5.2.9) asserts that ∇ is a **Riemannian connection** and equation (5.2.10) asserts that ∇ is **torsion-free**. Thus Lemma 5.2.7 can be restated as asserting that the **Levi-Civita connection** is the unique torsion-free Riemannian connection on the tangent bundle.

Exercise 5.2.16 Extend the notion of a connection to a general vector bundle E, both as a collection of linear operators $\nabla_X : \Omega^0(M, E) \to \Omega^0(M, E)$, one for every vector field $X \in \text{Vect}(M)$, and as a linear operator

$$\nabla : \Omega^0(M, E) \to \Omega^1(M, E)$$

satisfying the analogue of equation (5.2.8). Interpret this equation as a Leibniz rule for the product of a function on M with a section of E. Show that $\nabla^\perp$ is a connection on $TM^\perp$. Extend the notion of curvature to connections on general vector bundles.

Exercise 5.2.17 Show that the field which assigns to each $p \in M$ the multi-linear map $R_p^\perp : T_pM \times T_pM \to \mathcal{L}(T_pM^\perp, T_pM^\perp)$ characterized by

$$R^\perp(\partial_s \gamma, \partial_t \gamma)Y = \nabla_s^\perp \nabla_t^\perp Y - \nabla_t^\perp \nabla_s^\perp Y$$

for $\gamma : \mathbb{R}^2 \to M$ and $Y \in \text{Vect}^\perp(\gamma)$ satisfies the equation

$$R_p^\perp(u, v) = h_p(u)h_p(v)^* - h_p(v)h_p(u)^*$$

for $p \in M$ and $u, v \in T_pM$.

5.2.5   Riemannian Metrics on Lie Groups

We begin with a calculation of the Riemann curvature tensor on a Lie subgroup of the orthogonal group $O(n)$ with the Riemannian metric inherited from the standard inner product

$$\langle v, w \rangle := \text{trace}(v^\mathsf{T} w) \tag{5.2.20}$$

on the ambient space $\mathfrak{gl}(n, \mathbb{R}) = \mathbb{R}^{n \times n}$. This fits into the extrinsic setting used throughout most of this book. Note that every Lie subgroup of $O(n)$ is a closed subset of $O(n)$ by Theorem 2.5.27 and hence is compact.

Example 5.2.18   Let $G \subset O(n)$ be a Lie subgroup and let

$$\mathfrak{g} := \text{Lie}(G) = T_1 G$$

be the Lie algebra of G. Consider the Riemannian metric on G induced by the inner product (5.2.20) on $\mathbb{R}^{n \times n}$. Then the Riemann curvature tensor on G can be expressed in terms of the Lie bracket (see item (d) below).

(a) The maps $g \mapsto ag$, $g \mapsto ga$, $g \mapsto g^{-1}$ are isometries of G for every $a \in G$.
(b) Let $\gamma : \mathbb{R} \to G$ be a smooth curve and $X \in \text{Vect}(\gamma)$ be a smooth vector field along γ. Then the covariant derivative of X is given by

$$\gamma(t)^{-1} \nabla X(t) = \frac{d}{dt}\gamma(t)^{-1} X(t) + \frac{1}{2}\left[\gamma(t)^{-1}\dot{\gamma}(t), \gamma(t)^{-1}X(t)\right]. \tag{5.2.21}$$

(**Exercise:** Prove equation (5.2.21). **Hint:** Since $\mathfrak{g} \subset \mathfrak{o}(n)$ we have the identity $\text{trace}((\xi\eta + \eta\xi)\zeta) = 0$ for all $\xi, \eta, \zeta \in \mathfrak{g}$.)
(c) A smooth map $\gamma : \mathbb{R} \to G$ is a geodesic if and only if there exist matrices $g \in G$ and $\xi \in \mathfrak{g}$ such that

$$\gamma(t) = g \exp(t\xi). \tag{5.2.22}$$

For $G = O(n)$ we have seen this in Example 4.3.12 and in the general case this follows from equation (5.2.21) with $X = \dot{\gamma}$. Hence the exponential map $\exp : \mathfrak{g} \to G$ defined by the exponential matrix (as in Sect. 2.5) agrees with the time-1-map of the geodesic flow (as in Sect. 4.3).
(d) The Riemann curvature tensor on G is given by

$$g^{-1} R_g(u, v)w = -\frac{1}{4}[[g^{-1}u, g^{-1}v], g^{-1}w] \tag{5.2.23}$$

for $g \in G$ and $u, v, w \in T_g G$. Note that the first Bianchi identity is equivalent to the Jacobi identity. (**Exercise:** Prove equation (5.2.23).)

Definition 5.2.19 (Bi-invariant Riemannian metric) Let G be a Lie subgroup of $GL(n, \mathbb{R})$ and let $\mathfrak{g} = \mathrm{Lie}(G) = T_{\mathbb{1}}G$ be its Lie algebra. A Riemannian metric on G is called **bi-invariant** iff it has the form

$$\langle v, w \rangle_g := \langle vg^{-1}, wg^{-1} \rangle \tag{5.2.24}$$

for $g \in G$ and $v, w \in T_g G$, where $\langle \cdot, \cdot \rangle$ is an inner product on the Lie algebra $\mathfrak{g}$ that is **invariant under conjugation**, i.e. it satisfies the equation

$$\langle \xi, \eta \rangle = \langle g\xi g^{-1}, g\eta g^{-1} \rangle. \tag{5.2.25}$$

for all $\xi, \eta \in \mathfrak{g}$ and all $g \in G$.

Exercise 5.2.20 Prove that the Riemannian metric induced by (5.2.20) on any Lie subgroup $G \subset O(n)$ is bi-invariant.

Exercise 5.2.21 Use the Haar measure ([66, Chapter 8]) to prove that every compact Lie group admits a bi-invariant Riemannian metric.

Exercise 5.2.22 Prove that all the assertions in Example 5.2.18 carry over verbatim to any Lie group equipped with a bi-invariant Riemann metric.

Exercise 5.2.23 (Invariant inner product) Prove that, if an inner product on the Lie algebra $\mathfrak{g}$ of a Lie group G is invariant under conjugation, then it satisfies the equation

$$\langle [\xi, \eta], \zeta \rangle = \langle \xi, [\eta, \zeta] \rangle \tag{5.2.26}$$

for all $\xi, \eta, \zeta \in \mathfrak{g}$. If G is connected, prove that, conversely, equation (5.2.26) implies (5.2.25). An inner product on an arbitrary Lie algebra $\mathfrak{g}$ is called **invariant** iff it satisfies equation (5.2.26).

Exercise 5.2.24 (Commutant) Let $\mathfrak{g}$ be a finite-dimensional Lie algebra. The linear subspace spanned by all vectors of the form $[\xi, \eta]$ is called the **commutant of** $\mathfrak{g}$ and is denoted by $[\mathfrak{g}, \mathfrak{g}] := \mathrm{span}\{[\xi, \eta] \,|\, \xi, \eta \in \mathfrak{g}\}$. If $\mathfrak{g}$ is equipped with an invariant inner product, prove that $[\mathfrak{g}, \mathfrak{g}]^{\perp} = Z(\mathfrak{g})$ is the center of $\mathfrak{g}$ (Exercise 2.5.34) and hence $\mathfrak{g} = [\mathfrak{g}, \mathfrak{g}] \oplus Z(\mathfrak{g})$. Prove that the Heisenberg algebra $\mathfrak{h}$ in Exercise 2.5.15 satisfies $[\mathfrak{h}, \mathfrak{h}] = Z(\mathfrak{h})$ and hence does not admit an invariant inner product.

Example 5.2.25 (Killing form) Every finite-dimensional Lie algebra $\mathfrak{g}$ admits a natural symmetric bilinear form $\kappa : \mathfrak{g} \times \mathfrak{g} \to \mathbb{R}$ that satisfies equation (5.2.26). It is called the **Killing form** and is defined by

$$\kappa(\xi, \eta) := \mathrm{trace}\big(\mathrm{ad}(\xi)\mathrm{ad}(\eta)\big), \qquad \xi, \eta \in \mathfrak{g}, \tag{5.2.27}$$

where $\mathrm{ad} : \mathfrak{g} \to \mathrm{Der}(\mathfrak{g})$ is the adjoint representation in Example 2.5.23. The Killing form may have a kernel (which always contains the center of $\mathfrak{g}$), and even if it is nondegenerate, it may be indefinite.

Exercise 5.2.26 Prove that $\kappa([\xi, \eta], \zeta) = \kappa(\xi, [\eta, \zeta])$ for all $\xi, \eta, \zeta \in \mathfrak{g}$.

Exercise 5.2.27 Assume that $\mathfrak{g}$ admits an invariant inner product. For each $\xi \in \mathfrak{g}$ prove that the derivation $\mathrm{ad}(\xi)$ is skew-adjoint with respect to this inner product and deduce that $\kappa(\xi, \xi) = -\mathrm{trace}(\mathrm{ad}(\xi)^*\mathrm{ad}(\xi)) = -|\mathrm{ad}(\xi)|^2$. Deduce that the Killing form is nondegenerate whenever $\mathfrak{g}$ has a trivial center and admits an invariant inner product.

Example 5.2.28 (Right invariant Riemannian metric) Let G be any Lie subgroup of $GL(n, \mathbb{R})$ (not necessarily contained in $O(n)$), and let $\mathfrak{g} := \mathrm{Lie}(G) = T_{\mathbb{1}}G$ be its Lie algebra. Fix any inner product on the Lie algebra $\mathfrak{g}$ (not necessarily invariant under conjugation) and consider the Riemannian metric on G defined by

$$\langle v, w \rangle_g := \langle vg^{-1}, wg^{-1} \rangle \tag{5.2.28}$$

for $v, w \in T_gG$. This metric is called **right invariant**.

(a) The map $g \mapsto ga$ is an isometry of G for every $a \in G$.
(b) Define the linear map $A : \mathfrak{g} \to \mathrm{End}(\mathfrak{g})$ by

$$\langle A(\xi)\eta, \zeta \rangle = \frac{1}{2}\Big(\langle \xi, [\eta, \zeta] \rangle - \langle \eta, [\zeta, \xi] \rangle - \langle \zeta, [\xi, \eta] \rangle \Big) \tag{5.2.29}$$

for $\xi, \eta, \zeta \in \mathfrak{g}$. Then A is the unique linear map that satisfies

$$A(\xi) + A(\xi)^* = 0, \qquad A(\eta)\xi - A(\xi)\eta = [\xi, \eta]$$

for all $\xi, \eta \in \mathfrak{g}$, where $A(\xi)^*$ is the adjoint operator with respect to the inner product on $\mathfrak{g}$. Let $\gamma : \mathbb{R} \to G$ be a smooth curve and $X \in \mathrm{Vect}(\gamma)$ be a smooth vector field along γ. Then the covariant derivative of X is given by

$$\nabla X = \left(\frac{d}{dt}(X\gamma^{-1}) + A(\dot{\gamma}\gamma^{-1})X\gamma^{-1} \right)\gamma. \tag{5.2.30}$$

(**Exercise:** Prove this. Moreover, if the inner produt on $\mathfrak{g}$ is invariant, prove that $A(\xi)\eta = -\frac{1}{2}[\xi, \eta]$ for all $\xi, \eta \in \mathfrak{g}$.)
(c) A smooth curve $\gamma : \mathbb{R} \to G$ is a geodesic if and only if it satisfies

$$\frac{d}{dt}(\dot{\gamma}\gamma^{-1}) + A(\dot{\gamma}\gamma^{-1})\dot{\gamma}\gamma^{-1} = 0. \tag{5.2.31}$$

(**Exercise:** G is complete.)
(d) The Riemann curvature tensor on G is given by

$$\big(R_g(\xi g, \eta g)\zeta g \big)g^{-1} = \Big(A([\xi, \eta]) + [A(\xi), A(\eta)] \Big)\zeta \tag{5.2.32}$$

for $g \in G$ and $\xi, \eta, \zeta \in \mathfrak{g}$. (**Exercise:** Prove this.)

5.3 Generalized Theorema Egregium

We will now show that geodesics, covariant differentiation, parallel transport, and the Riemann curvature tensor are all intrinsic, i.e. they are intertwined by isometries. In the extrinsic setting these results are somewhat surprising since these objects are all defined using the second fundamental form, whereas isometries need not preserve the second fundamental form in any sense but only the first fundamental form.

Below we shall give a formula expressing the Gaußian curvature of a surface M^2 in $\mathbb{R}^3$ in terms of the Riemann curvature tensor and the first fundamental form. It follows that the Gaußian curvature is also intrinsic. This fact was called by Gauß the "Theorema Egregium" which explains the title of this section.

5.3.1 Pushforward

We assume throughout this section that $M \subset \mathbb{R}^n$ and $M' \subset \mathbb{R}^{n'}$ are smooth submanifolds of the same dimension m. As in Sect. 5.1 we denote objects on M' by the same letters as objects in M with primes affixed. In particular, g' denotes the first fundamental form on M' and R' denotes the Riemann curvature tensor on M'.

Let $\phi : M \to M'$ be a diffeomorphism. Using ϕ we can move objects on M to M'. For example the pushforward of a smooth curve $\gamma : I \to M$ is the curve

$$\phi_*\gamma := \phi \circ \gamma : I \to M',$$

the pushforward of a smooth function $f : M \to \mathbb{R}$ is the function

$$\phi_*f := f \circ \phi^{-1} : M' \to \mathbb{R},$$

the pushforward of a vector field $X \in \text{Vect}(\gamma)$ along a curve $\gamma : I \to M$ is the vector field $\phi_*X \in \text{Vect}(\phi_*\gamma)$ defined by

$$(\phi_*X)(t) := d\phi(\gamma(t))X(t)$$

for $t \in I$, and the pushforward of a global vector field $X \in \text{Vect}(M)$ is the vector field $\phi_*X \in \text{Vect}(M')$ defined by

$$(\phi_*X)(\phi(p)) := d\phi(p)X(p)$$

for $p \in M$. Recall that the first fundamental form on M is the Riemannian metric g defined as the restriction of the Euclidean inner product on the ambient space to each tangent space of M. It assigns to each $p \in M$ the bilinear map $g_p : T_pM \times T_pM \to \mathbb{R}$ given by

$$g_p(u, v) = \langle u, v \rangle, \qquad u, v \in T_pM.$$

Its pushforward is the Riemannian metric which assigns to each $p' \in M'$ the inner product $(\phi_* g)_{p'} : T_{p'} M' \times T_{p'} M' \to \mathbb{R}$ defined by

$$(\phi_* g)_{\phi(p)}(d\phi(p)u, d\phi(p)v) := g_p(u, v)$$

for $p := \phi^{-1}(p') \in M$ and $u, v \in T_p M$. The pushforward of the Riemann curvature tensor is the tensor which assigns to each $p' \in M'$ the bilinear map $(\phi_* R)_{p'} : T_{p'} M' \times T_{p'} M' \to \mathcal{L}(T_{p'} M', T_{p'} M')$, defined by

$$(\phi_* R)_{\phi(p)}(d\phi(p)u, d\phi(p)v) := d\phi(p) R_p(u, v) d\phi(p)^{-1}$$

for $p := \phi^{-1}(p') \in M$ and $u, v \in T_p M$.

5.3.2 Theorema Egregium

Theorem 5.3.1 (Theorema Egregium) *The first fundamental form, covariant differentiation, geodesics, parallel transport, and the Riemann curvature tensor are intrinsic. This means that for every isometry $\phi : M \to M'$ the following holds.*

(i) $\phi_* g = g'$.
(ii) *If $X \in \mathrm{Vect}(\gamma)$ is a vector field along a smooth curve $\gamma : I \to M$, then*

$$\nabla'(\phi_* X) = \phi_* \nabla X, \tag{5.3.1}$$

and if $X, Y \in \mathrm{Vect}(M)$ are global vector fields, then

$$\nabla'_{\phi_* X} \phi_* Y = \phi_*(\nabla_X Y). \tag{5.3.2}$$

(iii) *If $\gamma : I \to M$ is a geodesic, then $\phi \circ \gamma : I \to M'$ is a geodesic.*
(iv) *If $\gamma : I \to M$ is a smooth curve, then for all $s, t \in I$, we have*

$$\Phi'_{\phi \circ \gamma}(t, s) d\phi(\gamma(s)) = d\phi(\gamma(t)) \Phi_\gamma(t, s). \tag{5.3.3}$$

(v) $\phi_* R = R'$.

Proof Assertion (i) is simply a restatement of Theorem 5.1.1. To prove (ii) we choose a local smooth parametrization $\psi : \Omega \to U$ of an open set $U \subset M$, defined on an open set $\Omega \subset \mathbb{R}^m$, so that $\psi^{-1} : U \to \Omega$ is a coordinate chart. Suppose without loss of generality that $\gamma(t) \in U$ for all $t \in I$ and define $c : I \to \Omega$ and $\xi : I \to \mathbb{R}^m$ by

$$\gamma(t) = \psi(c(t)), \qquad X(t) = \sum_{i=1}^{m} \xi^i(t) \frac{\partial \psi}{\partial x^i}(c(t)).$$

Recall from equations (3.6.6) and (3.6.7) that

$$\nabla X(t) = \sum_{k=1}^{m} \left(\dot{\xi}^k(t) + \sum_{i,j=1}^{m} \Gamma_{ij}^k(c(t))\dot{c}^i(t)\xi^j(t) \right) \frac{\partial \psi}{\partial x^k}(c(t)),$$

where the Christoffel symbols $\Gamma_{ij}^k : \Omega \to \mathbb{R}$ are defined by

$$\Pi(\psi)\frac{\partial^2 \psi}{\partial x^i \partial x^j} = \sum_{k=1}^{m} \Gamma_{ij}^k \frac{\partial \psi}{\partial x^k}.$$

Now consider the same formula for $\phi_* X$ using the parametrization

$$\psi' := \phi \circ \psi : \Omega \to U' := \phi(U) \subset M'.$$

The Christoffel symbols $\Gamma_{ij}'^k : \Omega \to \mathbb{R}$ associated to this parametrization of U' are defined by the same formula as the Γ_{ij}^k with ψ replaced by ψ'. But the metric tensor for ψ agrees with the metric tensor for ψ':

$$g_{ij} = \left\langle \frac{\partial \psi}{\partial x^i}, \frac{\partial \psi}{\partial x^j} \right\rangle = \left\langle \frac{\partial \psi'}{\partial x^i}, \frac{\partial \psi'}{\partial x^j} \right\rangle.$$

Hence it follows from Lemma 3.6.5 that $\Gamma_{ij}'^k = \Gamma_{ij}^k$ for all i, j, k. This implies that the covariant derivative of $\phi_* X$ is given by

$$\nabla'(\phi_* X) = \sum_{k=1}^{m} \left(\dot{\xi}^k + \sum_{i,j=1}^{m} \Gamma_{ij}^k(c)\dot{c}^i \xi^j \right) \frac{\partial \psi'}{\partial x^k}(c)$$

$$= d\phi(\psi(c)) \sum_{k=1}^{m} \left(\dot{\xi}^k + \sum_{i,j=1}^{m} \Gamma_{ij}^k(c)\dot{c}^i \xi^j \right) \frac{\partial \psi}{\partial x^k}(c)$$

$$= \phi_* \nabla X.$$

This proves (5.3.1). Equation (5.3.2) follows immediately from (5.3.1) and Remark 5.2.4.

Here is a second proof of (ii). For every vector field $X \in \text{Vect}(M)$ we define the operator $D_X : \text{Vect}(M) \to \text{Vect}(M)$ by

$$D_X Y := \phi^*(\nabla_{\phi_* X} \phi_* Y).$$

Then, for all $X, Y \in \text{Vect}(M)$, we have

$$D_Y X - D_X Y = \phi^*(\nabla_{\phi_* Y} \phi_* X - \nabla_{\phi_* X} \phi_* Y) = \phi^*[\phi_* X, \phi_* Y] = [X, Y].$$

Moreover, it follows from (i) that

$$
\begin{aligned}
\phi_* \mathcal{L}_X \langle Y, Z \rangle &= \mathcal{L}_{\phi_* X} \langle \phi_* Y, \phi_* Z \rangle \\
&= \langle \nabla_{\phi_* X} \phi_* Y, \phi_* Z \rangle + \langle \phi_* Y, \nabla_{\phi_* X} \phi_* Z \rangle \\
&= \langle \phi_* D_X Y, \phi_* Z \rangle + \langle \phi_* Y, \phi_* D_X Z \rangle \\
&= \phi_* \big(\langle D_X Y, Z \rangle + \langle Y, D_X Z \rangle \big).
\end{aligned}
$$

and hence $\mathcal{L}_X \langle Y, Z \rangle = \langle D_X Y, Z \rangle + \langle Y, D_X Z \rangle$ for all $X, Y, Z \in \mathrm{Vect}(M)$. Thus the operator $X \mapsto D_X$ satisfies equations (5.2.9) and (5.2.10) and, by Lemma 5.2.7, it follows that $D_X Y = \nabla_X Y$ for all $X, Y \in \mathrm{Vect}(M)$. This completes the second proof of (ii).

We prove (iii). Since ϕ preserves the first fundamental form it also preserves the energy of curves, namely

$$
E(\phi \circ \gamma) = E(\gamma)
$$

for every smooth map $\gamma : [0, 1] \to M$. Hence γ is a critical point of the energy functional if and only if $\phi \circ \gamma$ is a critical point of the energy functional. Alternatively it follows from (ii) that

$$
\nabla' \left(\frac{d}{dt} \phi \circ \gamma \right) = \nabla' \phi_* \dot{\gamma} = \phi_* \nabla \dot{\gamma}
$$

for every smooth curve $\gamma : I \to M$. If γ is a geodesic, the last term vanishes and hence $\phi \circ \gamma$ is a geodesic as well. As a third proof we can deduce (iii) from the formula $\phi(\exp_p(v)) = \exp_{\phi(p)}(d\phi(p)v)$ in the proof of Theorem 5.1.1.

We prove (iv). For $t_0 \in I$ and $v_0 \in T_{\gamma(t_0)} M$ define

$$
X(t) := \Phi_\gamma(t, t_0) v_0, \qquad X'(t) := \Phi'_{\phi \circ \gamma}(t, t_0) d\phi(\gamma(t_0)) v_0.
$$

By (ii) the vector fields X' and $\phi_* X$ along $\phi \circ \gamma$ are both parallel and they agree at $t = t_0$. Hence $X'(t) = \phi_* X(t)$ for all $t \in I$ and this proves (5.3.3).

We prove (v). Fix a smooth map $\gamma : \mathbb{R}^2 \to M$ and a smooth vector field Z along γ, and define

$$
\gamma' = \phi \circ \gamma : \mathbb{R}^2 \to M', \qquad Z' := \phi_* Z \in \mathrm{Vect}(\gamma').
$$

Then it follows from (ii) that

$$
\begin{aligned}
R'(\partial_s \gamma', \partial_t \gamma') Z' &= \nabla'_s \nabla'_t Z' - \nabla'_t \nabla'_s Z' \\
&= \phi_* (\nabla_s \nabla_t Z - \nabla_t \nabla_s Z) \\
&= d\phi(\gamma) R(\partial_s \gamma, \partial_t \gamma) Z \\
&= (\phi_* R)(\partial_s \gamma', \partial_t \gamma') Z'.
\end{aligned}
$$

This proves (v) and Theorem 5.3.1. $\qquad\square$

The assertions of Theorem 5.3.1 carry over in slightly modified form to local isometries $\phi : M \to M'$. In particular, the pushforward of a vector field on M under ϕ is only defined when ϕ is a diffeomorphism while the pushforward of a vector field along a curve is defined for any smooth map ϕ. Also, the pushforward of the Riemann curvature tensor under a local isometry is only defined locally, and local isometries satisfy the local analogue of the first assertion in Theorema Egregium by definition.

Corollary 5.3.2 (Theorema Egregium for Local Isometries) *Every local isometry $\phi : M \to M'$ has the following properties.*

(i)   *Every vector field X along a smooth curve $\gamma : I \to M$ satisfies (5.3.1).*
(ii) *If $\gamma : I \to M$ is a geodesic, then so is $\phi \circ \gamma : I \to M'$.*
(iii) *Parallel transport along a smooth curve $\gamma : I \to M$ satisfies (5.3.3).*
(iv) *The curvature tensors R of M and R' of M' are related by the formula*

$$R'_{\phi(p)}(d\phi(p)u, d\phi(p)v) = d\phi(p)R_p(u,v)d\phi(p)^{-1} \tag{5.3.4}$$

for all $p \in M$ and all $u, v \in T_p M$.

Proof Let $p_0 \in M$. Then, by the Inverse Function Theorem 2.2.17, there exists an open neighborhood $U \subset M$ of p_0 such that $U' := \phi(U)$ is an open subset of M' and the restriction $\phi|_U : U \to U'$ is a diffeomorphism. This restriction is an isometry by Theorem 5.1.1. Hence, by Theorem 5.3.1 the assertions (i) and (ii) hold for the restriction of γ to $I_0 := \gamma^{-1}(U)$ (a union of subintervals of I) and (iv) holds for all $p \in U$. Since these are local statements and p_0 was chosen arbitrary, this proves (i), (ii), and (iv). Part (iii) follows directly from (i) as in the proof of Theorem 5.3.1 and this proves Corollary 5.3.2.   □

The next corollary spells out a useful consequence of Corollary 5.3.2. For sufficiently small tangent vectors equation (5.3.5) below already appeared in the proof of Theorem 5.1.1 and was used in Lemma 5.1.10 and Exercise 5.1.11. When M is not complete, recall the notation $V_p \subset T_p M$ for the domain of the exponential map of M at a point p (Definition 4.3.5). For $p' \in M'$ denote the domain of the exponential map by $V'_{p'} \subset T_{p'} M'$.

Corollary 5.3.3 *Let $\phi : M \to M'$ be a local isometry and let $p \in M$. Then $d\phi(p)V_p \subset V'_{\phi(p)}$ and, for every $v \in V_p$,*

$$\phi(\exp_p(v)) = \exp'_{\phi(p)}(d\phi(p)v). \tag{5.3.5}$$

Proof Let $v \in V_p \subset T_p M$ and define $\gamma(t) := \exp_p(tv)$ for $0 \le t \le 1$. Then $\gamma : [0, 1] \to M$ is a geodesic by Lemma 4.3.6, and hence so is the curve $\gamma' := \phi \circ \gamma : [0, 1] \to M'$ by Corollary 5.3.2. Moreover,

$$\gamma'(0) = \phi(\gamma(0)) = \phi(p), \qquad \dot{\gamma}'(0) = d\phi(\gamma(0))\dot{\gamma}(0) = d\phi(p)v$$

by the chain rule. Hence it follows from the definition of the exponential map (Definition 4.3.5) that $d\phi(p)v \in V'_{\phi(p)}$ and

$$\exp'_{\phi(p)}(d\phi(p)v) = \gamma'(1) = \phi(\gamma(1)) = \phi(\exp_p(v)).$$

This proves Corollary 5.3.3.   □

5.3.3 Gaußian Curvature

As a special case we shall now consider a **hypersurface** $M \subset \mathbb{R}^{m+1}$, i.e. a smooth submanifold of codimension one. We assume that there exists a smooth map $\nu :$ $M \to \mathbb{R}^{m+1}$ such that, for every $p \in M$, we have

$$\nu(p) \perp T_p M, \qquad |\nu(p)| = 1.$$

Such a map always exists locally (see Example 3.1.3). Note that $\nu(p)$ is an element of the unit sphere in $\mathbb{R}^{m+1}$ for every $p \in M$ and hence we can regard ν as a map from M to S^m, i.e. $\nu : M \to S^m$. Such a map is called a **Gauß map** for M. Note that if $\nu : M \to S^m$ is a Gauß map, so is $-\nu$, but this is the only ambiguity when M is connected. Differentiating ν at $p \in M$ we obtain a linear map

$$d\nu(p) : T_p M \to T_{\nu(p)} S^m = T_p M$$

Here we use the fact that $T_{\nu(p)} S^m = \nu(p)^\perp$ and, by definition of the Gauß map ν, the tangent space of M at p is also equal to $\nu(p)^\perp$. Thus $d\nu(p)$ is a linear map from the tangent space of M at p to itself.

Definition 5.3.4 The **Gaußian curvature** of the hypersurface M is the real valued function $K : M \to \mathbb{R}$ defined by

$$K(p) := \det\big(d\nu(p) : T_p M \to T_p M\big)$$

for $p \in M$. (Replacing ν by $-\nu$ has the effect of replacing K by $(-1)^m K$; so K is independent of the choice of the Gauß map when m is even.)

Remark 5.3.5 Given a subset $B \subset M$, the set $\nu(B) \subset S^m$ is often called the **spherical image** of B. If ν is a diffeomorphism on a neighborhood of B, the change of variables formula for an integral gives

$$\int_{\nu(B)} \mu_S = \int_B |K|\mu_M.$$

Here μ_M and μ_S denote the volume elements on M and S^m, respectively. Introducing the notation $\mathrm{Area}_M(B) := \int_B \mu_M$ we obtain the formula

$$|K(p)| = \lim_{B \to p} \frac{\mathrm{Area}_S(\nu(B))}{\mathrm{Area}_M(B)}.$$

This says that the curvature at p is roughly the ratio of the (m-dimensional) area of the spherical image $\nu(B)$ to the area of B where B is a very small open neighborhood of p in M. The sign of $K(p)$ is positive when the linear map $d\nu(p) : T_pM \to T_pM$ preserves orientation and negative when it reverses orientation.

Remark 5.3.6 We see that the Gaußian curvature is a natural generalization of **Euler's curvature** for a plane curve. Indeed if $M \subset \mathbb{R}^2$ is a 1-manifold and $p \in M$, we can choose a curve $\gamma = (x, y) : (-\varepsilon, \varepsilon) \to M$ such that $\gamma(0) = p$ and $|\dot{\gamma}(s)| = 1$ for every s. This curve parametrizes M by the arclength and the unit normal vector pointing to the right with respect to the orientation of γ is $\nu(x, y) = (\dot{y}, -\dot{x})$. This is a local Gauß map and its derivative $(\ddot{y}, -\ddot{x})$ is tangent to the curve. The inner product of the latter with the unit tangent vector $\dot{\gamma} = (\dot{x}, \dot{y})$ is the Gaußian curvature. Thus

$$K := \frac{dx}{ds}\frac{d^2y}{ds^2} - \frac{dy}{ds}\frac{d^2x}{ds^2} = \frac{d\theta}{ds}$$

where s is the arclength parameter and θ is the angle made by the normal (or the tangent) with some constant line. With this convention K is positive at a left turn and negative at a right turn.

Exercise 5.3.7 The Gaußian curvature of an m-dimensional sphere of radius r is constant and has the value r^{-m} (with respect to an outward pointing Gauß map when m is odd).

Exercise 5.3.8 Show that the Gaußian curvature of the surface $z = x^2 - y^2$ is -4 at the origin.

We now restrict to the case of **surfaces**, i.e. of 2-dimensional submanifolds of $\mathbb{R}^3$. Figure 5.1 illustrates the difference between positive and negative Gaußian curvature in dimension two.

Fig. 5.1 Positive and negative Gaußian curvature

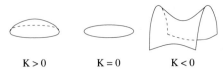

K > 0   K = 0   K < 0

Theorem 5.3.9 (Gaußian curvature) *Let $M \subset \mathbb{R}^3$ be a surface and fix a point $p \in M$. If $u, v \in T_p M$ is a basis, then*

$$K(p) = \frac{\langle R(u, v)v, u \rangle}{|u|^2 |v|^2 - \langle u, v \rangle^2}. \tag{5.3.6}$$

Moreover,

$$R(u, v)w = -K(p)\langle v(p), u \times v \rangle v(p) \times w \tag{5.3.7}$$

for all $u, v, w \in T_p M$.

Proof The orthogonal projection of $\mathbb{R}^3$ onto the tangent space $T_p M = v(p)^\perp$ is given by the 3×3-matrix

$$\Pi(p) = \mathbb{1} - v(p)v(p)^\mathsf{T}.$$

Hence

$$d\Pi(p)u = -v(p)(dv(p)u)^\mathsf{T} - (dv(p)u)v(p)^\mathsf{T}.$$

Here the first summand is the second fundamental form, which maps $T_p M$ to $T_p M^\perp$, and the second summand is its dual, which maps $T_p M^\perp$ to $T_p M$. Thus

$$h_p(v) = -v(p)(dv(p)v)^\mathsf{T} : T_p M \to T_p M^\perp,$$
$$h_p(u)^* = -(dv(p)u)v(p)^\mathsf{T} : T_p M^\perp \to T_p M.$$

By the Gauß–Codazzi formula this implies

$$\begin{aligned}
R_p(u, v)w &= h_p(u)^* h_p(v)w - h_p(v)^* h_p(u)w \\
&= (dv(p)u)(dv(p)v)^\mathsf{T} w - (dv(p)v)(dv(p)u)^\mathsf{T} w \\
&= \langle dv(p)v, w \rangle dv(p)u - \langle dv(p)u, w \rangle dv(p)v
\end{aligned}$$

and hence

$$\langle R_p(u, v)w, z \rangle = \langle dv(p)u, z \rangle \langle dv(p)v, w \rangle - \langle dv(p)u, w \rangle \langle dv(p)v, z \rangle. \tag{5.3.8}$$

Now fix four tangent vectors $u, v, w, z \in T_p M$ and consider the composition

$$\mathbb{R}^3 \xrightarrow{A} \mathbb{R}^3 \xrightarrow{B} \mathbb{R}^3 \xrightarrow{C} \mathbb{R}^3$$

of the linear maps

$$A\xi := \xi^1 v(p) + \xi^2 u + \xi^3 v,$$

$$B\eta := \begin{cases} dv(p)\eta, & \text{if } \eta \perp v(p), \\ \eta, & \text{if } \eta \in \mathbb{R}v(p), \end{cases}$$

$$C\zeta := \begin{pmatrix} \langle \zeta, v(p) \rangle \\ \langle \zeta, z \rangle \\ \langle \zeta, w \rangle \end{pmatrix}.$$

This composition is represented by the matrix

$$CBA = \begin{pmatrix} 1 & 0 & 0 \\ 0 & \langle dv(p)u, z \rangle & \langle dv(p)v, z \rangle \\ 0 & \langle dv(p)u, w \rangle & \langle dv(p)v, w \rangle \end{pmatrix}.$$

Hence, by (5.3.8), we have

$$\begin{aligned} \langle R_p(u, v)w, z \rangle &= \det(CBA) \\ &= \det(A)\det(B)\det(C) \\ &= \langle v(p), u \times v \rangle K(p) \langle v(p), z \times w \rangle \\ &= -K(p)\langle v(p), u \times v \rangle \langle v(p) \times w, z \rangle. \end{aligned}$$

This implies (5.3.7) and

$$\begin{aligned} \langle R_p(u, v)v, u \rangle &= K(p)\langle v(p), u \times v \rangle^2 \\ &= K(p)|u \times v|^2 \\ &= K(p)\Big(|u|^2|v|^2 - \langle u, v \rangle^2\Big). \end{aligned}$$

This proves Theorem 5.3.9. $\square$

Remark 5.3.10 Equation (5.3.6) implies

$$\langle R_p(u, v)w, z \rangle = K(p)\Big(\langle u, z \rangle \langle v, w \rangle - \langle u, w \rangle \langle v, z \rangle\Big) \tag{5.3.9}$$

for all $p \in M$ and all $u, v, w, z \in T_pM$. This is proved in Theorem 6.4.8 below.
Exercise: Deduce this formula from (5.3.7).

Corollary 5.3.11 (Theorema Egregium of Gauß) *The Gaußian curvature is intrinsic, i.e. if*

$$\phi : M \to M'$$

is an isometry of surfaces in $\mathbb{R}^3$, *then*

$$K = K' \circ \phi : M \to \mathbb{R}.$$

Proof Theorem 5.3.1 and Theorem 5.3.9. $\square$

Exercise 5.3.12 For $m = 1$ the Gaußian curvature is clearly *not* intrinsic as any two curves are locally isometric (parameterized by arclength). Show that the curvature $K(p)$ is intrinsic for even m while its absolute value $|K(p)|$ is intrinsic for odd $m \geq 3$. **Hint:** We still have the equation (5.3.8) which, for $z = u$ and $v = w$, can be written in the form

$$\langle R_p(u, v)v, u \rangle = \det \begin{pmatrix} \langle dv(p)u, u \rangle & \langle dv(p)u, v \rangle \\ \langle dv(p)v, u \rangle & \langle dv(p)v, v \rangle \end{pmatrix}.$$

Thus, for every orthonormal basis $v_1, \ldots, v_m$ of $T_p M$, the 2×2 minors of the matrix

$$\left(\langle dv(p)v_i, v_j \rangle \right)_{i,j=1,\ldots,m}$$

are intrinsic. Hence everything reduces to the following assertion.

Lemma *The determinant of an $m \times m$ matrix is an expression in its 2×2 minors if m is even; the absolute value of the determinant is an expression in the 2×2 minors if m is odd and greater than or equal to 3.*

The lemma is proved by induction on m. For the absolute value, note the formula

$$\det(A)^m = \det(\det(A)\mathbb{1}_m) = \det(AB) = \det(A)\det(B)$$

for an $m \times m$-matrix A where B is the transposed matrix of cofactors.

5.4 Curvature in Local Coordinates*

5.4.1 Riemann

Let $M \subset \mathbb{R}^k$ be an m-dimensional manifold and let $\phi = \psi^{-1} : U \to \Omega$ be a local coordinate chart on an open set $U \subset M$ with values in an open set $\Omega \subset \mathbb{R}^m$. Define the vector fields $E_1, \ldots, E_m$ along ψ by

$$E_i(x) := \frac{\partial \psi}{\partial x^i}(x) \in T_{\psi(x)}M.$$

These vector fields form a basis of $T_{\psi(x)}M$ for every $x \in \Omega$ and the coefficients $g_{ij} : \Omega \to \mathbb{R}$ of the first fundamental form are $g_{ij} = \langle E_i, E_j \rangle$. Recall from Lemma 3.6.5 that the Christoffel $\Gamma_{ij}^k : \Omega \to \mathbb{R}$ are the coefficients of the Levi-Civita connection, defined by

$$\nabla_i E_j = \sum_{k=1}^{m} \Gamma_{ij}^k E_k$$

and that they are given by the formula

$$\Gamma_{ij}^k := \sum_{\ell=1}^m g^{k\ell} \frac{1}{2} \left(\partial_i g_{j\ell} + \partial_j g_{i\ell} - \partial_\ell g_{ij} \right).$$

Define the coefficients $R_{ijk}^\ell : \Omega \to \mathbb{R}$ and $R_{ijk\ell} : \Omega \to \mathbb{R}$ of the Riemann curvature tensor by

$$R(E_i, E_j)E_k = \sum_{\ell=1}^m R_{ijk}^\ell E_\ell, \tag{5.4.1}$$

$$R_{ijk\ell} := \langle R(E_i, E_j)E_k, E_\ell \rangle = \sum_{\nu=1}^m R_{ijk}^\nu g_{\nu\ell}. \tag{5.4.2}$$

These coefficients are given by

$$R_{ijk}^\ell = \partial_i \Gamma_{jk}^\ell - \partial_j \Gamma_{ik}^\ell + \sum_{\nu=1}^m \left(\Gamma_{i\nu}^\ell \Gamma_{jk}^\nu - \Gamma_{j\nu}^\ell \Gamma_{ik}^\nu \right). \tag{5.4.3}$$

The coefficients of the Riemann curvature tensor have the symmetries

$$R_{ijk\ell} = -R_{jik\ell} = -R_{ij\ell k} = R_{k\ell ij} \tag{5.4.4}$$

and the **first Bianchi identity** has the form

$$R_{ijk}^\ell + R_{jki}^\ell + R_{kij}^\ell = 0, \qquad R_{ijk\ell} + R_{jki\ell} + R_{kij\ell} = 0. \tag{5.4.5}$$

Warning: Care must be taken with the ordering of the indices. Some authors use the notation R_{kij}^ℓ for what we call R_{ijk}^ℓ and $R_{\ell kij}$ for what we call $R_{ijk\ell}$.

Exercise 5.4.1 Prove equations (5.4.3), (5.4.4), and (5.4.5). Use (5.4.3) to give an alternative proof of Theorem 5.3.1.

5.4.2 Gauß

If $M \subset \mathbb{R}^n$ is a 2-manifold (not necessarily embedded in $\mathbb{R}^3$), we can use equation (5.3.6) as the definition of the Gaußian curvature $K : M \to \mathbb{R}$. Let $\psi : \Omega \to U$ be a local parametrization of an open set $U \subset M$ defined on an open set $\Omega \subset \mathbb{R}^2$. Denote the coordinates in $\mathbb{R}^2$ by (x, y) and define the functions $E, F, G : \Omega \to \mathbb{R}$ by

$$E := |\partial_x \psi|^2, \qquad F := \langle \partial_x \psi, \partial_y \psi \rangle, \qquad G := |\partial_y \psi|^2.$$

We abbreviate $D := EG - F^2$. Then the composition of the Gaußian curvature $K : M \to \mathbb{R}$ with the parametrization ψ is given by

$$
K \circ \psi = \frac{1}{D^2} \det \begin{pmatrix} E & F & \partial_y F - \frac{1}{2}\partial_x G \\ F & G & \frac{1}{2}\partial_y G \\ \frac{1}{2}\partial_x E & \partial_x F - \frac{1}{2}\partial_y E & -\frac{1}{2}\partial_y^2 E + \partial_x \partial_y F - \frac{1}{2}\partial_x^2 G \end{pmatrix}
$$

$$
- \frac{1}{D^2} \det \begin{pmatrix} E & F & \frac{1}{2}\partial_y E \\ F & G & \frac{1}{2}\partial_x G \\ \frac{1}{2}\partial_y E & \frac{1}{2}\partial_x G & 0 \end{pmatrix}
$$

$$
= -\frac{1}{2\sqrt{D}} \frac{\partial}{\partial x}\left(\frac{E\partial_x G - F\partial_y E}{E\sqrt{D}} \right)
$$

$$
+ \frac{1}{2\sqrt{D}} \frac{\partial}{\partial y}\left(\frac{2E\partial_x F - F\partial_x E - E\partial_y E}{E\sqrt{D}} \right)
$$

$$
= -\frac{1}{2\sqrt{D}} \frac{\partial}{\partial x}\left(\frac{\partial_x G - \partial_y F}{\sqrt{D}} \right) - \frac{1}{2\sqrt{D}} \frac{\partial}{\partial y}\left(\frac{\partial_y E - \partial_x F}{\sqrt{D}} \right)
$$

$$
- \frac{1}{4D^2} \det \begin{pmatrix} E & \partial_x E & \partial_y E \\ F & \partial_x F & \partial_y F \\ G & \partial_x G & \partial_y G \end{pmatrix}.
$$

This expression simplifies dramatically when $F = 0$ and we get

$$
K \circ \psi = -\frac{1}{2\sqrt{EG}}\left(\frac{\partial}{\partial x}\frac{\partial_x G}{\sqrt{EG}} + \frac{\partial}{\partial y}\frac{\partial_y E}{\sqrt{EG}} \right). \tag{5.4.6}
$$

Exercise 5.4.2 Prove that the Riemannian metric

$$
E = G = \frac{4}{(1 + x^2 + y^2)^2}, \qquad F = 0,
$$

on $\mathbb{R}^2$ has constant constant curvature $K = 1$ and the Riemannian metric

$$
E = G = \frac{4}{(1 - x^2 - y^2)^2}, \qquad F = 0,
$$

on the open unit disc has constant curvature $K = -1$.

Geometry and Topology

6

In this chapter we address what might be called the "fundamental problem of intrinsic differential geometry": when are two manifolds isometric? The central tool for addressing this question is the Cartan–Ambrose–Hicks Theorem (Sect. 6.1). In the subsequent sections we will use this result to examine flat spaces (Sect. 6.2), symmetric spaces (Sect. 6.3), and constant sectional curvature manifolds (Sect. 6.4). The chapter then examines manifolds of nonpositive sectional curvature and includes a proof of the Cartan Fixed Point Theorem (Sect. 6.5). The last three sections introduce the Ricci tensor and show that complete manifolds with uniformly positive Ricci tensor are compact (Sect. 6.6) and discuss the scalar curvature (Sect. 6.7) and the Weyl tensor (Sect. 6.8).

6.1 The Cartan–Ambrose–Hicks Theorem

The Cartan–Ambrose–Hicks Theorem answers the question (at least locally) when two manifolds are isometric. In general the equivalent conditions given there are probably more difficult to verify in most examples than the condition that there exist an isometry. However, under additional assumptions it has many important consequences. The section starts with some basic observations about homotopy and simple connectivity.

6.1.1 Homotopy

Definition 6.1.1 Let M be a manifold and let $I = [a, b]$ be a compact interval. A **(smooth) homotopy** of maps from I to M is a smooth map $\gamma : [0, 1] \times I \to M$. We often write $\gamma_\lambda(t) = \gamma(\lambda, t)$ for $\lambda \in [0, 1]$ and $t \in I$ and call γ a **(smooth) homotopy between** γ_0 **and** γ_1. We say the homotopy has **fixed endpoints** if $\gamma_\lambda(a) = \gamma_0(a)$ and $\gamma_\lambda(b) = \gamma_0(b)$ for all $\lambda \in [0, 1]$. (See Fig. 6.1.)

J.W. Robbin, D.A. Salamon, *Introduction to Differential Geometry*,
Springer Studium Mathematik (Master), https://doi.org/10.1007/978-3-662-64340-2_6

Fig. 6.1 A homotopy with
fixed endpoints

We remark that a homotopy and a variation are essentially the same thing,
namely a curve of maps (curves). The difference is pedagogical. We used the word
"variation" to describe a curve of maps through a given map; when we use this
word we are going to differentiate the curve to find a tangent vector (field) to the
given map. The word "homotopy" is used to describe a curve joining two maps; it
is a global rather than a local (infinitesimal) concept.

Definition 6.1.2 A manifold M is called **simply connected** iff for any two
curves $\gamma_0, \gamma_1 : [a, b] \to M$ with $\gamma_0(a) = \gamma_1(a)$ and $\gamma_0(b) = \gamma_1(b)$ there exists a
homotopy from γ_0 to γ_1 with endpoints fixed. (The idea is that the space $\Omega_{p,q}$ of
curves from p to q is connected.)

Remark 6.1.3 Two smooth maps $\gamma_0, \gamma_1 : [a, b] \to M$ with the same endpoints can
be joined by a continuous homotopy if and only if they can be joined by a smooth
homotopy. This follows from the Weierstrass approximation theorem.

Remark 6.1.4 Assume M is a connected smooth manifold. Then the topological
space $\Omega_{p,q}$ of all smooth curves in M with the endpoints p and q is connected for
some pair of points $p, q \in M$ if and only if it is connected for every pair of points
$p, q \in M$. (Prove this!)

Example 6.1.5 The Euclidean space $\mathbb{R}^m$ is simply connected; any two curves
$\gamma_0, \gamma_1 : [a, b] \to \mathbb{R}^m$ with the same endpoints can be joined by the homotopy $\gamma_\lambda(t) :=$
$\gamma_0(t) + \lambda(\gamma_1(t) - \gamma_0(t))$.

Example 6.1.6 The punctured plane $\mathbb{C} \setminus \{0\}$ is not simply connected; two curves
of the form

$$\gamma_n(t) := e^{2\pi i n t}, \qquad 0 \le t \le 1,$$

are not homotopic with fixed endpoints for distinct integers n.

Exercise 6.1.7 Prove that the m-sphere S^m is simply connected for $m \neq 1$.

6.1.2 The Global C-A-H Theorem

Theorem 6.1.8 (Global C-A-H Theorem) *Let $M \subset \mathbb{R}^n$ and $M' \subset \mathbb{R}^{n'}$ be nonempty, connected, simply connected, complete m-manifolds. Fix two elements $p_0 \in M$ and $p_0' \in M'$ and let $\Phi_0 : T_{p_0}M \to T_{p_0'}M'$ be an orthogonal linear isomorphism. Then the following are equivalent.*

(i) *There exists an isometry $\phi : M \to M'$ satisfying*

$$\phi(p_0) = p_0', \qquad d\phi(p_0) = \Phi_0. \tag{6.1.1}$$

(ii) *If (Φ, γ, γ') is a development satisfying the initial condition*

$$\gamma(0) = p_0, \qquad \gamma'(0) = p_0', \qquad \Phi(0) = \Phi_0, \tag{6.1.2}$$

then

$$\gamma(1) = p_0 \qquad \Longrightarrow \qquad \gamma'(1) = p_0', \quad \Phi(1) = \Phi_0$$

(iii) *If $(\Phi_0, \gamma_0, \gamma_0')$ and $(\Phi_1, \gamma_1, \gamma_1')$ are developments satisfying (6.1.2), then*

$$\gamma_0(1) = \gamma_1(1) \qquad \Longrightarrow \qquad \gamma_0'(1) = \gamma_1'(1).$$

(iv) *If (Φ, γ, γ') is a development satisfying (6.1.2), then $\Phi_* R_\gamma = R_{\gamma'}'$.*

Example 6.1.9 Before giving the proof let us interpret the conditions in case M and M' are two-dimensional spheres of radius r and r' respectively in three-dimensional Euclidean space $\mathbb{R}^3$. Imagine that the spheres are tangent at $p_0 = p_0'$. Clearly the spheres will be isometric exactly when $r = r'$.

Condition (ii) says that if the spheres are rolled along one another without sliding or twisting, then the endpoint $\gamma'(1)$ of one curve of contact depends only on the endpoint $\gamma(1)$ of the other and not on the intervening curve $\gamma(t)$. This condition is violated in the case $r \neq r'$ (see Fig. 6.2).

By Theorem 5.3.9 the Riemann curvature of a 2-manifold at p is determined by the Gaußian curvature $K(p)$; and for spheres we have $K(p) = 1/r^2$.

Fig. 6.2 Diagram for Example 6.1.9

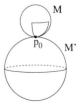

Exercise 6.1.10 Let γ be the closed curve which bounds an octant as shown in the diagram for Example 6.1.9. Find γ'.

Exercise 6.1.11 Show that in case M is two-dimensional, the condition $\Phi(1) = \Phi_0$ in Theorem 6.1.8 may be dropped from (ii).

Lemma 6.1.12 *Let $\phi : M \to M'$ be a local isometry and let $\gamma : I \to M$ be a smooth curve on an interval I. Fix an element $t_0 \in I$ and define*

$$p_0 := \gamma(t_0), \qquad q_0 := \phi(p_0), \qquad \Phi_0 := d\phi(p_0). \tag{6.1.3}$$

Then there exists a unique development (Φ, γ, γ') of M along M' on the entire interval I satisfying the initial conditions

$$\gamma'(t_0) = q_0, \qquad \Phi(t_0) = \Phi_0. \tag{6.1.4}$$

This development is given by

$$\gamma'(t) = \phi(\gamma(t)), \qquad \Phi(t) = d\phi(\gamma(t)) \tag{6.1.5}$$

for $t \in I$.

Proof Define

$$\gamma'(t) := \phi(\gamma(t)), \qquad \Phi(t) := d\phi(\gamma(t))$$

for $t \in I$. Then $\dot{\gamma}'(t) = \Phi(t)\dot{\gamma}(t)$ for all $t \in I$ by the chain rule, and every vector field X along γ satisfies $\Phi \nabla X = \nabla'(\Phi X)$ by Corollary 5.3.2. Hence (Φ, γ, γ') is a development by Lemma 3.5.19. By (6.1.3) this development satisfies the initial condition (6.1.4). Hence the assertion follows from the uniqueness result for developments in Theorem 3.5.21. This proves Lemma 6.1.12. $\square$

Proof of Theorem 6.1.8 We first prove a slightly different theorem. Namely, we weaken condition (i) to assert that ϕ is a local isometry (i.e. not necessarily bijective), and prove that this weaker condition is equivalent to (ii), (iii), and (iv) whenever M is connected and simply connected and M' is complete. Thus we drop the hypotheses that M be complete and M' be connected and simply connected.

We prove that (i) implies (ii). Given a development as in (ii) we have, by Lemma 6.1.12,

$$\gamma'(1) = \phi(\gamma(1)) = \phi(p_0) = p_0', \qquad \Phi(1) = d\phi(\gamma(1)) = d\phi(p_0) = \Phi_0,$$

as required.

We prove that (ii) implies (iii) when M' is complete. Choose developments $(\Phi_i, \gamma_i, \gamma_i')$ for $i = 0, 1$ as in (iii). Define a curve $\gamma : [0, 1] \to M$ by "composition", i.e.

$$\gamma(t) := \begin{cases} \gamma_0(2t), & 0 \le t \le 1/2, \\ \gamma_1(2 - 2t), & 1/2 \le t \le 1, \end{cases}$$

so that γ is continuous and piecewise smooth and $\gamma(1) = p_0$. By Theorem 3.5.21 there exists a development (Φ, γ, γ') on the interval $[0, 1]$ satisfying (6.1.2) (because M' is complete). Since $\gamma(1) = p_0$ it follows from (ii) that $\gamma'(1) = p_0'$ and $\Phi(1) = \Phi_0$. By the uniqueness of developments and the invariance under reparametrization, we have

$$(\Phi(t), \gamma(t), \gamma'(t)) = \begin{cases} (\Phi_0(2t), \gamma_0(2t), \gamma_0'(2t)), & 0 \le t \le 1/2, \\ (\Phi_1(2 - 2t), \gamma_1(2 - 2t), \gamma_1'(2 - 2t)), & 1/2 \le t \le 1. \end{cases}$$

Hence $\gamma_0'(1) = \gamma'(1/2) = \gamma_1'(1)$ as required.

We prove that (iii) implies (i) when M' is complete and M is connected. Define the map $\phi : M \to M'$ as follows. Fix an element $p \in M$. Since M is connected, there exists a smooth curve $\gamma : [0, 1] \to M$ such that $\gamma(0) = p_0$ and $\gamma(1) = p$. Since M' is complete, there exists a development (Φ, γ, γ') with $\gamma'(0) = p_0'$ and $\Phi(0) = \Phi_0$ (Theorem 3.5.21). Now define $\phi(p) := \gamma'(1)$. By (iii) the endpoint $p' := \gamma'(1)$ is independent of the choice of the curve γ, and so ϕ is well-defined. We prove that this map ϕ satisfies the following

(a) If (Φ, γ, γ') is a development satisfying $\gamma(0) = p_0$, $\gamma'(0) = p_0'$, $\Phi(0) = \Phi_0$, then $\phi(\gamma(t)) = \gamma'(t)$ for $0 \le t \le 1$.

(b) If $p, q \in M$ satisfy $0 < d(p, q) < \mathrm{inj}(p; M)$ and $d(p, q) < \mathrm{inj}(\phi(p); M')$, then $d'(\phi(p), \phi(q)) = d(p, q)$.

That ϕ satisfies (a) follows directly from the definition and the fact that the triple $(\Phi_t, \gamma_t, \gamma_t')$ defined by $\Phi_t(s) := \Phi(st)$, $\gamma_t(s) := \gamma(st)$, $\gamma_t'(s) := \gamma'(st)$ for $0 \le s \le 1$ is a development. To prove (b), choose $v \in T_p M$ such that

$$|v| = d(p, q) \qquad \exp_p(v) = q$$

(Theorem 4.4.4) and let $\gamma : [0, 1] \to M$ be a smooth curve with

$$\gamma(0) = p_0, \qquad \gamma(t) = \exp_p((2t - 1)v)$$

for $\frac{1}{2} \le t \le 1$. Let (Φ, γ, γ') be the unique development of M along M' satisfying $\gamma'(0) = p_0'$ and $\Phi(0) = \Phi_0$ (Theorem 3.5.21). Then, by (a),

$$\gamma'(\tfrac{1}{2}) = \phi(p), \qquad \gamma'(1) = \phi(q).$$

Also, by part (ii) of Lemma 3.5.19 with $X = \dot\gamma$, the restriction of γ' to the interval $[\frac{1}{2}, 1]$ is a geodesic. Thus $\gamma'(t) = \exp_{\phi(p)}((2t - 1)v')$ for $\frac{1}{2} \le t \le 1$, where the tangent vector $v' \in T_{\phi(p)}M'$ is given by $v' := \dot\gamma'(\tfrac{1}{2}) = \Phi(\tfrac{1}{2})v$ and hence satisfies $|v'| = |v| = d(p, q) < \mathrm{inj}(\phi(p); M')$. Thus it follows from Theorem 4.4.4 that $d'(\phi(p), \phi(q)) = d'(\phi(p), \exp_{\phi(p)}'(v')) = |v'| = d(p, q)$ and this proves (b). It follows from (b) and Theorem 5.1.1 that ϕ is a local isometry.

We prove that (i) implies (iv). Given a development as in (ii) we have

$$\gamma'(t) = \phi(\gamma(t)), \qquad \Phi(t) = d\phi(\gamma(t))$$

for every t, by Lemma 6.1.12. Hence it follows from part (iv) of Corollary 5.3.2 (Theorema Egregium for local isometries) that

$$\Phi(t)_* R_{\gamma(t)} = (\phi_* R)_{\gamma'(t)} = R'_{\gamma'(t)}$$

for all t as required.

We prove that (iv) implies (iii) when M' is complete and M is simply connected. Choose developments $(\Phi_i, \gamma_i, \gamma_i')$ for $i = 0, 1$ as in (iii). Since M is simply connected there exists a homotopy

$$[0, 1] \times [0, 1] \to M : (\lambda, t) \mapsto \gamma(\lambda, t) = \gamma_\lambda(t)$$

from γ_0 to γ_1 with endpoints fixed. By Theorem 3.5.21 there is, for each λ, a development $(\Phi_\lambda, \gamma_\lambda, \gamma_\lambda')$ on the interval $[0, 1]$ with initial conditions

$$\gamma_\lambda'(0) = p_0', \qquad \Phi_\lambda(0) = \Phi_0$$

(because M' is complete). The proof of Theorem 3.5.21 also shows that $\gamma_\lambda(t)$ and $\Phi_\lambda(t)$ depend smoothly on both t and λ. We must prove that

$$\gamma_1'(1) = \gamma_0'(1).$$

To see this we will show that, for each fixed t, the curve

$$\lambda \mapsto (\Phi_\lambda(t), \gamma_\lambda(t), \gamma_\lambda'(t))$$

is a development; then by the definition of development we have that the curve $\lambda \mapsto \gamma_\lambda'(1)$ is smooth and

$$\partial_\lambda \gamma_\lambda'(1) = \Phi_\lambda(1) \partial_\lambda \gamma_\lambda(1) = 0$$

as required.

First choose a basis $e_1, \ldots, e_m$ of $T_{p_0} M$ and extend it to obtain vector fields $E_i \in \mathrm{Vect}(\gamma)$ along the homotopy γ by imposing the conditions that the vector fields $t \mapsto E_i(\lambda, t)$ be parallel, i.e.

$$\nabla_t E_i(\lambda, t) = 0, \qquad E_i(\lambda, 0) = e_i. \tag{6.1.6}$$

Then the vectors $E_1(\lambda, t), \ldots E_m(\lambda, t)$ form a basis of $T_{\gamma_\lambda(t)} M$ for all λ and t. Second, define the vector fields E_i' along γ' by

$$E_i'(\lambda, t) := \Phi_\lambda(t) E_i(\lambda, t) \tag{6.1.7}$$

so that $\nabla_t' E_i' = 0$. Third, define the functions $\xi^1, \ldots, \xi^m : [0, 1]^2 \to \mathbb{R}$ by

$$\partial_t \gamma =: \sum_{i=1}^m \xi^i E_i, \qquad \partial_t \gamma' = \sum_{i=1}^m \xi^i E_i'. \tag{6.1.8}$$

Here the second equation follows from (6.1.7) and the fact that $\Phi_\lambda \partial_t \gamma = \partial_t \gamma'$.

Now consider the vector fields

$$X' := \partial_\lambda \gamma', \qquad Y_i' := \nabla_\lambda' E_i' \tag{6.1.9}$$

along γ'. They satisfy the equations

$$\nabla_t' X' = \nabla_t' \partial_\lambda \gamma' = \nabla_\lambda' \partial_t \gamma' = \nabla_\lambda' \left(\sum_{i=1}^m \xi^i E_i' \right) = \sum_{i=1}^m (\partial_\lambda \xi^i E_i' + \xi^i Y_i')$$

and

$$\nabla_t' Y_i' = \nabla_t' \nabla_\lambda' E_i' - \nabla_\lambda' \nabla_t' E_i'$$
$$= R'(\partial_t \gamma', \partial_\lambda \gamma') E_i'.$$

To sum up we have $X'(\lambda, 0) = Y_i'(\lambda, 0) = 0$ and

$$\nabla_t' X' = \sum_{i=1}^m (\partial_\lambda \xi^i E_i' + \xi^i Y_i'), \qquad \nabla_t' Y_i' = R'(\partial_t \gamma', X') E_i'. \tag{6.1.10}$$

On the other hand, the vector fields

$$X' := \Phi_\lambda \partial_\lambda \gamma, \qquad Y_i' := \Phi_\lambda \nabla_\lambda E_i \tag{6.1.11}$$

along γ' satisfy the same equations, namely

$$\nabla_t' X' = \Phi_\lambda \nabla_t \partial_\lambda \gamma = \Phi_\lambda \nabla_\lambda \partial_t \gamma = \Phi_\lambda \nabla_\lambda \left(\sum_{i=1}^m \xi^i E_i \right)$$
$$= \Phi_\lambda \sum_{i=1}^m (\partial_\lambda \xi^i E_i + \xi^i \nabla_\lambda E_i) = \sum_{i=1}^m (\partial_\lambda \xi^i E_i' + \xi^i Y_i'),$$
$$\nabla_t' Y_i' = \Phi_\lambda (\nabla_t \nabla_\lambda E_i - \nabla_\lambda \nabla_t E_i) = \Phi_\lambda R(\partial_t \gamma, \partial_\lambda \gamma) E_i$$
$$= R'(\Phi_\lambda \partial_t \gamma, \Phi_\lambda \partial_\lambda \gamma) \Phi_\lambda E_i = R'(\partial_t \gamma', X') E_i'.$$

Here the last but one equation follows from (iv).

Since the tuples (6.1.9) and (6.1.11) satisfy the same differential equation (6.1.10) and vanish at $t = 0$ they must agree. Hence

$$\partial_\lambda \gamma' = \Phi_\lambda \partial_\lambda \gamma, \qquad \nabla_\lambda' E_i' = \Phi_\lambda \nabla_\lambda E_i$$

for $i = 1, \ldots, m$. This says that $\lambda \mapsto (\Phi_\lambda(t), \gamma_\lambda(t), \gamma'_\lambda(t))$ is a development. For $t = 1$ we obtain $\partial_\lambda \gamma'(\lambda, 1) = 0$ as required.

Now the modified theorem (where ϕ is a local isometry) is proved. The original theorem follows immediately. Condition (iv) is symmetric in M and M'. Thus, if we assume (iv), there are local isometries $\phi : M \to M'$ and $\psi : M' \to M$ satisfying $\phi(p_0) = p'_0$, $d\phi(p_0) = \Phi_0$ and $\psi(p'_0) = p_0$, $d\psi(p'_0) = \Phi_0^{-1}$. But then $\psi \circ \phi$ is a local isometry with $\psi \circ \phi(p_0) = p_0$ and $d(\psi \circ \phi)(p_0) = \mathrm{id}$. Hence $\psi \circ \phi$ is the identity. Similarly $\phi \circ \psi$ is the identity so ϕ is bijective (and $\psi = \phi^{-1}$) as required. This proves Theorem 6.1.8.   □

Remark 6.1.13   The proof of Theorem 6.1.8 shows that the various implications in the **weak version** of the theorem (where ϕ is only a local isometry) require the following conditions on M and M':

- (i) always implies (ii), (iii), and (iv);
- (ii) implies (iii) whenever M' is complete;
- (iii) implies (i) whenever M' is complete and M is connected;
- (iv) implies (iii) whenever M' is complete and M is simply connected.

Remark 6.1.14   The proof that (iii) implies (i) in Theorem 6.1.8 can be slightly shortened by using the following observation. *Let $\phi : M \to M'$ be a map between smooth manifolds. Assume that $\phi \circ \gamma$ is smooth for every smooth curve $\gamma : [0, 1] \to M$. Then ϕ is smooth.*

Corollary 6.1.15   *Let M and M' be nonempty, connected, simply connected, complete Riemannian manifolds and let $\phi : M \to M'$ be a local isometry. Then ϕ is bijective and hence is an isometry.*

Proof   This follows by combining the weak and strong versions of the global C-A-H Theorem 6.1.8. Let $p_0 \in M$ and define $p'_0 := \phi(p_0)$ and $\Phi_0 := d\phi(p_0)$. Then the tuple M, M', p_0, p'_0, Φ_0 satisfies condition (i) of the weak version of Theorem 6.1.8. Hence this tuple also satisfies condition (iv) of Theorem 6.1.8. Since M and M' are connected, simply connected, and complete we may apply the strong version of Theorem 6.1.8 to obtain an isometry $\psi : M \to M'$ satisfying $\psi(p_0) = p'_0$ and $d\psi(p_0) = \Phi_0$. Since every isometry is also a local isometry and M is connected it follows from Lemma 5.1.10 that $\phi(p) = \psi(p)$ for all $p \in M$. Hence ϕ is an isometry, as required.   □

Remark 6.1.16   Refining the argument in the proof of Corollary 6.1.15 one can show that a local isometry $\phi : M \to M'$ must be surjective whenever M is complete and M' is connected. None of these assumptions can be removed. (Take an isometric embedding of a disc in the plane or an embedding of a complete space M into a space with two components, one of which is isometric to M.)

Likewise, one can show that a local isometry $\phi : M \to M'$ must be injective whenever M is complete and connected and M' is simply connected. Again none of these asumptions can be removed. (Take a covering $\mathbb{R} \to S^1$, or a covering of a disjoint union of two isometric complete simply connected spaces onto one copy of this space, or some noninjective immersion of a disc into the plane and choose the pullback metric on the disc.)

6.1.3 The Local C-A-H Theorem

Theorem 6.1.17 (Local C-A-H Theorem) *Let M and M' be smooth m-manifolds, let $p_0 \in M$ and $p_0' \in M'$, and let $\Phi_0 : T_{p_0} M \to T_{p_0'} M'$ be an orthogonal linear isomorphism. Let $r > 0$ be smaller than the injectvity radii of M at p_0 and of M' at p_0' and define*

$$U_r := \{p \in M \mid d(p_0, p) < r\}, \qquad U_r' := \{p' \in M' \mid d'(p_0', p') < r\}.$$

Then the following are equivalent.

(i) *There exists an isometry $\phi : U_r \to U_r'$ satisfying (6.1.1).*
(ii) *If (Φ, γ, γ') is a development on an interval $I \subset \mathbb{R}$ with $0 \in I$, satisfying the initial condition (6.1.2) as well as $\gamma(I) \subset U_r$ and $\gamma'(I) \subset U_r'$, then*

$$\gamma(1) = p_0 \quad \Longrightarrow \quad \gamma'(1) = p_0', \quad \Phi(1) = \Phi_0.$$

(iii) *If $(\Phi_0, \gamma_0, \gamma_0')$ and $(\Phi_1, \gamma_1, \gamma_1')$ are developments as in (ii), then*

$$\gamma_0(1) = \gamma_1(1) \quad \Longrightarrow \quad \gamma_0'(1) = \gamma_1'(1).$$

(iv) *If $v \in T_{p_0} M$ with $|v| < r$ and*

$$\gamma(t) := \exp_{p_0}(tv), \quad \gamma'(t) := \exp_{p_0'}(t\Phi_0 v), \quad \Phi(t) := \Phi_{\gamma'}'(t, 0)\Phi_0\Phi_\gamma(0, t),$$

then $\Phi(t)_* R_{\gamma(t)} = R_{\gamma'(t)}'$ for $0 \leq t \leq 1$.

If these equivalent conditions are satisfied, then

$$\phi(\exp_{p_0}(v)) = \exp_{p_0'}(\Phi_0 v)$$

for all $v \in T_{p_0} M$ with $|v| < r$.

The proof is based on the following lemma.

Lemma 6.1.18 *Let $p \in M$ and $v, w \in T_p M$. For $0 \leq t \leq 1$ define*

$$\gamma(t) := \exp(tv), \qquad X(t) := \frac{\partial}{\partial\lambda}\bigg|_{\lambda=0} \exp_p\big(t(v + \lambda w)\big) \in T_{\gamma(t)} M.$$

Then

$$\nabla_t \nabla_t X + R(X, \dot{\gamma})\dot{\gamma} = 0, \qquad X(0) = 0, \qquad \nabla_t X(0) = w. \qquad (6.1.12)$$

*A vector field along γ satisfying the first equation in (6.1.12) is called a **Jacobi field** along γ.*

Proof   Define

$$\gamma(\lambda, t) := \exp_p(t(v + \lambda w)), \qquad X(\lambda, t) := \partial_\lambda \gamma(\lambda, t)$$

for all λ and t. Since $\gamma(\lambda, 0) = p$ for all λ we have $X(\lambda, 0) = 0$ and

$$\nabla_t X(\lambda, 0) = \nabla_t \partial_\lambda \gamma(\lambda, 0) = \nabla_\lambda \partial_t \gamma(\lambda, 0) = \frac{d}{d\lambda}(v + \lambda w) = w.$$

Moreover, $\nabla_t \partial_t \gamma = 0$ and hence

$$\begin{aligned}
\nabla_t \nabla_t X &= \nabla_t \nabla_t \partial_\lambda \gamma \\
&= \nabla_t \nabla_\lambda \partial_t \gamma - \nabla_\lambda \nabla_t \partial_t \gamma \\
&= R(\partial_t \gamma, \partial_\lambda \gamma) \partial_t \gamma \\
&= R(\partial_t \gamma, X) \partial_t \gamma.
\end{aligned}$$

This proves Lemma 6.1.18. $\square$

Proof of Theorem 6.1.17   The proofs (i) $\Longrightarrow$ (ii) $\Longrightarrow$ (iii) $\Longrightarrow$ (i) $\Longrightarrow$ (iv) are as before; the reader might note that when $L(\gamma) \leq r$ we also have $L(\gamma') \leq r$ for any development so that there are plenty of developments with $\gamma : [0, 1] \to U_r$ and $\gamma' : [0, 1] \to U_r'$. The proof that (iv) implies (i) is a little different since (iv) here is somewhat weaker than (iv) of the global theorem: the equation $\Phi_* R = R'$ is only assumed for certain developments.

Hence assume (iv) and define $\phi : U_r \to U_r'$ by

$$\phi := \exp_{p_0'}' \circ \Phi_0 \circ \exp_{p_0}^{-1} : U_r \to U_r'.$$

We must prove that ϕ is an isometry. Thus we fix a point $q \in U_r$ and a tangent vector $u \in T_q M$ and choose $v, w \in T_p M$ with $|v| < r$ such that

$$\exp_{p_0}(v) = q, \qquad d \exp_{p_0}(v)w = u. \qquad (6.1.13)$$

Define $\gamma : [0, 1] \to U_r$, $\gamma' : [0, 1] \to U_r'$, $X \in \text{Vect}(\gamma)$, and $X' \in \text{Vect}(\gamma')$ by

$$\gamma(t) = \exp_{p_0}(tv), \qquad X(t) := \left. \frac{\partial}{\partial\lambda} \right|_{\lambda=0} \exp_{p_0}(t(v + \lambda w)),$$

$$\gamma'(t) = \exp_{p_0'}'(t\Phi_0 v), \qquad X'(t) := \left. \frac{\partial}{\partial\lambda} \right|_{\lambda=0} \exp_{p_0'}'(t(\Phi_0 v + \lambda \Phi_0 w)).$$

Then, by definition of ϕ, we have

$$\gamma' := \phi \circ \gamma, \qquad d\phi(\gamma)X = X'. \tag{6.1.14}$$

By Lemma 6.1.18, X is a solution of (6.1.12) and X' is a solution of

$$\nabla_t \nabla_t X' = R'(\partial_t \gamma', X')\partial_t \gamma', \quad X'(\lambda, 0) = 0, \quad \nabla_t X'(\lambda, 0) = \Phi_0 w. \tag{6.1.15}$$

Now define $\Phi(t) : T_{\gamma(t)} M \to T_{\gamma'(t)} M'$ by

$$\Phi(t) := \Phi'_{\gamma'}(t, 0)\Phi_0 \Phi_\gamma(0, t).$$

Then Φ intertwines covariant differentiation. Since $\dot{\gamma}$ and $\dot{\gamma}'$ are parallel vector fields with $\dot{\gamma}'(0) = \Phi_0 v = \Phi(0)\dot{\gamma}(0)$, we have

$$\Phi(t)\dot{\gamma}(t) = \dot{\gamma}'(t)$$

for every t. Moreover, it follows from (iv) that $\Phi_* R_\gamma = R'_{\gamma'}$. Combining this with (6.1.12) we obtain

$$\nabla'_t \nabla'_t(\Phi X) = \Phi \nabla_t \nabla_t X = R'(\Phi \dot{\gamma}, \Phi X)\Phi \dot{\gamma} = R'(\dot{\gamma}', \Phi X)\dot{\gamma}'.$$

Hence the vector field ΦX along γ' also satisfies the initial value problem (6.1.15) and thus

$$\Phi X = X' = d\phi(\gamma)X.$$

Here we have also used (6.1.14). Using (6.1.13) we find

$$\gamma(1) = \exp_{p_0}(v) = q, \qquad X(1) = d\exp_{p_0}(v)w = u,$$

and so

$$d\phi(q)u = d\phi(\gamma(1))X(1) = X'(1) = \Phi(1)u.$$

Since $\Phi(1) : T_{\gamma(1)} M \to T_{\gamma'(1)} M'$ is an orthogonal transformation this gives

$$|d\phi(q)u| = |\Phi(1)u| = |u|.$$

Hence ϕ is an isometry as claimed. This proves Theorem 6.1.17. $\square$

6.2 Flat Spaces

Our aim in the next few sections is to give applications of the Cartan-Ambrose-Hicks Theorem. It is clear that the hypothesis $\Phi_* R = R'$ for *all* developments will be difficult to verify without drastic hypotheses on the curvature. The most drastic such hypothesis is that the curvature vanishes identically.

Definition 6.2.1   A Riemannian manifold M is called **flat** iff the Riemann curvature tensor R vanishes identically.

Theorem 6.2.2   *Let $M \subset \mathbb{R}^n$ be a smooth m-manifold.*

(i)   *M is flat if and only if every point has a neighborhood which is isometric to an open subset of $\mathbb{R}^m$, i.e. at each point $p \in M$ there exist local coordinates $x^1, \ldots, x^m$ such that the coordinate vectorfields $E_i = \partial/\partial x^i$ are orthonormal.*

(ii)   *Assume M is connected, simply connected, and complete. Then M is flat if and only if there is an isometry $\phi : M \to \mathbb{R}^m$ onto Euclidean space.*

Proof   Assertion (i) follows immediately from Theorem 6.1.17 and (ii) follows immediately from Theorem 6.1.8.   □

Exercise 6.2.3   Carry over the Cartan–Ambrose–Hicks theorem and Theorem 6.2.2 to the intrinsic setting.

Exercise 6.2.4   A one-dimensional manifold is always flat.

Exercise 6.2.5   If M_1 and M_2 are flat, so is $M = M_1 \times M_2$.

Example 6.2.6   By Exercises 6.2.4 and 6.2.5 the standard torus

$$\mathbb{T}^m = \big\{ z = (z_1, \ldots, z_m) \in \mathbb{C}^m \,\big|\, |z_1| = \cdots = |z_m| = 1 \big\}$$

is flat.

Exercise 6.2.7   For $a, b > 0$ and $c \geq 0$ define $M \subset \mathbb{C}^3$ by

$$M := M(a, b, c) := \Big\{ (u, v, w) \in \mathbb{C}^3 \,\Big|\, |u| = a, \ |v| = b, \ w = c \frac{u\,v}{a\,b} \Big\}.$$

Then M is diffeomorphic to a torus (a product of two circles) and M is flat. If $a', b' > 0$ and $c' \geq 0$, prove that there is an isometry ϕ from $M' = M(a', b', c')$ to $M = M(a, b, c)$ if and only if the triples (a', b', c') and (a, b, c) are related by a permutation.

Hint: Show first that an isometry $\phi : M' \to M$ that satisfies the condition $\phi(a', b', c') = (a, b, c)$ must have the form

$$\phi(u', v', w') = \left(a \left(\frac{u'}{a'}\right)^\alpha \left(\frac{v'}{b'}\right)^\beta, \ b \left(\frac{u'}{a'}\right)^\gamma \left(\frac{v'}{b'}\right)^\delta, \ c \left(\frac{u'}{a'}\right)^{\alpha+\gamma} \left(\frac{v'}{b'}\right)^{\beta+\delta} \right)$$

for integers $\alpha, \beta, \gamma, \delta$ that satisfy $\alpha\delta - \beta\gamma = \pm 1$. Show that this map ϕ is an isometry if and only if

$$a'^2 + c'^2 = \alpha^2 a^2 + \gamma^2 b^2 + (\alpha + \gamma)^2 c^2,$$
$$c'^2 = \alpha\beta a^2 + \gamma\delta b^2 + (\alpha + \gamma)(\beta + \delta)c^2,$$
$$b'^2 + c'^2 = \beta^2 a^2 + \delta^2 b^2 + (\beta + \delta)^2 c^2.$$

Exercise 6.2.8 (Developable manifolds) Let $n = m + 1$ and let $E(t)$ be a one-parameter family of hyperplanes in $\mathbb{R}^n$. Then there exists a smooth map $u : \mathbb{R} \to \mathbb{R}^n$ such that

$$E(t) = u(t)^\perp, \qquad |u(t)| = 1, \qquad (6.2.1)$$

for every t. We assume that $\dot{u}(t) \neq 0$ for every t so that $u(t)$ and $\dot{u}(t)$ are linearly independent. Show that

$$L(t) := u(t)^\perp \cap \dot{u}(t)^\perp = \lim_{s \to t} E(t) \cap E(s). \qquad (6.2.2)$$

Thus $L(t)$ is a linear subspace of dimension $m - 1$. Now let $\gamma : \mathbb{R} \to \mathbb{R}^n$ be a smooth curve such that

$$\langle \dot{\gamma}(t), u(t) \rangle = 0, \qquad \langle \dot{\gamma}(t), \dot{u}(t) \rangle \neq 0 \qquad (6.2.3)$$

for all t. This means that $\dot{\gamma}(t) \in E(t)$ and $\dot{\gamma}(t) \notin L(t)$; thus $E(t)$ is spanned by $L(t)$ and $\dot{\gamma}(t)$. For $t \in \mathbb{R}$ and $\varepsilon > 0$ define

$$L(t)_\varepsilon := \{v \in L(t) \mid |v| < \varepsilon\}.$$

Let $I \subset \mathbb{R}$ be a bounded open interval such that the restriction of γ to the closure of I is injective. Prove that, for $\varepsilon > 0$ sufficiently small, the set

$$M_0 := \bigcup_{t \in I} \left(\gamma(t) + L(t)_\varepsilon \right)$$

is a smooth manifold of dimension $m = n - 1$. A manifold which arises this way is called **developable**. Show that the tangent spaces of M_0 are the original subspaces $E(t)$, i.e.

$$T_p M_0 = E(t) \qquad \text{for} \qquad p \in \gamma(t) + L(t)_\varepsilon.$$

(One therefore calls M_0 the *"envelope"* of the hyperplanes $\gamma(t) + E(t)$.) Show that M_0 is flat. (**Hint:** use Gauß–Codazzi.) If (Φ, γ, γ') is a development of M_0 along $\mathbb{R}^m$, show that the map $\phi : M_0 \to \mathbb{R}^m$, defined by

$$\phi(\gamma(t) + v) := \gamma'(t) + \Phi(t)v$$

for $v \in L(t)_\varepsilon$, is an isometry onto an open set $M_0' \subset \mathbb{R}^m$. Thus a development *"unrolls"* M_0 onto the Euclidean space $\mathbb{R}^m$. When $n = 3$ and $m = 2$ one can visualize M_0 as a twisted sheet of paper (see Fig. 6.3).

Remark 6.2.9   Given a codimension-1 submanifold

$$M \subset \mathbb{R}^{m+1}$$

and a curve $\gamma : \mathbb{R} \to M$ we may form the **osculating developable** M_0 to M along γ by taking

$$E(t) := T_{\gamma(t)} M.$$

This developable has common affine tangent spaces with M along γ as

$$T_{\gamma(t)} M_0 = E(t) = T_{\gamma(t)} M$$

for every t. This gives a nice interpretation of parallel transport: M_0 may be un-rolled onto a hyperplane where parallel transport has an obvious meaning and the identification of the tangent spaces thereby defines parallel transport in M. (See Remark 3.5.16.)

Exercise 6.2.10   Each of the following is a developable surface in $\mathbb{R}^3$.

(i)   A **cone on a plane curve** $\Gamma \subset H$, i.e.

$$M = \{ tp + (1-t)q \mid t > 0, q \in \Gamma \}$$

where $H \subset \mathbb{R}^3$ is an affine hyperplane, $p \in \mathbb{R}^3 \setminus H$, and $\Gamma \subset H$ is a 1-manifold.

(ii) A **cylinder on a plane curve** Γ, i.e.

$$M = \{ q + tv \mid q \in \Gamma, t \in \mathbb{R} \}$$

where H and Γ are as in (i) and v is a fixed vector not parallel to H. (This is a cone with the cone point p at infinity.)

(iii) The **tangent developable** to a space curve $\gamma : \mathbb{R} \to \mathbb{R}^3$, i.e.

$$M = \{ \gamma(t) + s\dot\gamma(t) \mid |t - t_0| < \varepsilon,\ 0 < s < \varepsilon \},$$

where $\dot\gamma(t_0)$ and $\ddot\gamma(t_0)$ are linearly independent and $\varepsilon > 0$ is sufficiently small.

(iv) The **paper model of a Möbius strip** (see Fig. 6.3).

Fig. 6.3 Developable surfaces

 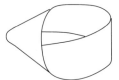

Fig. 6.4 A circular one-
sheeted hyperboloid

Remark 6.2.11 A 2-dimensional submanifold $M \subset \mathbb{R}^3$ is called a **ruled surface** iff there is a straight line in M through every point. Every developable surface is ruled, however, there are ruled surfaces that are not developable. An example is the manifold $M = \{\gamma(t) + s\ddot{\gamma}(t) \mid |t - t_0| < \varepsilon, |s| < \varepsilon\}$ where $\gamma : \mathbb{R} \to \mathbb{R}^3$ is a smooth curve with $|\dot{\gamma}| \equiv 1$ and $\ddot{\gamma}(t_0) \neq 0$, and $\varepsilon > 0$ is sufficiently small; this surface is not developable in general. Other examples are the **elliptic hyperboloid of one sheet**

$$M := \left\{ (x, y, z) \in \mathbb{R}^3 \,\middle|\, \frac{x^2}{a^2} + \frac{y^2}{b^2} - \frac{z^2}{c^2} = 1 \right\} \tag{6.2.4}$$

depicted in Fig. 6.4, the **hyperbolic paraboloid**

$$M := \left\{ (x, y, z) \in \mathbb{R}^3 \,\middle|\, z = \frac{x^2}{a^2} - \frac{y^2}{b^2} \right\}. \tag{6.2.5}$$

(both with two straight lines through every point in M), **Plücker's conoid**

$$M := \left\{ (x, y, z) \in \mathbb{R}^3 \,\middle|\, x^2 + y^2 \neq 0, \ z = \frac{2xy}{x^2 + y^2} \right\}, \tag{6.2.6}$$

the **helicoid**

$$M := \left\{ (x, y, z) \in \mathbb{R}^3 \,\middle|\, \frac{x + iy}{\sqrt{x^2 + y^2}} = e^{i\alpha z} \right\}, \tag{6.2.7}$$

and the **Möbius strip**

$$M := \left\{ \begin{pmatrix} \cos(s) \\ \sin(s) \\ 0 \end{pmatrix} + \frac{t}{2} \begin{pmatrix} \cos(s/2)\cos(s) \\ \cos(s/2)\sin(s) \\ \sin(s/2) \end{pmatrix} \,\middle|\, \begin{matrix} s \in \mathbb{R} \text{ and} \\ -1 < t < 1 \end{matrix} \right\}. \tag{6.2.8}$$

These five surfaces have negative Gaußian curvature. The Möbius strip in (6.2.8) is not developable, while the paper model of the Möbius strip is. The helicoid in (6.2.7) is a **minimal surface**, i.e. its **mean curvature** (the trace of the second fundamental form) vanishes. A minimal surface which is not ruled is the **catenoid**

$$M := \{ (x, y, z) \in \mathbb{R}^3 \mid x^2 + y^2 = c^2 \cosh(z/c) \}.$$

(**Exercise:** Prove all this.)

6.3 Symmetric Spaces

In the last section we applied the Cartan-Ambrose-Hicks Theorem in the flat case; the hypothesis $\Phi_* R = R'$ was easy to verify since both sides vanish. To find more general situations where we can verify this hypothesis note that for any development (Φ, γ, γ') satisfying the initial conditions

$$\gamma(0) = p_0, \qquad \gamma'(0) = p_0', \qquad \Phi(0) = \Phi_0,$$

we have

$$\Phi(t) = \Phi_{\gamma'}'(t, 0)\Phi_0\Phi_\gamma(0, t)$$

so that the hypothesis $\Phi_* R = R'$ is certainly implied by the three hypotheses

$$\Phi_\gamma(t, 0)_* R_{p_0} = R_{\gamma(t)}$$
$$\Phi_{\gamma'}'(t, 0)_* R_{p_0'}' = R_{\gamma'(t)}'$$
$$(\Phi_0)_* R_{p_0} = R_{p_0'}'.$$

The last hypothesis is a condition on the initial linear isomorphism

$$\Phi_0 : T_{p_0} M \to T_{p_0'} M'$$

while the former hypotheses are conditions on M and M' respectively, namely, that the Riemann curvature tensor is invariant by parallel transport. It is rather amazing that this condition is equivalent to a simple geometric condition as we now show.

6.3.1 Symmetric Spaces

Definition 6.3.1 A Riemannian manifold M is called **symmetric about the point** $p \in M$ iff there exists a (necessarily unique) isometry

$$\phi : M \to M$$

satisfying

$$\phi(p) = p, \qquad d\phi(p) = -\mathrm{id}; \tag{6.3.1}$$

M is called a **symmetric space** iff it is symmetric about each of its points. A Riemannian manifold M is called **locally symmetric about the point** $p \in M$ iff, for $r > 0$ sufficiently small, there exists an isometry

$$\phi : U_r(p, M) \to U_r(p, M), \qquad U_r(p, M) := \{q \in M \mid d(p, q) < r\},$$

satisfying (6.3.1); M is called a **locally symmetric space** iff it is locally symmetric about each of its points.

Remark 6.3.2 The proof of Theorem 6.3.4 below will show that, if M is locally symmetric, the isometry $\phi : U_r(p, M) \to U_r(p, M)$ with $\phi(p) = p$ and $d\phi(p) = -\mathrm{id}$ exists whenever $0 < r \le \mathrm{inj}(p)$.

Exercise 6.3.3 Every symmetric space is complete. **Hint:** If $\gamma : I \to M$ is a geodesic and $\phi : M \to M$ is a symmetry about the point $\gamma(t_0)$ for $t_0 \in I$, then $\phi(\gamma(t_0 + t)) = \gamma(t_0 - t)$ for all $t \in \mathbb{R}$ with $t_0 + t, t_0 - t \in I$.

Theorem 6.3.4 *Let $M \subset \mathbb{R}^n$ be an m-dimensional submanifold. Then the following are equivalent.*

(i) *M is locally symmetric.*
(ii) *The covariant derivative ∇R (defined below) vanishes identically, i.e.*

$$(\nabla_v R)_p(v_1, v_2)w = 0$$

for all $p \in M$ and $v, v_1, v_2, w \in T_p M$.
(iii) *The curvature tensor R is invariant under parallel transport, i.e.*

$$\Phi_\gamma(t, s)_* R_{\gamma(s)} = R_{\gamma(t)} \qquad (6.3.2)$$

for every smooth curve $\gamma : \mathbb{R} \to M$ and all $s, t \in \mathbb{R}$.

Proof See Sect. 6.3.2. □

Corollary 6.3.5 *Let M and M' be locally symmetric spaces and fix two points $p_0 \in M$ and $p_0' \in M'$, and let $\Phi_0 : T_{p_0} M \to T_{p_0'} M'$ be an orthogonal linear isomorphism. Let $r > 0$ be less than the injectivity radius of M at p_0 and the injectivity radius of M' at p_0'. Then the following holds.*

(i) *There exists an isometry $\phi : U_r(p_0, M) \to U_r(p_0', M')$ with $\phi(p_0) = p_0'$ and $d\phi(p_0) = \Phi_0$ if and only if Φ_0 intertwines R and R', i.e.*

$$(\Phi_0)_* R_{p_0} = R'_{p_0'}. \qquad (6.3.3)$$

(ii) *Assume M and M' are connected, simply connected, and complete. Then there exists an isometry $\phi : M \to M'$ with $\phi(p_0) = p_0'$ and $d\phi(p_0) = \Phi_0$ if and only if Φ_0 satisfies (6.3.3).*

Proof In (i) and (ii) the "only if" statement follows from Theorem 5.3.1 (Theorema Egregium) with $\Phi_0 := d\phi(p_0)$. To prove the "if" statement, let (Φ, γ, γ') be a development satisfying $\gamma(0) = p_0$, $\gamma'(0) = p_0'$, and $\Phi(0) = \Phi_0$. Since R and R' are invariant under parallel transport, by Theorem 6.3.4, it follows from the discussion in the beginning of this section that $\Phi_* R = R'$. Hence assertion (i) follows from the local C-A-H Theorem 6.1.17 and assertion (ii) follows from the global C-A-H Theorem 6.1.8. □

Corollary 6.3.6 *A connected, simply connected, complete, locally symmetric space is symmetric.*

Proof Corollary 6.3.5 (ii) with $M' = M$, $p'_0 = p_0$, and $\Phi_0 = -\text{id}$. □

Corollary 6.3.7 *A connected symmetric space M is **homogeneous**; i.e. given $p, q \in M$ there exists an isometry $\phi : M \to M$ with $\phi(p) = q$.*

Proof If M is simply connected, the assertion follows from part (ii) of Corollary 6.3.5 with $M = M'$, $p_0 = p$, $p'_0 = q$, and $\Phi_0 = \Phi_\gamma(1, 0) : T_p M \to T_q M$, where $\gamma : [0, 1] \to M$ is a curve from p to q. If M is not simply connected, we can argue as follows. There is an equivalence relation on M defined by

$$p \sim q \quad :\Longleftrightarrow \quad \exists \text{ isometry } \phi : M \to M \ni \phi(p) = q.$$

Let $p, q \in M$ and suppose that $d(p, q) < \text{inj}(p)$. By Theorem 4.4.4 there is a unique shortest geodesic $\gamma : [0, 1] \to M$ connecting p to q. Since M is symmetric there is an isometry $\phi : M \to M$ such that $\phi(\gamma(1/2)) = \gamma(1/2)$ and $d\phi(\gamma(1/2)) = -\text{id}$. This isometry satisfies $\phi(\gamma(t)) = \gamma(1 - t)$ and hence $\phi(p) = q$. Thus $p \sim q$ whenever $d(p, q) < \text{inj}(p)$. This shows that each equivalence class is open, hence each equivalence class is also closed, and hence there is only one equivalence class because M is connected. This proves Corollary 6.3.7. □

6.3.2   Covariant Derivative of the Curvature

For two vector spaces V, W and an integer $k \geq 1$ we denote by $\mathcal{L}^k(V, W)$ the vector space of all multi-linear maps from $V^k = V \times \cdots \times V$ to W. Thus $\mathcal{L}^1(V, W) = \mathcal{L}(V, W)$ is the space of all linear maps from V to W.

Definition 6.3.8 The **covariant derivative of the Riemann curvature tensor** assigns to every $p \in M$ a linear map

$$(\nabla R)_p : T_p M \to \mathcal{L}^2(T_p M, \mathcal{L}(T_p M, T_p M))$$

such that

$$(\nabla R)(X)(X_1, X_2)Y = \nabla_X\big(R(X_1, X_2)Y\big) - R(\nabla_X X_1, X_2)Y$$
$$- R(X_1, \nabla_X X_2)Y - R(X_1, X_2)\nabla_X Y \tag{6.3.4}$$

for all $X, X_1, X_2, Y \in \text{Vect}(M)$. We also use the notation

$$(\nabla_v R)_p := (\nabla R)_p(v)$$

for $p \in M$ and $v \in T_p M$ so that

$$(\nabla_X R)(X_1, X_2)Y := (\nabla R)(X)(X_1, X_2)Y$$

for all $X, X_1, X_2, Y \in \text{Vect}(M)$.

Remark 6.3.9 One verifies easily that the map

$$\mathrm{Vect}(M)^4 \to \mathrm{Vect}(M) : (X, X_1, X_2, Y) \mapsto (\nabla_X R)(X_1, X_2)Y,$$

defined by the right hand side of equation (6.3.4), is multi-linear over the ring of functions $\mathcal{F}(M)$. Hence it follows as in Remark 5.2.13 that ∇R is well defined, i.e. that the right hand side of (6.3.4) at $p \in M$ depends only on the tangent vectors $X(p), X_1(p), X_2(p), Y(p)$.

Remark 6.3.10 Let $\gamma : I \to M$ be a smooth curve on an interval $I \subset \mathbb{R}$ and

$$X_1, X_2, Y \in \mathrm{Vect}(\gamma)$$

be smooth vector fields along γ. Then equation (6.3.4) continues to hold with X replaced by $\dot{\gamma}$ and each ∇_X on the right hand side replaced by the covariant derivative of the respective vector field along γ:

$$\begin{aligned}
(\nabla_{\dot{\gamma}} R)(X_1, X_2)Y &= \nabla(R(X_1, X_2)Y) - R(\nabla X_1, X_2)Y \\
&\quad - R(X_1, \nabla X_2)Y - R(X_1, X_2)\nabla Y.
\end{aligned} \tag{6.3.5}$$

Theorem 6.3.11

(i) *If $\gamma : \mathbb{R} \to M$ is a smooth curve such that $\gamma(0) = p$ and $\dot{\gamma}(0) = v$, then*

$$(\nabla_v R)_p = \frac{d}{dt}\bigg|_{t=0} \Phi_\gamma(0, t)_* R_{\gamma(t)} \tag{6.3.6}$$

(ii) *The covariant derivative of the Riemann curvature tensor satisfies the **second Bianchi identity***

$$(\nabla_X R)(Y, Z) + (\nabla_Y R)(Z, X) + (\nabla_Z R)(X, Y) = 0. \tag{6.3.7}$$

Proof We prove (i). Let $v_1, v_2, w \in T_p M$ and choose parallel vector fields $X_1, X_2, Y \in \mathrm{Vect}(\gamma)$ along γ satisfying the initial conditions $X_1(0) = v_1$, $X_2(0) = v_2$, $Y(0) = w$. Thus

$$X_1(t) = \Phi_\gamma(t, 0)v_1, \qquad X_2(t) = \Phi_\gamma(t, 0)v_2, \qquad Y(t) = \Phi_\gamma(t, 0)w.$$

Then the last three terms on the right vanish in equation (6.3.5) and hence

$$\begin{aligned}
(\nabla_v R)(v_1, v_2)w &= \nabla(R(X_1, X_2)Y)(0) \\
&= \frac{d}{dt}\bigg|_{t=0} \Phi_{\gamma(0,t)} R_{\gamma(t)}(X_1(t), X_2(t))Y(t) \\
&= \frac{d}{dt}\bigg|_{t=0} \Phi_{\gamma(0,t)} R_{\gamma(t)}(\Phi_\gamma(t, 0)v_1, \Phi_\gamma(t, 0)v_2)\Phi_\gamma(t, 0)w \\
&= \frac{d}{dt}\bigg|_{t=0} \left(\Phi_\gamma(0, t)_* R_{\gamma(t)}\right)(v_1, v_2)w.
\end{aligned}$$

Here the second equation follows from Theorem 3.3.6. This proves (i).

We prove (ii). Choose a smooth function $\gamma : \mathbb{R}^3 \to M$ and denote by (r, s, t) the coordinates on $\mathbb{R}^3$. If Y is a vector field along γ, we have

$$
\begin{aligned}
(\nabla_{\partial_r \gamma} R)(\partial_s \gamma, \partial_t \gamma)Y &= \nabla_r \big(R(\partial_s \gamma, \partial_t \gamma)Y \big) - R(\partial_s \gamma, \partial_t \gamma)\nabla_r Y \\
&\quad - R(\nabla_r \partial_s \gamma, \partial_t \gamma)Y - R(\partial_s \gamma, \nabla_r \partial_t \gamma)Y \\
&= \nabla_r (\nabla_s \nabla_t Y - \nabla_t \nabla_s Y) - (\nabla_s \nabla_t - \nabla_t \nabla_s)\nabla_r Y \\
&\quad + R(\partial_t \gamma, \nabla_r \partial_s \gamma)Y - R(\partial_s \gamma, \nabla_t \partial_r \gamma)Y.
\end{aligned}
$$

Permuting the variables r, s, t cyclically and taking the sum of the resulting three equations we obtain

$$
\begin{aligned}
(\nabla_{\partial_r \gamma} R)&(\partial_s \gamma, \partial_t \gamma)Y + (\nabla_{\partial_s \gamma} R)(\partial_t \gamma, \partial_r \gamma)Y + (\nabla_{\partial_t \gamma} R)(\partial_r \gamma, \partial_s \gamma)Y \\
&= \nabla_r (\nabla_s \nabla_t Y - \nabla_t \nabla_s Y) - (\nabla_s \nabla_t - \nabla_t \nabla_s)\nabla_r Y \\
&\quad + \nabla_s (\nabla_t \nabla_r Y - \nabla_r \nabla_t Y) - (\nabla_t \nabla_r - \nabla_r \nabla_t)\nabla_s Y \\
&\quad + \nabla_t (\nabla_r \nabla_s Y - \nabla_s \nabla_r Y) - (\nabla_r \nabla_s - \nabla_s \nabla_r)\nabla_t Y.
\end{aligned}
$$

The terms on the right cancel out. This proves Theorem 6.3.11.   □

Proof of Theorem 6.3.4   We prove that (iii) implies (i). This follows from the local Cartan–Ambrose–Hicks Theorem 6.1.17 with

$$
p_0' = p_0 = p, \qquad \Phi_0 = -\mathrm{id} : T_p M \to T_p M.
$$

This isomorphism satisfies

$$
(\Phi_0)_* R_p = R_p.
$$

Hence it follows from the discussion in the beginning of this section that

$$
\Phi_* R = R'
$$

for every development (Φ, γ, γ') of M along itself satisfying

$$
\gamma(0) = \gamma'(0) = p, \qquad \Phi(0) = -\mathrm{id}.
$$

Hence, by the local C-A-H Theorem 6.1.17, there is an isometry

$$
\phi : U_r(p, M) \to U_r(p, M)
$$

satisfying

$$
\phi(p) = p, \qquad d\phi(p) = -\mathrm{id}
$$

whenever $0 < r < \mathrm{inj}(p; M)$.

We prove that (i) implies (ii). By Theorem 5.3.1 (Theorema Egregium), every isometry $\phi : M \to M'$ preserves the Riemann curvature tensor and covariant differentiation, and hence also the covariant derivative of the Riemann curvature tensor, i.e.

$$\phi_*(\nabla R) = \nabla' R'.$$

Applying this to the local isometry $\phi : U_r(p, M) \to U_r(p, M)$ we obtain

$$\left(\nabla_{d\phi(p)v} R\right)_{\phi(p)}(d\phi(p)v_1, d\phi(p)v_2) = d\phi(p)(\nabla_v R)(v_1, v_2)d\phi(p)^{-1}$$

for all $v, v_1, v_2 \in T_p M$. Since

$$d\phi(p) = -\mathrm{id}$$

this shows that ∇R vanishes at p.

We prove that (ii) imlies (iii). If ∇R vanishes, then equation (6.3.6) in Theorem 6.3.11 shows that the function

$$s \mapsto \Phi_\gamma(t, s)_* R_{\gamma(s)} = \Phi_\gamma(t, 0)_* \Phi_\gamma(0, s)_* R_{\gamma(s)}$$

is constant and hence is everywhere equal to $R_{\gamma(t)}$. This implies (6.3.2) and completes the proof of Theorem 6.3.4.   □

6.3.3 Covariant Derivative of the Curvature in Local Coordinates

Let $\phi : U \to \Omega$ be a local coordinate chart on M with values in an open set $\Omega \subset \mathbb{R}^m$, denote its inverse by $\psi := \phi^{-1} : \Omega \to U$, and let

$$E_i(x) := \frac{\partial \psi}{\partial x^i}(x) \in T_{\psi(x)} M, \qquad x \in \Omega, \qquad i = 1, \ldots, m,$$

be the local frame of the tangent bundle determined by this coordinate chart. Let $\Gamma_{ij}^k : \Omega \to \mathbb{R}$ denote the Christoffel symbols and $R_{ijk}^\ell : \Omega \to \mathbb{R}$ the coefficients of the Riemann curvature tensor so that

$$\nabla_i E_j = \sum_k \Gamma_{ij}^k E_k, \qquad R(E_i, E_j)E_k = \sum_\ell R_{ijk}^\ell E_\ell.$$

Given $i, j, k, \ell \in \{1, \ldots, m\}$ we can express the vector field $(\nabla_{E_i} R)(E_j, E_k)E_\ell$ along ψ for each $x \in \Omega$ as a linear combination of the basis vectors $E_i(x)$. This gives rise to functions

$$\nabla_i R_{jk\ell}^\nu : \Omega \to \mathbb{R}$$

defined by

$$(\nabla_{E_i} R)(E_j, E_k)E_\ell =: \sum_\nu \nabla_i R^\nu_{jk\ell} E_\nu. \tag{6.3.8}$$

These functions are given by

$$\nabla_i R^\nu_{jk\ell} = \partial_i R^\nu_{jk\ell} + \sum_\mu \Gamma^\nu_{i\mu} R^\mu_{jk\ell}$$
$$- \sum_\mu \Gamma^\mu_{ij} R^\nu_{\mu k\ell} - \sum_\mu \Gamma^\mu_{ik} R^\nu_{j\mu\ell} - \sum_\mu \Gamma^\mu_{i\ell} R^\nu_{jk\mu}. \tag{6.3.9}$$

The second Bianchi identity has the form

$$\nabla_i R^\nu_{jk\ell} + \nabla_j R^\nu_{ki\ell} + \nabla_k R^\nu_{ij\ell} = 0. \tag{6.3.10}$$

Exercise: Prove equations (6.3.9) and (6.3.10). **Warning:** As in Sect. 5.4, care must be taken with the ordering of the indices. Some authors use the notation $\nabla_i R^\nu_{\ell jk}$ for what we call $\nabla_i R^\nu_{jk\ell}$.

6.3.4   Examples and Exercises

Example 6.3.12   Every flat manifold is locally symmetric.

Example 6.3.13   If M_1 and M_2 are (locally) symmetric, so is $M_1 \times M_2$.

Example 6.3.14 $M = \mathbb{R}^m$ with the standard metric is a symmetric space. Recall that the isometry group $\mathcal{I}(\mathbb{R}^m)$ consists of all affine transformations of the form

$$\phi(x) = Ax + b, \qquad A \in O(m), \qquad b \in \mathbb{R}^m.$$

(See Exercise 5.1.4.) The isometry with fixed point $p \in \mathbb{R}^m$ and $d\phi(p) = -\mathrm{id}$ is given by $\phi(x) = 2p - x$ for $x \in \mathbb{R}^m$.

Example 6.3.15   The flat tori of Exercise 6.2.7 in the previous section are symmetric (but not simply connected). This shows that the hypothesis of simple connectivity cannot be dropped in part (ii) of Corollary 6.3.5.

Example 6.3.16   Below we define manifolds of constant curvature and show that they are locally symmetric. The simplest example, after a flat space, is the unit sphere $S^m = \{x \in \mathbb{R}^{m+1} \mid |x| = 1\}$. The symmetry ϕ of the sphere about a point $p \in M$ is given by

$$\phi(x) := -x + 2\langle p, x\rangle p$$

for $x \in S^m$. This extends to an orthogonal linear transformation of the ambient space. In fact the group of isometries of S^m is the group $O(m + 1)$ of orthogonal linear transformations of $\mathbb{R}^{m+1}$ (see Example 6.4.16 below). In accordance with Corollary 6.3.7 this group acts transitively on S^m.

Example 6.3.17 A compact two-dimensional manifold of constant negative curvature is locally symmetric (as its universal cover is symmetric) but not homogeneous (as closed geodesics of a given period are isolated). Hence it is not symmetric. This shows that the hypothesis that M be simply connected cannot be dropped in Corollary 6.3.6.

Example 6.3.18 The real projective space $\mathbb{R}P^n$ with the metric inherited from S^n is a symmetric space and the orthogonal group $O(n + 1)$ acts on it by isometries. The complex projective space $\mathbb{C}P^n$ with the Fubini–Study metric in Example 3.7.5 is a symmetric space and the unitary group $U(n + 1)$ acts on it by isometries. The complex Grassmannian $G_k(\mathbb{C}^n)$ in Example 3.7.6 is a symmetric space and the unitary group $U(n)$ acts on it by isometries. (**Exercise:** Prove this.)

Example 6.3.19 The simplest example of a symmetric space which is not of constant curvature is the orthogonal group $O(n) = \{g \in \mathbb{R}^{n \times n} \mid g^{\mathsf{T}}g = \mathbb{1}\}$ with the Riemannian metric (5.2.20) of Example 5.2.18. The symmetry ϕ about the point $a \in O(n)$ is given by $\phi(g) = ag^{-1}a$. This discussion extends to every Lie subgroup $G \subset O(n)$. (**Exercise:** Prove this.)

6.4 Constant Curvature

In the Sect. 5.3 we saw that the Gaußian curvature of a two-dimensional surface is intrinsic: we gave a formula for it in terms of the Riemann curvature tensor and the first fundamental form. We may use this formula to define the Gaußian curvature for *any* two-dimensional manifold (even if its codimension is greater than one). We make a slightly more general definition.

6.4.1 Sectional Curvature

Definition 6.4.1 Let $M \subset \mathbb{R}^n$ be a smooth m-dimensional submanifold. Let $p \in M$ and let $E \subset T_pM$ be a 2-dimensional linear subspace of the tangent space. The **sectional curvature** of M at (p, E) is the number

$$K(p, E) = \frac{\langle R_p(u, v)v, u \rangle}{|u|^2|v|^2 - \langle u, v \rangle^2}, \tag{6.4.1}$$

where $u, v \in E$ are linearly independent (and hence form a basis of E).

The right hand side of (6.4.1) remains unchanged if we multiply u or v by a nonzero real number or add to one of the vectors a real multiple of the other; hence it depends only on the linear subspace spanned by u ad v.

Example 6.4.2 If $M \subset \mathbb{R}^3$ is a 2-manifold, then by Theorem 5.3.9 the sectional curvature $K(p, T_p M) = K(p)$ is the Gaußian curvature of M at p. More generally, for any 2-manifold $M \subset \mathbb{R}^n$ (whether or not it has codimension one) we *define* the **Gaußian curvature** of M at p by

$$K(p) := K(p, T_p M). \tag{6.4.2}$$

Example 6.4.3 If $M \subset \mathbb{R}^{m+1}$ is a submanifold of codimension one and $v : M \to S^m$ is a Gauß map, then the sectional curvature of a 2-dimensional subspace $E \subset T_p M$ spanned by two linearly independent tangent vectors $u, v \in T_p M$ is given by

$$K(p, E) = \frac{\langle u, dv(p)u \rangle \langle v, dv(p)v \rangle - \langle u, dv(p)v \rangle^2}{|u|^2 |v|^2 - \langle u, v \rangle^2}. \tag{6.4.3}$$

This follows from equation (5.3.8) in the proof of Theorem 5.3.9 which holds in all dimensions. In particular, when $M = S^m$, we have $v(p) = p$ and hence $K(p, E) = 1$ for all p and E. For a sphere of radius r we have $v(p) = p/r$ and hence $K(p, E) = 1/r^2$.

Example 6.4.4 Let $G \subset O(n)$ be a Lie subgroup equipped with the Riemannian metric

$$\langle v, w \rangle := \operatorname{trace}(v^\mathsf{T} w)$$

for $v, w \in T_g G \subset \mathbb{R}^{n \times n}$. Then, by Example 5.2.18, the sectional curvature of G at the identity matrix $\mathbb{1}$ is given by

$$K(\mathbb{1}, E) = \frac{1}{4} \|[\xi, \eta]\|^2$$

for every 2-dimensional linear subspace $E \subset \mathfrak{g} = \operatorname{Lie}(G) = T_{\mathbb{1}} G$ with an orthonormal basis ξ, η.

Exercise 6.4.5 Let $E \subset T_p M$ be a 2-dimensional linear subspace, let $r > 0$ be smaller than the injectivity radius of M at p, and let $N \subset M$ be the 2-dimensional submanifold given by

$$N := \exp_p(\{v \in E \mid |v| < r\}).$$

Show that the sectional curvature $K(p, E)$ of M at (p, E) agrees with the Gaußian curvature of N at p.

Exercise 6.4.6 Let $p \in M \subset \mathbb{R}^n$ and let $E \subset T_p M$ be a 2-dimensional linear subspace. For $r > 0$ let L denote the ball of radius r in the $(n - m + 2)$-dimensional affine subspace of $\mathbb{R}^n$ through p and parallel to the vector subspace $E + T_p M^{\perp}$:

$$L = \left\{ p + v + w \mid v \in E, \ w \in T_p M^{\perp}, \ |v|^2 + |w|^2 < r^2 \right\}.$$

Show that, for r sufficiently small, $L \cap M$ is a 2-dimensional manifold with Gaußian curvature $K_{L \cap M}(p)$ at p given by $K_{L \cap M}(p) = K(p, E)$.

6.4.2 Constant Sectional Curvature

Definition 6.4.7 Let $k \in \mathbb{R}$ and $m \geq 2$ be an integer. An m-manifold $M \subset \mathbb{R}^n$ is said to have **constant sectional curvature** k iff $K(p, E) = k$ for every $p \in M$ and every 2-dimensional linear subspace $E \subset T_p M$.

Theorem 6.4.8 *Let $M \subset \mathbb{R}^n$ be an m-manifold and fix an element $p \in M$ and a real number k. Then the following are equivalent.*

(i) $K(p, E) = k$ *for every 2-dimensional linear subspace $E \subset T_p M$.*
(ii) *The Riemann curvature tensor of M at p is given by*

$$\langle R_p(v_1, v_2) v_3, v_4 \rangle = k \Big(\langle v_1, v_4 \rangle \langle v_2, v_3 \rangle - \langle v_1, v_3 \rangle \langle v_2, v_4 \rangle \Big) \qquad (6.4.4)$$

for all $v_1, v_2, v_3, v_4 \in T_p M$.

Proof That (ii) implies (i) follows directly from the definition of the sectional curvature in (6.4.1) by taking $v_1 = v_4 = u$ and $v_2 = v_3 = v$ in (6.4.4). Conversely, assume (i) and define the multi-linear map $Q : T_p M^4 \to \mathbb{R}$ by

$$Q(v_1, v_2, v_3, v_4) := \langle R_p(v_1, v_2) v_3, v_4 \rangle - k \Big(\langle v_1, v_4 \rangle \langle v_2, v_3 \rangle - \langle v_1, v_3 \rangle \langle v_2, v_4 \rangle \Big).$$

Then, for all $u, v, v_1, v_2, v_3, v_4 \in T_p M$, the map Q satisfies the equations

$$Q(v_1, v_2, v_3, v_4) + Q(v_2, v_1, v_3, v_4) = 0, \qquad (6.4.5)$$
$$Q(v_1, v_2, v_3, v_4) + Q(v_2, v_3, v_1, v_4) + Q(v_3, v_1, v_2, v_4) = 0, \qquad (6.4.6)$$
$$Q(v_1, v_2, v_3, v_4) - Q(v_3, v_4, v_1, v_2) = 0, \qquad (6.4.7)$$
$$Q(u, v, u, v) = 0. \qquad (6.4.8)$$

Here the first three equations follow from Theorem 5.2.14 and the last follows from the definition of Q and the hypothesis that the sectional curvature is $K(p, E) = k$ for every 2-dimensional linear subspace $E \subset T_p M$.

We must prove that Q vanishes. Using (6.4.7) and (6.4.8) we find

$$
\begin{aligned}
0 = Q(u, v_1 + v_2, u, v_1 + v_2) \\
= Q(u, v_1, u, v_2) + Q(u, v_2, u, v_1) \\
= 2Q(u, v_1, u, v_2)
\end{aligned}
$$

for all $u, v_1, v_2 \in T_p M$. This implies

$$
\begin{aligned}
0 = Q(u_1 + u_2, v_1, u_1 + u_2, v_2) \\
= Q(u_1, v_1, u_2, v_2) + Q(u_2, v_1, u_1, v_2)
\end{aligned}
$$

for all $u_1, u_2, v_1, v_2 \in T_p M$. Hence

$$
\begin{aligned}
Q(v_1, v_2, v_3, v_4) = -Q(v_3, v_2, v_1, v_4) \\
= Q(v_2, v_3, v_1, v_4) \\
= -Q(v_3, v_1, v_2, v_4) - Q(v_1, v_2, v_3, v_4).
\end{aligned}
$$

Here the second equation follows from (6.4.5) and the last from (6.4.6). Thus

$$
Q(v_1, v_2, v_3, v_4) = -\frac{1}{2} Q(v_3, v_1, v_2, v_4) = \frac{1}{2} Q(v_1, v_3, v_2, v_4)
$$

for all $v_1, v_2, v_3, v_4 \in T_p M$ and, repeating this argument,

$$
Q(v_1, v_2, v_3, v_4) = \frac{1}{4} Q(v_1, v_2, v_3, v_4).
$$

Hence $Q \equiv 0$ as claimed. This proves Theorem 6.4.8.   □

Remark 6.4.9 The symmetric group S_4 on four symbols acts naturally on the space $\mathcal{L}^4(T_p M, \mathbb{R})$ of multi-linear maps from $T_p M^4$ to $\mathbb{R}$. The conditions (6.4.5), (6.4.6), (6.4.7), and (6.4.8) say that the four elements

$$
\begin{aligned}
a = \mathrm{id} + (12), \\
c = \mathrm{id} + (123) + (132), \\
b = \mathrm{id} - (34), \\
d = \mathrm{id} + (13) + (24) + (13)(24)
\end{aligned}
$$

of the group ring of S_4 annihilate Q. This suggests an alternate proof of Theorem 6.4.8. A representation of a finite group is completely reducible so one can prove that $Q = 0$ by showing that any vector in any irreducible representation of S_4 which is annihilated by the four elements a, b, c and d must necessarily be zero. This can be checked case by case for each irreducible representation. (The group S_4 has 5 irreducible representations: two of dimension 1, two of dimension 3, and one of dimension 2.)

If M and M' are two m-dimensional manifolds with constant curvature k, then every orthogonal isomorphism $\Phi : T_p M \to T_{p'} M'$ intertwines the Riemann curvature tensors by Theorem 6.4.8. Hence by the appropriate version (local or global) of the C-A-H Theorem we have the following corollaries.

Corollary 6.4.10 *Every Riemannian manifold with constant sectional curvature is locally symmetric.*

Proof Theorem 6.3.4 and Theorem 6.4.8.   □

Corollary 6.4.11 *Let M and M' be m-dimensional Riemannian manifolds with constant curvature k and let $p \in M$ and $p' \in M'$. If $r > 0$ is smaller than the injectivity radii of M at p and of M' at p', then for every orthogonal isomorphism*

$$\Phi : T_p M \to T_{p'} M'$$

there exists an isometry

$$\phi : U_r(p, M) \to U_r(p', M')$$

such that

$$\phi(p) = p', \qquad d\phi(p) = \Phi.$$

Proof This follows from Corollary 6.3.5 and Corollary 6.4.10. Alternatively one can use Theorem 6.4.8 and the local C-A-H Theorem 6.1.17.   □

Corollary 6.4.12 *Any two connected, simply connected, complete Riemannian manifolds with the same constant sectional curvature and the same dimension are isometric.*

Proof Theorem 6.4.8 and the global C-A-H Theorem 6.1.8.   □

Corollary 6.4.13 *Let $M \subset \mathbb{R}^n$ be a connected, simply connected, complete manifold. Then the following are equivalent.*

(i)   *M has constant sectional curvature.*
(ii) *For every pair of points $p, q \in M$ and every orthogonal linear isomorphism $\Phi : T_p M \to T_q M$ there exists an isometry $\phi : M \to M$ such that*

$$\phi(p) = q, \qquad d\phi(p) = \Phi.$$

Proof That (i) implies (ii) follows immediately from Theorem 6.4.8 and the global C-A-H Theorem 6.1.8. Conversely assume (ii). Then, for every pair of points $p, q \in M$ and every orthogonal linear isomorphism

$$\Phi : T_p M \to T_q M,$$

it follows from Theorem 5.3.1 (Theorema Egregium) that

$$\Phi_* R_p = R_q$$

and so

$$K(p, E) = K(q, \Phi E)$$

for every 2-dimensional linear subspace $E \subset T_p M$. Since, for every pair of points $p, q \in M$ and of 2-dimensional linear subspaces

$$E \subset T_p M, \qquad F \subset T_q M,$$

we can find an orthogonal linear isomorphism $\Phi : T_p M \to T_q M$ such that

$$\Phi E = F,$$

this implies (i).   □

Corollary 6.4.13 asserts that a connected, simply connected, complete Riemannian m-manifold M has constant sectional curvature if and only if the isometry group $\mathcal{I}(M)$ acts transitively on its orthonormal frame bundle $\mathcal{O}(M)$. Note that, by Lemma 5.1.10, this group action is also free.

6.4.3   Examples and Exercises

Example 6.4.14   Any flat Riemannian manifold has constant sectional curvature $k = 0$.

Example 6.4.15   The manifold

$$M = \mathbb{R}^m$$

with its standard metric is, up to isometry, the unique connected, simply connected, complete Riemannian m-manifold with constant sectional curvature

$$k = 0.$$

Example 6.4.16   For $m \geq 2$ the unit sphere

$$M = S^m$$

with its standard metric is, up to isometry, the unique connected, simply connected, complete Riemannian m-manifold with constant sectional curvature

$$k = 1.$$

Hence, by Corollary 6.4.12, every connected simply connected, complete Riemannian manifold with positive sectional curvature $k = 1$ is compact. Moreover, by Corollary 6.4.13, the isometry group $\mathcal{I}(S^m)$ is isomorphic to the group $O(m + 1)$ of orthogonal linear transformations of $\mathbb{R}^{m+1}$. Thus, by Corollary 6.4.13, the orthonormal frame bundle $\mathcal{O}(S^m)$ is diffeomorphic to $O(m + 1)$. This follows also from the fact that, if

$$v_1, \ldots, v_m$$

is an orthonormal basis of $T_p S^m = p^\perp$ then

$$p, v_1, \ldots, v_m$$

is an orthonormal basis of $\mathbb{R}^{m+1}$.

Example 6.4.17 A product of spheres is *not* a space of constant sectional curvature, but it *is* a symmetric space. **Exercise: Prove this.**

Example 6.4.18 For $n \geq 4$ the orthogonal group $O(n)$ is not a space of constant sectional curvature, but it is a symmetric space and has nonnegative sectional curvature (see Example 6.4.4).

6.4.4 Hyperbolic Space

Fix an integer $m \geq 2$. The **hyperbolic space** $\mathbb{H}^m$ is, up to isometry, the unique connected, simply connected, complete Riemannian m-manifold with constant sectional curvature

$$k = -1.$$

A model for $\mathbb{H}^m$ can be constructed as follows. A point in $\mathbb{R}^{m+1}$ will be denoted by

$$p = (x_0, x), \qquad x_0 \in \mathbb{R}, \qquad x = (x_1, \ldots, x_m) \in \mathbb{R}^m.$$

Let $Q : \mathbb{R}^{m+1} \times \mathbb{R}^{m+1} \to \mathbb{R}$ denote the symmetric bilinear form given by

$$Q(p, q) := -x_0 y_0 + x_1 y_1 + \cdots + x_m y_m \tag{6.4.9}$$

for $p = (x_0, x), q = (y_0, y) \in \mathbb{R}^{m+1}$. Since Q is nondegenerate the space

$$\mathbb{H}^m := \{p = (x_0, x) \in \mathbb{R}^{m+1} \mid Q(p, p) = -1, \ x_0 > 0\}$$

is a smooth m-dimensional submanifold of $\mathbb{R}^{m+1}$ and the tangent space of $\mathbb{H}^m$ at p is given by

$$T_p \mathbb{H}^m = \{v \in \mathbb{R}^{m+1} \mid Q(p, v) = 0\}.$$

For $p = (x_0, x) \in \mathbb{R}^{m+1}$ and $v = (\xi_0, \xi) \in \mathbb{R}^{m+1}$ we have

$$p \in \mathbb{H}^m \iff x_0 = \sqrt{1 + |x|^2},$$

$$v \in T_p\mathbb{H}^m \iff \xi_0 = \frac{\langle \xi, x \rangle}{\sqrt{1 + |x|^2}}.$$

Now let us define a Riemannian metric on $\mathbb{H}^m$ by

$$\begin{aligned}
g_p(v, w) &:= Q(v, w) \\
&= \langle \xi, \eta \rangle - \xi_0 \eta_0 \\
&= \langle \xi, \eta \rangle - \frac{\langle \xi, x \rangle \langle \eta, x \rangle}{1 + |x|^2}
\end{aligned} \tag{6.4.10}$$

for $v = (\xi_0, \xi) \in T_p\mathbb{H}^m$ and $w = (\eta_0, \eta) \in T_p\mathbb{H}^m$.

Theorem 6.4.19 $\mathbb{H}^m$ *is a connected, simply connected, complete Riemannian m-manifold with constant sectional curvature* $k = -1$.

We remark that the manifold $\mathbb{H}^m$ does not quite fit into the extrinsic framework of most of this book as it is not exhibited as a submanifold of Euclidean space but rather of "pseudo-Euclidean space": the positive definite inner product $\langle v, w \rangle$ of the ambient space $\mathbb{R}^{m+1}$ is replaced by a nondegenerate symmetric bilinear form $Q(v, w)$. However, all the theory developed thus far goes through (reading $Q(v, w)$ for $\langle v, w \rangle$) provided we impose the additional hypothesis (true in the example $M = \mathbb{H}^m$) that the first fundamental form $g_p = Q|_{T_p M}$ is positive definite. For then $Q|_{T_p M}$ is nondegenerate and we may define the orthogonal projection $\Pi(p)$ onto $T_p M$ as before. The next lemma summarizes the basic observations; the proof is an exercise in linear algebra.

Lemma 6.4.20 *Let* Q *be a symmetric bilinear form on a vector space* V *and for each subspace* E *of* V *define its orthogonal complement by*

$$E^{\perp \varrho} := \{u \in V \mid Q(u, v) = 0 \; \forall v \in E\}.$$

Assume Q *is nondegenerate, i.e.* $V^{\perp \varrho} = \{0\}$. *Then, for every linear subspace* $E \subset V$, *we have*

$$V = E \oplus E^{\perp \varrho} \qquad \iff \qquad E \cap E^{\perp \varrho} = \{0\},$$

i.e. $E^{\perp \varrho}$ *is a vector space complement of* E *if and only if the restriction of* Q *to* E *is nondegenerate.*

Proof of Theorem 6.4.19 The proofs of the various properties of $\mathbb{H}^m$ are entirely analogous to the corresponding proofs for S^m. Thus the unit normal field to $\mathbb{H}^m$ is given by

$$\nu(p) = p$$

for $p \in \mathbb{H}^m$ although the "square of its length" is

$$Q(p, p) = -1.$$

For $p \in \mathbb{H}^m$ we introduce the Q-orthogonal projection $\Pi(p)$ of $\mathbb{R}^{m+1}$ onto the subspace $T_p\mathbb{H}^m$. It is characterized by the conditions

$$\Pi(p)^2 = \Pi(p), \qquad \ker \Pi(p) \perp_Q \operatorname{im}\Pi(p), \qquad \operatorname{im}\Pi(p) = T_p\mathbb{H}^m,$$

and is given by the explicit formula

$$\Pi(p)v = v + Q(v, p)p$$

for $v \in \mathbb{R}^{m+1}$.

The covariant derivative of a vector field $X \in \operatorname{Vect}(\gamma)$ along a smooth curve $\gamma : \mathbb{R} \to \mathbb{H}^m$ is given by

$$\begin{aligned}
\nabla X(t) &= \Pi(\gamma(t))\dot{X}(t) \\
&= \dot{X}(t) + Q(\dot{X}(t), \gamma(t))\gamma(t) \\
&= \dot{X}(t) - Q(X(t), \dot{\gamma}(t))\gamma(t).
\end{aligned}$$

The last identity follows by differentiating the equation $Q(X, \gamma) \equiv 0$. This can be interpreted as the hyperbolic Gauß–Weingarten formula as follows. For $p \in \mathbb{H}^m$ and $u \in T_p\mathbb{H}^m$ we introduce, as before, the second fundamental form

$$h_p(u) : T_p\mathbb{H}^m \to (T_p\mathbb{H}^m)^{\perp_Q}$$

via

$$h_p(u)v := \big(d\Pi(p)u\big)v$$

and denote its Q-adjoint by

$$h_p(u)^* : (T_p\mathbb{H}^m)^{\perp_Q} \to T_p\mathbb{H}^m.$$

For all $p \in \mathbb{H}^m$, $u \in T_p\mathbb{H}^m$, and $v \in \mathbb{R}^{m+1}$ we have

$$\begin{aligned}
\big(d\Pi(p)u\big)v &= \frac{d}{dt}\bigg|_{t=0} \big(v + Q(v, p + tu)(p + tu)\big) \\
&= Q(v, p)u + Q(v, u)p,
\end{aligned}$$

where the first summand on the right is tangent to $\mathbb{H}^m$ and the second summand is Q-orthogonal to $T_p\mathbb{H}^m$. Hence

$$h_p(u)v = Q(v, u)p, \qquad h_p(u)^*w = Q(w, p)u \qquad (6.4.11)$$

for $v \in T_p\mathbb{H}^m$ and $w \in (T_p\mathbb{H}^m)^{\perp_Q}$.

With this understood, the Gauß-Weingarten formula

$$\dot{X} = \nabla X + h_\gamma(\dot{\gamma})X$$

extends to the present setting. The reader may verify that the operators

$$\nabla : \text{Vect}(\gamma) \to \text{Vect}(\gamma)$$

thus defined satisfy the axioms of Theorem 3.7.8 and hence define the Levi-Civita connection on $\mathbb{H}^m$.

Now a smooth curve $\gamma : I \to \mathbb{H}^m$ is a geodesic if and only if it satisfies the equivalent conditions

$$\nabla\dot{\gamma} \equiv 0 \quad \Longleftrightarrow \quad \ddot{\gamma}(t) \perp_Q T_{\gamma(t)}\mathbb{H}^m \ \forall \, t \in I \quad \Longleftrightarrow \quad \ddot{\gamma} = Q(\dot{\gamma}, \dot{\gamma})\gamma.$$

A geodesic must satisfy the equation

$$\frac{d}{dt}Q(\dot{\gamma}, \dot{\gamma}) = 2Q(\ddot{\gamma}, \dot{\gamma}) = 0$$

because $\ddot{\gamma}$ is a scalar multiple of γ, and hence $Q(\dot{\gamma}, \dot{\gamma})$ is constant. Fix an element $p \in \mathbb{H}^m$ and a tangent vector $v \in T_p\mathbb{H}^m$ such that

$$Q(v, v) = 1.$$

Then the geodesic $\gamma : \mathbb{R} \to \mathbb{H}^m$ with $\gamma(0) = p$ and $\dot{\gamma}(0) = v$ is given by

$$\gamma(t) = \cosh(t)p + \sinh(t)v, \qquad (6.4.12)$$

where

$$\cosh(t) := \frac{e^t + e^{-t}}{2}, \qquad \sinh(t) := \frac{e^t - e^{-t}}{2}.$$

In fact we have $\ddot{\gamma}(t) = \gamma(t) \perp_Q T_{\gamma(t)}\mathbb{H}^m$. It follows that the geodesics exist for all time and hence $\mathbb{H}^m$ is geodesically complete. Moreover, being diffeomorphic to Euclidean space, $\mathbb{H}^m$ is connected and simply connected.

It remains to prove that $\mathbb{H}^m$ has constant sectional curvature $k = -1$. To see this we use the Gauß–Codazzi formula in the hyperbolic setting, i.e.

$$R_p(u, v) = h_p(u)^*h_p(v) - h_p(v)^*h_p(u). \qquad (6.4.13)$$

By equation (6.4.11), this gives

$$\begin{aligned}
\langle R_p(u,v)v,u \rangle &= Q(h_p(u)u, h_p(v)v) - Q(h_p(v)u, h_p(u)v) \\
&= Q(Q(u,u)p, Q(v,v)p) - Q(Q(u,v)p, Q(u,v)p) \\
&= -Q(u,u)Q(v,v) + Q(u,v)^2 \\
&= -g_p(u,u)g_p(v,v) + g_p(u,v)^2
\end{aligned}$$

for all $u, v \in T_p\mathbb{H}^m$. Hence, for every $p \in M$ and every 2-dimensional linear subspace $E \subset T_pM$ with a basis $u, v \in E$ we have

$$K(p, E) = \frac{\langle R_p(u,v)v,u \rangle}{g_p(u,u)g_p(v,v) - g_p(u,v)^2} = -1.$$

This proves Theorem 6.4.19.   □

Exercise 6.4.21 Prove that the pullback of the metric on $\mathbb{H}^m$ under the diffeomorphism

$$\mathbb{R}^m \to \mathbb{H}^m : x \mapsto \left(\sqrt{1 + |x|^2}, x \right)$$

is given by

$$|\widehat{x}|_x = \sqrt{|\widehat{x}|^2 - \frac{\langle x, \xi \rangle^2}{1 + |x|^2}}$$

for $x \in \mathbb{R}^m$ and $\widehat{x} \in \mathbb{R}^m = T_x\mathbb{R}^m$. Thus the metric tensor is given by

$$g_{ij}(x) = \delta_{ij} - \frac{x_i x_j}{1 + |x|^2} \tag{6.4.14}$$

for $x = (x_1, \ldots, x_m) \in \mathbb{R}^m$.

Exercise 6.4.22 The **Poincaré model** of hyperbolic space is the open unit disc $\mathbb{D}^m \subset \mathbb{R}^m$ equipped with the **Poincaré metric**

$$|\widehat{y}|_y = \frac{2|\widehat{y}|}{1 - |y|^2}$$

for $y \in \mathbb{D}^m$ and $\widehat{y} \in \mathbb{R}^m = T_y\mathbb{D}^m$. Thus the metric tensor is given by

$$g_{ij}(y) = \frac{4\delta_{ij}}{\left(1 - |y|^2\right)^2}, \qquad y \in \mathbb{D}^m. \tag{6.4.15}$$

Prove that the diffeomorphism

$$\mathbb{D}^m \to \mathbb{H}^m : y \mapsto \left(\frac{1 + |y|^2}{1 - |y|^2}, \frac{2y}{1 - |y|^2} \right)$$

is an isometry with the inverse

$$\mathbb{H}^m \to \mathbb{D}^m : (x_0, x) \mapsto \frac{x}{1 + x_0}.$$

Interpret this map as a stereographic projection from the *south pole* $(-1, 0)$.

Exercise 6.4.23   The composition of the isometries in Exercise 6.4.21 and Exercise 6.4.22 is the diffeomorphism $\mathbb{R}^m \to \mathbb{D}^m : x \mapsto y$ given by

$$y = \frac{x}{\sqrt{1 + |x|^2} + 1}, \qquad x = \frac{2y}{1 - |y|^2}, \qquad \sqrt{1 + |x|^2} = \frac{1 + |y|^2}{1 - |y|^2}.$$

Prove that this is an isometry intertwining the Riemannian metrics (6.4.14) and (6.4.15).

Exercise 6.4.24   This exercise shows that every nonconstant geodesic in the Poincaré model $\mathbb{D}^m$ of hyperbolic space in Exercise 6.4.22 converges to two points on the boundary $S^{m-1} = \partial\mathbb{D}^m$ in forward and backward time, and that any two distinct points on the boundary are the asymptotic limits of a unique geodesic in $\mathbb{D}^m$ up to reparametrization.

Fix an element $y \in \mathbb{D}^m$ and a tangent vector $\widehat{y} \in T_y\mathbb{D}^m = \mathbb{R}^m$ of norm one in the hyperbolic metric, i.e.

$$\lambda|\widehat{y}| = 1, \qquad \lambda := \frac{2}{1 - |y|^2}. \tag{6.4.16}$$

Let $\gamma : \mathbb{R} \to \mathbb{D}^m$ be the unique geodesic satisfying $\gamma(0) = y$ and $\dot{\gamma}(0) = \widehat{y}$. Prove the following.

(a) The geodesic γ is given by the explicit formula

$$\gamma(t) = \frac{\cosh(t)\lambda y + \sinh(t)\left(\lambda\widehat{y} + \langle \lambda y, \lambda\widehat{y} \rangle y\right)}{1 + \cosh(t)(\lambda - 1) + \sinh(t)\langle \lambda y, \lambda\widehat{y} \rangle} \tag{6.4.17}$$

for $t \in \mathbb{R}$. **Hint:** Use (6.4.12) and the isometries in Exercise 6.4.22.
(b) The limits

$$y_\pm := \lim_{t \to \pm\infty} \gamma(t) \in S^{m-1}$$

exist and are given by

$$y_+ = \frac{\lambda y + \lambda\widehat{y} + \langle \lambda y, \lambda\widehat{y} \rangle y}{\lambda - 1 + \langle \lambda y, \lambda\widehat{y} \rangle}, \qquad y_- = \frac{\lambda y - \lambda\widehat{y} - \langle \lambda y, \lambda\widehat{y} \rangle y}{\lambda - 1 - \langle \lambda y, \lambda\widehat{y} \rangle}. \tag{6.4.18}$$

(c) Assume $\widehat{y} \notin \mathbb{R}y$. Then there is a unique circle in $\mathbb{R}^m$ through y_- and y_+ that is orthogonal to S^{m-1} at $y_\pm$. The center $c \in \mathbb{R}^m$ and the radius r of this circle are given by

$$c = \frac{y_+ + y_-}{1 + \langle y_+, y_- \rangle} = \frac{(\lambda^2 - \lambda - \langle \lambda y, \lambda \widehat{y} \rangle^2)y - \langle \lambda y, \lambda \widehat{y} \rangle \lambda \widehat{y}}{\lambda^2 - 2\lambda - \langle \lambda y, \lambda \widehat{y} \rangle^2},$$

$$r^2 = \frac{1 - \langle y_+, y_- \rangle}{1 + \langle y_+, y_- \rangle} = \frac{1}{\lambda^2 - 2\lambda - \langle \lambda y, \lambda \widehat{y} \rangle^2}. \qquad (6.4.19)$$

(d) Let c and r be as in (c). Then the geodesic γ in (a) satisfies $|\gamma(t) - c| = r$ for all t. **Hint:** It suffices to verify this equation for $t = 0$.

(e) If $\widehat{y} \in \mathbb{R}y$, then $y_- + y_+ = 0$ and the geodesic γ traverses a segment of a straight line through the origin.

(f) Fix two distinct points y_- and y_+ on the unit sphere S^{m-1}. Then there exists a geodesic $\gamma : \mathbb{R} \to \mathbb{D}^m$ such that $\lim_{t \to \pm\infty} \gamma(t) = y_\pm$. If $\gamma' : \mathbb{R} \to \mathbb{D}^m$ is any other geodesic satisfying $\lim_{t \to \pm\infty} \gamma'(t) = y_\pm$, then there exist real numbers a, b such that $a > 0$ and $\gamma'(t) = \gamma(at + b)$ for all $t \in \mathbb{R}$.

Exercise 6.4.25 Prove that the isometry group of $\mathbb{H}^m$ is the pseudo-orthogonal group

$$\mathcal{I}(\mathbb{H}^m) = \mathrm{O}(m, 1) := \left\{ g \in \mathrm{GL}(m+1) \,\middle|\, \begin{array}{l} Q(gv, gw) = Q(v, w) \\ \text{for all } v, w \in \mathbb{R}^{m+1} \end{array} \right\}.$$

Thus, by Corollary 6.4.13, the orthonormal frame bundle $\mathcal{O}(\mathbb{H}^m)$ is diffeomorphic to $\mathrm{O}(m, 1)$.

Exercise 6.4.26 Prove that the exponential map $\exp_p : T_p\mathbb{H}^m \to \mathbb{H}^m$ is given by

$$\exp_p(v) = \cosh\left(\sqrt{Q(v, v)}\right) p + \frac{\sinh\left(\sqrt{Q(v, v)}\right)}{\sqrt{Q(v, v)}} v \qquad (6.4.20)$$

for $v \in T_p\mathbb{H}^m = p^{\perp_Q}$. Prove that this map is a diffeomorphism for every $p \in \mathbb{H}^m$. Thus any two points in $\mathbb{H}^m$ are connected by a unique geodesic. Prove that the intrinsic distance function on hyperbolic space is given by

$$d(p, q) = \cosh^{-1}(Q(p, q)) \qquad (6.4.21)$$

for $p, q \in \mathbb{H}^m$. Compare this with Example 4.3.11.

Exercise 6.4.27 In the case $m = 2$ the Poincaré model of hyperbolic space in Exercise 6.4.22 is the open unit disc $\mathbb{D} \subset \mathbb{C}$ in the complex plane. It can be identified with the upper half plane

$$\mathbb{H} = \{z \in \mathbb{C} \mid \mathrm{Im}(z) > 0\}$$

via the diffeomorphism

$$\mathbb{H} \to \mathbb{D} : z \mapsto \frac{i - z}{i + z}.$$

Show that the pullback of the Poincaré metric on $\mathbb{D}$ under this diffeomorphism is the Riemannian metric on $\mathbb{H}$ given by

$$|\widehat{z}|_z = \frac{|\widehat{z}|}{y}$$

for $z = x + iy \in \mathbb{H}$ and $\widehat{z} \in T_z\mathbb{H} = \mathbb{C}$. Show that the isometries of $\mathbb{H}$ (in the identity component) have the form

$$\phi(z) = \frac{az + b}{cz + d}, \qquad \begin{pmatrix} a & b \\ c & d \end{pmatrix} \in \mathrm{SL}(2, \mathbb{R}),$$

and deduce that the Lie group $\mathrm{PSL}(2, \mathbb{R}) := \mathrm{SL}(2, \mathbb{R})/\{\pm 1\}$ is isomorphic to the identity component of $\mathrm{O}(2, 1)$. Prove that every nonconstant geodesic in $\mathbb{H}$ traverses either a vertical half line or a semicircle centered at a point on the boundary $\partial\mathbb{H} = \mathbb{R}$.

6.5 Nonpositive Sectional Curvature

In the previous section we have seen that any two points in a connected, simply connected, complete manifold M of constant negative curvature are joined by a unique geodesic (Exercise 6.4.26). Thus the entire manifold M is geodesically convex and its injectivity radius is infinity. This continues to hold in much greater generality for manifolds with nonpositive sectional curvature. It is convenient, at this point, to extend the discussion to Riemannian manifolds in the intrinsic setting. In particular, at some point in the proof of the main theorem of this section and in our main example, we shall work with a Riemannian metric that does not arise (in any obvious way) from an embedding.

Definition 6.5.1 A Riemannian manifold M is said to have **nonpositive sectional curvature** iff $K(p, E) \leq 0$ for every $p \in M$ and every 2-dimensional linear subspace $E \subset T_pM$ or, equivalently, $\langle R_p(u, v)v, u \rangle \leq 0$ for all $p \in M$ and all $u, v \in T_pM$. A nonempty, connected, simply connected, complete Riemannian manifold with nonpositive sectional curvature is called a **Hadamard manifold**.

6.5.1 The Cartan–Hadamard Theorem

The next theorem shows that every Hadamard manifold is diffeomorphic to Euclidean space and has infinite injectivity radius. This is in sharp contrast to positive curvature manifolds as the example $M = S^m$ shows.

Theorem 6.5.2 (Cartan–Hadamard) *Let M be a connected, simply connected, complete Riemannan manifold. Then the following are equivalent.*

(i)   *M has nonpositive sectional curvature.*
(ii) *The derivative of each exponential map is length increasing, i.e.*

$$\left| d \exp_p(v)\widehat{v} \right| \geq |\widehat{v}|$$

for all $p \in M$ and all $v, \widehat{v} \in T_p M$.
(iii) *Each exponential map is **distance increasing**, i.e.*

$$d(\exp_p(v_0), \exp_p(v_1)) \geq |v_0 - v_1|$$

for all $p \in M$ and all $v_0, v_1 \in T_p M$.

Moreover, if these equivalent conditions are satisfied, then the exponential map $\exp_p : T_p M \to M$ is a diffeomorphism for every $p \in M$. Thus any two points in M are joined by a unique geodesic.

The proof makes use of the following two exercises.

Exercise 6.5.3 Let $\xi : [0, \infty) \to \mathbb{R}^n$ be a smooth function such that

$$\xi(0) = 0, \qquad \dot{\xi}(0) \neq 0, \qquad \xi(t) \neq 0 \quad \forall\, t > 0.$$

Prove that the function $f : [0, \infty) \to \mathbb{R}$ given by $f(t) := |\xi(t)|$ is smooth. **Hint:** The function $\eta : [0, \infty) \to \mathbb{R}^n$ defined by

$$\eta(t) := \begin{cases} t^{-1}\xi(t), & \text{for } t > 0, \\ \dot{\xi}(0), & \text{for } t = 0, \end{cases}$$

is smooth. Show that f is differentiable and $\dot{f} = |\eta|^{-1}\langle \eta, \dot{\xi} \rangle$.

Exercise 6.5.4 Let $\xi : \mathbb{R} \to \mathbb{R}^n$ be a smooth function such that

$$\xi(0) = 0, \qquad \ddot{\xi}(0) = 0.$$

Prove that there exist constants $\varepsilon > 0$ and $c > 0$ such that, for all $t \in \mathbb{R}$,

$$|t| < \varepsilon \qquad \Longrightarrow \qquad |\xi(t)|^2 |\dot{\xi}(t)|^2 - \langle \xi(t), \dot{\xi}(t) \rangle^2 \leq c|t|^6.$$

Hint: Write

$$\xi(t) = tv + \eta(t), \qquad \dot{\xi}(t) = v + \dot{\eta}(t)$$

with $\eta(t) = O(t^3)$ and $\dot{\eta}(t) = O(t^2)$. Show that the terms of order 2 and 4 cancel in the Taylor expansion at $t = 0$.

Proof of Theorem 6.5.2 We prove that (i) implies (ii). Fix a point $p \in M$ and two tangent vectors $v, \widehat{v} \in T_p M$. Assume without loss of generality that $\widehat{v} \neq 0$ and define the curve $\gamma : \mathbb{R} \to M$ and the vector field $X \in \text{Vect}(\gamma)$ along γ by

$$\gamma(t) := \exp_p(tv), \qquad X(t) := \frac{\partial}{\partial s}\Big|_{s=0} \exp_p(t(v + s\widehat{v})) \in T_{\gamma(t)} M \qquad (6.5.1)$$

for $t \in \mathbb{R}$. Then

$$X(0) = 0, \qquad X(t) = d\exp_p(tv)t\widehat{v}, \qquad \nabla X(0) = \widehat{v} \neq 0. \qquad (6.5.2)$$

To see this, define the map $\beta : \mathbb{R}^2 \to M$ by $\beta(s,t) := \exp_p(t(v + s\widehat{v}))$. It satisfies $\beta(0,t) = \gamma(t)$, $\partial_s \beta(0,t) = X(t)$, $\beta(s,0) = p$, and $\partial_t \beta(s,0) = v + s\widehat{v}$ for all $s, t \in \mathbb{R}$. Hence $\nabla X(0) = \nabla_t \partial_s \beta(0,0) = \nabla_s \partial_t \beta(0,0) = \widehat{v}$. Moreover, the curve $\beta(s, \cdot)$ is a geodesic for every s, and hence Lemma 6.1.18 asserts that $X = \partial_s \beta(0, \cdot)$ is a Jacobi field along γ, i.e.

$$\nabla\nabla X + R(X, \dot{\gamma})\dot{\gamma} = 0. \qquad (6.5.3)$$

It follows from Exercise 6.5.3 with $\xi(t) := \Phi_\gamma(0,t)X(t)$ that the function

$$[0, \infty) \to \mathbb{R} : t \mapsto |X(t)|$$

is smooth and

$$\frac{d}{dt}\Big|_{t=0} |X(t)| = |\nabla X(0)| = |\widehat{v}|.$$

Moreover, for $t > 0$, we have

$$\begin{aligned}
\frac{d^2}{dt^2}|X| &= \frac{d}{dt}\frac{\langle X, \nabla X\rangle}{|X|} \\
&= \frac{|\nabla X|^2 + \langle X, \nabla\nabla X\rangle}{|X|} - \frac{\langle X, \nabla X\rangle^2}{|X|^3} \\
&= \frac{|X|^2|\nabla X|^2 - \langle X, \nabla X\rangle^2}{|X|^3} + \frac{\langle X, R(\dot{\gamma}, X)\dot{\gamma}\rangle}{|X|} \\
&\geq 0.
\end{aligned} \qquad (6.5.4)$$

Here the third equality follows from the fact that X is a Jacobi field along γ, and the inequality follows from the nonpositive sectional curvature condition in (i) and from the Cauchy–Schwarz inequality. Thus the second derivative of the function $[0, \infty) \to \mathbb{R} : t \mapsto |X(t)| - t|\widehat{v}|$ is nonnegative; so its first derivative is nondecreasing and it vanishes at $t = 0$; thus

$$|X(t)| - t|\widehat{v}| \geq 0$$

for every $t \geq 0$. In particular, for $t = 1$ we obtain

$$\left| d \exp_p(v)\widehat{v} \right| = |X(1)| \geq |\widehat{v}|.$$

as claimed. Thus we have proved that (i) implies (ii).

We prove that (ii) implies (i). Assume, by contradiction, that (ii) holds but there exists a point $p \in M$ and a pair of vectors $v, \widehat{v} \in T_p M$ such that

$$\langle R_p(v, \widehat{v})v, \widehat{v} \rangle < 0. \tag{6.5.5}$$

Define $\gamma : \mathbb{R} \to M$ and $X \in \mathrm{Vect}(\gamma)$ by (6.5.1) so that (6.5.2) and (6.5.3) are satisfied. Thus X is a Jacobi field with

$$X(0) = 0, \qquad \nabla X(0) = \widehat{v} \neq 0.$$

Hence it follows from Exercise 6.5.4 with $\xi(t) := \Phi_\gamma(0, t)X(t)$ that there is a constant $c > 0$ such that, for $t > 0$ sufficiently small, we have the inequality

$$|X(t)|^2 |\nabla X(t)|^2 - \langle X(t), \nabla X(t) \rangle^2 \leq ct^6.$$

Moreover, by (6.5.1) and (6.5.2), $\lim_{t \searrow 0} \dot{\gamma}(t) = v$ and $\lim_{t \searrow 0} t^{-1}X(t) = \widehat{v}$. Hence, by (6.5.5) there exist constants $\delta > 0$ and $\varepsilon > 0$ such that

$$|X(t)| \geq \delta t, \qquad \langle X(t), R(\dot{\gamma}(t), X(t))\dot{\gamma}(t) \rangle \leq -\varepsilon t^2,$$

for $t > 0$ sufficiently small. By (6.5.4) this implies

$$\frac{d^2}{dt^2}|X| = \frac{|X|^2 |\nabla X|^2 - \langle X, \nabla X \rangle^2}{|X|^3} + \frac{\langle X, R(\dot{\gamma}, X)\dot{\gamma} \rangle}{|X|} \leq \frac{ct^3}{\delta^3} - \frac{\varepsilon t}{\delta}.$$

Integrate this inequality over an interval $[0, t]$ with $ct^2 < \varepsilon\delta^2$ to obtain

$$\frac{d}{dt}|X(t)| < \left.\frac{d}{dt}\right|_{t=0} |X(t)| = |\nabla X(0)|$$

Integrating this inequality again gives $|X(t)| < t|\nabla X(0)|$ for small positive t. Hence it follows from (6.5.2) that

$$\left| d \exp_p(tv)t\widehat{v} \right| = |X(t)| < t|\nabla X(0)| = t|\widehat{v}|$$

for $t > 0$ sufficiently small. This contradicts (ii).

We prove that (ii) implies that the exponential map $\exp_p : T_p M \to M$ is a diffeomorphism for every $p \in M$. By (ii) $\exp_p$ is a *local diffeomorphism*, i.e. its derivative $d \exp_p(v) : T_p M \to T_{\exp_p(v)} M$ is bijective for every $v \in T_p M$. Hence we can define a Riemannian metric on $M' := T_p M$ by pulling back the metric on M under the exponential map. To make this more explicit we choose a basis $e_1, \ldots, e_m$ of $T_p M$ and define the map $\psi : \mathbb{R}^m \to M$ by

$$\psi(x) := \exp_p \left(\sum_{i=1}^m x^i e_i \right)$$

for $x = (x^1, \ldots, x^m) \in \mathbb{R}^m$. Define the metric tensor by

$$g_{ij}(x) := \left\langle \frac{\partial \psi}{\partial x^i}(x), \frac{\partial \psi}{\partial x^j}(x) \right\rangle, \qquad i, j = 1, \ldots, m.$$

Then $(\mathbb{R}^m, g)$ is a Riemannian manifold (covered by a single coordinate chart) and $\psi : (\mathbb{R}^m, g) \to M$ is a local isometry, by definition of g. The manifold $(\mathbb{R}^m, g)$ is clearly connected and simply connected. Moreover, for every $\xi = (\xi^1, \ldots, \xi^n) \in \mathbb{R}^m = T_0 \mathbb{R}^m$, the curve $\mathbb{R} \to \mathbb{R}^m : t \mapsto t\xi$ is a geodesic with respect to g (because ψ is a local isometry and the image of the curve under ψ is a geodesic in M). Hence it follows from Theorem 4.6.5 that $(\mathbb{R}^m, g)$ is complete.

Since both $(\mathbb{R}^m, g)$ and M are connected, simply connected, and complete, it follows from Corollary 6.1.15 that the local isometry ψ is bijective. Thus the exponential map $\exp_p : T_p M \to M$ is a diffeomorphism as claimed. It follows that any two points in M are joined by a unique geodesic.

We prove that (ii) implies (iii). Fix a point $p \in M$ and two tangent vectors $v_0, v_1 \in T_p M$. Let $\gamma : [0, 1] \to M$ be the geodesic with the endpoints

$$\gamma(0) = \exp_p(v_0), \qquad \gamma(1) = \exp_p(v_1)$$

and let $v : [0, 1] \to T_p M$ be the unique curve satisfying $\exp_p(v(t)) = \gamma(t)$ for all t. Then $v(0) = v_0$, $v(1) = v_1$, and

$$d(\exp_p(v_0), \exp_p(v_1)) = L(\gamma)$$

$$= \int_0^1 \left| d \exp_p(v(t)) \dot{v}(t) \right| dt$$

$$\geq \int_0^1 |\dot{v}(t)| \, dt$$

$$\geq \left| \int_0^1 \dot{v}(t) \, dt \right|$$

$$= |v_1 - v_0|.$$

Here the third inequality follows from (ii). This shows that (ii) implies (iii).

We prove that (iii) implies (ii). Fix a point $p \in M$ and a tangent vector $v \in T_p M$ and denote $q := \exp_p(v)$. By (iii) the exponential map $\exp_q : T_q M \to M$ is injective and, since M is complete, it is bijective (see Theorem 4.6.6). Hence there exists a unique geodesic from q to any other point in M and therefore, by Theorem 4.4.4, we have

$$|w| = d(q, \exp_q(w)) \tag{6.5.6}$$

for every $w \in T_q M$. Now define $\phi := \exp_q^{-1} \circ \exp_p : T_p M \to T_q M$. This map satisfies $\phi(v) = 0$. Moreover, it is differentiable in a neighborhood of v and, by the chain rule, $d\phi(v) = d \exp_p(v) : T_p M \to T_q M$. Now choose $w := \phi(v + \widehat{v})$ with $\widehat{v} \in T_p M$. Then $\exp_q(w) = \exp_q(\phi(v + \widehat{v})) = \exp_p(v + \widehat{v})$ and hence it follows from (6.5.6) and part (iii) that

$$|\phi(v + \widehat{v}))| = |w| = d(q, \exp_q(w)) = d(\exp_p(v), \exp_p(v + \widehat{v})) \geq |\widehat{v}|.$$

This gives

$$\left| d \exp_p(v)\widehat{v} \right| = |d\phi(v)\widehat{v}| = \lim_{t \to 0} \frac{|\phi(v + t\widehat{v})|}{t} \geq \lim_{t \to 0} \frac{|t\widehat{v}|}{t} = |\widehat{v}|.$$

Thus we have proved that (iii) implies (ii) and this proves Theorem 6.5.2. $\qquad\square$

The next lemma establishes a useful inequality for Hadamard manifolds that amplifies the expanding property of the exponential map.

Lemma 6.5.5 *Let M be a Hadamard manifold. Fix an element $p \in M$ and two tangent vectors $v_0, v_1 \in T_p M$. Then, for $0 < t \leq T$,*

$$|v_0 - v_1| \leq \frac{d(\exp_p(tv_0), \exp_p(tv_1))}{t} \leq \frac{d(\exp_p(Tv_0), \exp_p(Tv_1))}{T}. \tag{6.5.7}$$

Proof The first inequality in (6.5.7) is part (iii) of Theorem 6.5.2. To prove the second inequality, assume $v_0 \neq v_1$ and define

$$\gamma_0(t) := \exp_p(tv_0), \qquad \gamma_1(t) := \exp_p(tv_1)$$

for $t \in \mathbb{R}$. For each $t \in \mathbb{R}$ let the curve $[0, 1] \to M : s \mapsto \gamma(s, t)$ be the unique geodesic with the endpoints $\gamma(0, t) = \gamma_0(t)$ and $\gamma(1, t) = \gamma_1(t)$. Then

$$\rho(t) := d(\gamma_0(t), \gamma_1(t)) = \int_0^1 |\partial_s \gamma| \, ds = |\partial_s \gamma(s, t)|$$

for all s and t and hence

$$\begin{aligned}
\dot{\rho}(t) &= \int_0^1 \frac{\langle \partial_s \gamma, \nabla_t \partial_s \gamma \rangle}{|\partial_s \gamma|} \, ds \\
&= \frac{\langle \partial_s \gamma(1, t), \partial_t \gamma(1, t) \rangle - \langle \partial_s \gamma(0, t), \partial_t \gamma(0, t) \rangle}{\rho(t)}.
\end{aligned} \tag{6.5.8}$$

Since $\frac{d}{dt}(\rho\dot\rho) = \rho\ddot\rho + \dot\rho^2$ and γ_0 and γ_1 are geodesics, this implies

$$
\begin{aligned}
\rho(t)\ddot\rho(t) + \dot\rho(t)^2 &= \frac{d}{dt}\Big(\langle\partial_s\gamma(1,t),\partial_t\gamma(1,t)\rangle - \langle\partial_s\gamma(0,t),\partial_t\gamma(0,t)\rangle\Big) \\
&= \langle\nabla_t\partial_s\gamma(1,t),\partial_t\gamma(1,t)\rangle - \langle\nabla_t\partial_s\gamma(0,t),\partial_t\gamma(0,t)\rangle \\
&= \int_0^1 \frac{\partial}{\partial s}\langle\nabla_t\partial_s\gamma,\partial_t\gamma\rangle\, ds \\
&= \int_0^1 \Big(|\nabla_t\partial_s\gamma|^2 + \langle\nabla_s\nabla_t\partial_s\gamma,\partial_t\gamma\rangle\Big)\, ds \\
&= \int_0^1 \Big(|\nabla_t\partial_s\gamma|^2 + \langle R(\partial_s\gamma,\partial_t\gamma)\partial_s\gamma,\partial_t\gamma\rangle\Big)\, ds \\
&\geq \dot\rho(t)^2.
\end{aligned}
$$

Here the last step follows from (6.5.8), the Cauchy–Schwarz inequality, and the nonpositive sectional curvature assumption. Thus $\rho : [0,T] \to \mathbb{R}$ is a convex function satisfying $\rho(0) = 0$ and hence $\rho(t) \leq t\rho(T)/T$ for $0 \leq t \leq T$. This proves Lemma 6.5.5.   □

6.5.2 Cartan's Fixed Point Theorem

Recall from Definition 6.5.1 that a Hadamard manifold is a nonempty, connected, simply connected, complete Riemannian manifold of nonpositive sectional curvature.

Theorem 6.5.6 (Cartan) *Let M be a Hadamard manifold and let G be a compact topological group that acts on M by isometries. Then there exists a point $p \in M$ such that $gp = p$ for every $g \in G$.*

The proof follows the argument given by Bill Casselmann in [16] and requires the following two lemmas. The first lemma asserts that every complete, connected, simply connected Riemannian manifold of nonpositive sectional curvature is a semi-hyperbolic space in the sense of Alexandrov [3].

Lemma 6.5.7 (Alexandrov) *Let M be a Hadamard manifold, let $m \in M$ and $v \in T_m M$, and define*

$$p_0 := \exp_m(-v), \qquad p_1 := \exp_m(v).$$

Then

$$2d(m,q)^2 + \frac{d(p_0,p_1)^2}{2} \leq d(p_0,q)^2 + d(p_1,q)^2 \tag{6.5.9}$$

for every $q \in M$ (see Fig. 6.5).

Fig. 6.5 Alexandrov semi-hyperbolic space

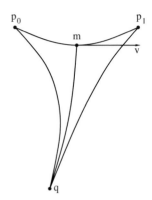

Proof By Theorem 6.5.2 the exponential map $\exp_m : T_m M \to M$ is a diffeomorphism. Hence $d(p_0, p_1) = 2|v|$. Now let $q \in M$. Then there is a unique tangent vector $w \in T_m M$ such that

$$q = \exp_m(w), \qquad d(m, q) = |w|.$$

Since the exponential map is expanding, by Theorem 6.5.2, we have

$$d(p_0, q) \geq |w + v|, \qquad d(p_1, q) \geq |w - v|.$$

Hence

$$
\begin{aligned}
d(m, q)^2 &= |w|^2 \\
&= \frac{|w + v|^2 + |w - v|^2}{2} - |v|^2 \\
&\leq \frac{d(p_0, q)^2 + d(p_1, q)^2}{2} - \frac{d(p_0, p_1)^2}{4}.
\end{aligned}
$$

This proves Lemma 6.5.7. □

Exercise 6.5.8 Equality holds in (6.5.9) whenever M is flat.

The next lemma is **Serre's Uniqueness Theorem** for the *circumcentre* of a bounded set in a *semi-hyperbolic space*.

Lemma 6.5.9 (Serre) *Let M be a Hadamard manifold and, for $p \in M$ and $r \geq 0$, denote by $B(p, r) \subset M$ the closed ball of radius r centered at p. Let $\Omega \subset M$ be a nonempty bounded set and define*

$$r_\Omega := \inf\{r > 0 \mid \text{there exists a } p \in M \text{ such that } \Omega \subset B(p, r)\}.$$

Then there exists a unique point $p_\Omega \in M$ such that $\Omega \subset B(p_\Omega, r_\Omega)$ (see Fig. 6.6).

Fig. 6.6 The circumcenter of
a bounded set

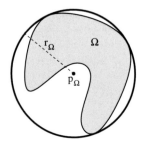

Proof We prove existence. Choose sequences $r_i > r_\Omega$ and $p_i \in M$ such that

$$\Omega \subset B(p_i, r_i), \qquad \lim_{i \to \infty} r_i = r_\Omega.$$

Choose $q \in \Omega$. Then $d(q, p_i) \le r_i$ for every i. Since the sequence r_i is bounded and M is complete, it follows that p_i has a convergent subsequence, still denoted by p_i. Its limit

$$p_\Omega := \lim_{i \to \infty} p_i$$

satisfies $\Omega \subset B(p_\Omega, r_\Omega)$.

We prove uniqueness. Let $p_0, p_1 \in M$ such that

$$\Omega \subset B(p_0, r_\Omega) \cap B(p_1, r_\Omega).$$

Since the exponential map $\exp_p : T_p M \to M$ is a diffeomorphism (by Theorem 6.5.2), there exists a unique vector $v_0 \in T_{p_0} M$ such that $p_1 = \exp_{p_0}(v_0)$. Denote the midpoint between p_0 and p_1 by

$$m := \exp_{p_0}\left(\tfrac{1}{2} v_0\right).$$

Then it follows from Lemma 6.5.7 that

$$d(m, q)^2 \le \frac{d(p_0, q)^2 + d(p_1, q)^2}{2} - \frac{d(p_0, p_1)^2}{4}$$
$$\le r_\Omega^2 - \frac{d(p_0, p_1)^2}{4}$$

for every $q \in \Omega$. Since $\sup_{q \in \Omega} d(m, q) \ge r_\Omega$ (by definition of r_Ω), it follows that $d(p_0, p_1) = 0$ and hence $p_0 = p_1$. This proves Lemma 6.5.9. $\qquad\square$

Proof of Theorem 6.5.6 Let $q \in M$ and consider the group orbit

$$\Omega := \{gq \mid g \in G\}.$$

Since G is compact, this set is bounded. Let r_Ω and p_Ω be as in Lemma 6.5.9. Then $\Omega \subset B(p_\Omega, r_\Omega)$. Since G acts on M by isometries, this implies

$$\Omega = g\Omega \subset B(gp_\Omega, r_\Omega)$$

for all $g \in G$. Hence it follows from the uniqueness statement in Lemma 6.5.9 that $gp_\Omega = p_\Omega$ for every $g \in G$. This proves Theorem 6.5.6. □

6.5.3 Positive Definite Symmetric Matrices

We close this section with an example of a nonpositive sectional curvature manifold which plays a key role in Donaldson's approach to Lie algebra theory [17] (see Sect. 7.5.2). Let m be a positive integer and consider the space

$$\mathcal{P} := \mathcal{P}(\mathbb{R}^m) := \left\{ P \in \mathbb{R}^{m \times m} \,\middle|\, P^\mathsf{T} = P > 0 \right\} \tag{6.5.10}$$

of positive definite symmetric $m \times m$-matrices. (Here the notation "$P > 0$" means $\langle x, Px \rangle > 0$ for every nonzero vector $x \in \mathbb{R}^m$.) Thus $\mathcal{P}$ is an open subset of the vector space $\mathcal{S} := \{ S \in \mathbb{R}^{m \times m} \mid S^\mathsf{T} = S \}$ of symmetric matrices and hence the tangent space of $\mathcal{P}$ is $T_P\mathcal{P} = \mathcal{S}$ for every $P \in \mathcal{P}$. However, we do not use the metric inherited from the inclusion into $\mathcal{S}$ but define a Riemannian metric by

$$\langle \widehat{P}_1, \widehat{P}_2 \rangle_P := \mathrm{trace}\left(\widehat{P}_1 P^{-1} \widehat{P}_2 P^{-1} \right) \tag{6.5.11}$$

for $P \in \mathcal{P}$ and $\widehat{P}_1, \widehat{P}_2 \in T_P\mathcal{P} = \mathcal{S}$.

Theorem 6.5.10 *The space $\mathcal{P}$ with the Riemannian metric (6.5.11) is a connected, simply connected, complete Riemannian manifold with nonpositive sectional curvature, and the distance function on $\mathcal{P}$ is given by*

$$d(P, Q) = \sqrt{\mathrm{trace}\left(\left(\log(P^{-1/2} Q P^{-1/2}) \right)^2 \right)} \tag{6.5.12}$$

for $P, Q \in \mathcal{P}$. Moreover, $\mathcal{P}$ is a symmetric space and the group $\mathrm{GL}(m, \mathbb{R})$ of non-singular $m \times m$-matrices acts on $\mathcal{P}$ by isometries via $P \mapsto gPg^\mathsf{T}$ for $g \in \mathrm{GL}(m, \mathbb{R})$.

Proof See below. □

Remark 6.5.11 Let V be an m-dimensional vector space and $\mathcal{H} \subset S^2 V^*$ be the set of inner products on V. Define a Riemannian metric on $\mathcal{H}$ by

$$\langle \widehat{h}_1, \widehat{h}_2 \rangle_h := \mathrm{trace}(S_1 S_2), \qquad h(\cdot, S_i \cdot) := \widehat{h}_i, \tag{6.5.13}$$

for $h \in \mathcal{H}$ and $\widehat{h}_1, \widehat{h}_2 \in T_h\mathcal{H} = S^2 V^*$. Then every vector space isomorphism $\alpha : \mathbb{R}^m \to V$ determines a diffeomorphism $\phi_\alpha : \mathcal{H} \to \mathcal{P}$ via

$$\phi_\alpha(h) = P \qquad \Longleftrightarrow \qquad h(\alpha\xi, \alpha\eta) = \langle \xi, P^{-1}\eta \rangle_{\mathbb{R}^m}. \qquad (6.5.14)$$

The derivative of ϕ_α at h in the direction $\widehat{h} \in T_h\mathcal{H}$ is given by

$$d\phi_\alpha(h)\widehat{h} = \widehat{P} \qquad \Longleftrightarrow \qquad \widehat{P}P^{-1} = -\alpha^{-1}S\alpha, \quad h(\cdot, S\cdot) := \widehat{h}. \qquad (6.5.15)$$

Thus ϕ_α is an isometry with respect to the Riemannian metrics (6.5.13) on $\mathcal{H}$ and (6.5.11) on $\mathcal{P}$. The ϕ_α form an atlas on $\mathcal{H}$ with the transition maps $\phi_{\beta\alpha}(P) := \phi_\beta \circ \phi_\alpha^{-1}(P) = g_{\beta\alpha} P g_{\beta\alpha}^{\mathsf{T}}$, where $g_{\beta\alpha} := \beta^{-1}\alpha \in \mathrm{GL}(m, \mathbb{R})$.

Remark 6.5.12   The submanifold

$$\mathcal{P}_0 := \mathcal{P}_0(\mathbb{R}^m) := \{ P \in \mathcal{P} \mid \det(P) = 1 \} \qquad (6.5.16)$$

of positive definite symmetric $m \times m$-matrices with determinant one is totally geodesic (see Remark 6.5.13 below). Hence all the assertions of Theorem 6.5.10 (with $\mathrm{GL}(m, \mathbb{R})$ replaced by $\mathrm{SL}(m, \mathbb{R})$) remain valid for $\mathcal{P}_0$.

Remark 6.5.13   Let M be a Riemannian manifold and $L \subset M$ be a submanifold. Then the following are equivalent.

(i)   If $\gamma : I \to M$ is a geodesic on an open interval I such that $0 \in I$ and

$$\gamma(0) \in L, \qquad \dot{\gamma}(0) \in T_{\gamma(0)}L,$$

then there is a constant $\varepsilon > 0$ such that $\gamma(t) \in L$ for $|t| < \varepsilon$.
(ii)   If $\gamma : I \to L$ is a smooth curve on an open interval I and Φ_γ denotes parallel transport along γ in M, then

$$\Phi_\gamma(t, s)T_{\gamma(s)}L = T_{\gamma(t)}L \qquad \forall\, s, t \in I.$$

(iii)   If $\gamma : I \to L$ is a smooth curve on an open interval I and $X \in \mathrm{Vect}(\gamma)$ is a vector field along γ (with values in TM), then

$$X(t) \in T_{\gamma(t)}L \quad \forall\, t \in I \qquad \Longrightarrow \qquad \nabla X(t) \in T_{\gamma(t)}L \quad \forall\, t \in I.$$

A submanifold that satisfies these equivalent conditions is called **totally geodesic**.

Exercise 6.5.14   Prove the equivalence of (i), (ii), (iii) in Remark 6.5.13. **Hint:** Choose suitable coordinates and translate each of the three assertions into conditions on the Christoffel symbols.

Exercise 6.5.15 Prove that $\mathcal{P}_0$ is a totally geodesic submanifold of $\mathcal{P}$. Prove that $\mathcal{P}$ is diffeomorphic to the quotient $GL(m, \mathbb{R})/O(m)$ via polar decomposition and that $\mathcal{P}_0$ is diffeomorphic to the quotient $SL(m, \mathbb{R})/SO(m)$. **Hint:** Consider the map $GL(m, \mathbb{R}) \to \mathcal{P} : g \mapsto \sqrt{gg^\mathsf{T}}$.

Exercise 6.5.16 In the case $m = 2$ prove that $\mathcal{P}_0$ is isometric to the hyperbolic space $\mathbb{H}^2$.

The proof of Theorem 6.5.10 is based on the calculation of the Levi-Civita connection and the formulas for geodesics and the Riemann curvature tensor in the following three lemmas.

Lemma 6.5.17 *Let $I \to \mathcal{P} : t \mapsto P(t)$ be a smooth path in $\mathcal{P}$ on an interval $I \subset \mathbb{R}$ and let $I \to \mathcal{S} : t \mapsto S(t)$ be a vector field along P. Then the covariant derivative of S is given by*

$$\nabla S = \dot{S} - \frac{1}{2} S P^{-1} \dot{P} - \frac{1}{2} \dot{P} P^{-1} S. \tag{6.5.17}$$

Proof The formula (6.5.17) determines a family of linear operators on the spaces of vector fields along paths that satisfy the torsion-free condition

$$\nabla_s \partial_t P = \nabla_t \partial_s P$$

for every smooth map $\mathbb{R}^2 \to \mathcal{P} : (s, t) \mapsto P(s, t)$ and the Leibniz rule

$$\nabla \langle S_1, S_2 \rangle_P = \langle \nabla S_1, S_2 \rangle_P + \langle S_1, \nabla S_2 \rangle_P$$

for any two vector fields S_1 and S_2 along P. These two conditions determine the covariant derivative uniquely (see Lemma 3.6.5 and Theorem 3.7.8). This proves Lemma 6.5.17. $\qquad \square$

Lemma 6.5.18 *The geodesics in $\mathcal{P}$ are given by*

$$\begin{aligned}
\gamma(t) &= P \exp\left(t P^{-1} \widehat{P}\right) \\
&= \exp\left(t \widehat{P} P^{-1}\right) P \\
&= P^{1/2} \exp\left(t P^{-1/2} \widehat{P} P^{-1/2}\right) P^{1/2}
\end{aligned} \tag{6.5.18}$$

for $P \in \mathcal{P}$, $\widehat{P} \in T_P \mathcal{P} = \mathcal{S}$, and $t \in \mathbb{R}$. In particular, $\mathcal{P}$ is complete.

Proof The curve $\gamma : \mathbb{R} \to \mathcal{P}$ defined by (6.5.18) satisfies

$$\dot{\gamma}(t) = \widehat{P} \exp\left(t P^{-1} \widehat{P}\right) = \widehat{P} P^{-1} \gamma(t).$$

Hence it follows from Lemma 6.5.17 that

$$\nabla \dot{\gamma}(t) = \ddot{\gamma}(t) - \dot{\gamma}(t)\gamma(t)^{-1}\dot{\gamma}(t)$$
$$= \ddot{\gamma}(t) - \widehat{P} P^{-1}\dot{\gamma}(t)$$
$$= 0$$

for every $t \in \mathbb{R}$. Hence γ is a geodesic. Since the curve $\gamma : \mathbb{R} \to \mathcal{P}$ in (6.5.18) satisfies $\gamma(0) = P$ and $\dot{\gamma}(0) = \widehat{P}$, this proves Lemma 6.5.18. $\square$

Lemma 6.5.19 *For $P \in \mathcal{P}$, $S, T, A \in \mathcal{S}$ the curvature tensor on $\mathcal{P}$ is*

$$R_P(S,T)A = -\frac{1}{4}SP^{-1}TP^{-1}A - \frac{1}{4}AP^{-1}TP^{-1}S$$
$$+ \frac{1}{4}TP^{-1}SP^{-1}A + \frac{1}{4}AP^{-1}SP^{-1}T \qquad (6.5.19)$$
$$= -\frac{1}{4}\big[[SP^{-1}, TP^{-1}], AP^{-1}\big]P.$$

Proof Choose smooth maps $P : \mathbb{R}^2 \to \mathcal{P}$ and $A : \mathbb{R}^2 \to \mathcal{S}$ and define $S := \partial_s P$ and $T := \partial_t P$. Then $\partial_s T = \partial_t S$ and $R_P(S,T)A = \nabla_s \nabla_t A - \nabla_t \nabla_s A$. Moreover, by Lemma 6.5.17 we have $\nabla_s A = \partial_s A - \frac{1}{2}AP^{-1}S - \frac{1}{2}SP^{-1}A$ and $\nabla_t A = \partial_t A - \frac{1}{2}AP^{-1}T - \frac{1}{2}TP^{-1}A$. Hence

$$R_P(S,T)A = \partial_s \nabla_t A - \frac{1}{2}(\nabla_t A)P^{-1}S - \frac{1}{2}SP^{-1}(\nabla_t A)$$
$$- \partial_t \nabla_s A + \frac{1}{2}(\nabla_s A)P^{-1}T + \frac{1}{2}TP^{-1}(\nabla_s A)$$
$$= \partial_s \Big(\partial_t A - \frac{1}{2}AP^{-1}T - \frac{1}{2}TP^{-1}A\Big)$$
$$- \frac{1}{2}\Big(\partial_t A - \frac{1}{2}AP^{-1}T - \frac{1}{2}TP^{-1}A\Big)P^{-1}S$$
$$- \frac{1}{2}SP^{-1}\Big(\partial_t A - \frac{1}{2}AP^{-1}T - \frac{1}{2}TP^{-1}A\Big)$$
$$- \partial_t \Big(\partial_s A - \frac{1}{2}AP^{-1}S - \frac{1}{2}SP^{-1}A\Big)$$
$$+ \frac{1}{2}\Big(\partial_s A - \frac{1}{2}AP^{-1}S - \frac{1}{2}SP^{-1}A\Big)P^{-1}T$$
$$+ \frac{1}{2}TP^{-1}\Big(\partial_s A - \frac{1}{2}AP^{-1}S - \frac{1}{2}SP^{-1}A\Big).$$

A term by term inspection shows that the partial derivatives of A, S, T cancel because $\partial_s T = \partial_t S$. Hence we are left with the drivatives of P, so

$$
\begin{aligned}
R_P&(S,T)A \\
&= \frac{1}{2}AP^{-1}(\partial_s P)P^{-1}T + \frac{1}{2}TP^{-1}(\partial_s P)P^{-1}A \\
&\quad + \frac{1}{4}AP^{-1}TP^{-1}S + \frac{1}{4}TP^{-1}AP^{-1}S + \frac{1}{4}SP^{-1}AP^{-1}T + \frac{1}{4}SP^{-1}TP^{-1}A \\
&\quad - \frac{1}{2}AP^{-1}(\partial_t P)P^{-1}S - \frac{1}{2}SP^{-1}(\partial_t P)P^{-1}A \\
&\quad - \frac{1}{4}AP^{-1}SP^{-1}T - \frac{1}{4}SP^{-1}AP^{-1}T - \frac{1}{4}TP^{-1}AP^{-1}S - \frac{1}{4}TP^{-1}SP^{-1}A.
\end{aligned}
$$

Insert $\partial_s P = S$, $\partial_t P = T$ to obtain (6.5.19). This proves Lemma 6.5.19.   □

Proof of Theorem 6.5.10   The manifold $\mathcal{P}$ is obviously connected and simply connected as it is a convex open subset of a finite-dimensional vector space. That the map $\mathrm{GL}(m, \mathbb{R}) \times \mathcal{P} \to \mathcal{P} : (g, P) \mapsto gPg^{\mathsf{T}}$ defines a group action of $\mathrm{GL}(m, \mathbb{R})$ on $\mathcal{P}$ by isometries follows directly from the definitions. The remaining assertions will be proved in three steps.

Step 1 *The manifold $\mathcal{P}$ has nonpositive sectional curvature.*

By Lemma 6.5.19 with $A = T$ and equation (6.5.11) we have

$$
\begin{aligned}
\langle S, &R_P(S,T)T\rangle_P \\
&= \mathrm{trace}\big(SP^{-1}(R_P(S,T)T)P^{-1}\big) \\
&= -\frac{1}{4}\mathrm{trace}\big(SP^{-1}\big[[SP^{-1}, TP^{-1}], TP^{-1}\big]\big) \\
&= \frac{1}{2}\mathrm{trace}\big(SP^{-1}TP^{-1}SP^{-1}TP^{-1}\big) - \frac{1}{2}\mathrm{trace}\big(SP^{-1}TP^{-1}TP^{-1}SP^{-1}\big) \\
&= \frac{1}{2}\mathrm{trace}\big(X^2\big) - \frac{1}{2}\mathrm{trace}\big(X^{\mathsf{T}}X\big),
\end{aligned}
$$

where $X := P^{-1/2}SP^{-1}TP^{-1/2} \in \mathbb{R}^{m \times m}$. Write $X =: (x_{ij})_{i,j=1,\dots,m}$. Then, by the Cauchy–Schwarz inequality, we have

$$
\mathrm{trace}(X^2) = \sum_{i,j} x_{ij} x_{ji} \le \sum_{i,j} x_{ij}^2 = \mathrm{trace}(X^{\mathsf{T}}X).
$$

Thus $\langle S, R_P(S,T)T\rangle_P \le 0$ for all $P \in \mathcal{P}$ and $S, T \in \mathcal{S}$. This proves Step 1.

Step 2 *$\mathcal{P}$ is a symmetric space.*

Fix an element $A \in \mathcal{P}$ and define the map $\phi : \mathcal{P} \to \mathcal{P}$ by $\phi(P) := AP^{-1}A$ for $P \in \mathcal{P}$. This map is a diffeomorphism, fixes the matrix $A = \phi(A)$, and its derivative at $P \in \mathcal{P}$ is given by $d\phi(P)\widehat{P} = -AP^{-1}\widehat{P}P^{-1}A$ for $\widehat{P} \in T_P\mathcal{P}$. Hence $d\phi(A) = -\mathrm{id}$. Moreover, $(d\phi(P)\widehat{P})\phi(P)^{-1} = -AP^{-1}\widehat{P}A^{-1}$ and so

$$|d\phi(P)\widehat{P}|^2_{\phi(P)} = \mathrm{trace}\left((AP^{-1}\widehat{P}A^{-1})^2\right) = \mathrm{trace}\left((P^{-1}\widehat{P})^2\right) = |\widehat{P}|^2_P$$

for all $P \in \mathcal{P}$ and $\widehat{P} \in T_P\mathcal{P}$. Thus ϕ is an isometry and this proves Step 2.

Step 3 *The distance function on $\mathcal{P}$ is given by* (6.5.12).

Let $P, Q \in \mathcal{P}$. Then, since $\mathcal{P}$ is a Hadamard manifold by Step 1 and Lemma 6.5.18, there exists a unique matrix $\widehat{P} \in \mathcal{S}$ such that $\exp_P(\widehat{P}) = Q$. By Lemma 6.5.18 this equation reads $P^{1/2}\exp(P^{-1/2}\widehat{P}P^{-1/2})P^{1/2} = Q$. Thus $S := P^{-1/2}\widehat{P}P^{-1/2} = \log(P^{-1/2}QP^{-1/2})$ and

$$d(P, Q)^2 = |\widehat{P}|^2_P = \mathrm{trace}\left(\widehat{P}P^{-1}\widehat{P}P^{-1}\right) = \mathrm{trace}\left(S^2\right).$$

This proves Step 3 and Theorem 6.5.10.   □

Remark 6.5.20 Theorem 6.5.10 carries over to the complex setting as follows. Replace $\mathcal{P}$ by the space

$$\mathcal{Q} := \{Q \in \mathbb{C}^{m\times m} \mid Q^* = Q > 0\} \tag{6.5.20}$$

of positive definite Hermitian matrices. Here Q^* denotes the conjugate transposed matrix of $Q \in \mathbb{C}^{m\times m}$ and the notation "$Q > 0$" means $z^*Qz > 0$ for every nonzero vector $z \in \mathbb{C}^m$. Thus $\mathcal{Q}$ is an open subset of the vector space of Hermitian $m \times m$-matrices. Define the Riemannian metric on $\mathcal{Q}$ by

$$\langle H_1, H_2 \rangle_Q := \mathrm{Re}\left(\mathrm{trace}(H_1Q^{-1}H_2Q^{-1})\right)$$

for $Q \in \mathcal{Q}$ and $H_1, H_2 \in T_Q\mathcal{Q}$. Then all the assertions of Theorem 6.5.10 (with $\mathrm{GL}(m, \mathbb{R})$ replaced by $\mathrm{GL}(m, \mathbb{C})$) carry over to $\mathcal{Q}$. The proof is verbatim the same, with the transposed matric replaced by the conjugate tramsposed matrix and the trace replaced by the real part of the trace.

Remark 6.5.21 The set $\mathcal{Q}_0 := \{Q \in \mathcal{Q} \mid \det(Q) = 1\}$ of positive definite Hermitian matrices with determinant one is a totally geodesic submanifold of $\mathcal{Q}$. Hence all the assertions of Theorem 6.5.10 (with $\mathrm{GL}(m, \mathbb{R})$ replaced by $\mathrm{SL}(m, \mathbb{C})$) remain valid for $\mathcal{Q}_0$.

Exercise 6.5.22 Show that Theorem 6.5.10 remains valid for $\mathcal{Q}$, and $\mathcal{Q}_0$ is a totally geodesic submanifold of $\mathcal{Q}$. Prove that $\mathcal{Q}$ is diffeomorphic to the quotient $\mathrm{GL}(m, \mathbb{C})/\mathrm{U}(m)$ and $\mathcal{Q}_0$ is diffeomorphic to $\mathrm{SL}(m, \mathbb{C})/\mathrm{SU}(m)$. **Hint:** Consider the map $\mathrm{SL}(m, \mathbb{C}) \to \mathcal{Q}_0 : g \mapsto \sqrt{gg^*}$. Show that the pullback metric on $\mathrm{SL}(m, \mathbb{C})/\mathrm{SU}(m)$ is given by the norm of the Hermitian part of the matrix $g^{-1}\widehat{g}$ for $\widehat{g} \in T_g\mathrm{SL}(m, \mathbb{C})$.

Exercise 6.5.23 In the case $m = 2$ prove that $\mathfrak{Q}_0$ is isometric to the hyperbolic space $\mathbb{H}^3$.

The space $\mathfrak{Q}_0 \cong \mathrm{SL}(m, \mathbb{C})/\mathrm{SU}(m)$ (with nonpositive sectional curvature) can be viewed as a kind of dual space of the Lie group $\mathrm{SU}(m)$ (with nonnegative sectional curvature). They have the same dimension and in both cases the Riemann curvature tensor is given by the Lie bracket (see equation (5.2.23) for $\mathrm{SU}(m)$ and equation (6.5.19) for $\mathfrak{Q}_0$). One can think of the noncompact Lie group $G^c := \mathrm{SL}(m, \mathbb{C})$ as the *complexification* of the compact Lie group $G := \mathrm{SU}(m)$. It can be written in the form

$$G^c = \{\exp(i\eta)u \mid u \in G, \eta \in \mathfrak{g}\}, \tag{6.5.21}$$

the Lie algebra of G^c is the complexification $\mathfrak{g}^c = \mathfrak{g} \oplus i\mathfrak{g}$ of the Lie algebra of G, and the quotient G^c/G is a Hadamard manifold. These observations carry over to all Lie subgroups $G \subset \mathrm{SU}(m)$. For an exposition see [20].

Exercise 6.5.24 (Siegel upper half space)

(i)   The standard symplectic form ω_0 on $\mathbb{R}^{2n} = \mathbb{R}^n \times \mathbb{R}^n$ is given by

$$\omega_0(z, \zeta) := (J_0 z)^{\mathsf{T}} \zeta, \qquad J_0 := \begin{pmatrix} 0 & -\mathbb{1} \\ \mathbb{1} & 0 \end{pmatrix},$$

for $z, \zeta \in \mathbb{R}^{2n}$ and the space of ω_0-compatible linear complex structures is the $(n^2 + n)$-dimensional manifold

$$\mathcal{J}(\mathbb{R}^{2n}, \omega_0) := \left\{ J \in \mathbb{R}^{2n \times 2n} \left| \begin{array}{l} J^2 = -\mathbb{1}, \ JJ_0 + J_0 J^{\mathsf{T}} = 0, \\ \omega_0(\zeta, J\zeta) > 0 \text{ for } 0 \neq \zeta \in \mathbb{R}^{2n} \end{array} \right. \right\}. \tag{6.5.22}$$

Define a Riemannian metric on $\mathcal{J}(M, \omega_0)$ by

$$\langle \widehat{J}_1, \widehat{J}_2 \rangle := \mathrm{trace}\big(\widehat{J}_1 \widehat{J}_2\big) \tag{6.5.23}$$

for $\widehat{J}_1, \widehat{J}_2 \in T_J \mathcal{J}(\mathbb{R}^{2n}, \omega_0)$. Show that the symplectic linear group $\mathrm{Sp}(2n)$ (Exercise 2.5.5) acts on the space $\mathcal{J}(\mathbb{R}^{2n}, \omega_0)$ by isometries $J \mapsto gJg^{-1}$. If $J \in \mathcal{J}_0(\mathbb{R}^{2n}, \omega_0)$ and $P := -JJ_0 = P^{\mathsf{T}}$, show that $\omega_0(\cdot, J\cdot) = \langle \cdot, P^{-1} \cdot \rangle$. Show that the map $\mathcal{J}(\mathbb{R}^{2n}, \omega_0) \to \mathcal{P}_0(\mathbb{R}^{2n}) : J \mapsto -JJ_0$ is an $\mathrm{Sp}(2n)$-equivariant isometric embedding, whose image is a totally geodesic sub-manifold of $\mathcal{P}_0(\mathbb{R}^{2n})$. Deduce that $\mathcal{J}(\mathbb{R}^{2n}, \omega_0)$ is a Hadamard manifold and a symmetric space. For every $J \in \mathcal{J}(\mathbb{R}^{2n}, \omega_0)$ show that the map $J' \mapsto -JJ'J$ is an isometry fixing J whose derivative at J is $-\mathrm{id}$.

(ii) **Siegel upper half space** is the manifold $S_n \subset \mathbb{C}^{n \times n}$ of symmetric complex $n \times n$-matrices with positive definite imaginary part [71]. The symplectic linear group $\mathrm{Sp}(2n)$ acts on this space via

$$g_* Z := (AZ + B)(CZ + D)^{-1}, \qquad g = \begin{pmatrix} A & B \\ C & D \end{pmatrix}, \tag{6.5.24}$$

$$A^\mathsf{T} C = C^\mathsf{T} A, \qquad B^\mathsf{T} D = D^\mathsf{T} B, \qquad A^\mathsf{T} D - C^\mathsf{T} B = \mathbb{1},$$

for $g \in \mathrm{Sp}(2n)$ and $Z \in S_n$. Show that this is a well-defined group action. (For hints see [49, page 72].) Show that there is a unique $\mathrm{Sp}(2n)$-equivariant diffeomorphism from S_n to $\mathcal{J}(\mathbb{R}^{2n}, \omega_0)$ that sends $i\mathbb{1}$ to J_0. Show that this map is given by the explicit formula

$$J(Z) = \begin{pmatrix} XY^{-1} & -Y - XY^{-1}X \\ Y^{-1} & -Y^{-1}X \end{pmatrix} \in \mathcal{J}(\mathbb{R}^{2n}, \omega_0), \qquad Z = X + iY \in S_n.$$

Show that the diffeomorphism $S_n \to \mathcal{J}(\mathbb{R}^{2n}, \omega_0) : Z \mapsto J(Z)$ is an isometry with respect to the Riemannian metric on S_n given by

$$|\widehat{Z}|_Z^2 = 2\mathrm{trace}\big((Y^{-1}\widehat{X})^2 + (Y^{-1}\widehat{Y})^2\big) \tag{6.5.25}$$

for $Z = X + iY \in S_n$ and $\widehat{Z} = \widehat{X} + i\widehat{Y} \in T_Z S_n$.

6.6   Positive Ricci Curvature*

In this section we prove that every complete connected manifold $M \subset \mathbb{R}^n$ whose Ricci curvature satisfies a uniform positive lower bound is necessarily compact. If the sectional curvature is constant and positive, this follows from Corollary 6.4.12 as was noted in Example 6.4.16.

Definition 6.6.1 (Ricci tensor) Let $M \subset \mathbb{R}^n$ be an m-dimensional submanifold and fix an element $p \in M$. The **Ricci tensor** of M at p is the symmetric bilinear form

$$\mathrm{Ric}_p : T_p M \times T_p M \to \mathbb{R}$$

defined by

$$\mathrm{Ric}_p(u, v) := \sum_{i=1}^m \langle R_p(e_i, u)v, e_i \rangle, \tag{6.6.1}$$

where $e_1, \ldots, e_m$ is an orthonormal basis of $T_p M$. The Ricci tensor is independent of the choice of this orthonormal frame and is symmetric by equations (5.2.17) and (5.2.19) in Theorem 5.2.14.

6.6.1   The Ricci Tensor in Local Coordinates

Let $\phi : U \to \Omega$ be a local coordinate chart on an open set $U \subset M$ with values in an open set $\Omega \subset \mathbb{R}^m$, denote its inverse by $\psi := \phi^{-1} : \Omega \to U$, and let

$$E_i(x) := \frac{\partial \psi}{\partial x^i}(x) \in T_{\psi(x)} M, \qquad x \in \Omega, \qquad i = 1, \ldots, m,$$

be the local frame of the tangent bundle determined by this coordinate chart. Denote the coefficients of the first fundamental form by $g_{ij} := \langle E_i, E_j \rangle$ and the coefficients of the Riemann curvature tensor by $R^\ell_{ijk} : \Omega \to \mathbb{R}$ so that

$$R(E_i, E_j) E_k = \sum_\ell R^\ell_{ijk} E_\ell.$$

(see Section 5.4). Then

$$\mathrm{Ric}_{ij} := \mathrm{Ric}(E_i, E_j) = \sum_{\nu=1}^m R^\nu_{\nu ij} = \sum_{\mu,\nu=1}^m R_{\nu ij\mu} g^{\mu\nu}. \qquad (6.6.2)$$

(**Exercise:** Prove this.)

6.6.2   The Bonnet–Myers Theorem

Theorem 6.6.2 (Bonnet–Myers)   *Let $M \subset \mathbb{R}^n$ be a complete, connected manifold of dimension $m \geq 2$ and suppose that there exists a $\delta > 0$ such that*

$$\mathrm{Ric}_p(v, v) \geq (m - 1)\delta |v|^2 \qquad (6.6.3)$$

for every $p \in M$ and every $v \in T_p M$. Then $d(p, q) \leq \pi/\sqrt{\delta}$ for all $p, q \in M$ and hence M is compact.

The proof is based on the following lemma.

Lemma 6.6.3   *Let $\mathbb{R} \times [0, 1] \to M : (s, t) \mapsto \gamma_s(t)$ be a smooth map such that $\gamma := \gamma_0 : [0, 1] \to M$ is a geodesic and $\gamma_s(0) = \gamma(0)$ and $\gamma_s(1) = \gamma(1)$ for all $s \in \mathbb{R}$. Define the vector field X along γ by $X(t) := \frac{\partial}{\partial s}\big|_{s=0} \gamma_s(t)$ for $0 \leq t \leq 1$. Then*

$$\frac{d^2}{ds^2}\bigg|_{s=0} E(\gamma_s) = -\int_0^1 \langle \nabla\nabla X + R(X, \dot\gamma)\dot\gamma, X \rangle \, dt$$

$$= \int_0^1 \left(|\nabla X|^2 - \langle R(X, \dot\gamma)\dot\gamma, X \rangle \right) dt. \qquad (6.6.4)$$

Proof In the proof of Theorem 4.1.4 we have seen that

$$\frac{d}{ds} E(\gamma_s) = -\int_0^1 \langle \nabla_t \partial_t \gamma_s(t), \partial_s \gamma_s(t) \rangle \, dt$$

for all $s \in \mathbb{R}$ (see equation (4.1.10)). Differentiate this equation again with respect to s and use the identity $\nabla_s \partial_t = \nabla_t \partial_s$ to obtain

$$\frac{d^2}{ds^2} E(\gamma_s) = -\frac{d}{ds}\Big|_{s=0} \int_0^1 \langle \nabla_t \partial_t \gamma_s, \partial_s \gamma_s \rangle \, dt$$

$$= -\int_0^1 \langle \nabla_s \nabla_t \partial_t \gamma_s, \partial_s \gamma_s \rangle \, dt - \int_0^1 \langle \nabla_t \partial_t \gamma_s, \nabla_s \partial_s \gamma_s \rangle \, dt$$

$$= -\int_0^1 \langle \nabla_t \nabla_t \partial_s \gamma_s + R(\partial_s \gamma_s, \partial_t \gamma_s)\partial_t \gamma_s, \partial_s \gamma_s \rangle \, dt$$

$$- \int_0^1 \langle \nabla_t \partial_t \gamma_s, \nabla_s \partial_s \gamma_s \rangle \, dt.$$

Now take $s = 0$. Then $\nabla_t \partial_t \gamma = 0$ because γ is a geodesic and hence

$$\frac{d^2}{ds^2}\Big|_{s=0} E(\gamma_s) = -\int_0^1 \langle \nabla_t \nabla_t X + R(X, \dot\gamma)\dot\gamma, X \rangle \, dt$$

This proves the first equality in (6.6.4). To prove the second equality in (6.6.4) use integration by parts and the fact that $X(0) = 0$ and $X(1) = 0$. This proves Lemma 6.6.3.   □

Proof of Theorem 6.6.2 Let $p, q \in M$. By the Hopf–Rinow Theorem 4.6.6 there exists a geodesic $\gamma : [0, 1] \to M$ such that

$$\gamma(0) = p, \qquad \gamma(1) = q, \qquad L(\gamma) = d(p, q).$$

Let $X \in \mathrm{Vect}(\gamma)$ be a vector field along γ such that $X(0) = 0$ and $X(1) = 0$ and define $\gamma_s(t) := \exp_{\gamma(t)}(sX(t))$ for $s \in \mathbb{R}$ and $0 \le t \le 1$. Then $s = 0$ is the absolute minimum of the function $\mathbb{R} \to \mathbb{R} : s \mapsto E(\gamma_s)$ by Lemma 4.4.1. Hence $\frac{d^2}{ds^2}\big|_{s=0} E(\gamma_s) \ge 0$ and by Lemma 6.6.3 this implies

$$\int_0^1 \langle R(X, \dot\gamma)\dot\gamma, X \rangle \, dt \le \int_0^1 |\nabla_t X(t)|^2 \, dt. \tag{6.6.5}$$

Now assume $p \neq q$ and choose an orthonormal frame $E_1, \ldots, E_m$ along γ such that $E_1 = \dot{\gamma}/|\dot{\gamma}|$ and $\nabla_t E_i \equiv 0$ for $i = 1, \ldots m$. Define

$$X_i(t) := \sin(\pi t) E_i(t)$$

for $i = 1, \ldots m$ and $0 \leq t \leq 1$. Then $|\nabla_t X_i(t)| = \pi \cos(\pi t)$ for all i and t and

$$\delta(m-1)|\dot{\gamma}(t)|^2 \leq \text{Ric}_{\gamma(t)}(\dot{\gamma}(t), \dot{\gamma}(t)) = \sum_{i=2}^{m} \langle R(E_i(t), \dot{\gamma}(t))\dot{\gamma}(t), E_i(t) \rangle$$

for $0 \leq t \leq 1$. Multiply this inequality by $\sin^2(\pi t)$, integrate over the unit interval, and use the identities $|\dot{\gamma}(t)| = d(p, q)$ and $\int_0^1 \sin^2(\pi t)\, dt = 1/2$ to obtain the estimate

$$\frac{\delta(m-1)}{2} d(p, q)^2 = \int_0^1 \delta(m-1)\sin^2(\pi t)|\dot{\gamma}(t)|^2\, dt$$

$$\leq \sum_{i=2}^{m} \int_0^1 \langle R(X_i(t), \dot{\gamma}(t))\dot{\gamma}(t), X_i(t) \rangle\, dt$$

$$\leq \sum_{i=2}^{m} \int_0^1 |\nabla_t X_i(t)|^2\, dt$$

$$= (m-1) \int_0^1 \pi^2 \cos^2(\pi t)\, dt$$

$$= \frac{\pi^2(m-1)}{2}.$$

Here the third step uses (6.6.5). Since $m \geq 2$ it follows from this estimate that $d(p, q)^2 \leq \pi^2/\delta$ and this proves Theorem 6.6.2. □

A direct consequence of Theorem 6.6.2 is that every compact manifold with positive Ricci curvature has a compact universal cover and hence has a finite fundamental group.

6.6.3 Positive Sectional Curvature

Corollary 6.6.4 *Let $M \subset \mathbb{R}^n$ be a complete, connected manifold of dimension $m \geq 2$ and suppose that there exists a $\delta > 0$ such that*

$$K(p, E) \geq \delta$$

for every $p \in M$ and every 2-dimensional linear subspace $E \subset T_pM$. Then

$$d(p,q) \leq \frac{\pi}{\sqrt{\delta}}$$

for all $p, q \in M$ and hence M is compact.

Proof   The condition $K \geq \delta$ implies (6.6.3) with $m := \dim(M)$ and hence the assertion follows from Theorem 6.6.2. This proves Corollary 6.6.4.   □

The example of the m-sphere shows that the estimate in Corollary 6.6.4 is sharp. Namely, $M := S^m$ has sectional curvature $K = 1$ and diameter π.

The paraboloid $M := (x, y, z) \in \mathbb{R}^3 \mid z = x^2 + y^2\}$ has positive Gaußian curvature and so positive Ricci curvature (Lemma 6.7.2) but is noncompact.

Remark 6.6.5 (Sphere Theorem)   The **Topological Sphere Theorem** asserts that every complete, connected, simply connected Riemannian m-manifold M whose sectional curvature satisfies the estimate

$$1/4 < K(p, E) \leq 1$$

for every $p \in M$ and every 2-dimensional linear subspace $E \subset T_pM$ must be homeomorphic to the m-sphere. The **Differentiable Sphere Theorem** asserts under the same assumptions that M is diffeomorphic to S^m.

The problem goes back to a question posed by Heinz Hopf [31, 32] in the 1920s. After intermediate results by Rauch [57] (with $1/4$ replaced by $3/4$) and others, the Toplogical Sphere Theorem was proved in 1961 by Berger [6] and Klingenberg [41]. The Differentiable Sphere Theorem was proved in 2007 by Brendle and Schoen [8–10]. They even weakened the assumption to $0 < \max_E K(p, E) < 4 \min_E K(p, E)$ for all $p \in M$, where the maximum and minimum are taken over all 2-dimensional linear subspaces $E \subset T_pM$.

The Topological Sphere Theorem is sharp, as the suitably scaled Fubini–Study metric on complex projective space satisfies $1/4 \leq K(p, E) \leq 1$ for all p and E. The Differentiable Sphere Theorem is a significant improvement, because in many dimensions there exist smooth m-manifolds that are homeomorphic, but not diffeomorphic, to S^m. These are the so-called *exotic spheres* and many of those do not even admit metrics of positive *scalar* curvature [30]. (For the definition of scalar curvature see Sect. 6.7.)

6.7   Scalar Curvature*

This section introduces the scalar curvature and explains how it is related to the Ricci tensor and the Riemann curvature tensor in dimensions 2 and 3. The section also includes a brief discussion of several problems in differential geometry in which the scalar curvature plays a central role.

6.7.1 Definition and Basic Properties

Let m be a positive integer and let $M \subset \mathbb{R}^n$ be an m-manifold. For each element $p \in M$ denote by $R_p : T_p M \times T_p M \to \mathrm{End}(T_p M)$ the Riemann curvature tensor and by $\mathrm{Ric}_p : T_p M \times T_p M \to \mathbb{R}$ the Ricci tensor of M at p.

Definition 6.7.1 (Scalar curvature) Fix an element $p \in M$. The **scalar curvature** $S(p)$ of M at p is the trace of the Ricci tensor and is given by

$$S(p) := \sum_{i=1}^{m} \mathrm{Ric}_p(e_i, e_i) = \sum_{i,j=1}^{m} \langle R_p(e_i, e_j)e_j, e_i \rangle, \tag{6.7.1}$$

where $e_1, \ldots, e_m$ is an orthonormal basis of $T_p M$. The scalar curvature is independent of the choice of this orthonormal frame.

Lemma 6.7.2 *Assume* $m = 2$ *and let* $K : M \to \mathbb{R}$ *be the Gaußian curvature of* M *in* (6.4.2). *Then*

$$S(p) = 2K(p), \qquad \mathrm{Ric}_p(u, v) = K(p)\langle u, v \rangle, \tag{6.7.2}$$

$$\langle R_p(u, v)w, z \rangle = K(p)\big(\langle u, z\rangle\langle v, w\rangle - \langle u, w\rangle\langle v, z\rangle\big) \tag{6.7.3}$$

for all $p \in M$ *and* $u, v, w, z \in T_p M$.

Proof By definition, the Gaußian curvature is given by

$$K(p) = \frac{\langle R_p(u, v)v, u \rangle}{|u|^2 |v|^2 - \langle u, v \rangle^2} \tag{6.7.4}$$

for every pair of linearly independent tangent vectors $u, v \in T_p M$. Take u to be a unit vector orthogonal to v to obtain $\mathrm{Ric}_p(v, v) = K(p)|v|^2$ for every $v \in T_p M$. Since $\mathrm{Ric}_p : T_p M \times T_p M \to \mathbb{R}$ is a symmetric bilinear form, this implies the second equality in (6.7.2). With this understood the first equality in (6.7.2) follows directly from the definition of the scalar curvature in (6.7.1). Equation (6.7.3) follows from (6.7.4) by Theorem 6.4.8. This proves Lemma 6.7.2. $\square$

Lemma 6.7.3 *Assume* $m = 3$. *Then*

$$\begin{aligned}
\langle R_p(u, v)w, z \rangle = {} & \mathrm{Ric}_p(v, w)\langle u, z\rangle - \mathrm{Ric}_p(u, w)\langle v, z\rangle \\
& + \mathrm{Ric}_p(u, z)\langle v, w\rangle - \mathrm{Ric}_p(v, z)\langle u, w\rangle \\
& - \frac{S(p)}{2}\big(\langle u, z\rangle\langle v, w\rangle - \langle u, w\rangle\langle v, z\rangle\big)
\end{aligned} \tag{6.7.5}$$

for all $p \in M$ *and all* $u, v, w, z \in T_p M$.

Proof Choose an orthonormal basis e_1, e_2, e_3 of the tangent space T_pM and define the endomorphism

$$Q_p : T_pM \to T_pM$$

by

$$Q_p u := \sum_i R_p(u, e_i)e_i$$

for $u \in T_pM$. The right hand side of this equation is independent of the choice of the orthonormal basis and the endomorphism Q_p satisfies

$$\text{trace}(Q_p) = S(p)$$

and

$$\langle Q_p u, v \rangle = \text{Ric}_p(u, v)$$

for all $u, v \in T_pM$. With this notation equation (6.7.5) takes the form

$$R_p(u, v)w = \text{Ric}_p(v, w)u + \langle v, w \rangle \left(Q_p u - \frac{S(p)}{2}u \right)$$
$$- \text{Ric}_p(u, w)v - \langle u, w \rangle \left(Q_p v - \frac{S(p)}{2}v \right). \tag{6.7.6}$$

It suffices to verify equation (6.7.6) in the following three cases.

(a) u, v are linarly dependent.
(b) u, v, w are orthonormal.
(c) u, v are orthonormal and $w = v$.

In the case (a) both sides of equation (6.7.6) vanish. In the case (b) we have

$$R_p(u, v)w = \langle R_p(u, v)w, u \rangle u + \langle R_p(u, v)w, v \rangle v$$
$$= \text{Ric}_p(v, w)u - \text{Ric}_p(u, w)v,$$

and this is equivalent to (6.7.6).

In the case (c) equation (6.7.6) takes the form

$$R_p(u, v)v = \text{Ric}_p(v, v)u - \text{Ric}_p(u, v)v + Q_p u - \frac{S(p)}{2}u. \tag{6.7.7}$$

To verify this formula, choose a unit vector w that is orthogonal to u and v. Then it follows from the definition of $S(p)$ and Q_p that

$$\frac{S(p)}{2} = \langle R_p(u, v)v, u \rangle + \langle R_p(w, v)v, w \rangle + \langle R_p(u, w)w, u \rangle$$
$$= \text{Ric}_p(v, v) + \langle R_p(u, w)w, u \rangle,$$

and

$$Q_p u = R_p(u, v)v + R_p(u, w)w$$
$$= R_p(u, v)v + \langle R_p(u, w)w, v \rangle v + \langle R_p(u, w)w, u \rangle u$$
$$= R_p(u, v)v + \text{Ric}_p(u, v)v + \frac{S(p)}{2}u - \text{Ric}_p(v, v)u.$$

This proves (6.7.7) and Lemma 6.7.3. □

6.7.2 Scalar Curvature in Local Coordinates

Let $\phi : U \to \Omega$ be a local coordinate chart on an open set $U \subset M$ with values in an open set $\Omega \subset \mathbb{R}^m$, denote its inverse by

$$\psi := \phi^{-1} : \Omega \to U,$$

and denote by

$$E_i(x) := \partial_i \psi(x) \in T_{\psi(x)}M$$

for $x \in \Omega$ and $i = 1, \ldots, m$ the local frame of the tangent bundle determined by this coordinate chart. Denote the coefficients of the first fundamental form by $g_{ij} := \langle E_i, E_j \rangle$, of the Ricci tensor by

$$\text{Ric}_{ij} := \text{Ric}(E_i, E_j),$$

and of the Riemann curvature tensor by $R_{ijk\ell}$ and R_{ijk}^ℓ so that

$$R_{ijk\ell} = \langle R(E_i, E_j)E_k, E_\ell \rangle, \qquad R(E_i, E_j)E_k = \sum_{\ell=1}^m R_{ijk}^\ell E_\ell.$$

Then the scalar curvature is the function $S : \Omega \to \mathbb{R}$ given by

$$S = \sum_{i,j=1}^m \text{Ric}_{ij} g^{ij} = \sum_{i,j,\nu=1}^m R_{\nu ij}^\nu g^{ij} = \sum_{i,j,k,\ell=1}^m R_{ijk\ell} g^{jk} g^{i\ell}. \qquad (6.7.8)$$

(**Exercise:** Prove this.)

6.7.3 Positive Scalar Curvature

An important question for a compact smooth manifold M is whether or not it admits a Riemannian metric of positive scalar curvature. A theorem of Lichnerowicz [46] asserts that, if M is a compact spin manifold of dimension $m = 4n$ that admits a Riemannian metric of positive scalar curvature, then a certain characteristic class of this manifold (the $\widehat{A}$-genus) must vanish. The definitions of the terms that appear in this sentence (spin structure and $\widehat{A}$-genus) as well as in the proof, which involves the Dirac operator, the Atiyah–Singer index theorem, and the Weitzenböck formula, go beyond the scope of the present book. For an exposition see [65, Theorem 6.30].

A nonlinear variant of Lichnerowicz' theorem asserts that a compact oriented smooth 4-manifold with $b_2^+ - b_1$ odd and $b_2^+ > 1$ that admits a Riemannian metric of positive scalar curvature has vanishing Seiberg-Witten invariants (see [65, Proposition 7.32]).

In another direction Gromov and Lawson [22] proved that if M_1 and M_2 are two compact manifolds of dimension $m \geq 3$ that admit Riemannian metrics of positive scalar curvature, then so does their connected sum $M_1 \# M_2$ (see also [65, Theorem 2.18]).

In the late 1970's Schoen and Yau [70] proved, using minimal surfaces, that the torus $\mathbb{T}^m = \mathbb{R}^m / \mathbb{Z}^m$ does not admit a metric of positive scalar curvature for $m \leq 7$. We remark that the $\widehat{A}$-genus of the torus vanishes and so Lichnerowicz' theorem does not apply. In [22] Gromov and Lawson refined the techniques of Lichnerowicz to prove that, for any m, the m-torus does not admit a metric of positive scalar curvature. In fact, they proved that for any compact spin manifold M of dimension m the connected sum $N := M \# \mathbb{T}^m$ does not admit a metric of positive scalar curvature. Moreover, they proved that if N admits a metric of nonnegative scalar curvature, then this metric must be flat and N must be the standard m-torus.

6.7.4   Constant Scalar Curvature

An interesting class of Riemannian metrics consists of those that have constant scalar curvature. In dimension $m = 2$ the existence of such a metric is the content of the **uniformisation theorem**. The proof involves the solution of the Kazdan-Warner equation and goes beyond the scope of this book. For an exposition see [65, Theorem 2.20 and Theorem D.1].

In dimension two Lemma 6.7.2 shows that the constant scalar curvature condition is equivalent to constant sectional curvature. However, in higher dimensions the constant scalar curvature condition is more general. By Corollary 6.4.12 every compact simply connected m-manifold with constant sectional curvature is diffeomorphic to the m-sphere, while constant scalar curvature metrics exist on every compact manifold. Examples are the Fubini-Study metric on complex projective space (Example 2.8.5 and Example 3.7.5), locally symmetric spaces (Theorem 6.3.4), and products of Riemannian manifolds with constant scalar curvature.

Definition 6.7.4 Let M be a Riemannian manifold with the metric g. A Riemannian metric g' on M is called **conformally equivalent** to g iff there exists a smooth function $\lambda : M \to (0, \infty)$ such that $g' = \lambda g$. The set of all such Riemannian metrics is called the **conformal class** of g.

Remark 6.7.5 (Yamabe problem) The **Yamabe problem** asserts that *the conformal class of every Riemannian metric on a compact m-manifold M of dimension $m \geq 2$ contains a metric of constant scalar curvature.*

This problem was formulated in 1960 by Hidehiko Yamabe and was eventually settled in the affirmative in 1984 by the combined work of Hidehiko Yamabe [78], Thierry Aubin [5], Neil Trudinger [75], and Richard Schoen [69]. The proof for a compact manifold M of dimension $m > 2$ relies on finding a positive function $f : M \to \mathbb{R}$ and a real number c that satisfy the **Yamabe equation**

$$\frac{4(m-1)}{m-2} \Delta_g f + S_g f = c f^{(m+2)/(m-2)}. \tag{6.7.9}$$

Here $S_g : M \to \mathbb{R}$ denotes the scalar curvature of the Riemannian metric g and Δ_g denotes its **Laplace–Beltrami operator**. In local coordinates this operator is given by the formula

$$\Delta_g = \frac{1}{\sqrt{\det(g)}} \sum_{i,j} \frac{\partial}{\partial x^i} g^{ij} \sqrt{\det(g)} \frac{\partial}{\partial x^j}. \tag{6.7.10}$$

If $f : M \to (0, \infty)$ is a positive solution of (6.7.9), then the Riemannian metric $f^{4/(m-2)} g$ has the constant scalar curvature c. **Exercise**: Prove this. **Hint**: Show first that

$$S_{u^2 g} = u^{-2} S_g + 2(m-1)u^{-3} \Delta_g u - (m-1)(m-4)u^{-4} |du|_g^2. \tag{6.7.11}$$

Then take $f := u^{m/2-1}$ and use the identities

$$|du^n|_g^2 = n^2 u^{2n-2} |du|_g^2, \tag{6.7.12}$$

$$\Delta_g u^n = n u^{n-1} \Delta_g u + n(n-1)u^{n-2} |du|_g^2 \tag{6.7.13}$$

for a smooth function $u : M \to (0, \infty)$ and a real number $n > 0$.

Examples of constant scalar curvature metrics arise from the Einstein condition (see Lemma 6.7.7 below).

Definition 6.7.6 (Einstein metric) A Riemannian manifold M is called an **Einstein manifold** iff its Ricci tensor is a scalar multiple of the first fundamental form, i.e. there exists a smooth function $\lambda : M \to \mathbb{R}$

$$\mathrm{Ric}_p(u, v) = \lambda(p)\langle u, v \rangle \tag{6.7.14}$$

for all $p \in M$ and all $u, v \in T_p M$. It follows from the definitions that the factor λ in (6.7.14) is related to the scalar curvature S by

$$\lambda = \frac{S}{m}. \tag{6.7.15}$$

Lemma 6.7.2 shows that every Riemannian metric on a 2-manifold is an Einstein metric and the factor $\lambda = K$ is the Gaußian curvature.

Lemma 6.7.7 *Let M be an Einstein manifold of dimension $m \geq 3$. Then the scalar curvature of M is locally constant.*

Proof Let $p \in M$. Then there exists a local orthonormal frame of the tangent bundle $E_1, \ldots, E_m \in \mathrm{Vect}(U)$ in a neighborhood U of p such that the covariant derivatives ∇E_i all vanish at p. (**Exercise:** Prove this.) Denote the deriviative of a function f at p in the direction $E_i(p)$ by $\partial_i f$. Then

$$
\begin{aligned}
0 &= \sum_{j,k} \langle (\nabla_{E_i} R)(E_j, E_k) E_k, E_j \rangle \\
&\quad + \sum_{j,k} \langle (\nabla_{E_j} R)(E_k, E_i) E_k, E_j \rangle + \sum_{j,k} \langle (\nabla_{E_k} R)(E_i, E_j) E_k, E_j \rangle \\
&= \partial_i \sum_{j,k} \langle R(E_j, E_k) E_k, E_j \rangle \\
&\quad + \sum_j \partial_j \sum_k \langle R(E_k, E_i) E_k, E_j \rangle + \sum_k \partial_k \sum_j \langle R(E_i, E_j) E_k, E_j \rangle \\
&= \partial_i S - \sum_j \partial_j \mathrm{Ric}(E_i, E_j) - \sum_k \partial_k \mathrm{Ric}(E_i, E_k) \\
&= \frac{m-2}{m} \partial_i S.
\end{aligned}
$$

Here the the first equality follows from the second Bianchi identity (6.3.7), the second holds at p because $\nabla E_i(p) = 0$, the third follows from the definitions of Ric and S, and the last uses the identity $\mathrm{Ric}(E_i, E_j) = \delta_{ij} S/m$, which holds by (6.7.14) and (6.7.15). Since $m \geq 3$ it follows that the derivative of S vanishes everywhere, and this proves Lemma 6.7.7.   □

Examples of Einstein metrics include all constant sectional curvature metrics by Theorem 6.4.8, the Fubini-Study metric on complex projective space, and all metrics with vanishing Ricci tensor. Examples of the latter are Calabi–Yau metrics on complex manifolds and G_2-structures on 7-manifolds. These are again subjects that go far beyond the scope of this book. In general, the construction of Einstein metrics and the question of their existence is a highly nontrivial problem in differential geometry. The study of this problem has a long history and there are many deep theorems and interesting open questions about this subject.

6.8   The Weyl Tensor*

This section introduces the Weyl tensor and explains some of its basic properties. The section closes with brief discussions of locally conformally flat metrics and self-dual four-manifolds.

6.8.1 Definition and Basic Properties

Let m be a positive integer and let $M \subset \mathbb{R}^n$ be an m-manifold. For each element $p \in M$ denote by $R_p : T_pM \times T_pM \to \text{End}(T_pM)$ the Riemann curvature tensor and by $\text{Ric}_p : T_pM \times T_pM \to \mathbb{R}$ the Ricci tensor of M at p. Also let $S : M \to \mathbb{R}$ be the scalar curvature in Definition 6.7.1.

Definition 6.8.1 (Weyl tensor) Assume $m \geq 3$. The **Weyl tensor** of M at an element $p \in M$ is the bilinear map

$$W_p : T_pM \times T_pM \to \text{End}(T_pM)$$

defined by

$$
\begin{aligned}
\langle W_p(u,v)w, z \rangle := {}& \langle R_p(u,v)w, z \rangle \\
& - \frac{1}{m-2}\Big(\text{Ric}_p(v,w)\langle u,z \rangle - \text{Ric}_p(u,w)\langle v,z \rangle \Big) \\
& - \frac{1}{m-2}\Big(\text{Ric}_p(u,z)\langle v,w \rangle - \text{Ric}_p(v,z)\langle u,w \rangle \Big) \\
& + \frac{S(p)}{(m-1)(m-2)}\Big(\langle u,z \rangle\langle v,w \rangle - \langle v,z \rangle\langle u,w \rangle \Big)
\end{aligned}
\tag{6.8.1}
$$

for $u, v, w, z \in T_pM$.

Lemma 6.7.3 shows that the Weyl tensor vanishes in dimension three. In higher dimensions the Weyl tensor may be nonzero. The next lemma summarizes the basic algebraic properties of the Weyl tensor.

Lemma 6.8.2 *The Weyl tensor $W_p : T_pM \times T_pM \to \text{End}(T_pM)$ at an element $p \in M$ is a skew-symmetric bilinear map with values in the space of skew-adjoint endomorphisms of T_pM and, for all $u, v, w, z \in T_pM$ and every orthonormal basis $e_1, \dots, e_m$ of T_pM, it satisfies*

$$\langle W_p(u,v)w, z \rangle = \langle W_p(w,z)u, v \rangle, \tag{6.8.2}$$

$$W_p(u,v)w + W_p(v,w)u + W_p(w,u)v = 0, \tag{6.8.3}$$

$$\sum_{i=1}^{m} \langle W_p(e_i, u)v, e_i \rangle = 0. \tag{6.8.4}$$

Proof The skew-symmetry of the Weyl tensor and equation (6.8.2) follow directly from the definition and Theorem 5.2.14. It then follows from (6.8.2) that $W_p(u,v)$ is a skew-adjoint endomorphism of T_pM for all $u, v \in T_pM$. The verification of the Bianchi identity (6.8.3) is a straight forward computation which we leave as an exercise. To prove (6.8.4), let $u, v \in T_pM$ and choose an orthonormal basis $e_1, \ldots, e_m$ of T_pM. Then

$$
\sum_{i=1}^m \langle W_p(e_i, u)v, e_i \rangle = \sum_{i=1}^m \langle R_p(e_i, u)v, e_i \rangle
$$

$$
- \frac{1}{m-2} \sum_{i=1}^m \Big(\mathrm{Ric}_p(u, v)\langle e_i, e_i \rangle - \mathrm{Ric}_p(e_i, v)\langle u, e_i \rangle \Big)
$$

$$
- \frac{1}{m-2} \sum_{i=1}^m \Big(\mathrm{Ric}_p(e_i, e_i)\langle u, v \rangle - \mathrm{Ric}_p(u, e_i)\langle e_i, v \rangle \Big)
$$

$$
+ \frac{S(p)}{(m-1)(m-2)} \sum_{i=1}^m \Big(\langle e_i, e_i \rangle \langle u, v \rangle - \langle u, e_i \rangle \langle e_i, v \rangle \Big)
$$

$$
= \mathrm{Ric}_p(u, v)
$$

$$
- \frac{m-1}{m-2}\mathrm{Ric}_p(u, v)
$$

$$
- \frac{S(p)}{m-2}\langle u, v \rangle + \frac{1}{m-2}\mathrm{Ric}_p(u, v)
$$

$$
+ \frac{S(p)}{m-2}\langle u, v \rangle
$$

$$
= 0.
$$

This proves (6.8.4) and Lemma 6.8.2.   □

6.8.2   The Weyl Tensor in Local Coordinates

Let $\phi : U \to \Omega$ be a local coordinate chart on an open set $U \subset M$ with values in an open set $\Omega \subset \mathbb{R}^m$, denote its inverse by $\psi := \phi^{-1} : \Omega \to U$ and let $E_i(x) := \partial_i \psi(x) \in T_{\psi(x)}M$ for $x \in \Omega$ and $i = 1, \ldots, m$ be the be the local frame of the tangent bundle determined by this coordinate chart. Denote the coefficients of the first fundamental form by $g_{ij} := \langle E_i, E_j \rangle$, of the Ricci tensor by $\mathrm{Ric}_{ij} := \mathrm{Ric}(E_i, E_j)$, and of the Riemann curvature tensor by R^ℓ_{ijk}. Let $S = \sum_{i,j} \mathrm{Ric}_{ij} g^{ij} = \sum_{i,j,\nu}^m R^\nu_{\nu ij} g^{ij}$ be the scalar curvature in local coordinates. Then the cofficients $W^\ell_{ijk} : \Omega \to \mathbb{R}$ of the Weyl tensor are defined by $W(E_i, E_j)E_k = \sum_\ell W^\ell_{ijk} E_\ell$. and they can be expresses in the form

$$
W_{ijk\ell} := \langle W(E_i, E_j)E_k, E_\ell \rangle = \sum_\nu W^\nu_{ijk} g_{\nu\ell}
$$

$$
= \sum_\nu R^\nu_{ijk} g_{\nu\ell} + \frac{S}{(m-1)(m-2)}\Big(g_{i\ell}g_{jk} - g_{ik}g_{j\ell} \Big) \tag{6.8.5}
$$

$$
- \frac{1}{m-2}\Big(\mathrm{Ric}_{jk}g_{i\ell} - \mathrm{Ric}_{ik}g_{j\ell} + \mathrm{Ric}_{i\ell}g_{jk} - \mathrm{Ric}_{j\ell}g_{ik} \Big).
$$

6.8.3 Conformal Invariance

Definition 6.8.3 (Locally conformally flat metric) Let M be a Riemannian manifold. The metric g on M is called **locally conformally flat**, iff for each $p \in M$ there exists a Riemannian metric on M that is conformally equivalent to g (see Definition 6.7.4) and flat in a neighborhood of p.

By the local C-A-H Theorem 6.1.17 a Riemannian m-manifold M is locally conformally flat if and only if each $p \in M$ has an open neighborhood U that is **conformally diffeomorphic** to an open subset $\Omega \subset \mathbb{R}^m$, i.e. there exists a coordinate chart $\phi : U \to \Omega$ and a smooth function $\lambda : U \to (0, \infty)$ such that $|v| = \lambda(p)|d\phi(p)v|_{\mathbb{R}^m}$ for all $p \in U$ and all $v \in T_pM$.

A remarkable property of the Weyl tensor is that it remains unchanged under multiplication of the Riemannian metric by a positive function and so is an invariant of the conformal class of the metric. This is easy to see when the function is constant. In that case the Riemann curvature tensor and the Ricci tensor remain unchanged, the scalar curvature gets multiplied by the inverse, and so the Weyl tensor remains unchanged. As Remark 6.7.5 shows, the situation is more complicated when instead of multiplying the metric by a constant, we multiply it by a *nonconstant* positive function. The conformal invariance of the Weyl tensor can then be proved by a somewhat cumbersome calculation in local coordinates.

Remark 6.8.4 This discussion shows that the Weyl tensor vanishes for every Riemannian metric that is locally conformally flat. In fact, it turns out that in dimension $m \geq 4$ the Weyl tensor of M vanishes if and only if the Riemannian metric on M is locally conformally flat (see [43]). Moreover, a **theorem of Kuiper** [33, 42] asserts that every compact, connected, simply connected Riemannian m-manifold that is locally conformally flat is conformally diffeomorphic to the m-sphere with its constant sectional curvature metric. Thus every compact, connected, simply connected Riemannian manifold of dimension $m \geq 4$ that is not diffeomorphic to S^m must have a nonvanishing Weyl tensor.

6.8.4 Self-Dual Four-Manifolds

The lowest dimension in which the study of the Weyl tensor is interesting is $m = 4$. To explain this, it is useful to consider the notion of an **oriented manifold**.

Definition 6.8.5 (Orientation) An **orientation** of an m-manifold M is a collection of orientations of the tangent spaces T_pM, one for each $p \in M$, that depend continuously on p, i.e. if $E_1, \ldots, E_m$ are pointwise linearly independent vector fields in a connected open neighborhood $U \subset M$ of p and the vectors $E_1(p), \ldots, E_m(p)$ form a positive basis of T_pM, then for every $q \in U$ the vectors $E_1(q), \ldots, E_m(q)$ form a positive basis of T_qM. In the intrinsic language an oriented manifold is one equipped with an atlas such that all the transition maps are orientation preserving diffeomorphisms.

Definition 6.8.6 (2-Form) Let M be a smooth manifold. A 2-**form** on M is a collection of skew-symmetric bilinear maps $\omega_p : T_p M \times T_p M \to \mathbb{R}$, one for each $p \in M$, which is smooth in the sense that for every pair of smooth vector fields X, Y on M the assignment $p \mapsto \omega_p(X(p), Y(p))$ defines a smooth function on M. The space of all 2-forms on M is denoted by $\Omega^2(M)$.

In the similar vein the Weyl tensor of a Riemannian m-manifold M can be thought of as 2-form with values in the endomorphism bundle $\mathrm{End}(TM)$. By the symmetry properties in Lemma 6.8.2 the Weyl tensor induces a linear map $\mathcal{W} : \Omega^2(M) \to \Omega^2(M)$ via the formula

$$(\mathcal{W}\omega)_p(u, v) := \sum_{1 \le i < j \le m} \langle W_p(u, v) e_i, e_j \rangle \omega_p(e_i, e_j) \qquad (6.8.6)$$

for $\omega \in \Omega^2(M)$, $p \in M$, and $u, v \in T_p M$, where $e_1, \dots, e_m$ is an orthonormal basis of $T_p M$. The right hand side of equation (6.8.6) is independent of the choice of this orthonormal basis and is a 2-form by Lemma 6.8.2.

Now let M be an oriented Riemannian 4-manifold. Then a 2-form ω on M is called **self-dual** iff it satisfies the condition

$$\omega_p(e_0, e_1) = \omega_p(e_2, e_3) \qquad (6.8.7)$$

for every $p \in M$ and every positive orthonormal basis e_0, e_1, e_2, e_3 of $T_p M$. It is called **anti-self-dual** iff it satisfies (6.8.7) for every $p \in M$ and every negative orthonormal basis e_0, e_1, e_2, e_3 of $T_p M$. Thus ω is anti-self-dual if and only if it is self-dual for the opposite orientation. Denote the space of self-dual 2-forms by $\Omega^{2,+}(M)$ and the space of anti-self-dual 2-forms by $\Omega^{2,-}(M)$. Then there is a direct sum decomposition

$$\Omega^2(M) = \Omega^{2,+}(M) \oplus \Omega^{2,-}(M).$$

Lemma 6.8.7 *Let M be an oriented Riemannian 4-manifold. Then the linear operator $\mathcal{W} : \Omega^2(M) \to \Omega^2(M)$ in (6.8.6) preserves the subspace of self-dual 2-forms and the subspace of anti-self-dual 2-forms.*

Proof Fix any orthonormal basis e_0, e_1, e_2, e_3 of $T_p M$ and abbreviate

$$w_{ijk\ell} := \langle W_p(e_i, e_j) e_k, e_\ell \rangle$$

for $i, j, k, \ell \in \{0, 1, 2, 3\}$. Then by Lemma 6.8.2 we have

$$w_{ijk\ell} = -w_{jik\ell} = -w_{ij\ell k} = w_{k\ell ij}, \qquad w_{ijk\ell} + w_{jki\ell} + w_{kij\ell} = 0, \qquad (6.8.8)$$

$$w_{0ij0} + w_{1ij1} + w_{2ij2} + w_{3ij3} = 0 \qquad (6.8.9)$$

for all i, j, k, ℓ. It follows from (6.8.8) and (6.8.9) that

$$
\begin{aligned}
w_{0102} &= w_{2331}, & w_{0103} &= w_{2312}, & w_{0203} &= w_{3112}, \\
w_{2302} &= w_{0131}, & w_{2303} &= w_{0112}, & w_{3103} &= w_{0212}.
\end{aligned}
\tag{6.8.10}
$$

Namely, by (6.8.8) each of these six identities is equivalent to an equation of the form $\sum_{i=0}^{3} w_{ijki} = 0$ with $j \neq k$ and this holds by (6.8.9). Use (6.8.8) and (6.8.9) again to obtain

$$
\begin{aligned}
& w_{0101} - w_{2323} - w_{0202} + w_{3131} \\
&= w_{0220} + w_{1221} + w_{3223} - w_{0110} - w_{2112} - w_{3113} \\
&= 0
\end{aligned}
$$

and hence, by cyclic permutation of $1, 2, 3$,

$$
\varepsilon := w_{0101} - w_{2323} = w_{0202} - w_{3131} = w_{0303} - w_{1212}.
$$

Since $w_{0101} + w_{0202} + w_{0303} = 0$ by (6.8.9), this implies

$$
3\varepsilon = w_{1221} + w_{2332} + w_{1331} = w_{1221} - w_{0330} = w_{0303} - w_{1212} = \varepsilon.
$$

Thus $\varepsilon = 0$ and so

$$
w_{0101} = w_{2323}, \qquad w_{0202} = w_{3131}, \qquad w_{0303} = w_{1212}. \tag{6.8.11}
$$

Now assume that τ is a nonzero self-dual 2-form and fix an element $p \in M$. Then there exists a positive orthonormal basis e_0, e_1, e_2, e_3 of $T_p M$ and a real number $\lambda \neq 0$ such that $\tau(e_0, e_1) = \tau(e_2, e_3) = \lambda$ and $\tau(e_i, e_j) = 0$ for all other pairs i, j. Hence $(\mathcal{W}\tau)(e_i, e_j) = \lambda w_{01ij} + \lambda w_{23ij}$ for all i and j. Take $\lambda = 1$ and use the equations (6.8.8), (6.8.10), and (6.8.11) to obtain

$$
\begin{aligned}
(\mathcal{W}\tau)(e_0, e_1) &= w_{0101} + w_{2301} = w_{0123} + w_{2323} = (\mathcal{W}\tau)(e_2, e_3), \\
(\mathcal{W}\tau)(e_0, e_2) &= w_{0102} + w_{2302} = w_{0131} + w_{2331} = (\mathcal{W}\tau)(e_3, e_1), \\
(\mathcal{W}\tau)(e_0, e_3) &= w_{0103} + w_{2303} = w_{0112} + w_{2312} = (\mathcal{W}\tau)(e_1, e_2).
\end{aligned}
$$

Thus $\mathcal{W}\tau$ is self-dual. The anti-self-dual case follows by reversing the orientation. This proves Lemma 6.8.7. □

Lemma 6.8.8 *Let M be an oriented Riemannian 4-manifold and denote by $\mathcal{W}$: $\Omega^2(M) \to \Omega^2(M)$ the linear operator determined by the Weyl tensor via (6.8.6). Then the following are equivalent.*

(i) *If $\tau \in \Omega^2(M)$ is anti-self-dual, then $\mathcal{W}\tau = 0$.*

(ii) *The Weyl tensor W satisfies satisfies the equation*

$$
W_p(e_0, e_1) = W_p(e_2, e_3) \tag{6.8.12}
$$

for every $p \in M$ and every positive orthonormal basis e_0, e_1, e_2, e_3 of $T_p M$.

Proof We prove that (i) implies (ii). Thus assume (i) and choose a positive orthonormal basis e_0, e_1, e_2, e_3 of $T_p M$. Let τ be the 2-form defined by

$$\tau(e_0, e_1) := 1, \qquad \tau(e_2, e_3) := -1$$

and $\tau(e_i, e_j) := 0$ for all other pairs i, j. Then τ is anti-self-dual and hence $\mathcal{W}\tau = 0$ by (i). Thus it follows from (6.8.2) and (6.8.6) that

$$\langle W(e_0, e_1)u - W(e_2, e_3)u, v \rangle = (\mathcal{W}\tau)(u, v) = 0$$

for all $u, v \in T_p M$ and this proves (ii).

Conversely, suppose that (ii) holds, choose a positive orthonormal basis e_0, e_1, e_2, e_3 of $T_p M$, and let τ be an anti-self-dual 2-form. Then

$$\lambda_1 := \tau(e_0, e_1) = -\tau(e_2, e_3),$$
$$\lambda_2 := \tau(e_0, e_2) = -\tau(e_3, e_1),$$
$$\lambda_3 := \tau(e_0, e_3) = -\tau(e_1, e_2)$$

and hence

$$\mathcal{W}\tau = \lambda_1 \langle (W(e_0, e_1) - W(e_2, e_3)) \cdot, \cdot \rangle$$
$$+ \lambda_2 \langle (W(e_0, e_2) - W(e_3, e_1)) \cdot, \cdot \rangle$$
$$+ \lambda_3 \langle (W(e_0, e_3) - W(e_1, e_2)) \cdot, \cdot \rangle = 0$$

by (ii). This proves Lemma 6.8.8. $\qquad\qquad\qquad\qquad\qquad\qquad\qquad\square$

Definition 6.8.9 An oriented Riemannian 4-manifold is called **self-dual** iff its Weyl tensor satisfies the equivalent conditions of Lemma 6.8.8. It is called **anti-self-dual** iff it is self-dual for the opposite orientation.

Examples of self-dual 4-manifolds are all 4-manifolds with constant sectional curvature by Theorem 6.4.8, or more generally all locally conformally flat 4-manifolds such as $S^1 \times S^3$. Other examples are the complex projective plane with its Fubini–Study metric (Examples 2.8.5 and 3.7.5), and Ricci flat Kähler surfaces with the opposite of the complex orientation (the K3-surface and the Enriques surface). Compact simply connected smooth 4-manifolds with signature zero that are not diffeomorphic to the 4-sphere, such as $S^2 \times S^2$ or the one-point blowup of the projective plane, do not admit any self-dual metrics by Kuiper's theorem (Remark 6.8.4), because every self-dual metric on such a manifold is locally conformally flat. For a survey of these basic examples see Kalafat [34].

The study of self-dual 4-manifolds was initiated by Penrose [56] and Atiyah–Hitchin–Singer [4] and was later taken up by many authors including Taubes [73], LeBrun [45], and Donaldson [18].

In [18] Donaldson proposes to study the moduli space of self-dual metrics on a compact oriented smooth 4-manifold (without boundary) modulo the action of the group of diffeomorphisms. This is a very difficult problem. The self-duality equation is a system of nonlinear partial differential equations and they do give rise to a *"finite-dimensional moduli space"*. However, this space may be highly singular and it is noncompact unless it is empty. It would need to be *"compactified"* in a suitable way, one would need to find a way to understand the singularities, and one would have to assign to it some kind of *"virtual fundamental class"* to obtain numerical invariants by pairing this class with suitable cohomology classes in the ambient space.

Topics in Geometry

7

This chapter explores various topics in differential geometry that are central to the subject and accessible with the material covered in this book, but go beyond the scope of a one semester lecture course. The first section builds on the material in Chap. 4 about geodesics and can be read directly after that chapter. It introduces conjugate points on geodesics and proves the Morse Index Theorem. This result is then used to show that every geodesic without conjugate points minimizes the length *locally (and strictly)* in the space of all curves joining the same endpoints and, conversely, that locally minimizing geodesics have no interior conjugate points (Sect. 7.1). This result in turn plays an essential role in the proof of the Continuity Theorem for the injectivity radius (Sect. 7.2). The next section examines the group of isometries of a connected Riemannian manifold and contains a proof of the Myers–Steenrod Theorem, which asserts that the isometry group always admits the natural structure of a finite–dimensional Lie group (Sect. 7.3). The proof given here has several parallels to the proof of the Closed Subgroup Theorem. This section is based on the study of isometries and the Riemann curvature tensor in Chap. 5 and can be read directly after that chapter. The present chapter then deals with the specific example of the isometry group of a compact connected Lie group equipped with a bi-invariant Riemannian metric (Sect. 7.4). The last two sections are devoted to Donaldson's differential geometric approach to Lie algebra theory [17]. They build on the material in Chap. 6 and include Donaldson's existence theorems for critical points of convex functions on Hadamard manifolds (Sect. 7.5) and his existence proof for symmetric inner products on simple Lie algebras (Sect. 7.6). Corollaries include the uniqueness of maximal compact subgroups, and Cartan's theorem about the compact real form of a semisimple complex Lie algebra.

7.1 Conjugate Points and the Morse Index*

This section introduces conjugate points on geodesics and contains a proof of the Morse Index Theorem. As an application we prove that every geodesic without conjugate points minimizes the length and the energy *locally (and strictly)* among

J.W. Robbin, D.A. Salamon, *Introduction to Differential Geometry*, Springer Studium Mathematik (Master), https://doi.org/10.1007/978-3-662-64340-2_7

all nearby curves joining the same endpoints. Conversely, locally minimizing geodesics have no interior conjugate points. The results of this section will be used in Sect. 7.2 to prove the Continuity Theorem for the injectivity radius. Assume throughout that $M \subset \mathbb{R}^n$ is a smooth m-manifold and recall the notation $\Omega_{p,q}$ for the space of all smooth paths $\gamma : [0, 1] \to M$ with the endpoints $\gamma(0) = p$ and $\gamma(1) = q$ (Sect. 4.1.2).

7.1.1 Conjugate Points

We have seen in Theorem 4.4.4 and Corollary 4.4.6 that geodesics minimize the length on short time intervals. A natural question arising from this observation is how long the time interval can be chosen on which a geodesic minimizes the length at least locally in some neighborhood in the space of paths. An answer to this question is closely related to the Hessian of the energy functional $E : \Omega_{p,q} \to \mathbb{R}$ in (4.1.2). It was established in Lemma 6.6.3 that the Hessian of the energy functional E at a geodesic $\gamma : [0, 1] \to M$ is the linear operator $\mathcal{H}_\gamma : \mathrm{Vect}_0(\gamma) \to \mathrm{Vect}(\gamma)$ defined by

$$\mathcal{H}_\gamma X := -\nabla\nabla X - R(X, \dot\gamma)\dot\gamma. \tag{7.1.1}$$

The domain of this operator is the space $\mathrm{Vect}_0(\gamma)$ of all vector fields X along γ that vanish at the endoints, i.e. $X(0) = 0$ and $X(1) = 0$. Recall that a Jacobi field along a geodesic is a solution of the equation

$$\nabla\nabla X + R(X, \dot\gamma)\dot\gamma = 0, \tag{7.1.2}$$

and so the kernel of $\mathcal{H}_\gamma$ is the space of Jacobi fields along γ that vanish at the endpoints (Lemma 6.1.18).

Definition 7.1.1 Let $\gamma : [a, b] \to M$ be a geodesic. A **conjugate point** is a real number τ in the interval $a < \tau \le b$ such that there exists a nonvanishing Jacobi field X along the restriction $\gamma|_{[a,\tau]}$ that satisfies $X(a) = 0$ and $X(\tau) = 0$. The dimension of the space of solutions of this equation is called the **multiplicity** of the conjugate point τ and will be denoted by

$$m_\gamma(\tau) := \dim\left\{ X \in \mathrm{Vect}(\gamma|_{[a,\tau]}) \left| \begin{array}{l} \nabla\nabla X + R(X, \dot\gamma)\dot\gamma = 0, \\ X(a) = 0,\ X(\tau) = 0 \end{array} \right. \right\}. \tag{7.1.3}$$

Thus $m_\gamma(\tau) = 0$ whenever τ is not a conjugate point of γ.

In Sect. 4.4 we have addressed the question when a geodesic $\gamma : [0, 1] \to M$ with the endpoints $\gamma(0) = p$ and $\gamma(1) = q$ minimizes the lengths of curves with the same endpoints *globally*, i.e. when it satisfies $L(\gamma) = d(p, q)$. A weaker variant of this question is whether it minimizes the length *locally*, i.e. only among nearby curves in $\Omega_{p,q}$. Here the word *"nearby"* can have several meanings, depending on which topology on the space $\Omega_{p,q}$ is used. The relevant topologies in the present setting are the C^1 topology and the C^∞ topology. Since our manifold M is embedded in $\mathbb{R}^n$ these topologies are induced by the distance functions defined by

$$d_{C^1}(\gamma, \gamma') := \sup_{0 \leq t \leq 1} |\gamma(t) - \gamma'(t)| + \sup_{0 \leq t \leq 1} |\dot{\gamma}(t) - \dot{\gamma}'(t)|,$$

$$d_{C^\infty}(\gamma, \gamma') := \sum_{k=0}^{\infty} 2^{-k} \frac{\sup_{0 \leq t \leq 1} |\frac{d^k}{dt^k}(\gamma(t) - \gamma'(t))|}{1 + \sup_{0 \leq t \leq 1} |\frac{d^k}{dt^k}(\gamma(t) - \gamma'(t))|} \tag{7.1.4}$$

for $\gamma, \gamma' \in \Omega_{p,q}$. Note that, if $\gamma \in \Omega_{p,q}$ satisfies $\min_t |\dot{\gamma}(t)| > 0$, then so does every curve γ' in a sufficiently small C^1-neighborhood of γ in $\Omega_{p,q}$.

Theorem 7.1.2 *Let $\gamma : [0, 1] \to M$ be a nonconstant geodesic with the endpoints $\gamma(0) = p$ and $\gamma(1) = q$. Then the following holds.*

(i) *If there exists a C^∞-open neighborhood $\mathcal{U} \subset \Omega_{p,q}$ of γ such that every curve $\gamma' \in \mathcal{U}$ satisfies $L(\gamma') \geq L(\gamma)$, then γ has no conjugate points τ in the open interval $0 < \tau < 1$.*

(ii) *If γ has no conjugate points τ in the interval $0 < \tau \leq 1$, then there exists a C^1-open neighborhood $\mathcal{U} \subset \Omega_{p,q}$ of γ such that every curve $\gamma' \in \mathcal{U}$ satisfies $L(\gamma') \geq L(\gamma)$ with equality if and only if there exists a diffeomorphism $\rho : [0.1] \to [0, 1]$ such that $\rho(0) = 0$, $\rho(1) = 1$, and $\gamma' = \gamma \circ \rho$.*

The proof will be based on the Morse Index Theorem explained below.

Example 7.1.3 The archetypal example of a conjugate point is the endpoint of a geodesic $\gamma : [0, 1] \to S^2$ on the unit sphere $S^2 \subset \mathbb{R}^3$ of length π. This is the endpoint of an arc around half of a great circle, and at this point the geodesic seizes to be locally unique as there is a continuous family of geodesics joining north and south pole (the meridians).

Example 7.1.4 The absence of conjugate points does not signify that a geodesic $\gamma \in \Omega_{p,q}$ minimizes the length *globally*. The sinplest example is a geodesic around the unit circle $M = S^1$ of length $L(\gamma) > \pi$. Similar examples exist on the torus $M = \mathbb{T}^m$, on the cylinder $M = \mathbb{R} \times S^1$, and also on simply connected manifolds such as the sphere S^2 with a suitably chosen Riemannian metric (so that an open subset $U \subset S^2$ is isometric to a cylinder $(0, 1) \times S^1$, for example).

7.1.2 The Morse Index Theorem

Definition 7.1.5 (Morse index) Let $\gamma : [0, 1] \to M$ be a geodesic. The **Morse index** of γ is defined as the maximal dimension of a linear subspace $X \subset \mathrm{Vect}_0(\gamma)$, such that

$$\int_0^1 \left(|\nabla X|^2 - \langle R(X,\dot{\gamma})\dot{\gamma}, X \rangle \right) dt < 0 \qquad (7.1.5)$$

for every nonzero element $X \in \mathcal{X}$. It will be denoted by

$$\mu(\gamma) := \max\left\{ \dim(\mathcal{X}) \;\middle|\; \begin{array}{l} \mathcal{X} \text{ is a linear subspace of } \mathrm{Vect}_0(\gamma) \\ \text{and (7.1.5) holds for all } X \in \mathcal{X} \setminus \{0\} \end{array} \right\}. \qquad (7.1.6)$$

This is also the number of negative eigenvalues, counted with multiplicities, of the operator $\mathcal{H}_\gamma$ in (7.1.1).

Theorem 7.1.6 (Morse Index Theorem)   *Let $\gamma : [0,1] \to M$ be a geodesic. Then γ has only finitely many conjugate points τ, each with multiplicity $1 \le m_\gamma(\tau) \le m$, and the Morse index of γ is the number of conjugate points in the open interval $0 < \tau < 1$, counted with multiplicities, i.e.*

$$\mu(\gamma) = \sum_{0<\tau<1} m_\gamma(\tau). \qquad (7.1.7)$$

Exercise 7.1.7   Prove that geodesics on manifolds with nonpositive sectional curvature have no conjugate points. So their Morse indices are zero.

Exercise 7.1.8   Find geodesics on the unit sphere $M = S^2 \subset \mathbb{R}^3$ with arbitrarily large Morse index.

The proof of Theorem 7.1.6 requires some background in functional analysis which goes beyond the scope of this book and for which the interested reader is referred to [12]. Remark 7.1.9 below lists the main properties of the operator $\mathcal{H}_\gamma$ that are used in the proofs of Theorems 7.1.2 and 7.1.6. It will be convenient in some places to use the L^2 inner product

$$\langle X, Y \rangle_{L^2} := \int_0^1 \langle X(t), Y(t) \rangle \, dt$$

and the associated norm

$$\|X\|_{L^2} := \sqrt{\int_0^1 |X(t)|^2 \, dt}$$

for vector fields X, Y along γ.

Remark 7.1.9 (Properties of $\mathcal{H}_\gamma$) We begin with the properties that can be easily verified directly, without an appeal to Hilbert space theory.

(a) *Each eigenvalue of $\mathcal{H}_\gamma$ is a real number.*
   This assertion is a consequence of the fact that the operator $\mathcal{H}_\gamma$ is symmetric, i.e. $\langle \mathcal{H}_\gamma X, Y \rangle_{L^2} = \langle X, \mathcal{H}_\gamma Y \rangle_{L^2}$ for all $X, Y \in \mathrm{Vect}_0(\gamma)$. (**Exercise:** Verify assertion (a) by complexifying the space of vector fields along γ.)
(b) *Each eigenvalue of $\mathcal{H}_\gamma$ has multiplicity at most m.*
   The eigenspace of $\mathcal{H}_\gamma$ for an eigenvalue λ consists of solutions of the equation $-\nabla\nabla X - R(X, \dot{\gamma}) = \lambda X$ with $X(0) = 0$. Every solution is uniquely determined by $\nabla X(0)$ and is an eigenfunction for λ if and only if $X(1) = 0$.
(c) *Every $X \in \mathrm{Vect}_0(\gamma)$ satisfies the **Poincaré inequality***

$$\int_0^1 |X(t)|^2 \leq \frac{1}{\pi^2} \int_0^1 |\nabla X|^2 \, dt. \tag{7.1.8}$$

   To see this, choose a parallel orthonormal frame $E_1(t), \ldots, E_m(t) \in T_{\gamma(t)}M$ of the tangent bundle along γ and write $X(t) = \sum_{i=1}^m \xi_i(t) E_i(t)$. Then the inequality (7.1.8) takes the form $\pi^2 \int_0^1 |\xi|^2 \, dt \leq \int_0^1 |\dot{\xi}|^2 \, dt$ and this can be proved by writing ξ as a Fourier series $\xi(t) = \sum_{k=1}^\infty \sin(\pi k t)\xi_k$. (**Exercise:** Prove the estimate (7.1.8). Verify that it is sharp.)
(d) *Let $c_\gamma := \sup_t \sup_{|v|=1} \langle R(v, \dot{\gamma}(t))\dot{\gamma}(t), v \rangle$. Then, for all $X \in \mathrm{Vect}_0(\gamma)$,*

$$\int_0^1 \Big(|\nabla X|^2 - \langle R(X, \dot{\gamma})\dot{\gamma}, X \rangle \Big) \, dt \geq (\pi^2 - c_\gamma) \int_0^1 |X|^2 \, dt$$

   This follows directly from the Poincaré inequality in (c).
(e) *Let λ be an eigenvalue of $\mathcal{H}_\gamma$. Then $\lambda \geq \pi^2 - c_\gamma$.*
   There exists a nonzero element $X \in \mathrm{Vect}_0(\gamma)$ such that $\mathcal{H}_\gamma X = \lambda X$. By (d) this implies $\lambda \|X\|_{L^2}^2 = \langle X, \mathcal{H}_\gamma X \rangle_{L^2} \geq (\pi^2 - c_\gamma)\|X\|_{L^2}^2$ and so $\lambda \geq \pi^2 - c_\gamma$.
(f) *The set $\sigma(\mathcal{H}_\gamma)$ of eigenvalues of $\mathcal{H}_\gamma$ is a discrete subset of $\mathbb{R}$.*
   Let $\lambda \in \sigma(\mathcal{H}_\gamma)$ and let Π_λ be the orthogonal projection onto the eigenspace of λ. Then the operator $\mathcal{H}_\gamma - \lambda\mathrm{id} + \Pi_\lambda : \mathrm{Vect}_0(\gamma) \to \mathrm{Vect}(\gamma)$ is bijective, and hence so is the operator $\mathcal{H}_\gamma - \lambda'\mathrm{id} + \Pi_\lambda$ for λ' sufficently close to λ. Such a number λ' cannot be an eigenvalue of $\mathcal{H}_\gamma$. This argument uses the fact that $\mathcal{H}_\gamma$ is a Fredholm operator between appropriate Sobolev completions.
(g) *The smallest eigenvalue of $\mathcal{H}_\gamma$ is the supremum of all real numbers a that satisfy the inequality*

$$\int_0^1 \Big(|\nabla X|^2 - \langle R(X, \dot{\gamma})\dot{\gamma}, X \rangle \Big) \, dt \geq a \int_0^1 |X(t)|^2 \, dt \tag{7.1.9}$$

   for all $X \in \mathrm{Vect}_0(\gamma)$.

To prove this, define $a \in \mathbb{R}$ to be the infimum of the integrals on the left in (7.1.9) over all $X \in \mathrm{Vect}_0(\gamma)$ with $\|X\|_{L^2} = 1$. This infimum is finite by (d). Now choose a minimizing sequence $X_i \in \mathrm{Vect}_0(\gamma)$ with $\|X_i\|_{L^2} = 1$. This sequence satisfies a uniform upper bound on $\|\nabla X_i\|_{L^2}$. Hence, by the theorems of Banach–Alaoglu and Arzelà-Ascoli, there exists a subsequence that converges weakly in the Sobolev space $W^{1,2}$ and strongly in the supremum norm to a weak solution of the equation $-\nabla\nabla X - R(X, \dot\gamma)\dot\gamma = aX$ with zero boundary condition. Now one can verify by a bootstrapping argument that every weak solution is smooth. Hence (7.1.9) holds, a is an eigenvalue of $\mathcal{H}_\gamma$, and every eigenvalue λ of $\mathcal{H}_\gamma$ satisfies $\lambda \geq a$.

(h) *The Morse index of γ is finite.*

The operator $\mathcal{H}_\gamma$ has only finitely many negative eigenvalues by (e) and (f), and the direct sum X_γ of their eigenspaces is finite-dimensional by (b). Moreover, every nonzero element $X \in X_\gamma$ satisfies the inequality $\langle X, \mathcal{H}_\gamma X\rangle_{L^2} < 0$. One can then repeat the argument sketched in (g) for the L^2 orthogonal complement of X_γ to show that $\langle X, \mathcal{H}_\gamma X\rangle_{L^2} \geq 0$ for all $X \in X_\gamma^\perp$. Hence the dimension of X_γ is the Morse index of γ (Definition 7.1.5).

Proof of Theorem 7.1.6 Choose a parallel orthonormal frame

$$E_1(t), \ldots, E_m(t) \in T_{\gamma(t)}M$$

and, for $0 \leq t \leq 1$, define the symmetric matrix $S(t) = S(t)^\mathsf{T} \in \mathbb{R}^{m \times m}$ by

$$S(t) := (S_{ij}(t))_{i,j=1}^m, \qquad S_{ij}(t) := \langle R(E_i(t), \dot\gamma(t))\dot\gamma(t), E_j(t)\rangle,$$

Let $I := [0, 1]$ and define the operator $A : C_0^\infty(I, \mathbb{R}^m) \to C^\infty(I.\mathbb{R}^m)$ with the domain $C_0^\infty(I, \mathbb{R}^m) := \{\xi \in C^\infty(I, \mathbb{R}^m) \mid \xi(0) = \xi(1) = 0\}$, by

$$A\xi := -\ddot\xi - S\xi.$$

This operator is isomorphic to the operator (7.1.1) via the isomorphism that sends $\xi \in C_0^\infty(I, \mathbb{R}^m)$ to $X := \sum_i \xi_i E_i \in \mathrm{Vect}_0(\gamma)$. Now define a family of operators $A_\tau : C_0^\infty(I, \mathbb{R}^m) \to C^\infty(I.\mathbb{R}^m)$ by

$$(A_\tau\xi)(t) := -\frac{1}{\tau^2}\ddot\xi(t) - S(\tau t)\xi(t)$$

for $0 < \tau \leq 1$. Then A_τ is isomorphic to the operator (7.1.1) on the interval $[0, \tau]$ via the isomorphism that sends $\xi \in C_0^\infty(I, \mathbb{R}^m)$ to the vector field $X(t) := \sum_i \xi(\tau^{-1}t)E_i(t) \in T_{\gamma(t)}M$, $0 \leq t \leq \tau$, along the curve $\gamma|_{[0,\tau]}$. Thus τ is a conjugate point of γ if and only if A_τ has a nontrivial kernel, and the dimension of the kernel is the multiplicity $m_\gamma(\tau)$.

By Remark 7.1.9 A_τ has only positive eigenvalues for small positive τ. For $0 < \tau \leq 1$ define the operator $\dot A_\tau : C_0^\infty(I, \mathbb{R}^m) \to C^\infty(I, \mathbb{R}^m)$ by

$$(\dot A_\tau\xi)(t) := \frac{d}{d\tau}(A_\tau\xi)(t) = \frac{2}{\tau^3}\ddot\xi(t) - t\dot S(\tau t)\xi(t).$$

If τ is a conjugate point, we claim that every element $\xi \in \ker(A_\tau)$ satisfies

$$\Gamma_\tau(\xi) := \int_0^1 \langle \xi, \dot{A}_\tau \xi \rangle \, dt = -\frac{1}{\tau^3} |\dot{\xi}(1)|^2. \tag{7.1.10}$$

To see this, note that $\ddot{\xi}(t) + \tau^2 S(\tau t)\xi(t) = 0$ and $\frac{d}{dt}S(\tau t) = \tau \dot{S}(\tau t)$. Hence

$$\Gamma_\tau(\xi) = \int_0^1 \left(\frac{2}{\tau^3} \langle \xi(t), \ddot{\xi}(t) \rangle - \frac{t}{\tau} \langle \xi(t), \tau \dot{S}(\tau t)\xi(t) \rangle \right) dt$$

$$= \int_0^1 \left(\frac{2}{\tau^3} \langle \xi(t), \ddot{\xi}(t) \rangle + \frac{1}{\tau} \langle \xi(t), S(\tau t)\xi(t) \rangle + \frac{2t}{\tau} \langle \dot{\xi}(t), S(\tau t)\xi(t) \rangle \right) dt$$

$$= \int_0^1 \left(\frac{1}{\tau^3} \langle \xi(t), \ddot{\xi}(t) \rangle - \frac{2t}{\tau^3} \langle \dot{\xi}(t), \ddot{\xi}(t) \rangle \right) dt$$

$$= -\frac{1}{\tau^3} \int_0^1 \left(|\dot{\xi}(t)|^2 + 2t \langle \dot{\xi}(t), \ddot{\xi}(t) \rangle \right) dt$$

$$= -\frac{1}{\tau^3} \int_0^1 \frac{d}{dt} \left(t |\dot{\xi}(t)|^2 \right) dt$$

$$= -\frac{1}{\tau^3} |\dot{\xi}(1)|^2.$$

This proves (7.1.10). Note also that $\Gamma_\tau(\xi) < 0$ unless $\xi(t) = 0$ for all t.

With this understood, we appeal to the Kato Selection Theorem [35, Theorem II.5.4 and Theorem II.6.8]. It asserts in the case at hand that, near each point τ_0 with $k := \dim(\ker(A_{\tau_0})) > 0$, there exist k continuously differentiable functions $\lambda_i : (\tau_0 - \varepsilon, \tau_0 + \varepsilon) \to \mathbb{R}$ such that $\lambda_i(\tau_0) = 0$, the numbers $\lambda_1(\tau), \ldots, \lambda_k(\tau)$ are the eigenvalues of A_τ near zero, with multiplicities accounted for by repetitions, and the derivatives $\dot{\lambda}_i(\tau_0)$ are the eigenvalues of the crossing form $\Gamma_{\tau_0} : \ker A_{\tau_0} \to \mathbb{R}$ in (7.1.10), again with multiplicities accounted for by repetitions. The derivatives are all negative by (7.1.10). Hence there exists a $\delta > 0$ such that $\lambda_i(\tau) > 0$ for $\tau_0 - \delta < \tau < \tau_0$ and $\lambda_i(\tau) < 0$ for $\tau_0 < \tau < \tau_0 + \delta$. Hence conjugate points are isolated, and for $\tau_0 < \tau < \tau_0 + \delta$ the operator A_τ has precisely k more negative eigenvalues than for $\tau_0 - \delta < \tau < \tau_0$. Hence the number of the negative eigenvalues of A_1 with multiplicities is $\sum_{0 < \tau < 1} \dim(\ker(A_\tau))$. This proves Theorem 7.1.6. $\qquad \square$

For more detailed explanations and other closely related applications of the Kato Selection Theorem the reader is referred to [61, 76].

7.1.3 Locally Minimal Geodesics

Proof of Theorem 7.1.2   We prove part (i). Let $X \in \mathrm{Vect}_0(M)$ and define

$$\gamma_s(t) := \exp_{\gamma(t)}(sX(t))$$

for $0 \le t \le 1$ and $-\delta < s < \delta$. Here $\delta > 0$ is chosen so small that $sX(t)$ is contained in the domain $V_{\gamma(t)} \subset T_{\gamma(t)}M$ of the exponential map for all t and all s with $|s| < \delta$. Then $\lim_{s \to 0} d_{C^\infty}(\gamma, \gamma_s) = 0$ and so $\gamma_s \in \mathcal{U}$ for s sufficiently small. Hence the function $(-\delta, \delta) \to \mathbb{R} : s \mapsto L(\gamma_s)$ has a local minimum at $s = 0$, and since $E(\gamma) = L(\gamma)^2/2$ and $L(\gamma_s)^2/2 \le E(\gamma_s)$, so does the function $s \mapsto E(\gamma_s)$. By Lemma 6.6.3 this implies

$$0 \le \left.\frac{d^2}{ds^2}\right|_{s=0} E(\gamma_s) = \int_0^1 \left(|\nabla X|^2 - \langle R(X, \dot{\gamma})\dot{\gamma}, X \rangle \right) dt.$$

This inequality shows that the operator $\mathcal{H}_\gamma$ has no negative eigenvalues. Hence, by Theorem 7.1.6 the geodesic γ has no conjugate points τ in the interval $0 < \tau < 1$. This proves (i).

We prove part (ii) in six steps. Let $\gamma \in \Omega_{p,q}$ be a nonconstant geodesic without conjugate points.

Step 1 *There exist constants $\delta_1 > 0$ and $\varepsilon > 0$ such that*

$$\int_0^1 \left(|\nabla X|^2 - \langle R(X, \dot{\gamma}')\dot{\gamma}', X \rangle \right) dt \ge \varepsilon \int_0^1 \left(|\nabla X(t)|^2 + |X(t)|^2 \right) dt \qquad (7.1.11)$$

for every curve $\gamma' \in \Omega_{p,q}$ with $d_{C^1}(\gamma, \gamma') < \delta_1$ and every $X \in \mathrm{Vect}_0(\gamma')$.

By Theorem 7.1.6 the Hessian $\mathcal{H}_\gamma$ has no negative eigenvalues and, since 1 is not a conjugate point of γ, also zero is not an eigenvalue of $\mathcal{H}_\gamma$. Hence the smallest eigenvalue of $\mathcal{H}_\gamma$ is positive and so, by part (e) of Remark 7.1.9, the estimate (7.1.9) holds for all $X \in \mathrm{Vect}_0(\gamma)$ with a positive constant a. If the constant $b > 0$ is chosen such that $\langle R(v, \dot{\gamma}(t))\dot{\gamma}(t), v \rangle \le b|v|^2$ for all t and all $v \in T_{\gamma(t)}M$, we obtain the estimate (7.1.11) for $\gamma' = \gamma$ and $X \in \mathrm{Vect}_0(\gamma)$ with $\varepsilon := a/(a + b + 1)$. (**Exercise:** Verify this.) In a local coordinat chart on M the integrand on the left in (7.1.11) has the form

$$\sum_{k,\ell} \left(\dot{\xi}^k + \sum_{i,j} \Gamma_{ij}^k \xi^i \dot{c}^j \right) g_{k\ell} \left(\dot{\xi}^\ell + \sum_{\mu,\nu} \Gamma_{\mu\nu}^\ell \xi^\mu \dot{c}^\nu \right) - \sum_{i,j,k,\ell} R_{ijk\ell} \xi^i \dot{c}^j \dot{c}^k \xi^\ell,$$

where $c(t) := \phi(\gamma'(t))$ and $\xi(t) = d\phi(\gamma'(t))X(t)$ (and the coefficients g_{ij}, Γ_{ij}^k, and $R_{ijk\ell}$ are functions of $c(t)$). This expression depends continuously on the curve c and its derivative $\dot{c}$. Hence there exists a constant $\delta_1 > 0$ such that every curve $\gamma' \in \Omega_{p,q}$ with $d_{C^1}(\gamma, \gamma') < \delta_1$ and every $X \in \mathrm{Vect}_0(\gamma')$ satisfies the estimate (7.1.11) with $\varepsilon = a/2(a + b + 1)$. This proves Step 1.

Step 2 *Let $\delta_1 > 0$ be as in Step 1. Then there exists a constant $\delta_2 > 0$ with the following significance. If $Y \in \mathrm{Vect}_0(\gamma)$ satisfies*

$$\sup_{0 \leq t \leq 1} \left(|Y(t)| + |\nabla Y(t)| \right) < \delta_2, \tag{7.1.12}$$

then the curve $\gamma' := \exp_\gamma(Y) \in \Omega_{p,q}$ satisfies $d_{C^1}(\gamma, \gamma') < \delta_1$.

For $0 \leq t \leq 1$ define the map $f_t : V_{\gamma(t)} \times T_{\gamma(t)}M \to TM$ by

$$f_t(v, \widehat{v}) := \left(\exp(\gamma(t), v), d \exp(\gamma(t), v)(\dot{\gamma}(t), \widehat{v} + h_{\gamma(t)}(\dot{\gamma}(t), v)) \right) \tag{7.1.13}$$

for $v \in V_{\gamma(t)}$ and $\widehat{v} \in T_{\gamma(t)}M$. By Lemma 4.3.6 the map $(t, v, \widehat{v}) \mapsto f_t(v, \widehat{v})$ from the set $U := \{(t, v, \widehat{v}) \,|\, 0 \leq t \leq 1,\, v \in V_{\gamma(t)},\, \widehat{v} \in T_{\gamma(t)}M\}$ to TM is smooth and it satisfies $f_t(0, 0) = (\gamma(t), \dot{\gamma}(t))$. Thus there exist constants $r_1 > 0$ and $C_1 \geq 1$ such that all t and all $v, \widehat{v} \in T_{\gamma(t)}M$ with $|v| + |\widehat{v}| \leq r_1$ satisfy $v \in V_{\gamma(t)}$ and

$$|p - \gamma(t)| + |\widehat{p} - \dot{\gamma}(t)| \leq C_1\left(|v| + |\widehat{v}| \right), \qquad (p, \widehat{p}) := f_t(v, \widehat{v}). \tag{7.1.14}$$

Choose $Y \in \mathrm{Vect}_0(\gamma)$ such that $|Y(t)| + |\nabla Y(t)| \leq r_1$ for $0 \leq t \leq 1$ and define $\gamma'(t) := \exp_{\gamma(t)}(Y(t))$. Then it follows from the Gauß–Weingarten formula in Lemma 3.2.3 that $(\gamma'(t), \dot{\gamma}'(t)) = f_t(Y(t), \nabla Y(t))$ for all t. Hence the estimate (7.1.14) shows that Step 2 holds with $\delta_2 := \min\{r_1, \delta_1/C_1\}$.

Step 3 *Let $\delta_2 > 0$ be as in Step 2. Then there exists a constant $\delta_3 > 0$ with the following significance. If $\gamma' \in \Omega_{p,q}$ satisfies the inequality $d_{C^1}(\gamma, \gamma') < \delta_3$, then there exists a unique vector field $Y \in \mathrm{Vect}_0(\gamma)$ satisfying (7.1.12) and*

$$|Y(t)| < \mathrm{inj}(\gamma(t); M), \qquad \gamma'(t) = \exp_{\gamma(t)}(Y(t)) \tag{7.1.15}$$

for $0 \leq t \leq 1$.

Define $\rho := \inf_{0 \leq t \leq 1} \mathrm{inj}(\gamma(t); M) > 0$ and let f_t be the map in (7.1.13). Its derivative at the origin is given by $df_t(0, 0)(\eta, \widehat{\eta}) = (\eta, \widehat{\eta} + h_{\gamma(t)}(\dot{\gamma}(t), \eta))$ and so is invertible. Hence, by the implicit function theorem, there exist constants $r_2 > 0$ and $C_2 \geq 1$ such that, for $0 \leq t \leq 1$, the set

$$Q_t := \{(p, \widehat{p}) \in TM \,|\, |p - \gamma(t)| + |\widehat{p} - \dot{\gamma}(t)| \leq r_2\}$$

is compact, f_t restricts to a diffeomorphism from a compact neighborhood of the origin in $B_\rho(\gamma(t)) \times T_{\gamma(t)}M$ to Q_t, and every pair $(p, \widehat{p}) \in Q_t$ satisfies

$$|v| + |\widehat{v}| \leq C_2\left(|p - \gamma(t)| + |\widehat{p} - \dot{\gamma}(t)| \right), \qquad (v, \widehat{v}) := f_t^{-1}(p, \widehat{p}). \tag{7.1.16}$$

Choose $\gamma' \in \Omega_{p,q}$ such that $d_{C^1}(\gamma, \gamma') \le r_2$. Then $(\gamma'(t), \dot{\gamma}'(t)) \in Q_t$ for all t and hence there exists a unique vector field $Y \in \text{Vect}_0(\gamma)$ satisfying (7.1.15). By (7.1.13) it also satisfies $(Y(t), \nabla Y(t)) = f_t^{-1}(\gamma'(t), \dot{\gamma}'(t))$ for all t. Hence the estimate (7.1.16) shows that Step 3 holds with $\delta_3 := \min\{r_2, \delta_2/C_2\}$.

Step 4 *Let δ_3 be as in Step 3. Then there exists a constant $\delta_4 > 0$ with the following significance. If $\gamma' \in \Omega_{p,q}$ satisfies $d_{C^1}(\gamma, \gamma') < \delta_4$, then the map $\rho : [0,1] \to [0,1]$ defined by $\rho(t) := L(\gamma')^{-1} \int_0^t |\dot{\gamma}'(s)| \, ds$ for $0 \le t \le 1$ is a diffeomorphism and $d_{C^1}(\gamma, \gamma' \circ \rho^{-1}) < \delta_3$.*

Define $c := L(\gamma)$ and $C := \sup_t |\ddot{\gamma}(t)|$. Then $c > 0$ because γ is nonconstant. We claim that Step 4 holds for every constant $\delta_4 > 0$ that satisfies

$$\delta_4 < \frac{c}{2}, \qquad \left(32 + \frac{12C}{c}\right)\delta_4 < \delta_3. \tag{7.1.17}$$

To see this, fix a constant $\delta_4 > 0$ that satisfies (7.1.17) and let $\gamma' \in \Omega_{p,q}$ such that $d_{C^1}(\gamma, \gamma') < \delta_4$. Since the geodesic γ is parametrized by the arclength, it satisfies $|\dot{\gamma}(t)| = L(\gamma) = c$. Hence $||\dot{\gamma}'(t)| - c| < \delta_4$ for all t. Since $\delta_4 < c$ by (7.1.17), this implies that ρ is a diffeomorphism. It also implies the length inequality $|L(\gamma') - c| < \delta_4$ and hence

$$\frac{1}{3} < \frac{c - \delta_4}{c + \delta_4} < \dot{\rho}(t) = \frac{|\dot{\gamma}'(t)|}{L(\gamma')} < \frac{c + \delta_4}{c - \delta_4} < 3, \qquad |\dot{\rho}(t) - 1| < \frac{2\delta_4}{c - \delta_4} < \frac{4\delta_4}{c}.$$

Define $\beta := \rho^{-1}$ and $\gamma'' := \gamma' \circ \beta$. Then $\dot{\beta}(\rho(t)) = 1/\dot{\rho}(t)$ and hence

$$|\dot{\beta}(\rho(t)) - 1| = \frac{|1 - \dot{\rho}(t)|}{\dot{\rho}(t)} < \frac{12\delta_4}{c}.$$

This implies

$$|\dot{\beta}(t) - 1| < \frac{12\delta_4}{c}, \qquad |\beta(t) - t| < \frac{12\delta_4}{c}, \qquad \dot{\beta}(t) < 7.$$

Hence

$$\begin{aligned}
|\gamma(t) - \gamma''(t)| &\le |\gamma(t) - \gamma(\beta(t))| + |\gamma(\beta(t)) - \gamma'(\beta(t))| \\
&\le c|t - \beta(t)| + d_{C^1}(\gamma, \gamma') \\
&< 13\delta_4
\end{aligned}$$

and

$$\begin{aligned}
|\dot{\gamma}(t) - \dot{\gamma}''(t)| &\le |\dot{\gamma}(t) - \dot{\gamma}(\beta(t))| + |1 - \dot{\beta}(t)||\dot{\gamma}(\beta(t))| \\
&\quad + \dot{\beta}(t)|\dot{\gamma}(\beta(t)) - \dot{\gamma}'(\beta(t))| \\
&\le C|t - \beta(t)| + c|1 - \dot{\beta}(t)| + \dot{\beta}(t)d_{C^1}(\gamma, \gamma') \\
&< \left(\frac{12C}{c} + 12 + 7\right)\delta_4.
\end{aligned}$$

These inequalities imply $d_{C^1}(\gamma, \gamma'') < (12C/c + 32)\delta_4 < \delta_3$ by (7.1.17) and this proves Step 4.

Step 5 *Let $\delta_3 > 0$ be as in Step 3 and let $\gamma' \in \Omega_{p,q}$ such that $d_{C^1}(\gamma, \gamma') < \delta_3$. Then $E(\gamma') \geq E(\gamma)$ with equality if and only if $\gamma' = \gamma$.*

Assume $\gamma' \neq \gamma$. By Step 3 there exists a nonzero vector field $Y \in \mathrm{Vect}_0(\gamma)$ satisfying (7.1.12), (7.1.15), and $\gamma' = \exp_\gamma(Y)$. For $0 \leq s, t \leq 1$ define

$$\gamma_s(t) := \exp_{\gamma(t)}(sY(t)).$$

Then $\gamma_0 = \gamma$, $\gamma_1 = \gamma'$, and Step 2 asserts that

$$d_{C^1}(\gamma, \gamma_s) < \delta_1 \qquad \text{for } 0 \leq s \leq 1.$$

Hence it follows from the argument in the proof of Lemma 6.6.3 that

$$\frac{d^2}{ds^2} E(\gamma_s) = \int_0^1 \left(|\nabla_t \partial_s \gamma_s|^2 - \langle R(\partial_s \gamma_s, \partial_t \gamma_s) \partial_t \gamma_s, \partial_s \gamma_s \rangle \right) dt > 0$$

for all s. Here the equality uses the identity $\nabla_s \partial_s \gamma_s = 0$ and the inequality follows from Step 1 and the fact that $\partial_s \gamma_s = d\exp_\gamma(sY)Y$ is a nonzero vector field along γ_s. Thus the curve $[0, 1] \to \mathbb{R} : s \mapsto E(\gamma_s)$ is strictly convex. Since its derivative vanishes at $s = 0$, it follows that $E(\gamma_s) > E(\gamma)$ for $0 < s \leq 1$. Since $\gamma_1 = \gamma'$, this proves Step 5.

Step 6 *Let δ_4 be as in Step 4. Then the C^1-open set*

$$\mathcal{U} := \{\gamma' \in \Omega_{p,q} \mid d_{C^1}(\gamma, \gamma') < \delta_4\}$$

satisfies the requirements of part (ii) in Theorem 7.1.2.

Let $\gamma' \in \mathcal{U}$ and define the diffeomorphism $\rho : [0, 1] \to [0, 1]$ as in Step 4. Then, by Step 4 we have $d_{C^1}(\gamma, \gamma' \circ \rho^{-1}) < \delta_3$ and hence, by Step 5 this implies $E(\gamma' \circ \rho^{-1}) \geq E(\gamma)$ with equality if and only if $\gamma' \circ \rho^{-1} = \gamma$. Since γ and $\gamma' \circ \rho^{-1}$ are parametrized by the arclength, we find that

$$L(\gamma') = L(\gamma' \circ \rho^{-1}) = \sqrt{2E(\gamma' \circ \rho^{-1})} \geq \sqrt{2E(\gamma)} = L(\gamma),$$

and that equality holds if and only if $\gamma' \circ \rho^{-1} = \gamma$. Thus every curve $\gamma' \in \mathcal{U}$ that arises from γ by reparametrization has the same length as γ and every other curve in $\mathcal{U}$ is strictly longer. This proves Step 6 and Theorem 7.1.2. $\square$

We remark that there is a precise analogy between Theorem 7.1.2 about geodesics (extrema of the energy functional E) and extrema of a function

$$f : \mathbb{R}^n \to \mathbb{R}$$

on a finite-dimensional vector space. If the function f has a local minimum at a point $x_0 \in \mathbb{R}^n$, then the Hessian of f at x_0 has no negative eigenvalues and, conversely, if x_0 is a critical point of f and all the eigenvalues of the Hessian at x_0 are positive, then x_0 is a strict local minimum of f.

7.2 The Injectivity Radius*

Assume throughout that $M \subset \mathbb{R}^n$ is a smooth m-manifold. In this section we prove that the function which assigns to each point $p \in M$ its injectivity radius $\mathrm{inj}(p; M) \in (0, \infty]$ is continuous. Recall the concept of a local diffeomorphism as a smooth map between manifolds of the same dimension whose derivative at each point is a vector space isomorphism (Definition 2.2.20). Recall also the notation $V_p \subset T_p M$ for the domain of the exponential map at p and the notation $B_r(p) = \{v \in T_p M \mid |v| < r\}$ for $p \in M$ and $r > 0$.

Theorem 7.2.1 *The function $M \to (0, \infty] : p \mapsto \mathrm{inj}(p; M)$ is continuous.*

The proof of Theorem 7.2.1 is based on two lemmas.

Lemma 7.2.2 *The set $U_r := \{p \in M \mid r < \mathrm{inj}(p; M)\}$ is open for each real number $r > 0$.*

Proof Let $r > 0$ and $p_0 \in U_r$. We prove in three steps that there exists a $\delta > 0$ such that $r + \delta \le \mathrm{inj}(p; M)$ for every $p \in M$ with $d(p, p_0) < \delta$.

Step 1 *There exists a $\delta > 0$ such that $B_{r+\delta}(p) \subset V_p$ for every $p \in M$ with $d(p, p_0) < \delta$.*

Suppose, by contradiction, that such a constant δ does not exist. Then there exist sequences $p_i \in M$ and $v_i \in T_{p_i} M \setminus V_i$ such that $d(p_i, p_0) < 1/i$ and $|v_i| < r + 1/i$. Passing to a subsequence, if necessary, we may assume that v_i converges to a vector $v_0 \in T_{p_0} M$. Then $|v_0| \le r$ and so $(p_0, v_0) \in V$. Since $(p_i, v_i) \in TM \setminus V$ converges to (p_0, v_0), this contradicts the fact that V is open. This proves Step 1.

Step 2 *There exists a $\delta > 0$ such that the map $\exp_p : B_{r+\delta}(p) \to M$ is a local diffeomorphism for every $p \in M$ with $d(p, p_0) < \delta$.*

Suppose, by contradiction, that such a constant δ does not exist. Then there exist sequences $p_i \in M$ and $v_i \in T_{p_i} M$ such that $d(p_i, p_0) < 1/i$ and $|v_i| < r + 1/i$, and the derivative $d\exp_{p_i}(v_i) : T_{p_i} M \to T_{\exp_{p_i}(v_i)} M$ is not injective. Passing to a subsequence, if necessary, we may assume that v_i converges to a vector $v_0 \in T_{p_0} M$. Then, by smoothness of the exponential map, the derivative

$$d\exp_{p_0}(v_0) : T_{p_0} M \to T_{\exp_{p_0}(v_0)} M$$

is not injective. Since $|v_0| \le r$, this contradicts the fact that $r < \mathrm{inj}(p_0; M)$. This proves Step 2.

Step 3 *There exists a $\delta > 0$ such that the map* $\exp_p : B_{r+\delta}(p) \to M$ *is injective for every* $p \in M$ *with* $d(p, p_0) < \delta$.

Suppose, by contradiction, that such a constant δ does not exist. Then there exist sequences $p_i \in M$ and $u_i, v_i \in T_{p_i} M$ such that

$$d(p_i, p_0) < 1/i, \qquad |u_i| < r + 1/i, \qquad |v_i| < r + 1/i$$

and

$$u_i \ne v_i, \qquad \exp_{p_i}(u_i) = \exp_{p_i}(v_i) =: q_i.$$

Passing to a subsequence, if necessary, we may assume that the limits

$$u_0 = \lim_{i \to \infty} u_i \in T_{p_0} M, \qquad v_0 := \lim_{i \to \infty} v_i \in T_{p_0} M$$

exist. These limits satisfy

$$|u_0| \le r, \qquad |v_0| \le r, \qquad \exp_{p_0}(u_0) = \exp_{p_0}(v_0).$$

Since $r < \mathrm{inj}(p_0; M)$, this implies $u_0 = v_0$ and hence

$$\lim_{i \to \infty} u_i = \lim_{i \to \infty} v_i = v_0.$$

Now define

$$w_i := \frac{v_i - u_i}{\tau_i} \in T_{p_i} M, \qquad \tau_i := |v_i - u_i| > 0.$$

Passing to a further subsequence, we may assume that the limit

$$w_0 := \lim_{i \to \infty} w_i \in T_{p_0} M$$

exists. Then $|w_0| = 1$ and hence

$$\begin{aligned}
0 &= \lim_{i \to \infty} \frac{\exp_{p_i}(v_i) - \exp_{p_i}(u_i)}{\tau_i} \\
&= \lim_{i \to \infty} \int_0^1 d\exp_{p_i}(u_i + t(v_i - u_i)) \frac{v_i - u_i}{\tau_i} \, dt \\
&= d\exp_{p_0}(v_0) w_0 \\
&\ne 0.
\end{aligned}$$

This contradiction proves Step 3. Now choose a constant $\delta > 0$ that satisfies the requirements of Step 1, Step 2, and Step 3. Then, for every $p \in M$ with $d(p, p_0) < \delta$ we have $B_{r+\delta}(p) \subset V_p$ and the map $\exp_p : B_{r+\delta}(p) \to M$ is a diffeomorphism onto its image. Hence $\mathrm{inj}(p; M) \geq r + \delta > r$ for every element $p \in M$ with $d(p, p_0) < \delta$, and this proves Lemma 7.2.2. $\square$

Lemma 7.2.3 *The set $A_r := \{p \in M \mid r \leq \mathrm{inj}(p; M)\}$ is closed for each real number $r > 0$ and hence also for $r = \infty$.*

Proof Let $p_i \in A_r$ be a sequence that converges to an element $p \in M$. We prove in five steps that $r \leq \mathrm{inj}(p; M)$.

Step 1 $B_r(p) \subset V_p$.

Choose a tangent vector $v \in T_p M$ with $|v| < r$, choose a constant $\varepsilon > 0$ such that $|v| + \varepsilon < r$, and choose an integer i such that $d(p, p_i) \leq \varepsilon$. Define

$$K := \exp_{p_i} (\bar{B}_{|v|+\varepsilon}(p_i)) = \{q \in M \mid d(q, p_i) \leq |v| + \varepsilon\}.$$

Here the second equality follows from Theorem 4.4.4 and the fact that $|v| + \varepsilon < r \leq \mathrm{inj}(p_i; M)$. By definition, K is the image of a compact set under a continuous map, and so is a compact subset of M. Hence

$$\widetilde{K} := \{(q, w) \in TM \mid d(q, p_i) \leq |v| + \varepsilon, \; |w| = |v|\}$$

is a compact subset of TM. We claim that $v \in V_p$. Suppose, by contradiction, that this is not the case. Then $I_{p,v} \cap [0, \infty) = [0, T)$ with $0 < T \leq 1$. Denote by $\gamma : [0, T) \to M$ the geodesic $\gamma(t) := \exp_p(tv)$. Then

$$d(\gamma(t), p_i) \leq d(\exp_p(tv), p) + d(p, p_i) \leq |tv| + \varepsilon \leq |v| + \varepsilon$$

and $|\dot{\gamma}(t)| = |v|$, and hence $(\gamma(t), \dot{\gamma}(t)) \in \widetilde{K}$ for $0 \leq t < T$. By Lemma 4.3.3 and Corollary 2.4.15, this implies that there exists a constant $\delta > 0$ such that $[0, T + \delta) \subset I_{p,v}$, in contradiction to the definition of T. This contradiction shows that our assumption that v is not an element of V_p must have been wrong. Thus $v \in V_p$ and this proves Step 1.

Step 2 *Let $q \in M$ such that $d(p, q) < r$. Then there exists a vector $v \in T_p M$ such that $|v| = d(p, q)$ and $\exp_p(v) = q$.*

Since p_i converges to p, we have $\lim_{i \to \infty} d(p_i, q) = d(p, q) < r$ and so there exists an integer $i_0 \in \mathbb{N}$ such that $d(p_i, q) < r$ for all $i \geq i_0$. Then, for each $i \geq i_0$, there exists a tangent vector $v_i \in T_{p_i} M$ that satisfies the conditions

$\exp_{p_i}(v_i) = q$ and $|v_i| = d(p_i, q) < r$ (Theorem 4.4.4). Passing to a subsequence, if necessary, we may assume that the limit $v := \lim_{i\to\infty} v_i$ exists in $\mathbb{R}^n$. Then $v \in T_p M$, $|v| = \lim_{i\to\infty} |v_i| = \lim_{i\to\infty} d(p_i, q) = d(p, q)$, so $v \in B_r(p) \subset V_p$ and $\exp_p(v) = \lim_{i\to\infty} \exp_{p_i}(v_i) = q$. This proves Step 2.

Step 3 *Let* $v \in T_p M$ *such that* $|v| < r$. *Then* $d(p, \exp_p(v)) = |v|$.

Define $q := \exp_p(v)$, so $d(p, q) \leq |v| < r$. Choose a sequence $v_i \in T_{p_i} M$ such that $|v_i| = |v|$ for all i and $\lim_{i\to\infty} v_i = v$. Then the sequence $q_i := \exp_{p_i}(v_i)$ converges to $q = \exp_p(v)$ and satisfies $d(p_i, q_i) = |v_i|$ by Theorem 4.4.4. Hence $d(p, q) = \lim_{i\to\infty} d(p_i, q_i) = \lim_{i\to\infty} |v_i| = |v|$ and this proves Step 3.

Step 4 *The map* $\exp_p : B_r(p) \to M$ *is a local diffeomorphism.*

Let $v \in B_r(p)$ and let $\hat{v} \in T_p M$ be a nonzero vector. Choose any real number $1 < \lambda < r/|v|$ so that $|\lambda v| < r$ and define the geodesic $\gamma : [0, \lambda] \to M$ and the vector field $X \in \text{Vect}(\gamma)$ by

$$\gamma(t) := \exp_p(tv), \qquad X(t) := \frac{\partial}{\partial s}\bigg|_{s=0} \exp_p\big(t(v + s\hat{v})\big) = d\exp_p(tv)t\hat{v}$$

for $0 \leq t \leq \lambda$. By Lemma 6.1.18, X is a Jacobi field along γ and

$$X(0) = 0, \qquad \nabla X(0) = \hat{v} \neq 0, \qquad X(1) = d\exp_p(v)\hat{v}.$$

Since $L(\gamma) = d(p, \exp_p(\lambda v)) = d(\gamma(0), \gamma(\lambda))$ by Step 3, it follows from part (i) of Theorem 7.1.2 that the geodesic $\gamma : [0, \lambda] \to M$ has no conjugate points τ in the open interval $0 < \tau < \lambda$. In particular, $\tau = 1$ is not a conjugate point, and so $X(1) \neq 0$. Hence the derivative $d\exp_p(v)$ of the exponential map is bijective at every point $v \in B_r(p)$ and this proves Step 4.

Step 5 *The map* $\exp_p : B_r(p) \to M$ *is injective.*

This is a covering argument. Let $v_0, v_1 \in B_r(p)$ with $\exp_p(v_0) = \exp_p(v_1)$, choose a smooth path $v : [0, 1] \to B_r(p)$ such that $v(0) = v_0$ and $v(1) = v_1$, and define $\gamma(t) := \exp_p(v(t))$ for $0 \leq t \leq 1$, so $\gamma(0) = \gamma(1) = q := \exp_p(v_0)$. Choose $\rho < r$ and $i \in \mathbb{N}$ such that $|v(t)| < \rho$ for all t and $d(p_i, p) < r - \rho$. Then $d(p_i, \gamma(t)) \leq d(p_i, p) + d(p, \gamma(t)) < d(p_i, p) + \rho < r$ for all t. Define

$$\beta(s, t) := \exp_{p_i}\Big(s\exp_{p_i}^{-1}(q) + (1 - s)\exp_{p_i}^{-1}(\gamma(t))\Big), \qquad 0 \leq s, t \leq 1.$$

This map takes values in the set $U_r(p) = \{p' \in M \mid d(p, p') < r\}$ and satisfies

$$\beta(0, t) = \gamma(t) = \exp_p(v(t)), \qquad \beta(1, t) = q \qquad \text{for } 0 \leq t \leq 1,$$

and $\beta(s, 0) = \beta(s, 1) = q$ for all s. Since the map $\exp_p : B_r(p) \to U_r(p)$ is surjective by Step 2 and a local diffeomorphism by Step 4, a path lifting argument shows that there exists a smooth map $u : [0, 1]^2 \to B_r(p)$ such that

$$u(0, t) = v(t), \qquad \exp_p(u(s, t)) = \beta(s, t)$$

for $0 \le s, t \le 1$. This map satisfies $u(s, 0) = v_0$ and $u(s, 1) = v_1$ for all s and, moreover, the curve $t \mapsto u(1, t)$ must be constant. Hence $v_0 = v_1$. This proves.Step 5 and Lemma 7.2.3.   □

Proof of Theorem 7.2.1   The set $\{p \in M \mid a < \mathrm{inj}(p; M) < b\} = U_a \setminus A_b$ is open for all nunbers $0 < a < b \le \infty$ by Lemma 7.2.2 and Lemma 7.2.3. Hence the function $M \to (0, \infty] : p \mapsto \mathrm{inj}(p; M)$ is continuous.   □

7.3   The Group of Isometries*

This section is devoted to the Myers–Steenrod Theorem which asserts that the group $I(M)$ of isometries of a connected smooth Riemannian manifold M admits the natural structure of a finite-dimensional Lie group [52].

7.3.1   The Myers–Steenrod Theorem

Assume throughout that $M \subset \mathbb{R}^n$ is a nonempty connected smooth m-manifold. To state the main result, it is convenient to introduce the space

$$G := \left\{ (q, \Phi, p) \,\middle|\, \begin{array}{l} p, q \in M \text{ and } \Phi : T_p M \to T_q M \\ \text{is an orthogonal isomorphism} \end{array} \right\}. \tag{7.3.1}$$

This space is a **groupoid**, i.e. a category in which every morphism is an isomorphism. The space of objects is the manifold M, the morphisms from $p \in M$ to $q \in M$ are the triples of the form $(q, \Phi, p) \in G$, the identity morphism from p to itself is the triple $(p, \mathbb{1}, p)$, the inverse of $(q, \Phi, p) \in G$ is the triple (p, Φ^{-1}, q), and the composition of a morphism $(q, \Phi, p) \in G$ from p to q with a morphism $(r, \Psi, q) \in G$ from q to r is the triple $(r, \Psi\Phi, p)$.

The space G is a smooth manifold (in the intrinsic sense). To see this, consider the diagonal action of the orthogonal group $O(m)$ on the product of the orthonormal frame bundle $\mathcal{O}(M)$ with itself (Definition 3.4.3). This action is free and the map $\pi : \mathcal{O}(M) \times \mathcal{O}(M) \to G$ which sends the pair $((q, e'), (p, e))$ to the triple $(q, e' \circ e^{-1}, p) \in G$ descends to a bijection from the quotient $(\mathcal{O}(M) \times \mathcal{O}(M))/O(m)$ to G. By Theorem 2.9.14 there is a unique smooth structure on G such that the map

$\pi : \mathcal{O}(M) \times \mathcal{O}(M) \to G$ is a submersion. With this structure the maps $s, t : G \to M$ defined by

$$s(q, \Phi, p) := p, \qquad t(q, \Phi, p) := q$$

are smooth, the map $e : M \to G$ defined by

$$e(p) := (p, \mathbb{1}, p)$$

is an embedding, the inverse map $i : G \to G$ defined by

$$i(q, \Phi, p) := (p, \Phi^{-1}, q)$$

is a smooth involution, the map $s \times t : G \times G \to M \times M$ is a submersion, and the composition map $m : (s \times t)^{-1}(\Delta) \to G$ defined by

$$m((r, \Psi, q), (q, \Phi, p)) := (r, \Psi\Phi, p)$$

is smooth. (Here $\Delta := \{(q, q) \mid q \in M\}$ is the diagonal in $M \times M$.) These properties assert that the tuple (G, s, t, e, i, m) is a **smooth groupoid**.

For each $p \in M$ define

$$G_p := \left\{ (q, \Phi) \, \middle| \, \begin{array}{l} q \in M \text{ and } \Phi : T_p M \to T_q M \\ \text{is an orthogonal isomorphism} \end{array} \right\}. \tag{7.3.2}$$

This space is a submanifold of G because $s : G \to M$ is a submersion. The space G_p is also a smooth submanifold of $\mathbb{R}^n \times \mathcal{L}(T_p M, \mathbb{R}^n)$ and is diffeomorphic to the orthonormal frame bundle $\mathcal{O}(M) \subset \mathbb{R}^n \times \mathbb{R}^{n \times m}$ via the map $G_p \to \mathcal{O}(M) :$ $(q, \Phi) \mapsto (q, \Phi \circ e)$ for any orthonormal frame e of $T_p M$. Since M is connected, Lemma 5.1.10 asserts that the map

$$\iota_p : \mathcal{I}(M) \to G_p, \qquad \iota_p(\phi) := (\phi(p), d\phi(p)), \tag{7.3.3}$$

is injective for every $p \in M$. Denote the image of this map by

$$\mathcal{I}_p := \iota_p(\mathcal{I}(M)) = \{(\phi(p), d\phi(p)) \mid \phi \in \mathcal{I}(M)\} \subset G_p. \tag{7.3.4}$$

In the following theorem we do not assume that M is complete. For a space of smooth functions or maps on M we use the term C^∞ *topology* to mean the topology of uniform convergence with all derivatives on each compact subset of M. Likewise we use the term C^0 *topology* to mean the topology of uniform convergence on each compact subset of M. The latter is also called the **compact-open topology** (because a basis of the topology consists of sets of maps, one for each compact subset of the source and each open subset of the target, that send the given compact subset of the source into the given open subset of the target).

Theorem 7.3.1 (Myers–Steenrod) *Let $M \subset \mathbb{R}^n$ be a nonempty connected smooth m-manifold. Then the following holds.*

(i)   *There exists a unique smooth structure on the isometry group $\mathcal{I}(M)$ such that the map $\iota_p : \mathcal{I}(M) \to G_p$ in (7.3.3) is an embedding for every $p \in M$. The topology induced by this smooth structure agrees with the C^0 topology and with the C^∞ topology on $\mathcal{I}(M)$ and $\dim(\mathcal{I}(M)) \le m(m+1)/2$.*
(ii) *With the smooth structure in part (i) the maps*

$$\mathcal{I}(M) \times \mathcal{I}(M) \to \mathcal{I}(M) : (\psi, \phi) \mapsto \psi \circ \phi$$

and

$$\mathcal{I}(M) \to \mathcal{I}(M) : \phi \mapsto \phi^{-1}$$

are smooth. Thus $\mathcal{I}(M)$ is a finite-dimensional Lie group.
(iii) *The Lie algebra of $\mathcal{I}(M)$ is the space $\mathrm{Vect}_{K,c}(M)$ of complete Killing vector fields (defined below).*

Proof See Sect. 7.3.4.   □

7.3.2   The Topology on the Space of Isometries

The next lemma shows that for each $p \in M$ the set $\mathcal{I}_p$ in (7.3.4) is a closed subset of G_p and that the map $\iota_p : \mathcal{I}(M) \to \mathcal{I}_p$ in (7.3.3) is a homeomorphism with respect to the C^∞ topology on $\mathcal{I}(M)$.

Lemma 7.3.2 *Fix two elements $p_0, q_0 \in M$, let $\Phi_0 : T_{p_0}M \to T_{q_0}M$ be an orthogonal isomorphism, and let $\phi_i : M \to M$ be a sequence of isometries. Then the following are equivalent.*

(i)   *The sequence $\iota_{p_0}(\phi_i) \in \mathcal{I}_{p_0}$ converges to the pair $(q_0, \Phi_0) \in G_{p_0}$, i.e.*

$$\lim_{i \to \infty} \phi_i(p_0) = q_0, \qquad \lim_{i \to \infty} d\phi_i(p_0) = \Phi_0. \qquad (7.3.5)$$

(ii) *There exists an isometry $\phi : M \to M$ such that $\phi(p_0) = q_0$, $d\phi(p_0) = \Phi_0$, and ϕ_i converges to ϕ in the C^∞ topology.*

Proof That (ii) implies (i) follows directly from the definitions. We prove in three steps that (i) implies (ii). For $p \in M$ and $r > 0$ denote by $V_p \subset T_pM$ the domain of the exponential map of M at p (Definition 4.3.5), and define $U_r(p) := \{q \in M \mid d(p,q) < r\}$ and $B_r(p) := \{v \in T_pM \mid |v| < r\}$.

Step 1 *Assume* (7.3.5) *holds and choose a real number* $0 < r \leq \mathrm{inj}(p_0; M)$. *Then* $r \leq \mathrm{inj}(q_0; M)$ *and* ϕ_i *converges on the open set* $U_r(p_0)$ *in the* C^∞ *topology to the isometry* $\phi_0 := \exp_{q_0} \circ \, \Phi_0 \circ \exp_{p_0}^{-1} : U_r(p_0) \to U_r(q_0)$.

Since $r \leq \mathrm{inj}(p_0; M)$, Corollary 5.3.3 asserts that

$$B_r(\phi_i(p_0)) = d\phi_i(p_0) B_r(p_0) \subset d\phi_i(p_0) V_{p_0} \subset V_{\phi_i(p_0)}$$

and $\phi_i \circ \exp_{p_0} = \exp_{\phi_i(p_0)} \circ \, \Phi_i : B_r(p_0) \to U_r(\phi_i(p_0))$. Thus the map

$$\exp_{\phi_i(p_0)} = \phi_i \circ \exp_{p_0} \circ \, \Phi_i^{-1} : B_r(\phi_i(p_0)) \to U_r(\phi_i(p_0))$$

is a diffeomorphism and so $r \leq \mathrm{inj}(\phi_i(p_0); M)$ for each $i \in \mathbb{N}$. Since $\phi_i(p_0)$ converges to q_0, this implies $r \leq \mathrm{inj}(q_0; M)$ (Lemma 7.2.3). Hence the sequence of maps $\exp_{\phi_i(p_0)} \circ \, \Phi_i$ converges to the map $\exp_{q_0} \circ \, \Phi_0$ in the C^∞ topology on $B_r(p_0)$. Hence the sequence of isometries

$$\phi_i = \exp_{\phi_i(p_0)} \circ \, \Phi_i \circ \exp_{p_0}^{-1} : U_r(p_0) \to M$$

converges in the C^∞ topology to the diffeomorphism

$$\phi_0 := \exp_{q_0} \circ \, \Phi_0 \circ \exp_{p_0}^{-1} : U_r(p_0) \to U_r(q_0).$$

The map ϕ_0 satisfies

$$d(\phi_0(p), \phi_0(q)) = \lim_{i \to \infty} d(\phi_i(p), \phi_i(q)) = d(p, q)$$

for all $p, q \in U_r(p_0)$ and hence is an isometry (Theorem 5.1.1). This proves Step 1.

Step 2 *The sequence* ϕ_i *converges in the* C^∞ *topology on all of* M *to a smooth map* $\phi : M \to M$.

Define the set $M_0 := \{ p \in M \mid \text{the sequence } (\phi_i(p), d\phi_i(p)) \text{ converges} \}$. This set is nonempty because $p_0 \in M_0$. We prove that M_0 is open. Fix any element $p \in M_0$ and define $q := \lim_{i \to \infty} \phi_i(p)$ and $\Phi := \lim_{i \to \infty} d\phi_i(p)$. Choose a real number r such that $0 < r < \mathrm{inj}(p; M)$. Then Step 1 asserts that $r \leq \mathrm{inj}(q; M)$ and the sequence $\phi_i|_{U_r(p)}$ converges to the diffeomorphism $\exp_q \circ \, \Phi \circ \exp_p^{-1} : U_r(p) \to U_r(q)$ in the C^∞ topology on $U_r(p)$, and hence $U_r(p) \subset M_0$. This shows that M_0 is open, that $\phi_i|_{M_0}$ converges in the C^∞-topology to a smooth map $\phi : M_0 \to M$, and that $\phi(M_0)$ is an open subset of M.

We prove that M_0 is closed. Let $p_\nu \in M_0$ be a sequence that converges to an element $p \in M$. Then there exists a real number $r > 0$ such that $r < \mathrm{inj}(p_\nu; M)$ for all ν. Hence $B_r(p_\nu) \subset M_0$ for all ν by Step 1. Choose ν so large that $d(p_\nu, p) < r$ to obtain $p \in M_0$. This shows that M_0 is closed. Since M is connected, we deduce that $M_0 = M$. This proves Step 2.

Step 3 *The map $\phi : M \to M$ in Step 2 is an isometry.*

We claim that the triple p_0, q_0, Φ_0 in (7.3.5) satisfies

$$\lim_{i \to \infty} \phi_i^{-1}(q_0) = p_0, \qquad \lim_{i \to \infty} d\phi_i^{-1}(q_0) = \Phi_0^{-1}. \tag{7.3.6}$$

Namely, the sequence $d(\phi_i^{-1}(q_0), p_0) = d(q_0, \phi_i(p_0))$ converges to zero by assumption, and by Step 1 the derivatives $d\phi_i$ converge to $d\phi$ uniformly in some neighborhood $U_0 \subset M$ of p_0. Since $\phi_i^{-1}(q_0) \in U_0$ for i sufficiently large, this implies that the sequence $d\phi_i(\phi_i^{-1}(q_0))$ converges to $d\phi(p_0) = \Phi_0$ and hence the sequence $d\phi_i(\phi_i^{-1}(q_0))^{-1} = d\phi_i^{-1}(q_0)$ converges to Φ_0^{-1}. This proves (7.3.6). By (7.3.6) and Step 2 the sequence ϕ_i^{-1} converges in the C^∞ topology to a smooth map $\psi : M \to M$. By uniform convergence on compact subsets of M we have $\psi \circ \phi = \mathrm{id}$ and $\phi \circ \psi = \mathrm{id}$, so $\phi : M \to M$ is a diffeomorphism. Moreover, since ϕ_i is an isometry for each i and converges pointwise to ϕ, we have $d(\phi(p), \phi(q)) = \lim_{i \to \infty} d(\phi_i(p), \phi_i(q)) = d(p, q)$ for all $p, q \in M$, so ϕ is an isometry. This proves Step 3 and Lemma 7.3.2. $\square$

The next goal is to verify that the spaces $\mathcal{I}_p$ in (7.3.4) are diffeomorphic to each other. For $p_0, p_1 \in M$ define the map $\mathcal{F}_{p_1, p_0} : \mathcal{I}_{p_0} \to \mathcal{I}_{p_1}$ by

$$\mathcal{F}_{p_1, p_0}(\phi(p_0), d\phi(p_0)) := (\phi(p_1), d\phi(p_1)), \qquad \phi \in \mathcal{I}(M). \tag{7.3.7}$$

This map is well-defined by Lemma 5.1.10, because M is connected. Collectively, these maps give rise to a map $\mathcal{F} : M \times \mathcal{I} \to \mathcal{I}$ defined by

$$\begin{aligned}
\mathcal{I} &:= \big\{ (\phi(p), d\phi(p), p) \mid p \in M, \ \phi \in \mathcal{I}(M) \big\} \subset \mathcal{G}, \\
\mathcal{F}(p_1, (\phi(p_0), d\phi(p_0), p_0)) &:= (\phi(p_1), d\phi(p_1), p_1)
\end{aligned} \tag{7.3.8}$$

for $p_0, p_1 \in M$ and $\phi \in \mathcal{I}(M)$. The next lemma uses the concept of a smooth map on an arbitrary subset of Euclidean space as in the beginning of Sect. 2.1.

Lemma 7.3.3 *The map $\mathcal{F} : M \times \mathcal{I} \to \mathcal{I}$ is smooth. In particular, for each pair of points $p_0, p_1 \in M$, the map $\mathcal{F}_{p_1, p_0} : \mathcal{I}_{p_0} \to \mathcal{I}_{p_1}$ is a diffeomorphism.*

Proof Let $p_0, p_1 \in M$ and choose a smooth path $\gamma : I = [0, 1] \to M$ with the endpoints $\gamma(0) = p_0$ and $\gamma(1) = p_1$. Let $\mathcal{U}_\gamma \subset \mathcal{G}_{p_0}$ be the set of all pairs $(q_0, \Phi_0) \in \mathcal{G}_{p_0}$ such that there exists a development (Φ, γ, γ') on the interval I that satisfies

$$\gamma'(0) = q_0, \qquad \Phi(t) = \Phi_0. \tag{7.3.9}$$

By Remark 3.5.22, the set $\mathcal{U}_\gamma$ is open in $\mathcal{G}_{p_0}$ and the map $\mathcal{U}_\gamma \to \mathcal{G}$ that sends the pair $(q_0, \Phi_0) \in \mathcal{U}_\gamma$ to the pair $(\gamma'(1), \Phi(1)) \in \mathcal{G}_{p_1}$ is smooth. This shows that there is a unique diffeomorphism

$$\mathcal{F}_\gamma : \mathcal{U}_\gamma \to \mathcal{U}_{\gamma^{-1}}, \qquad \gamma^{-1}(t) := \gamma(1 - t), \qquad \mathcal{U}_\gamma \subset \mathcal{G}_{p_0}, \qquad \mathcal{U}_{\gamma^{-1}} \subset \mathcal{G}_{p_1}$$

that satisfies the condition

$$\mathcal{F}_\gamma(\gamma'(0), \Phi(0)) = (\gamma'(1), \Phi(1)) \tag{7.3.10}$$

for every development (Φ, γ, γ') of M along M on the interval I. The inverse of $\mathcal{F}_\gamma$ is the smooth map $\mathcal{F}_{\gamma^{-1}} : \mathcal{U}_{\gamma^{-1}} \to \mathcal{U}_\gamma$. If $\phi \in \mathcal{I}(M)$, then by Lemma 6.1.12 there exists a development (Φ, γ, γ') on I satisfying the initial conditions $\gamma'(0) = \phi(p_0)$ and $\Phi(0) = d\phi(p_0)$, and it is given by

$$\gamma'(t) = \phi(\gamma(t)), \qquad \Phi(t) = d\phi(\gamma(t)) \tag{7.3.11}$$

for $t \in I$. Hence $(\phi(p_0), d\phi(p_0)) \in \mathcal{U}_\gamma$ and by (7.3.10) and (7.3.11) we have

$$\mathcal{F}_\gamma(\phi(p_0), d\phi(p_0)) = (\phi(p_1), d\phi(p_1)) = \mathcal{F}_{p_1, p_0}(\phi(p_0), d\phi(p_0))$$

for every $\phi \in \mathcal{I}(M)$. Thus $\mathcal{I}_{p_0} \subset \mathcal{U}_\gamma$ and $\mathcal{I}_{p_1} \subset \mathcal{U}_{\gamma^{-1}}$ and $\mathcal{F}_\gamma|_{\mathcal{I}_{p_0}} = \mathcal{F}_{p_1, p_0}$. Hence $\mathcal{F}_{p_1, p_0}$ is smooth. The smoothness of $\mathcal{F}$ follows from the smooth dependence of the map $\mathcal{F}_\gamma$ on the curve γ, the verification of which we leave to the reader. This proves Lemma 7.3.3. $\qquad\square$

7.3.3 Killing Vector Fields

Killing vector fields are defined as those vector fields on M whose flows are one-parameter families of isometries. Assume throughout that M is a nonempty connected smooth m-dimensional submanifold of $\mathbb{R}^n$. Let X be a vector field on M and denote by

$$\mathbb{R} \times M \supset \mathcal{D} \to M : (t, p) \mapsto \phi(t, p) = \phi^t(p)$$

the flow of X (Definition 2.4.8). Then Theorem 2.4.9 asserts that $\mathcal{D}$ is an open subset of $\mathbb{R} \times M$ and ϕ is smooth. Thus, for every $t \in \mathbb{R}$, the set $\mathcal{D}_t := \{p \in M \mid (t, p) \in \mathcal{D}\}$ is open in M and the map $\phi^t : \mathcal{D}_t \to \mathcal{D}_{-t}$ is a diffeomorphism with the inverse $\phi^{-t} : \mathcal{D}_{-t} \to \mathcal{D}_t$.

Lemma 7.3.4 *In this situation the following are equivalent.*

(i) *For every $t \in \mathbb{R}$ the diffeomorphism $\phi^t : \mathcal{D}_t \to \mathcal{D}_{-t}$ is an isometry.*
(ii) *For every $p \in M$ and every pair of tangent vectors $u, v \in T_p M$, we have*

$$\langle \nabla_u X(p), v \rangle + \langle u, \nabla_v X(p) \rangle = 0. \tag{7.3.12}$$

Proof Let $p \in M$ and $v \in T_p M$. Choose a smooth curve $\alpha : \mathbb{R} \to M$ such that $\alpha(0) = p$ and $\dot\alpha(0) = v$, let $\Omega := \{(s, t) \in \mathbb{R}^2 \mid (s, \alpha(t)) \in \mathcal{D}\}$, and define the map $\gamma : \Omega \to M$ by $\gamma(s, t) := \phi^t(\alpha(s))$ for $(s, t) \in \Omega$. Then

$$\partial_s \gamma(0, t) = d\phi^t(p)v, \qquad \partial_t \gamma = X \circ \gamma, \qquad \nabla_s \partial_t \gamma = \nabla_{\partial_s \gamma} X(\gamma).$$

Thus the formula $\nabla_t \partial_s \gamma = \nabla_s \partial_t \gamma$ in Lemma 3.2.4 shows that

$$\nabla_t d\phi^t(p)v = \nabla_{d\phi^t(p)v} X(\phi^t(p)) \tag{7.3.13}$$

for all $t \in \mathbb{R}$ and this implies

$$\frac{d}{dt}\left|d\phi^t(p)v\right|^2 = 2\langle \nabla_{d\phi^t(p)v} X(\phi^t(p)), d\phi^t(p)v\rangle. \tag{7.3.14}$$

The right hand side vanishes for all p, v, t if and only if X satisfies (ii), and the left hand side vanishes for all p, v, t if and only if the flow of X satisfies (i). This proves Lemma 7.3.4. $\qquad\square$

Definition 7.3.5 A vector field $X \in \mathrm{Vect}(M)$ is called a **Killing vector field** (named after Wilhelm Karl Joseph Killing [39]) iff it satisfies equation (7.3.12) for all $p \in M$ and all $u, v \in T_p M$. The space of Killing vector fields will be denoted by $\mathrm{Vect}_K(M)$.

The space of Killing vector fields is a vector subspace of the Lie algebra of all vector fields on M. The next lemma shows that it is a finite-dimensional Lie subalgebra. Part (ii) is the linear counterpart of Lemma 5.1.10.

Lemma 7.3.6 *Let $M \subset \mathbb{R}^n$ be a nonempty connected smooth m-manifold. Then the following holds.*

(i) *If X is a Killing vector field on M and $\psi : M \to M$ is an isometry, then the pullback $\psi^* X$ is a Killing vector field. If X and Y are Killing vector fields on M, then so is their Lie bracket $[X, Y]$.*

(ii) *If X is a Killing vector field on M and there exists a $p_0 \in M$ such that*

$$X(p_0) = 0, \qquad \nabla X(p_0) = 0, \tag{7.3.15}$$

then $X(p) = 0$ for all $p \in M$.

(iii) *If M is complete and X is a Killing vector field, then X is complete.*

Proof We prove part (i). Let X be a Killing vector field, let $\phi^t : \mathcal{D}_t \to \mathcal{D}_{-t}$ be the flow of X, and let $\psi : M \to M$ be an isometry. Then

$$\psi^{-1} \circ \phi^t \circ \psi : \psi^{-1}(\mathcal{D}_t) \to \psi^{-1}(\mathcal{D}_{-t})$$

is the flow of $\psi^* X$. Hence, by Lemma 7.3.4, the flow of $\psi^* X$ is a one-parameter family of isometries and so $\psi^* X$ is a Killing vector field. Now assume that Y is another Killing vector field and denote its flow by ψ^t. Then each ψ^t is an isometry and so the pullback $(\psi^t)^* X$ is a Killing vector field for each t. Hence, by Lemma 2.4.18 and Exercise 5.2.9, the Lie bracket $[X, Y] = \frac{d}{dt}\big|_{t=0}(\psi^t)^* X$ is also a Killing vector field. This proves (i).

We prove part (ii). Let $\phi^t : \mathcal{D}_t \to \mathcal{D}_{-t}$ be the flow of a Killing vector field X and assume that there exists a $p_0 \in M$ such that (7.3.15) holds. Then it follows from (7.3.13) that $\phi^t(p_0) = p_0$ and $d\phi^t(p_0) = \mathrm{id}$ for all t. Hence, by Lemma 5.1.10 the isometry $\phi_t : \mathcal{D}_t \to \mathcal{D}_{-t}$ is the identity for each $t \in \mathbb{R}$. Thus $\mathcal{D}_t = \mathcal{D}_{-t} = M$ for all t and $X(p) = 0$ for all $p \in M$. This proves (ii).

We prove part (iii). Thus assume that M is complete and X is a Killing vector field. Let $\gamma : I \to M$ be an integral curve of X on its maximal existence interval $I \subset \mathbb{R}$ and denote $p := \gamma(0)$. Differentiate the equation $\dot{\gamma}(t) = X(\gamma(t))$ to obtain $\frac{d}{dt}|\dot{\gamma}(t)|^2 = 2\langle \nabla_{\dot{\gamma}(t)} X(\gamma(t)), \dot{\gamma}(t) \rangle = 0$ for all $t \in I$. Hence the function $I \to \mathbb{R} : t \mapsto |\dot{\gamma}(t)|$ is constant and this implies the inequality $d(p, \gamma(t)) \le \int_0^t |\dot{\gamma}(s)| \, ds = t|X(p)|$ for all $t \in I$. Since M is complete, the closed ball of radius R about p is compact for every $R > 0$ (Theorem 4.6.5). Hence the restriction of γ to any bounded subinterval of I is contained in a compact subset of M amd by Corollary 2.4.15 this implies $I = \mathbb{R}$. This proves (iii) and Lemma 7.3.6. $\qquad\square$

The proof of Theorem 7.3.1 is somewhat analogous to the proof of the Closed Subgroup Theorem 2.5.27. The first parallel is in part (i) of the next lemma, which asserts that the space of complete Killing vector fields is a Lie subalgebra of $\mathrm{Vect}(M)$ and can be viewed as an anlogue of Lemma 2.5.29. Part (ii) is the analogue of Lemma 2.5.28.

Lemma 7.3.7

(i)   *The set $\mathrm{Vect}_{K,c}(M)$ of complete Killing vector fields on M is a Lie subalgebra of $\mathrm{Vect}_K(M)$.*

(ii) *Let $\mathbb{R} \times M \to M : (t, p) \mapsto \psi_t(p)$ be a smooth map such that ψ_t is an isometry for every $t \in \mathbb{R}$ and define $X(p) := \frac{d}{dt}\big|_{t=0} \psi_t(p)$ for $p \in M$. Then X is a complete Killing vector field.*

Proof   The proof has three steps.

Step 1 *Let $U, V \subset M$ be nonempty open sets, let $\phi : U \to V$ be an isometry, and let $\phi_k : M \to M$ be a sequence of isometries that converges uniformly on every compact subset of U to ϕ. Then ϕ extends uniquely to an isometry from M to M and ϕ_k converges in the C^∞ topology to this extension.*

Fix an element $p_0 \in U$. Then by part (i) of Exercise 5.1.11 we have

$$\lim_{k \to \infty} \phi_k(p_0) = \phi(p_0), \qquad \lim_{k \to \infty} d\phi_k(p_0) = d\phi(p_0).$$

Hence the assertion of Step 1 follows from Lemma 7.3.2.

Step 2 *We prove part (ii).*

That X is a Killing vector field follows from the same argument as in Lemma 7.3.4 with time dependent vector fields. Denote by $\phi^t : \mathcal{D}_t \to \mathcal{D}_{-t}$ the flow of X. Then the sequence of isometries $\psi_{t/k}^k : M \to M$ converges to ϕ^t uniformly on every compact subset of $\mathcal{D}_t$ (see Exercise 7.3.10 below). Hence Step 1 asserts that ϕ^t extends to an isometry on all of M whenever $\mathcal{D}_t$ is nonempty, in particular for small t. The extended isometries still satisfy $\phi^{s+t} = \phi^s \circ \phi^t$ and $\partial_t \phi^t = X \circ \phi^t$. Thus $\mathcal{D}_t = M$ for small t and so for all t, because $\phi^{-s}(\mathcal{D}_{-s} \cap \mathcal{D}_t) \subset \mathcal{D}_{s+t}$ (Theorem 2.4.9). This proves Step 2.

Step 3 *We prove part (i).*

Let $X, Y \in \text{Vect}_{K,c}(M)$, let ϕ^t be the flow of X, and let ψ^t be the flow of Y. Then $\frac{d}{dt}\big|_{t=0} \phi^t \circ \psi^t(p) = X(p) + Y(p)$ for all $p \in M$, and so $X + Y$ is a complete Killing vector field by Step 2. Hence $\text{Vect}_{K,c}(M)$ is a vector space. It is finite-dimensional by Lemma 7.3.6. Moreover, $(\psi^{-t} \circ \phi^s \circ \psi^t)_{s \in \mathbb{R}}$ is the flow of the pullback vector field $(\psi^t)^* X$, hence $(\psi^t)^* X$ is a complete Killing vector field for each $t \in \mathbb{R}$, and hence so is the Lie bracket $[X, Y] = \frac{d}{dt}\big|_{t=0}(\psi^t)^* X$ by finite-dimensionality. This proves Step 3 and Lemma 7.3.7. $\square$

Exercise 7.3.8 Let $M \subset \mathbb{R}^n$ be a smooth manifold and let X be a vector field on M. Consider the following condition.

(K) *If $\gamma : I \to M$ is a geodesic, then $X \circ \gamma \in \text{Vect}(\gamma)$ is a Jacobi field along γ, i.e. it satisfies the differential equation $\nabla\nabla(X \circ \gamma) + R(X \circ \gamma, \dot{\gamma})\dot{\gamma} = 0$.*

Prove that every Killing vector field satisfies (K) and use this to give an alternative proof of part (ii) of Lemma 7.3.6. If M is compact, prove that a vector field satisfies (K) if and only if it is a Killing vector field. Find a vector field on a noncompact manifold that satisfies (K) but is not a Killing vector field. **Hint:** Differentiate the function $t \mapsto \langle X(\gamma(t)), \dot{\gamma}(t) \rangle$ twice.

Exercise 7.3.9 (Grönwall's inequality) *If $a, c \geq 0$ and $\sigma : [0, T] \to \mathbb{R}$ is a continuous function satisfying*

$$0 \leq \sigma(t) \leq a + c \int_0^t \alpha(s)\, ds \qquad \text{for } 0 \leq t \leq T, \qquad (7.3.16)$$

then $\sigma(t) \leq ae^{ct}$ for every real number $0 \leq t \leq T$. **Hint:** The function

$$\tau(t) := \int_0^t (ae^{cs} - \sigma(s))\, ds$$

satisfies $\dot{\tau}(t) \geq c\tau(t)$ for all t. Differentiate the function $t \mapsto e^{-ct}\tau(t)$ to show that τ is nonnegative and nondecreasing.

Exercise 7.3.10 Let T, c, ε be positive real numbers, let $M \subset \mathbb{R}^n$ be a smooth m-dimensional submanifold, and let $[0, T] \times M \to \mathbb{R}^n : (t, p) \mapsto X_t(p)$ and $[0, T] \times M \to \mathbb{R}^n : (t, p) \mapsto Y_t(p)$ be continuous maps that satisfy the conditions $X_t(p), Y_t(p) \in T_p M$ and

$$|X_t(p) - Y_t(p)|_{\mathbb{R}^n} \leq \varepsilon, \qquad |X_t(p) - X_t(q)|_{\mathbb{R}^n} \leq c|p - q|_{\mathbb{R}^n} \tag{7.3.17}$$

for all $p, q \in M$ and all $t \in [0, T]$. If the curves $\beta, \gamma : [0, T] \to M$ are solutions of the differential equations $\dot\beta(t) = X_t(\beta(t))$ and $\dot\gamma(t) = Y_t(\gamma(t))$, prove that

$$|\beta(t) - \gamma(t)|_{\mathbb{R}^n} \leq \left(|\beta(0) - \gamma(0)|_{\mathbb{R}^n} + T\varepsilon\right)e^{ct} \tag{7.3.18}$$

for $0 \leq t \leq T$. Relax the continuity requirement in t on the vector fields to piecewise continuity. **Hint:** Define the function $\sigma : [0, T] \to \mathbb{R}$ by

$$\sigma(t) := |\beta(t) - \gamma(t)|_{\mathbb{R}^n}.$$

Show that

$$\sigma(t) \leq \sigma(0) + t\varepsilon + c \int_0^t \sigma(s)\, ds$$

and use Grönwall's inequality.

7.3.4   Proof of the Myers–Steenrod Theorem

With these preparations we are ready to prove Theorem 7.3.1. The following lemma is the heart of the proof. It is the analogue of Lemma 2.5.30 in the proof of the Closed Subgroup Theorem 2.5.27.

Lemma 7.3.11 *Fix an element $p_0 \in M$, a tangent vector $v_0 \in T_{p_0} M$, and a linear map $A_0 : T_{p_0} M \to T_{p_0} M$. Let $\phi_i \in \mathcal{I}(M)$ be a sequence of isometries and let τ_i be a sequence of positive real numbers such that*

$$\lim_{i \to \infty} \tau_i = 0, \qquad \lim_{i \to \infty} \phi_i(p_0) = p_0, \qquad \lim_{i \to \infty} d\phi_i(p_0) = \mathbb{1}, \tag{7.3.19}$$

and, for all $v \in T_{p_0} M$,

$$\lim_{i \to \infty} \frac{\phi_i(p_0) - p_0}{\tau_i} = v_0, \qquad \lim_{i \to \infty} \frac{d\phi_i(p_0)v - v}{\tau_i} = A_0 v. \tag{7.3.20}$$

Then there exists a unique complete Killing vector field X on M such that

$$X(p_0) = v_0, \qquad dX(p_0) = A_0. \tag{7.3.21}$$

For every $p \in M$ and every $v \in T_p M$ this vector field satisfies

$$\lim_{i \to \infty} \frac{\phi_i(p) - p}{\tau_i} = X(p), \qquad \lim_{i \to \infty} \frac{d\phi_i(p)v - v}{\tau_i} = dX(p)v, \qquad (7.3.22)$$

and the convergence is uniform on compact subsets of TM.

Proof   By (7.3.19), (7.3.20), and Exercise 2.2.4 we have

$$(p_0, A_0) \in T_{(p_0, \mathbb{1})} G_{p_0}.$$

Let $p \in M$ and recall the definition of the map $\mathcal{F}_{p,p_0} : \mathcal{I}_{p_0} \to \mathcal{I}_p$ in (7.3.7). By Lemma 7.3.3 this map extends to a diffeomorphism $\mathcal{F}_\gamma$ from an open subset $\mathcal{U}_\gamma \subset G_{p_0}$ containing the set $\mathcal{I}_{p_0}$ to an open subset $\mathcal{U}_{\gamma^{-1}} \subset G_p$ containing the set $\mathcal{I}_p$ and it satisfies $\mathcal{F}_\gamma(p_0, \mathbb{1}) = (p, \mathbb{1})$. Define

$$(X(p), A(p)) := d\mathcal{F}_\gamma(p_0, \mathbb{1})(v_0, A_0) \in T_{(p,\mathbb{1})} G_p.$$

Since $\mathcal{F}_\gamma(\phi_i(p_0), d\phi_i(p_0)) = (\phi_i(p), d\phi_i(p))$ for all i, it follows from (7.3.20) and Exercise 2.2.16 that, for all $v \in T_p M$, we have

$$\lim_{i \to \infty} \frac{\phi_i(p) - p}{\tau_i} = X(p), \qquad \lim_{i \to \infty} \frac{d\phi_i(p)v - v}{\tau_i} = A(p)v. \qquad (7.3.23)$$

This formula shows that the pair $(X(p), A(p)) \in T_{(p,\mathbb{1})} G_p$ is independent of the choice of the extension $\mathcal{F}_\gamma$ used to define it. Moreover, it follows from the smoothness of the map $\mathcal{F}$ in Lemma 7.3.3 that the map $p \mapsto (X(p), A(p))$ is smooth and that the convergence in (7.3.23) is uniform as the pair (p, v) varies over any compact subset of TM.

Now let $v \in T_p M$ and define $\gamma(t) := \exp_p(tv)$ for $t \in I_{p,v}$. Then

$$
\begin{aligned}
X(\gamma(t)) - X(p) &= \lim_{i \to \infty} \frac{\phi_i(\gamma(t)) - \gamma(t)}{\tau_i} - \lim_{i \to \infty} \frac{\phi_i(p) - p}{\tau_i} \\
&= \lim_{i \to \infty} \frac{\phi_i(\gamma(t)) - \phi_i(\gamma(0)) - \gamma(t) + \gamma(0)}{\tau_i} \\
&= \lim_{i \to \infty} \int_0^t \frac{d\phi_i(\gamma(s))\dot{\gamma}(s) - \dot{\gamma}(s)}{\tau_i} \, ds \\
&= \int_0^t A(\gamma(s))\dot{\gamma}(s) \, ds.
\end{aligned}
$$

Here the last step uses uniform convergence on compact sets in (7.3.23). Divide by t and use the continuity of the curve $t \mapsto A(\gamma(t))\dot{\gamma}(t)$ to obtain

$$A(p)v = \lim_{t \to 0} \frac{1}{t} \int_0^t A(\gamma(s))\dot{\gamma}(s) \, ds = \lim_{t \to 0} \frac{X(\exp_p(tv)) - X(p)}{t}.$$

Hence, by Exercise 2.2.16, we deduce that, for all $p \in M$ and all $v \in T_p M$,

$$A(p)v = dX(p)v.$$

By (7.3.20) and (7.3.23) this shows that X satisfies (7.3.21) and (7.3.22). By (7.3.22) and Exercise 2.2.4 we have $(X(p), dX(p)) \in T_{(p,\mathbb{1})} \mathcal{G}_p$ and this implies $\langle v, dX(p)v \rangle = 0$ for all $v \in T_p M$. Here is a more direct proof of this crucial fact. Namely, by (7.3.22) we have

$$\lim_{i \to \infty} \frac{|d\phi_i(p)v - v|^2}{\tau_i^2} = |dX(p)v|^2.$$

Hence

$$
\begin{aligned}
\langle v, dX(p)v \rangle &= \lim_{i \to \infty} \frac{\langle v, d\phi_i(p)v - v \rangle}{\tau_i} \\
&= \lim_{i \to \infty} \frac{\langle v, d\phi_i(p)v \rangle - |v|^2}{\tau_i} \\
&= \lim_{i \to \infty} \frac{2\langle v, d\phi_i(p)v \rangle - |v|^2 - |d\phi_i(p)v|^2}{2\tau_i} \\
&= -\lim_{i \to \infty} \frac{|d\phi_i(p)v - v|^2}{\tau_i^2} \frac{\tau_i}{2} \\
&= 0
\end{aligned}
$$

for all $p \in M$ and all $v \in T_p M$. This shows that X is a Killing vector field and so it remains to prove that X is complete.

Let $\phi^t : \mathcal{D}_t \to \mathcal{D}_{-t}$ be the flow of X and, for $p \in M$, denote by

$$I_p := \{ t \in \mathbb{R} \mid p \in \mathcal{D}_t \}$$

the maximal existence interval for the solution γ of the initial value problem $\dot{\gamma}(t) = X(\gamma(t))$, $\gamma(0) = p$. We prove in three steps that X is complete.

Step 1 *Let $p \in M$ and let $T > 0$ such that $T|X(p)| < \mathrm{inj}(p; M)$. Then*

$$[-T, T] \subset I_p.$$

Define $\gamma(t) := \phi^t(p)$ for $t \in I_p$. Since X is a Killing vector field, the diffeomorphism $\phi^t : \mathcal{D}_t \to \mathcal{D}_{-t}$ is an isometry by Lemma 7.3.4, and hence

$$|\dot{\gamma}(t)| = |X(\phi^t(p))| = |d\phi^t(p)X(p)| = |X(p)|,$$

for all $t \in I_p$. This implies $d(p, \gamma(t)) \le t|X(p)|$ for all $t \in I_p$, and hence γ cannot leave the compact set $U_{T|X(p)|}(p) = \{ q \in M \mid d(p, q) \le T|X(p)| \}$ on any subinterval $I \subset [-T, T]$. Hence $[-T, T] \subset I_p$ by Corollary 2.4.15 and this proves Step 1.

Step 2 *Let $p \in M$. If $t \in \mathbb{R}$ satisfies $|t X(p)| < \text{inj}(p; M)$ and the sequence of integers $m_i \in \mathbb{Z}$ is chosen such that*

$$m_i \tau_i \leq t < (m_i + 1)\tau_i, \tag{7.3.24}$$

then $\phi^t(p) = \lim_{i \to \infty} \phi_i^{m_i}(p)$.

Let $T > 0$ such that $T|X(p)| < \text{inj}(p; M)$. Then $[0, T] \subset I_p$ by Step 1 and the set

$$K := \left\{ \phi^t(p) \,\middle|\, 0 \leq t \leq T \right\}$$

is compact and $|X(q)| = |X(p)|$ for all $q \in K$. By Lemma 4.2.7 there exists a constant $\delta > 0$ such that, for all $q, q_0, q_1 \in M$,

$$q \in K, \quad d(q, q_0) \leq \delta, \quad d(q, q_1) \leq \delta \qquad \Longrightarrow \qquad \frac{d(q_0, q_1)}{|q_0 - q_1|} \leq 2. \tag{7.3.25}$$

Fix a real number $0 < \varepsilon \leq 1$ and choose $i_0 \in \mathbb{N}$ such that, for all $i \geq i_0$,

$$\tau_i |X(p)| < \delta, \quad (T + \tau_i)|X(p)| < \text{inj}(p; M), \quad \sup_{q \in K} d(q, \phi_i(q)) < \delta, \tag{7.3.26}$$

$$\sup_{q \in K} \left| \frac{\phi_i(q) - q}{\tau_i} - X(q) \right| < \varepsilon, \qquad \sup_{q \in K} \left| \frac{\phi^{\tau_i}(q) - q}{\tau_i} - X(q) \right| < \varepsilon. \tag{7.3.27}$$

Then, for all $i \geq i_0$ and all $q \in K$, we have $d(q, \phi^{\tau_i}(q)) \leq \tau_i |X(p)| < \delta$ and $d(q, \phi_i(q)) < \delta$ by (7.3.26), and hence by (7.3.25) and (7.3.27),

$$\sup_{q \in K} d(\phi_i(q), \phi^{\tau_i}(q)) \leq 2 \sup_{q \in K} |\phi_i(q) - \phi^{\tau_i}(q)| \leq 4\tau_i \varepsilon. \tag{7.3.28}$$

Now choose a real number $0 \leq t \leq T$ and choose $m_i \in \mathbb{N}_0$ as in (7.3.24). Then, for $k = 1, \ldots, m_i - 1$ we have $\phi^{k\tau_i}(p) \in K$ and hence

$$d\left(\phi_i^{k+1}(p), \phi^{(k+1)\tau_i}(p) \right)$$
$$\leq d\left(\phi_i(\phi_i^k(p)), \phi_i(\phi^{k\tau_i}(p)) \right) + d\left(\phi_i(\phi^{k\tau_i}(p)), \phi^{\tau_i}(\phi^{k\tau_i}(p)) \right)$$
$$= d\left(\phi_i^k(p), \phi^{k\tau_i}(p) \right) + d\left(\phi_i(\phi^{k\tau_i}(p)), \phi^{\tau_i}(\phi^{k\tau_i}(p)) \right)$$
$$\leq d\left(\phi_i^k(p), \phi^{k\tau_i}(p) \right) + 4\tau_i \varepsilon.$$

Here the last inequality holds for $i \geq i_0$ by (7.3.28). By induction this implies

$$d(\phi_i^{m_i}(p), \phi^{m_i \tau_i}(p)) \leq 4 m_i \tau_i \varepsilon \leq 4T\varepsilon$$

for all $i \geq i_0$. Hence

$$\lim_{i \to \infty} \phi_i^{m_i}(p) = \lim_{i \to \infty} \phi^{m_i \tau_i}(p) = \phi^t(p)$$

and this proves Step 2 for $t \geq 0$. For $t \leq 0$ the argument is similar.

Step 3 $I_p = \mathbb{R}$ *for every* $p \in M$.

Fix an element $p \in M$ and real number $T > 0$ such that $T|X(p)| < \mathrm{inj}(p; M)$. Choose the sequence $m_i \in \mathbb{N}_0$ such that $\tau_i m_i \leq T < (m_i + 1)\tau_i$. Then Step 2 asserts that $\phi^T(p) = \lim_{i \to \infty} \phi_i^{m_i}(p)$. Since $\phi_i^{m_i} : M \to M$ is an isometry, it follows that $T|X(p)| < \mathrm{inj}(\phi_i^{m_i}(p); M)$ for all $i \in \mathbb{N}$ and hence

$$T|X(p)| < \mathrm{inj}\big(\phi^T(p); M\big).$$

By Step 1 this implies $[0, T] \subset I_{\phi^T(p)}$. Hence $[0, 2T] \subset I_p$ and, by Step 2,

$$\phi^{2T}(p) = \lim_{i \to \infty} \phi_i^{m_i}\big(\phi^T(p)\big).$$

Continue by induction to obtain for all $k \in \mathbb{N}$ that

$$T|X(p)| < \mathrm{inj}\big(\phi^{kT}(p); M\big)$$

and hence $[0, (k + 1)T] \subset I_p$ and

$$\phi^{(k+1)T}(p) = \lim_{i \to \infty} \phi_i^{m_i}\big(\phi^{kT}(p)\big).$$

Thus $[0, \infty) \subset I_p$ and the same argument shows that $(-\infty, 0] \subset I_p$. Hence X is complete. This proves Step 3 and Lemma 7.3.11. $\qquad\qquad\square$

The next lemma establishes the smooth structure on $\mathcal{I}(M)$, in analogy to the proof of the Closed Subgroup Theorem 2.5.27.

Lemma 7.3.12 *For every* $p \in M$ *the set* $\mathcal{I}_p$ *is a smooth submanifold of* G_p *and its tangent space at* $(q, \Phi) \in \mathcal{I}_p$ *is given by*

$$T_{(q,\Phi)}\mathcal{I}_p = \big\{(X(q), dX(q)\Phi) \,\big|\, X \in \mathrm{Vect}_{K,c}(M)\big\}. \qquad (7.3.29)$$

Proof By Lemma 3.4.5 with $\mathbb{R}^m$ replaced by $T_p M$ the tangent space of G_p at an element $(q, \Phi) \in G_p$ is given by

$$T_{(q,\Phi)}G_p = \left\{ (\widehat{q}, \widehat{\Phi}) \,\middle|\, \begin{array}{l} \widehat{q} \in T_q M, \ \widehat{\Phi} \in \mathcal{L}(T_p M, \mathbb{R}^n), \\ (\mathbb{1} - \Pi(q))\widehat{\Phi} = h_q(\widehat{q})\Phi, \text{ and} \\ \langle \Phi u, \widehat{\Phi} v \rangle + \langle \widehat{\Phi} u, \Phi v \rangle = 0 \ \forall \, u, v \in T_p M. \end{array} \right\}. \qquad (7.3.30)$$

If $(q, \Phi) \in G_p$ and $X \in \mathrm{Vect}(M)$, then $(\mathbb{1} - \Pi(q))dX(q)\Phi = h_q(X(q))\Phi$ by the Gauß–Weingarten formula in Remark 5.2.5. Moreover, if X is a Killing vector field, then $\langle \Phi u, dX(q)\Phi v \rangle + \langle dX(q)\Phi u, \Phi v \rangle = 0$ for all $u, v \in T_p M$, so it follows from (7.3.30) that the pair $(\widehat{q}, \widehat{\Phi})$ with $\widehat{q} = X(q)$ and $\widehat{\Phi} = dX(q)\Phi$ is a tangent vector of G_p at (q, Φ).

Now abbreviate

$$k := \dim\big(\mathrm{Vect}_{K,c}(M)\big) \leq \frac{m(m+1)}{2} = \dim(G_p) =: \ell.$$

Fix an isometry $\phi_0 : M \to M$ and define

$$(q_0, \Phi_0) := (\phi_0(p), d\phi_0(p)) = \iota_p(\phi_0) \in \mathcal{I}_p.$$

We must construct a coordinate chart on G_p in a neighborhood of the point $(q_0, \Phi_0) \in \mathcal{I}_p$ with values in an open set $\Omega \subset \mathbb{R}^\ell$ that sends $\mathcal{I}_p$ to the intersection of Ω with $\mathbb{R}^k \times \{0\}$. For this we first choose a basis $Y_1, \ldots, Y_k$ of $\mathrm{Vect}_{K,c}(M)$ and then a basis $\eta_1 = (\widehat{q}_1, \widehat{\Phi}_1), \ldots, \eta_\ell = (\widehat{q}_\ell, \widehat{\Phi}_\ell)$ of the tangent space $T_{(q_0, \Phi_0)} G_p$ such that

$$\eta_j = (\widehat{q}_j, \widehat{\Phi}_j) = \Big(Y_j(\phi_0(p)), dY_j(\phi_0(p))d\phi_0(p) \Big), \qquad j = 1, \ldots, k.$$

Next we choose any smooth map

$$\mathbb{R}^{\ell-k} \to G_p : (t^{k+1}, \ldots, t^\ell) \mapsto \iota(t^{k+1}, \ldots, t^\ell)$$

such that $\iota(0) = (q_0, \Phi_0)$ and

$$\frac{\partial \iota}{\partial t^j}(0, \ldots, 0) = \eta_j = (\widehat{q}_j, \widehat{\Phi}_j), \qquad j = k+1, \ldots, \ell.$$

For a complete Killing vector field $X \in \mathrm{Vect}_{K,c}(M)$ let $\psi_X \in \mathcal{I}(M)$ be the time-1 map of the flow of X and define the map $\Theta : \mathbb{R}^\ell \to G_p$ by

$$\Theta(t^1, \ldots, t^\ell) := \Big(\psi_X(q), d\psi_X(q)\Phi \Big),$$
$$X := t^1 Y_1 + \cdots + t^k Y_k, \qquad\qquad (7.3.31)$$
$$(q, \Phi) := \iota(t^{k+1}, \ldots, t^\ell)$$

for $(t^1, \ldots, t^\ell) \in \mathbb{R}^\ell$. Then

$$\Theta(0) = \iota(0) = (q_0, \Phi_0) = \iota_p(\phi_0) \in \mathcal{I}_p$$

and

$$d\Theta(0)(\widehat{t}^1, \ldots, \widehat{t}^\ell) = \widehat{t}^1 \eta_1 + \cdots + \widehat{t}^\ell \eta_\ell$$

for $(\hat{t}^1, \ldots, \hat{t}^\ell) \in \mathbb{R}^\ell$. Thus the derivative $d\Theta(0) : \mathbb{R}^\ell \to T_{(q_0, \Phi_0)} G_p$ is a bijective linear map and so the Inverse Function Theorem asserts that the map Θ restricts to a diffeomorphism from a sufficiently small open neighborhood $\Omega \subset \mathbb{R}^\ell$ of the origin onto its image $\iota(\Omega) \subset G_p$, which is an open neighborhood in G_p of the point $\Theta(0) = \iota_p(\phi_0) \in \mathcal{I}_p$. With this understood, the assertion that $\mathcal{I}_p$ is a smooth submanifold of G_p is a direct consequence of the following Claim.

Claim *There exists an open set* $\Omega_0 \subset \mathbb{R}^\ell$ *such that*

$$0 \in \Omega_0 \subset \Omega, \qquad \Theta\big(\Omega_0 \cap (\mathbb{R}^k \times \{0\})\big) = \mathcal{U}_0 \cap \mathcal{I}_p, \qquad \mathcal{U}_0 := \Theta(\Omega_0).$$
$$(7.3.32)$$

Suppose, by contradiction, that such an open set Ω_0 does not exist. Then there exist sequences $t_i = (t_i^1, \ldots, t_i^\ell) \in \mathbb{R}^\ell$ and $\phi_i \in \mathcal{I}(M)$ such that

$$\lim_{i \to \infty} t_i = 0, \qquad t_i \in \Omega \setminus (\mathbb{R}^k \times \{0\}), \qquad \Theta(t_i) = \iota_p(\phi_i) \in \mathcal{I}_p.$$

Define

$$X_i := t_i^1 Y_1 + \cdots + t_i^k Y_k, \qquad (q_i, \Phi_i) := \iota\big(t_i^{k+1}, \ldots, t_i^\ell\big).$$

Then $\big(\psi_{X_i}(q_i), d\psi_{X_i}(q_i)\Phi_i\big) = \Theta(t_i) = \iota_p(\phi_i) = \big(\phi_i(p), d\phi_i(p)\big)$ and hence

$$q_i = (\psi_{X_i}^{-1} \circ \phi_i)(p), \qquad \Phi_i = d\psi_{X_i}(q_i)^{-1} d\phi_i(p) = d(\psi_{X_i}^{-1} \circ \phi_i)(p).$$

Thus $\iota_p(\psi_{X_i}^{-1} \circ \phi_i) = (q_i, \Phi_i) = \Theta\big(0, \ldots, 0, t_i^{k+1}, \ldots, t_i^\ell\big)$ is still a sequence in the set $\Theta(\Omega \setminus (\mathbb{R}^k \times \{0\})) \cap \mathcal{I}_p$ that converges to $\iota_p(\phi_0)$. Thus we may assume without loss of generality that $t_i^1 = t_i^2 = \cdots = t_i^k = 0$ and so

$$\iota(t_i) = \iota_p(\phi_i), \qquad \lim_{i \to \infty} \iota_p(\phi_i) = \iota_p(\phi_0). \tag{7.3.33}$$

Then $\phi_i \circ \phi_0^{-1}$ converges to the identity in the C^∞ topology by Lemma 7.3.2. Since $\phi_i \neq \phi_0$, we have

$$\tau_i := \big|(\phi_i \circ \phi_0^{-1})(p) - p\big| + \sup_{0 \neq v \in T_p M} \frac{|d(\phi_i \circ \phi_0^{-1})(p)v - v|}{|v|} > 0 \tag{7.3.34}$$

for all i by Lemma 5.1.10, and $\lim_{i \to \infty} \tau_i = 0$. Hence, by Exercise 2.2.4 there exists a tangent vector $(v_0, A_0) \in T_{(p,\mathbb{1})} G_p$ and a subsequence, still denoted by ϕ_i, such that, for all $v \in T_p M$,

$$\lim_{i \to \infty} \frac{(\phi_i \circ \phi_0^{-1})(p) - p}{\tau_i} = v_0, \qquad \lim_{i \to \infty} \frac{d(\phi_i \circ \phi_0^{-1})(p)v - v}{\tau_i} = A_0 v. \tag{7.3.35}$$

It follows from (7.3.35) and Lemma 7.3.11 that there exists a complete Killing vector field $X \in \mathrm{Vect}_{K,c}(M)$ such that

$$X(p) = v_0, \qquad dX(p) = A_0.$$

This vector field is nonzero because $(v_0, A_0) \neq 0$ by (7.3.34). Moreover, by equation (7.3.22) in Lemma 7.3.11, with ϕ_i replaced by $\phi_i \circ \phi_0^{-1}$ and the pair (p, v) replaced by the pair $(\phi_0(p), d\phi_0(p)v)$, we obtain

$$X(\phi_0(p)) = \lim_{i \to \infty} \frac{\phi_i(p) - \phi_0(p)}{\tau_i},$$

$$dX(\phi_0(p))d\phi_0(p)v = \lim_{i \to \infty} \frac{d\phi_i(p)v - d\phi_0(p)v}{\tau_i}$$

for $v \in T_p M$. Since by (7.3.33) the sequence

$$(\phi_i(p), d\phi_i(p)) = \iota_p(\phi_i) = \iota(t_i)$$

converges to $(\phi_0(p), d\phi_0(p)) = \iota(0)$, the pair $(X(\phi_0(p)), dX(\phi_0(p))d\phi_0(p))$ belongs to the image of the derivative $d\iota(0)$ by Exercise 2.2.4, and hence is a nonzero linear combination of the vectors $\eta_{k+1}, \ldots, \eta_\ell$. It is also a linear combination of the vectors $\eta_1, \ldots, \eta_k$, because X is a complete Killing vector field and so is a linear combination of the vector fields $Y_1, \ldots, Y_k$. This is a contradiction, and this contradiction proves the Claim. Thus there does, after all, exist an open set $\Omega_0 \subset \mathbb{R}^\ell$ that satisfies (7.3.32), and the map $\Theta^{-1} : \mathcal{U}_0 \to \Omega_0$ is then a coordinate chart on G_p which satisfies

$$\Theta^{-1}(\mathcal{U}_0 \cap \mathcal{I}_p) = \Omega_0 \cap (\mathbb{R}^k \times \{0\}).$$

Hence $\mathcal{I}_p$ is a submanifold of G_p and this proves Lemma 7.3.12.   □

Proof of Theorem 7.3.1 By Lemma 7.3.3 and Lemma 7.3.12 there exists a unique smooth structure on $\mathcal{I}(M)$ such that the map $\iota_p : \mathcal{I}(M) \to G_p$ in (7.3.3) is an embedding for every $p \in M$. That the topology induced by this smooth structure agrees with the C^∞ topology was established in Lemma 7.3.2. That it also agrees with the compact open topology (i.e. with the C^0 topology of uniform convergence on conpact sets) is the content of Exercise 5.1.11. This proves part (i).

We prove part (ii). Recall the definition of the map

$$\mathcal{F} : M \times \mathcal{I} \to \mathcal{I}$$

in (7.3.8) and the target map $t : G \to M$, the inverse map $i : G \to G$, and the composition map $m : (s \times t)^{-1}(\Delta) \to G$ in the beginning of Sect. 7.3.1. These maps are all smooth and turn $\mathcal{I}$ into a smooth subgroupoid of G. For each $p \in M$ they endow the submanifold $\mathcal{I}_p \subset G_p$ with the structure of a Lie group as follows. The unit is the element

$$e_p := (p, \mathbb{1}) \in \mathcal{I}_p \tag{7.3.36}$$

The product is the map $m_p : \mathcal{I}_p \times \mathcal{I}_p \to \mathcal{I}_p$ defined by

$$(m_p(\eta, \xi), p) := m\Big(\mathcal{F}\big(t(\xi, p), (\eta, p)\big), (\xi, p)\Big) \tag{7.3.37}$$

for $\xi, \eta \in \mathcal{I}_p$, and the inverse map is the involution $i_p : \mathcal{I}_p \to \mathcal{I}_p$ defined by

$$(i_p(\xi), p) := \mathcal{F}\big(p, i(\xi, p)\big) \tag{7.3.38}$$

for $\xi \in \mathcal{I}_p$. The maps m_p in (7.3.37) and i_p in (7.3.38) are smooth by definition. To show that they define a group structure and that the map

$$\iota_p : \mathcal{I}(M) \to \mathcal{I}_p$$

in (7.3.3) is a group isomorphism, we must verify the identities

$$\iota_p(\mathrm{id}) = e_p,$$
$$m_p(\iota_p(\psi), \iota_p(\phi)) = \iota_p(\psi \circ \phi), \tag{7.3.39}$$
$$i_p(\iota_p(\phi)) = \iota_p(\phi^{-1})$$

for all $\phi, \psi \in \mathcal{I}(p)$. The first equation in (7.3.39) follows directly from the definitions. To prove the remaining equations, fix two isometries ϕ and ψ and define $\xi, \eta, \zeta \in \mathcal{I}_p$ by

$$\xi := \iota_p(\phi), \qquad \eta := \iota_p(\psi), \qquad \zeta := \iota_p(\psi \circ \phi).$$

Then, by the chain rule,

$$
\begin{aligned}
(\zeta, p) &= \Big(\psi(\phi(p)), d\psi(\phi(p))d\phi(p), p\Big) \\
&= m\Big(\big(\psi(\phi(p)), d\psi(\phi(p)), \phi(p)\big), (\phi(p), d\phi(p), p)\big)\Big) \\
&= m\Big(\mathcal{F}\big(\phi(p), (\psi(p), d\psi(p), p)\big), (\phi(p), d\phi(p), p)\big)\Big) \\
&= m\Big(\mathcal{F}\big(t(\xi, p), (\eta, p)\big), (\xi, p)\Big) \\
&= (m_p(\eta, \xi), p).
\end{aligned}
$$

Here the last equality follows from the definition of the map m_p in (7.3.37). This proves the second equation in (7.3.39).

To verify the third equation in (7.3.39), we compute

$$
\begin{aligned}
\left(\iota_p(\phi^{-1}), p\right) &= \left(\phi^{-1}(p), d\phi^{-1}(p), p\right) \\
&= \mathcal{F}\left(p, (\phi^{-1}(\phi(p)), d\phi^{-1}(\phi(p)), \phi(p))\right) \\
&= \mathcal{F}\left(p, (p, d\phi(p)^{-1}, \phi(p))\right) \\
&= \mathcal{F}\left(p, i(\phi(p), d\phi(p), p)\right) \\
&= \mathcal{F}\left(p, i(\xi, p)\right) \\
&= (i_p(\xi), p).
\end{aligned}
$$

Here the last equality follows from the definition of the map i_p in (7.3.38). This proves the third equation in (7.3.39) and part (ii).

That the Lie algebra of $\mathcal{I}_p$ is isomorphic to the space $\mathrm{Vect}_{K,c}(M)$ of complete Killing vector fields under the Lie algebra isomorphism

$$
\mathrm{Vect}_{K,c}(M) \to T_{e_p}\mathcal{I}_p : X \mapsto (X(p), dX(p))
$$

follows from Lemma 7.3.12. This proves part (iii) and Theorem 7.3.1. □

Corollary 7.3.13 *Let* $(p, e) \in \mathcal{O}(M)$. *Then there exists a constant* $\varepsilon > 0$ *with the following significance. If* $\phi : M \to M$ *is an isometry that satisfies*

$$
d(p, \phi(p)) < \varepsilon, \qquad |e - d\phi(p)e|_{\mathbb{R}^{n \times m}} < \varepsilon,
$$

then there exists a complete Killing vector field $X \in \mathrm{Vect}_{K,c}(M)$ *whose flow has the time-1-map* $\psi_X = \phi$.

Proof Use the construction of the map $\Theta : \Omega \to \mathcal{G}_p$ and the Claim in the proof of Lemma 7.3.12 with $\phi_0 = \mathrm{id}$ to obtain $\phi \in \Theta(\Omega \cap (\mathbb{R}^k \times \{0\}))$. □

7.3.5 Examples and Exercises

Throughout we denote by $\mathcal{I}_0(M) \subset \mathcal{I}(M)$ the connected component of the identity in the group of isometries of a Riemannian manifold M.

Example 7.3.14 The isometry group of $\mathbb{R}^m$ is the group of all affine transformations of $\mathbb{R}^m$ with orthogonal linear part (Exercise 5.1.4).

Example 7.3.15 We have seen in Example 6.4.16 that the isometry group of the m-sphere $S^m \subset \mathbb{R}^{m+1}$ is the group $\mathcal{I}(S^m) = O(m + 1)$ of orthogonal transformations of the ambient space. In the case $m = 2$ a theorem of Smale [72] asserts that the inclusion $O(3) = \mathcal{I}(S^2) \hookrightarrow \mathrm{Diff}(S^2)$ of the isometry group of the 2-sphere into the infinite-dimensional group of all diffeomorphisms of the 2-sphere is a homotopy equivalence.

Example 7.3.16   In Sect. 6.4.4 we have introduced the hyperbolic space $\mathbb{H}^m$. Its isometry group is the group $\mathcal{I}(\mathbb{H}^m) = O(m, 1)$ of all linear transformations of $\mathbb{R}^{m+1}$ that preserve the quadratic form Q in (6.4.9) (Exercise 6.4.25). In the case $m = 2$ the identity component of this group is isomorphic to the Lie group $\mathrm{PSL}(2, \mathbb{R}) = \mathrm{SL}(2, \mathbb{R})/\{\pm 1\}$. Geometrically this can be understood by examining the upper half space model of $\mathbb{H}^2$ (Exercise 6.4.27).

The preceding three examples are the constant sectional curvature manifolds discussed in Sect. 6.2 and Sect. 6.4. In all three cases the isometry group has the maximal dimension $\dim(\mathcal{I}(M)) = m(m + 1)/2$ and is diffeomorphic to the orthonormal frame bundle $\mathcal{O}(M)$, so these examples satisfy the condition $\mathcal{I}_p = \mathcal{G}_p$ in the notation of Sect. 7.3.1. By Corollary 6.4.13 a complete, connected, simply connected manifold M satisfies $\mathcal{I}_p = \mathcal{G}_p$ if and only if it has constant sectional curvature.

Exercise 7.3.17   Consider the incomplete 2-manifolds

$$M_0 := \mathbb{R}^2 \setminus \{(0, 0)\}, \qquad M_1 := \mathbb{R}^2 \setminus \mathbb{Z}^2.$$

Prove that for $i = 0, 1$ every isometry of M_i extends to an isometry of $\mathbb{R}^2$ and so $\mathcal{I}(M_i)$ is a subgroup of the Lie group $\mathcal{I}(\mathbb{R}^2)$ of affine maps with orthogonal linear part (Example 7.3.14). The isometry group of M_0 is isomorphic to the Lie group $O(2)$. The isometry group of M_1 is discrete and is an example of a so-called **wallpaper group** (of which there are 17). The Lie algebra of Killing vector fields in both cases has dimension 3. Which Killing vector fields are not complete? **Hint:** A Killing vector field on a connected open set $M \subset \mathbb{R}^2$ is a smooth map $M \to \mathbb{R}^2 : (x, y) \mapsto (u(x, y), v(x, y))$ that satisfies the equations $\partial_x u = \partial_y v = 0$ and $\partial_y u + \partial_x v = 0$. Deduce that the map (u, v) is affine and has a skew-symmetric linear part.

Example 7.3.18   The identity component of the isometry group of the complex projective space $\mathbb{C}P^n$ (Example 2.8.5) with the Fubini–Study metric (Example 3.7.5) is the group

$$\mathcal{I}_0(\mathbb{C}P^n) = \mathrm{PSU}(n + 1) = \mathrm{SU}(n + 1)/Z(\mathrm{SU}(n + 1)) \cong U(n + 1)/U(1).$$

Here $Z(\mathrm{SU}(n + 1)) = \{\lambda \mathbb{1} \mid \lambda \in S^1, \lambda^{n+1} = 1\} \cong \mathbb{Z}/(n + 1)\mathbb{Z}$ is the center of the group $\mathrm{SU}(n + 1)$. We emphasize that the dimension $n(n + 2)$ of the isometry group $\mathcal{I}(\mathbb{C}P^n)$ is smaller than the dimension $n(2n + 1)$ of the orthonormal frame bundle $\mathcal{O}(\mathbb{C}P^n)$ unless $n = 0$ or $n = 1$.

In the case $n = 1$ the projective line $\mathbb{C}P^1$ is isometric to the 2-sphere by stereographic projection (Exercise 2.8.13 and Example 3.7.5). Hence the identity component $\mathrm{PSU}(2)$ of the isometry group of $\mathbb{C}P^1$ is isomorphic to the identity component $\mathrm{SO}(3)$ of the isometry group of S^2. An explicit isomorphism is discussed in Exercise 2.5.22.

The full isometry group $\mathcal{I}(\mathbb{C}\mathrm{P}^n)$ has two connected components. In the case $n = 2$ it is an open question whether the inclusion $\mathcal{I}(\mathbb{C}\mathrm{P}^2) \hookrightarrow \mathrm{Diff}(\mathbb{C}\mathrm{P}^2)$ of the isometry group of $\mathbb{C}\mathrm{P}^2$ into the group of all diffeomorphisms of $\mathbb{C}\mathrm{P}^2$ is a homotopy equivalence. By deep results of Gromov [21] and Taubes [74] a positive answer to this question is equivalent to the assertion that the space of symplectic forms on $\mathbb{C}\mathrm{P}^2$ in a fixed cohomology class is contractible (the symplectic uniqueness conjecture for $\mathbb{C}\mathrm{P}^2$). It is not even known whether this space is connected or, equivalently, whether every diffeomorphism of $\mathbb{C}\mathrm{P}^2$ that induces the identity on cohomology is isotopic to the identity. For a more detailed discussion see [67, Example 3.4] and [49, Example 13.4.1].

Example 7.3.19 The identity component $\mathcal{I}_0(S^2 \times S^2)$ of the isometry group of the product manifold $M = S^2 \times S^2$ is the product group $\mathrm{SO}(3) \times \mathrm{SO}(3)$. In contrast to Smales' Theorem the inclusion $\mathcal{I}_0(S^2 \times S^2) \hookrightarrow \mathrm{Diff}_0(S^2 \times S^2)$ into the group of diffeomorphisms of $S^2 \times S^2$ that are isotopic to the identity is not a homotopy equivalence. For example, if $\phi_{x,\theta} : S^2 \to S^2$ denotes the rotation about the axis through $x \in S^2$ by the angle $\theta \in \mathbb{R}/2\pi\mathbb{Z}$ and the diffeomorphism $\psi_\theta : S^2 \times S^2 \to S^2 \times S^2$ is defined by

$$\psi_\theta(x, y) := (x, \phi_{x,\theta}(y)), \qquad x, y \in S^2,$$

then the loop $\mathbb{R}/2\pi\mathbb{Z} \to \mathrm{Diff}_0(S^2 \times S^2) : \theta \mapsto \psi_\theta$ is not contractible and neither is any of its iterates. Thus $\mathrm{Diff}_0(S^2 \times S^2)$ has an infinite fundamental group while the fundamental group of $\mathrm{SO}(3) \times \mathrm{SO}(3)$ is finite. For $S^2 \times S^2$ it is an open question whether every diffeomorphism that induces the identity on cohomology is isotopic to the identity. For a more detailed discussion see [67, Example 3.5] and [49, Example 13.4.2].

7.4   Isometries of Compact Lie Groups*

In the following theorem we denote by $\mathcal{I}_0(M)$ the connected component of the identity in the group of all isometries of a manifold M.

Theorem 7.4.1 *Let* $G \subset \mathrm{GL}(n, \mathbb{R})$ *be a compact connected Lie group equipped with a bi-invariant Riemannian metric. If* $\phi \in \mathcal{I}_0(G)$, *then there exist elements* $a, b \in G$ *such that*

$$\phi(g) = \phi_{a,b}(g) := agb^{-1} \qquad \textit{for all } g \in G. \tag{7.4.1}$$

The proof of Theorem 7.4.1 makes use of the Killing form introduced in Example 5.2.25. Recall the definition of the center $Z(\mathfrak{g})$ in Exercise 2.5.34 and of the commutant $[\mathfrak{g}, \mathfrak{g}] \subset \mathfrak{g}$ in Exercise 5.2.24. The heart of the proof is Lemma 7.4.3. The case where the Lie algebra has a nontrivial center is then dealt with in Lemma 7.4.5.

Lemma 7.4.2 *Let $\mathfrak{g}$ be a finite-dimensional Lie algebra that admits an invariant inner product and has a trivial center. Then the Killing form on $\mathfrak{g}$ is nondegenerate.*

Proof Exercise 5.2.27. □

Lemma 7.4.3 *Let $\mathfrak{g}$ be a finite-dimensional Lie algebra and assume that the Killing form on $\mathfrak{g}$ is nondegenerate. Then the following holds.*

(i) $Z(\mathfrak{g}) = \{0\}$ and $[\mathfrak{g}, \mathfrak{g}] = \mathfrak{g}$.
(ii) *The Lie algebra homomorphism* $\mathrm{ad} : \mathfrak{g} \to \mathrm{Der}(\mathfrak{g})$ *is bijective.*
(iii) *Every derivation* $\delta : \mathfrak{g} \to \mathfrak{g}$ *has trace zero.*
(iv) *Let* $\delta : \mathfrak{g} \to \mathfrak{g}$ *be a linear map such that, for all* $\xi, \eta, \zeta \in \mathfrak{g}$,

$$\delta[[\xi, \eta], \zeta] = [[\delta\xi, \eta], \zeta] + [[\xi, \delta\eta], \zeta] + [[\xi, \eta], \delta\zeta]. \tag{7.4.2}$$

Then there exists a unique element $\xi_\delta \in \mathfrak{g}$ *such that* $\delta\xi = [\xi_\delta, \xi]$ *for all* $\xi \in \mathfrak{g}$.

Proof We prove part (i). If $\xi \in Z(\mathfrak{g})$, then $\mathrm{ad}(\xi) = 0$, hence $\kappa(\xi, \eta) = 0$ for all $\eta \in \mathfrak{g}$, and hence $\xi = 0$ by nondegeneracy of the Killing form. To prove that $[\mathfrak{g}, \mathfrak{g}] = \mathfrak{g}$, assume that $\Lambda : \mathfrak{g} \to \mathbb{R}$ is any linear functional that vanishes on the commutant $[\mathfrak{g}, \mathfrak{g}]$. Since the Killing form is nondegenerate, there exists an element $\zeta \in \mathfrak{g}$ such that $\Lambda = \kappa(\zeta, \cdot)$. Hence $0 = \kappa(\zeta, [\xi, \eta]) = \kappa([\zeta, \xi], \eta)$ for all $\xi, \eta \in \mathfrak{g}$. Since the Killing form is nondegenerate, this implies $[\zeta, \xi] = 0$ for all $\xi \in \mathfrak{g}$, hence $\zeta \in Z(\mathfrak{g})$, hence $\zeta = 0$, and hence $\Lambda = 0$. This proves (i).

We prove part (ii). The map $\mathrm{ad} : \mathfrak{g} \to \mathrm{Der}(\mathfrak{g})$ is injective by part (i). Choose a basis $\xi_1, \dots, \xi_k$ of $\mathfrak{g}$. Since the Killing form is nondegenerate, there exists a dual basis $\eta_1, \dots, \eta_k$ of $\mathfrak{g}$ such that $\kappa(\xi_i, \eta_j) = \delta_{ij}$ for all i, j. These bases satisfy $\zeta = \sum_i \kappa(\eta_i, \zeta)\xi_i$ for all $\zeta \in \mathfrak{g}$. Now let $\delta \in \mathrm{Der}(\mathfrak{g})$ and define $\xi := \sum_i \mathrm{trace}\big(\delta\,\mathrm{ad}(\xi_i)\big)\eta_i \in \mathfrak{g}$. Then, for all $\zeta \in \mathfrak{g}$, we have

$$\mathrm{trace}\big(\delta\,\mathrm{ad}(\zeta)\big) = \sum_i \kappa(\eta_i, \zeta)\mathrm{trace}\big(\delta\,\mathrm{ad}(\xi_i)\big) = \kappa(\xi, \zeta) = \mathrm{trace}\big(\mathrm{ad}(\xi)\mathrm{ad}(\zeta)\big).$$

Thus the derivation $\varepsilon := \delta - \mathrm{ad}(\xi)$ satisfies $\mathrm{trace}(\varepsilon\,\mathrm{ad}(\zeta)) = 0$ for all $\zeta \in \mathfrak{g}$. Since $[\varepsilon, \mathrm{ad}(\eta)] = \mathrm{ad}(\varepsilon\eta)$, this implies

$$\kappa(\varepsilon\eta, \zeta) = \mathrm{trace}\big(\mathrm{ad}(\varepsilon\eta)\mathrm{ad}(\zeta)\big) = \mathrm{trace}\big([\varepsilon, \mathrm{ad}(\eta)]\mathrm{ad}(\zeta)\big)$$
$$= \mathrm{trace}\big(\varepsilon[\mathrm{ad}(\eta), \mathrm{ad}(\zeta)]\big) = \mathrm{trace}\big(\varepsilon\,\mathrm{ad}([\eta, \zeta])\big) = 0$$

for all $\eta, \zeta \in \mathfrak{g}$. Hence $\varepsilon = 0$ by nondegeneracy of the Killing form and hence $\delta = \mathrm{ad}(\xi)$. This proves (ii).

We prove part (iii). Since $\mathrm{trace}(\mathrm{ad}([\xi, \eta])) = \mathrm{trace}([\mathrm{ad}(\xi), \mathrm{ad}(\eta)]) = 0$ for all $\xi, \eta \in \mathfrak{g}$ and $[\mathfrak{g}, \mathfrak{g}] = \mathfrak{g}$ by part (i), we have $\mathrm{trace}(\mathrm{ad}(\xi)) = 0$ for all $\xi \in \mathfrak{g}$. Hence (iii) follows from part (ii).

We prove part (iv). Define the bilinear map $B_\delta : \mathfrak{g} \times \mathfrak{g} \to \mathfrak{g}$ by

$$B_\delta(\xi, \eta) := \delta[\xi, \eta] - [\delta\xi, \eta] - [\xi, \delta\eta] \tag{7.4.3}$$

for $\xi, \eta \in \mathfrak{g}$. Then equation (7.4.2) can be expressed in the form

$$B_\delta([\xi, \eta], \zeta) + [B_\delta(\xi, \eta), \zeta] = 0 \tag{7.4.4}$$

for $\xi, \eta, \zeta \in \mathfrak{g}$. By part (i) there exists a basis of $\mathfrak{g}$ consisting of vectors of the form $e_i = [\xi_i, \eta_i]$. Define the linear map $A_\delta : \mathfrak{g} \to \mathfrak{g}$ by $A_\delta e_i := -B_\delta(\xi_i, \eta_i)$. Then $B_\delta(e_i, \zeta) = [A_\delta e_i, \zeta]$ for all i and ζ by (7.4.4). Hence

$$B_\delta(\xi, \eta) = [A_\delta\xi, \eta] = [\xi, A_\delta\eta] \tag{7.4.5}$$

for all $\xi, \eta \in \mathfrak{g}$. Here the second equality holds by the skew-symmetry of B_δ. By the Jacobi identity and equations (7.4.4) and (7.4.5) we have

$$\begin{aligned}
2[B_\delta(\xi, \eta), \zeta] &= [[\xi, A_\delta\eta], \zeta] + [[A_\delta\xi, \eta], \zeta] \\
&= -[[\zeta, \xi], A_\delta\eta] - [[A_\delta\eta, \zeta], \xi] + [[A_\delta\xi, \eta], \zeta] \\
&= -B_\delta([\zeta, \xi], \eta) - [B_\delta(\eta, \zeta), \xi] + [[A_\delta\xi, \eta], \zeta] \\
&= [B_\delta(\zeta, \xi), \eta] + B_\delta([\eta, \zeta], \xi) + [[A_\delta\xi, \eta], \zeta] \\
&= [[\zeta, A_\delta\xi], \eta] + [[\eta, \zeta], A_\delta\xi] + [[A_\delta\xi, \eta], \zeta] = 0
\end{aligned}$$

for all $\xi, \eta, \zeta \in \mathfrak{g}$. Since $Z(\mathfrak{g}) = \{0\}$ by part (i), we find that $B_\delta(\xi, \eta) = 0$ for all $\xi, \eta \in \mathfrak{g}$, hence δ is a derivation, and hence δ is in the image of the map $\mathrm{ad} : \mathfrak{g} \to \mathrm{Der}(\mathfrak{g})$ by part (ii). This proves (iv) and Lemma 7.4.3. $\qquad\square$

Lemma 7.4.4 *The assertion of Theorem 7.4.1 holds under the additional assumption that the center of the Lie algebra $\mathfrak{g} = \mathrm{Lie}(G)$ is trivial.*

Proof   Let $\phi \in \mathcal{I}_0(G)$ and choose a smooth isotopy $[0, 1] \to \mathcal{I}_0(G) : t \mapsto \phi_t$ joining the identity $\phi_0 = \mathrm{id}$ to $\phi_1 = \phi$. For $0 \le t \le 1$ define the diffeomorphism $\psi_t : G \to G$ by $\psi_t(g) := \phi_t(\mathbb{1})^{-1}\phi_t(g)$. Then each ψ_t is an isometry and the path $[0, 1] \to \mathcal{I}_0(G) : t \mapsto \psi_t$ satisfies

$$\psi_0 = \mathrm{id}, \qquad \psi_t(\mathbb{1}) = \mathbb{1}$$

for all t. By Theorem 5.3.1 the derivatives $\Psi_t := d\psi_t(\mathbb{1}) : \mathfrak{g} \to \mathfrak{g}$ preserve the Riemann curvature tensor of G at $\mathbb{1}$ and by Example 5.2.18 (for the standard metric on Lie subgroups of $O(n)$) and Exercise 5.2.22 (for general bi-invariant Riemannian metrics) this translates into the condition

$$\Psi_t[[\xi, \eta], \zeta] = [[\Psi_t\xi, \Psi_t\eta], \Psi_t\zeta] \tag{7.4.6}$$

for all t and all $\xi, \eta, \zeta \in \mathfrak{g}$. Hence the endomorphism $\delta_t := \Psi_t^{-1}\frac{d}{dt}\Psi_t : \mathfrak{g} \to \mathfrak{g}$ satisfies (7.4.2) for each t. Moreover, by Lemma 7.4.2 the Killing form on $\mathfrak{g}$ is nondegenerate. Hence it follows from Lemma 7.4.3 that there exists a smooth path $[0, 1] \to \mathfrak{g} : t \mapsto \xi_t$ such that $\Psi_t^{-1}\frac{d}{dt}\Psi_t = \mathrm{ad}(\xi_t)$ for all t. Thus

$$\frac{d}{dt}\Psi_t = \Psi_t\mathrm{ad}(\xi_t), \qquad \Psi_0 = \mathbb{1}. \tag{7.4.7}$$

Now let $[0, 1] \to G : t \mapsto b_t$ be the solution of the differential equation

$$\frac{d}{dt}b_t = b_t\xi_t, \qquad b_0 = \mathbb{1}, \tag{7.4.8}$$

and define $\Phi_t := \mathrm{Ad}(b_t)$ (Example 2.5.23). Then $\frac{d}{dt}\Phi_t = \Phi_t\mathrm{ad}(\xi_t)$ for all t and $\Phi_0 = \mathbb{1}$. Thus $\Phi_t = \Psi_t$ and so

$$\Psi_t\eta = b_t\eta b_t^{-1} \tag{7.4.9}$$

for all t and η. Take $t = 1$, define $b := b_1$, and use Lemma 5.1.10 (uniqueness of local isometries) to obtain $\psi_1(g) = bgb^{-1}$ for all $g \in G$. Hence

$$\phi(g) = \phi(\mathbb{1})\psi_1(g) = \phi(\mathbb{1})bgb^{-1} = agb^{-1}$$

for all $g \in G$ with $a := \phi(\mathbb{1})b$. This proves Lemma 7.4.4. □

We will denote by $Z_0(G) \subset Z(G)$ the connected component of $\mathbb{1}$ in the center of G. Then $Z_0(G) \subset G$ is a compact connected abelian Lie subgroup of G (Exercise 2.5.34). Since G is connected, its Lie algebra is the center of $\mathfrak{g}$, i.e. $\mathrm{Lie}(Z_0(G)) = Z(\mathfrak{g})$.

Lemma 7.4.5 *Let* $G \subset GL(n, \mathbb{R})$ *be a compact connected Lie group with a bi-invariant Riemannian metric and let* $\phi \in \mathcal{I}_0(G)$. *Then*

$$\phi(hg) = \phi(gh) = \phi(g)h = h\phi(g) \tag{7.4.10}$$

for all $g \in G$ *and all* $h \in Z_0(G)$.

Proof The proof relies on the following basic observations.

(a) $Z_0(G)$ is a compact connected Lie subgroup of G and $\mathrm{Lie}(Z_0(G)) = Z(\mathfrak{g})$.
(b) $\exp(\xi + \eta) = \exp(\xi)\exp(\eta)$ for all $\xi, \eta \in Z(\mathfrak{g})$.
(c) The exponential map $\exp : Z(\mathfrak{g}) \to Z_0(G)$ is surjective.
(d) $\Lambda := \{\xi \in Z(\mathfrak{g}) \mid \exp(\xi) = \mathbb{1}\}$ is a discrete additive subgroup of $Z(\mathfrak{g})$ which spans $Z(\mathfrak{g})$.

Part (a) was noted above, part (b) follows from Exercise 2.5.39 because the Lie algebra $Z(\mathfrak{g})$ is commutative, and part (c) follows from the Hopf–Rinow Theorem 4.6.6. That the set Λ is an additive subgroup of $Z(\mathfrak{g})$ follows directly from (b). Moreover, by (a) and (b) the exponential map restricts to a local diffeomorphism $\exp : Z(\mathfrak{g}) \to Z_0(G)$. Hence Λ is discrete and by (c) the exponential map descends to a proper map from $Z(\mathfrak{g})/\Lambda$ onto $Z_0(G)$. Since $Z_0(G)$ is compact, the lattice Λ spans the vector space $Z(\mathfrak{g})$.

Now assume $\phi(\mathbb{1}) = \mathbb{1}$ and define $\Phi := d\phi(\mathbb{1}) : \mathfrak{g} \to \mathfrak{g}$. Then

$$\phi(\exp(\xi)) = \exp(\Phi\xi) \tag{7.4.11}$$

for all $\xi \in \mathfrak{g}$ by Corollary 5.3.3 and Example 5.2.18. Moreover, Φ is an orthogonal transformation of $\mathfrak{g}$ that preserves the Riemann curvature tensor (Theorem 5.3.1). Thus $\|[\Phi\xi, \Phi\eta]\| = \|[\xi, \eta]\|$ for all $\xi, \eta \in \mathfrak{g}$ by Example 5.2.18. So, if $\xi \in Z(\mathfrak{g})$, then $[\xi, \eta] = 0$ for all η, hence $[\Phi\xi, \Phi\eta] = 0$ for all η, and hence $\Phi\xi \in Z(\mathfrak{g})$. This shows that

$$\Phi Z(\mathfrak{g}) = Z(\mathfrak{g}). \tag{7.4.12}$$

By (7.4.11) and (7.4.12) we have $\xi \in \Lambda$ if and only if $\phi(\exp(\xi)) = \mathbb{1}$ if and only if $\exp(\Phi\xi) = \mathbb{1}$ if and only if $\Phi\xi \in \Lambda$, so that $\Phi\Lambda = \Lambda$. Since ϕ is isotopic to the identity through isometries, by assumption, there exists a smooth path of orthogonal transformations Φ_t of $\mathfrak{g}$ from $\Phi_0 = \mathbb{1}$ to $\Phi_1 = \Phi$ that satisfy $\Phi_t \Lambda = \Lambda$ for all t. Thus $\Phi\xi = \xi$ for all $\xi \in \Lambda$ and so for all $\xi \in Z(\mathfrak{g})$ because the lattice spans the subspace $Z(\mathfrak{g})$ by (d). Hence, by (7.4.11) and (c), we obtain $\phi(h) = h$ for all $h \in Z_0(G)$ whenever $\phi(\mathbb{1}) = \mathbb{1}$.

To prove equation (7.4.10) in general, fix an element $g \in G$ and define the diffeomorphism $\phi_g : G \to G$ by $\phi_g(h) := \phi(g)^{-1}\phi(gh)$ for $h \in G$. Then ϕ_g is an isometry and $\phi_g(\mathbb{1}) = \mathbb{1}$. Moreover, $\phi_g \in \mathcal{I}_0(G)$, because G is connected. Hence $\phi_g(h) = h$ for all $h \in Z_0(G)$ and this proves Lemma 7.4.5. $\square$

Proof of Theorem 7.4.1   Let $\phi \in \mathcal{I}_0(G)$. By Lemma 7.4.5, ϕ descends to a diffeomorphism $\bar{\phi} : \bar{G} \to \bar{G}$ of the quotient group

$$\bar{G} := G/Z_0(G).$$

The Lie algebra of $\bar{G}$ is the quotient space

$$\bar{\mathfrak{g}} := \mathfrak{g}/Z(\mathfrak{g}) \cong Z(\mathfrak{g})^{\perp}$$

and the invariant inner product on $\mathfrak{g}$ restricts to an invariant inner product on the orthogonal complement of $Z(\mathfrak{g})$. This defines a bi-invariant Riemannian metric on $\bar{G}$. The diffeomorphism $\bar{\phi}$ is an isometry with respect to this metric, because the derivative $d\phi(g) : T_g G \to T_{\phi(g)} G$ is an orthogonal transformation, which by (7.4.10) sends a tangent vector $g\eta \in T_g G$ with $\eta \in Z(\mathfrak{g})$ to $d\phi(g)g\eta = \phi(g)\eta \in T_{\phi(g)}G$, and hence it sends the subspace $g Z(\mathfrak{g})^{\perp} \subset T_g G$ to $\phi(g)Z(\mathfrak{g})^{\perp} \subset T_{\phi(g)}G$. Apply Lemma 7.4.4 to the isometry $\bar{\phi}$ to obtain elements $a, b \in G$ whose equivalence classes $\bar{a}, \bar{b} \in \bar{G}$ satisfy $\bar{\phi}(\bar{g}) = \bar{a}\bar{g}\bar{b}^{-1}$ for all $\bar{g} \in \bar{G}$. This implies

$$\alpha(g) := a^{-1}\phi(g)bg^{-1} \in Z_0(G) \tag{7.4.13}$$

for all $g \in G$. Moreover, it follows from Lemma 7.4.5 that

$$\alpha(gh) = \alpha(g) \tag{7.4.14}$$

for all $g \in G$ and all $h \in Z_0(\mathfrak{g})$. Now define the isometry $\psi : G \to G$ by

$$\psi(g) := a^{-1}\phi(g)b = \alpha(g)g \tag{7.4.15}$$

for $g \in G$. Fix an element $g \in G$ and define the linear maps $A, \Psi : \mathfrak{g} \to \mathfrak{g}$ by

$$A\xi := \alpha(g)^{-1}d\alpha(g)g\xi, \qquad \Psi\xi := \psi(g)^{-1}d\psi(g)g\xi \tag{7.4.16}$$

for $\xi \in \mathfrak{g}$. Then $\Psi = \mathrm{id} + A$ by (7.4.15) and the map A vanishes on $Z(\mathfrak{g})$ by (7.4.14) and takes values in $Z(\mathfrak{g})$ by (7.4.13). Since Ψ is an orthogonal transformation, this implies

$$\Psi = \mathrm{id}, \qquad A = 0.$$

Hence it follows from the definition of the linear map A in (7.4.16) that the map $\alpha : G \to Z_0(G)$ in (7.4.13) is constant. Thus

$$\phi(g) = \alpha(\mathbb{1})agb^{-1}$$

for all $g \in G$, and this proves Theorem 7.4.1.   □

Corollary 7.4.6 *Let* G *be a compact connected Lie group equipped with a bi-invariant Riemannian metric. Then the map* $(a,b) \mapsto \phi_{a,b}$ *in Theorem 7.4.1 descends to a Lie Group isomorphism*

$$\rho_G : (G \times G)/Z(G) \to \mathcal{I}_0(G). \tag{7.4.17}$$

Proof   The map $G \times G \to \mathcal{I}_0(G) : (a,b) \mapsto \phi_{a,b}$ is a group homomorphism by definition, is smooth by Theorem 7.3.1, and is surjective by Theorem 7.4.1. Moreover, $\phi_{a,b}$ is the identity if and only if $a = b \in Z(G)$. Hence the map ρ_G in (7.4.17) is a Lie group isomorphism, with the smooth structure on the quotient group $(G \times G)/Z(G)$ determined by Theorem 2.9.14. This proves Corollary 7.4.6.   □

Example 7.4.7   Theorem 7.4.1 establishes a one-to-one correspondence between isometries $\phi \in \mathcal{I}_0(G)$ that satisfy $\phi(\mathbb{1}) = \mathbb{1}$ (the case $a = b$) and Lie group automorphisms of G in the identity component. This correspondence does not extend to other connected components.

For example, in the case $G = \mathbb{T}^n = \mathbb{R}^n/\mathbb{Z}^n$ the group of automorphisms of $\mathbb{T}^n$ is the infinite group $\mathrm{Aut}(\mathbb{T}^n) = \mathrm{GL}(n, \mathbb{Z})$ of integer matrices with determinant ± 1, while the group of isometries that fix the origin is the finite subgroup $O(n, \mathbb{Z}) \subset \mathrm{GL}(n, \mathbb{Z})$ of orthogonal integer matrices. Also, if G is not abelian, then the isometry $G \to G : g \mapsto g^{-1}$ is not a Lie group automorphism, but a Lie group anti-automorphism.

Exercise 7.4.8 Examine the case $G = SU(2)$ (Example 2.5.21) and deduce that there exists a Lie group isomorphism

$$SO(4) \cong \big(SU(2) \times SU(2)\big)/\{\pm\mathbb{1}\}.$$

Exercise 7.4.9 Let G be a compact Lie group equipped with a bi-invariant Riemannian metric and let $\mathfrak{g} := \mathrm{Lie}(G)$. Show that the formula

$$[(\zeta, \delta), (\zeta', \delta')] := \Big(\delta\zeta' - \delta'\zeta, [\delta, \delta'] + \tfrac{1}{4}\mathrm{ad}\big([\zeta, \zeta']\big)\Big) \tag{7.4.18}$$

for $\zeta, \zeta' \in \mathfrak{g}$ and $\delta, \delta' \in \mathrm{Der}(\mathfrak{g})$ defines a Lie bracket on $\mathfrak{g} \times \mathrm{Der}(\mathfrak{g})$. Show that the map $\mathrm{Vect}_K(G) \to \mathfrak{g} \times \mathrm{Der}(\mathfrak{g}) : X \mapsto (X(\mathbb{1}), \nabla X(\mathbb{1}))$ identifies the Lie algebra $\mathrm{Vect}_K(G) = \mathrm{Lie}(\mathcal{I}(G))$ of Killing vector fields on G with the Lie algebra $\mathfrak{g} \times \mathrm{Der}(\mathfrak{g})$. Show that the homomorphism (7.4.17) induces the surjective Lie algebra homomorphism

$$\mathfrak{g} \times \mathfrak{g} \to \mathfrak{g} \times \mathrm{Der}(\mathfrak{g}) : (\xi, \eta) \mapsto \Big(\xi - \eta, \tfrac{1}{2}\mathrm{ad}(\xi + \eta)\Big), \tag{7.4.19}$$

whose kernel consist of all pairs $(\xi, \eta) \in \mathfrak{g} \times \mathfrak{g}$ that satisfy $\xi = \eta \in Z(\mathfrak{g})$. Show that $\dim(\mathrm{Der}(\mathfrak{g})) = \dim(\mathfrak{g}) - \dim(Z(\mathfrak{g}))$.

7.5 Convex Functions on Hadamard Manifolds*

The last two sections of this book are devoted to Donaldson's beautiful paper [17] in which he develops a differential geometric approach to Lie algebra theory. The results of [17] will be explained in reverse order. The first subsection examines the *sphere at infinity* of a Hadamard manifold M and contains a proof of [17, Theorem 4], which asserts that every convex function $f : M \to \mathbb{R}$ that is invariant under the action of a Lie group G by isometries must attain its minimum whenever the G-action has no fixed point at infinity (Sect. 7.5.1). The next subsection deals with the special case of [17, Theorem 3], where M is the manifold of inner products on a vector space V on which a Lie group $G \subset SL(V)$ acts irreducibly (Sect. 7.5.2). If G is the identity component of the isotropy subgroup of a nonzero vector $w \in W$ under a representation of the special linear group $\rho : SL(V) \to SL(W)$, then by [17, Theorem 2] there exists an inner product on V for which the Lie algebra $\mathfrak{g}$ of G is symmetric (Sect. 7.5.3). This is used in [17, Theorem 1] in the case where $V = \mathfrak{g}$ is a simple Lie algebra, w is the Lie bracket, and G is the identity component of the group of automorphisms of $\mathfrak{g}$, to establish the existence of symmetric inner products on $\mathfrak{g}$ and deduce various standard results in Lie algebra theory. These applications to Lie algebra theory are deferred to the next section.

7.5.1   Convex Functions and The Sphere at Infinity

Assume throughout that M is a Hadamard manifold, i.e. a nonempty, connected, simply connected, complete Riemannian manifold with nonpositive sectional curvature (Definition 6.5.1). For $p \in M$ denote the unit sphere in the tangent space $T_p M$ by

$$S_p := \{v \in T_p M \mid |v| = 1\}.$$

Their union determines a submanifold $SM := \{(p, v) \mid p \in M, \ v \in S_p\}$ of the tangent bundle, called the **unit sphere bundle**.

Definition 7.5.1   Define an equivalence relation $\sim$ on SM by

$$(p, v) \sim (q, w) \quad \overset{\text{def}}{\Longleftrightarrow} \quad \sup_{t \geq 0} d(\exp_p(tv), \exp_q(tw)) < \infty \qquad (7.5.1)$$

for $(p, v), (q, w) \in SM$. The equivalence class of a pair $(p, v) \in SM$ will be denoted by $[p, v] := \{(q, w) \in SM \mid (q, w) \sim (p, v)\}$ and the quotient space

$$S_\infty(M) := SM/\!\!\sim \ = \{[p, v] \mid (p, v) \in SM\}$$

is called the **sphere at infinity** of M.

The following lemma shows that the map $S_p \to S_\infty(M) : v \mapsto [p, v]$ is a homeomorphism with respect to the quotient topology on $S_\infty(M)$ for every $p \in M$ (see [17, Lemma 3]).

Lemma 7.5.2   *There exists a unique collection of maps $F_{q,p} : S_p \to S_q$, one for each pair of points $p, q \in M$, such that*

$$F_{q,p}(v) = \lim_{R \to \infty} \frac{\exp_q^{-1}(\exp_p(Rv))}{|\exp_q^{-1}(\exp_p(Rv))|} \qquad (7.5.2)$$

for all $p, q \in M$ and all $v \in S_p$. Moreover, the convergence in (7.5.2) is uniform on S_p, the maps $F_{q,p}$ are homeomorphisms, and they satisfy

$$w = F_{q,p}(v) \quad \Longleftrightarrow \quad \sup_{t \geq 0} d(\exp_p(tv), \exp_q(tw)) < \infty \qquad (7.5.3)$$

for all $(p, v), (q, w) \in SM$ and

$$F_{r,q} \circ F_{q,p} = F_{r,p}, \qquad F_{p,p} = \mathrm{id} \qquad (7.5.4)$$

for all $p, q, r \in M$.

Fig. 7.1 The sphere at infinity

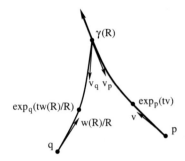

Proof Let $p, q \in M$ and define the maps $F_{R,q,p} : S_p \to S_q$ by

$$F_{R,q,p}(v) := \frac{\exp_q^{-1}(\exp_p(Rv))}{|\exp_q^{-1}(\exp_p(Rv))|} \tag{7.5.5}$$

for $R > 0$ and $v \in S_p$. We claim that, for all $R > d(p,q)$ and all $v \in S_p$,

$$\left| \frac{\partial}{\partial R} F_{R,q,p}(v) \right| \le \frac{d(p,q)}{R(R - d(p.q))}. \tag{7.5.6}$$

To see this, fix an element $v \in S_p$ and define the geodesic $\gamma : \mathbb{R} \to M$ by $\gamma(t) := \exp_p(tv)$. Define the curve $w : \mathbb{R} \to T_q M$ by $w(t) := \exp_q^{-1}(\gamma(t))$ for $t \in \mathbb{R}$. Then $F_{R,q,p}(v) = w(R)/|w(R)|$ and hence

$$\frac{\partial}{\partial R} F_{R,q,p}(v) = \frac{\dot{w}(R)}{|w(R)|} - \left\langle \frac{\dot{w}(R)}{|w(R)|}, \frac{w(R)}{|w(R)|} \right\rangle \frac{w(R)}{|w(R)|}.$$

This is the orthogonal projection of the vector $\dot{w}(R)/|w(R)|$ onto the orthogonal complement of $w(R)$. Hence its length connot decrease by adding to it a scalar multiple of $w(R)$, and so

$$\left| \frac{\partial}{\partial R} F_{R,q,p}(v) \right| \le \left| \frac{\dot{w}(R)}{|w(R)|} - \frac{w(R)}{R|w(R)|} \right| = \frac{|\dot{w}(R) - w(R)/R|}{|w(R)|}. \tag{7.5.7}$$

Next we use the expanding property of the exponential map in Theorem 6.5.2 twice, namely first the inequality for the derivative in part (ii) at the point q and then the inequality in part (iii) at the point $\gamma(R)$ (see Fig. 7.1). Define the tangent vectors $v_p, v_q \in T_{\gamma(R)}M$ by

$$v_p := -\dot{\gamma}(R) = -d\exp_q(w(R))\dot{w}(R),$$

$$v_q := -\frac{d}{dt}\bigg|_{t=R} \exp_q(tw(R)/R) = -d\exp_q(w(R))w(R)/R.$$

Then Theorem 6.5.2 yields the estimate

$$|\dot{w}(R) - w(R)/R| \leq |d\exp_q(w(R))(\dot{w}(R) - w(R)/R)| = |v_p - v_q|. \quad (7.5.8)$$

Also, $\exp_{\gamma(R)}(sv_p) = \exp_p((R-s)v)$, $\exp_{\gamma(R)}(sv_q) = \exp_q((R-s)w(R)/R)$. Take $s = R - t$ and use Lemma 6.5.5 to obtain, for $0 \leq t < R$,

$$|v_p - v_q| \leq \frac{d\left(\exp_p(tv), \exp_q(tw(R)/R)\right)}{R-t} \leq \frac{d(p,q)}{R}. \quad (7.5.9)$$

Since $d(p, \gamma(R)) = R$ and $d(q, \gamma(R)) = |w(R)|$, the triangle inequality yields

$$R - d(p,q) \leq |w(R)| \leq R + d(p,q). \quad (7.5.10)$$

Combinig the inequalities (7.5.7), (7.5.8), (7.5.9), and (7.5.10) we find that

$$\left|\frac{\partial}{\partial R} F_{R,q,p}(v)\right| \leq \frac{|\dot{w}(R) - w(R)/R|}{|w(R)|} \leq \frac{|v_p - v_q|}{|w(R)|} \leq \frac{d(p,q)}{R(R - d(p,q))}.$$

This proves the estimate (7.5.6).

It follows from (7.5.6) by integrating from a fixed number $R > d(p,q)$ to infinity that the maps $F_{R,q,p} : S_p \to S_q$ converge to a map $F_{q,p} : S_p \to S_q$ as R tends to infinity and that

$$\sup_{v \in S_p}\left|F_{q,p}(v) - F_{R,q,p}(v)\right| \leq \int_R^\infty \frac{d(p,q)}{r(r - d(p,q))}\,dr = \log\left(\frac{R}{R - d(p,q)}\right)$$

for all $R > d(p,q)$. Thus the convergence is uniform and hence the limit map $F_{q,p}$ is continuous.

Next we prove (7.5.3). Let $p, q \in M$ and $v \in S_p$. Then

$$w := F_{q,p}(v) = \lim_{R \to \infty} F_{R,q,p}(v) = \lim_{R \to \infty} \frac{w(R)}{|w(R)|} = \lim_{R \to \infty} \frac{w(R)}{R}.$$

Here the third equality follows from (7.5.5) and the definition of the vector $w(R) = \exp_q^{-1}(\exp_p(Rv))$, and the last equality follows from (7.5.10). Hence $\exp_q(tw) = \lim_{R \to \infty} \exp_q(tw(R)/R)$, and so it follows from (7.5.9) by taking the limit $R \to \infty$ that $d(\exp_p(tv), \exp_q(tw)) \leq d(p,q)$ for all $t \geq 0$. This proves " $\implies$ " in (7.5.3). The converse implication follows from the fact that, by Theorem 6.5.2, there can be at most one tangent vector $w \in S_q$ satisfying $\sup_{t>0} d(\exp_p(tv), \exp_p(tw)) < \infty$. Thus we have proved that the maps $F_{q,p} : S_p \to S_q$ in (7.5.2) satisfy (7.5.3). That they also satisfy (7.5.4) follows directly from (7.5.3) and the fact that (7.5.1) defines an equivalence relation on SM. Hence each map $F_{q,p} : S_p \to S_q$ is a homeomorphism with the inverse $F_{p,q} : S_q \to S_p$ and this proves Lemma 7.5.2.   □

Lemma 7.5.3 *If $\phi \in \mathcal{I}(M)$, $p, q \in M$, and $v \in S_p$, then*

$$F_{\phi(q),\phi(p)} \circ d\phi(p) = d\phi(q) \circ F_{q,p} : S_p \to S_{\phi(q)}. \tag{7.5.11}$$

Thus the group of isometries of M acts continuously on the sphere at infinity via $\mathcal{I}(M) \times S_\infty(M) \to S_\infty(M) : (\phi, [p, v]) \mapsto \phi_*[p, v] := [\phi(p), d\phi(p)v]$.

Proof Since ϕ is an isometry it satisfies $\phi \circ \exp_q = \exp_{\phi(q)} \circ d\phi(q)$ for all $q \in M$ by Corollary 5.3.3. Hence

$$d\phi(q)\Big(\exp_q^{-1}\big(\exp_p(Rv)\big)\Big) = \exp_{\phi(q)}^{-1}\Big(\phi\big(\exp_p(Rv)\big)\Big)$$
$$= \exp_{\phi(q)}^{-1}\Big(\exp_{\phi(p)}\big(R\,d\phi(p)v\big)\Big)$$

and so $d\phi(q) \circ F_{R,q,p} = F_{R,\phi(q),\phi(p)} \circ d\phi(p)$ for all $p, q \in M$ and all $R > 0$. Divide by the norm and take the limit $R \to \infty$ to obtain (7.5.11). This proves Lemma 7.5.3. $\qquad\qquad\square$

Definition 7.5.4 Let $G \subset \mathrm{GL}(n, \mathbb{R})$ be a Lie group. A **G-action on M by isometries** is a Lie group homomorphism

$$G \to \mathcal{I}(M) : g \mapsto \phi_g,$$

i.e. the map $G \times M \to M : (g, p) \mapsto \phi_g(p)$ is a smooth group action (Definition 2.5.40) and the map $\phi_g : M \to M$ is an isometry for each $g \in G$. In this situation we say that the G action has a **fixed point** iff there exists an element $p \in M$ such that

$$\phi_g(p) = p$$

for all $g \in G$. We say that the G-action has **a fixed point at infinity** iff the induced G-action on the sphere at infinity has a fixed point, i.e. there exist elements $p \in M$ and $v \in S_p$ such that

$$d\phi_g(p)v = F_{\phi_g(p),p}(v)$$

for all $g \in G$. If such a pair $(p, v) \in S(M)$ does not exist, we say that the G-action has **no fixed point at infinity**.

Definition 7.5.5 A smooth function $f : M \to \mathbb{R}$ is called **convex** iff the function $f \circ \gamma : \mathbb{R} \to \mathbb{R}$ is convex for every geodesic $\gamma : \mathbb{R} \to M$.

With these preparations in place we are ready to state the following existence theorem for critical points of a convex function (see [17, Theorem 4]).

Theorem 7.5.6 (Donaldson) *Let M be a Hadamard manifold equipped with a smooth action $G \to \mathcal{I}(M) : g \mapsto \phi_g$ of a Lie group G by isometries, let K be a compact subgroup of G, and let $f : M \to \mathbb{R}$ be a convex function such that*

$$f \circ \phi_g = f$$

for all $g \in G$. Assume that the G-action has no fixed point at infinity. Then there exists an element $p_0 \in M$ such that

$$f(p_0) = \inf_{p \in M} f(p), \qquad \phi_u(p_0) = p_0 \qquad \text{for all } u \in K. \tag{7.5.12}$$

As pointed out in [17], similar results can be found in the works of Bishop–O'Neill [7] and Bridson–Haefliger [11, Lemma 8.26]. We remark that the compactness of the subgroup $K \subset G$ is only needed for an appeal to Cartan's Fixed Point Theorem 6.5.6. If we assume instead that the action of K on M has a fixed point, compactness is not required. The proof of Theorem 7.5.6 is based in the following lemma (see [17, Lemma 5]).

Lemma 7.5.7 *Let p_i be a sequence in M such that $\lim_{i \to \infty} d(p, p_i) = \infty$ for some (and hence every) $p \in M$. Let $\mathcal{I} \subset \mathcal{I}(M)$ be a collection of isometries of M such that $\sup_i d(p_i, \phi(p_i)) < \infty$ for all $\phi \in \mathcal{I}$. Then the isometries in $\mathcal{I}$ have a common fixed point at infinity, i.e. there exists an element $[p, v] \in S_\infty(M)$ such that $\phi_*[p, v] = [p, v]$ for all $\phi \in \mathcal{I}$.*

Proof Fix an element $p \in M$ and define

$$v_i := \frac{\exp_p^{-1}(p_i)}{R_i}, \qquad R_i := |\exp_p^{-1}(p_i)| = d(p, p_i), \tag{7.5.13}$$

for each $i \in \mathbb{N}$ such that $p_i \neq p$. Passing to a subsequence, if necessary, we may assume that $p_i \neq p$ for all $i \in \mathbb{N}$ and that the limit $v := \lim_{i \to \infty} v_i \in S_p$ exists. We will prove that $d\phi(p)v = F_{\phi(p),p}(v)$ for all $\phi \in \mathcal{I}$. To see this, let $\phi \in \mathcal{I}$, choose $c > 0$ such that $d(p_i, \phi(p_i)) \leq c$ for all i, and define

$$v_i' := \frac{\exp_p^{-1}(\phi(p_i))}{R_i'}, \qquad R_i' := |\exp_p^{-1}(\phi(p_i))| = d(p, \phi(p_i)).$$

Since $\exp_p(R_i' v_i') = \phi(p_i)$ and $\exp_p(R_i v_i) = p_i$, Theorem 6.5.2 asserts that

$$|R_i v_i - R_i' v_i'| \leq d(p_i, \phi(p_i)) \leq c$$

for all i. Since $|R_i - R_i'| \leq c$ by the triangle inequality, it follows that

$$|v_i - v_i'| \leq \frac{2c}{R_i}$$

and hence

$$\lim_{i\to\infty} R_i' = \infty, \qquad \lim_{i\to\infty} v_i' = \lim_{i\to\infty} v_i = v.$$

Since $\exp_p(R_i' v_i') = \phi(p_i)$, this implies

$$F_{q,p}(v) = \lim_{i\to\infty} \frac{\exp_q^{-1}(\exp_p(R_i' v_i'))}{|\exp_q^{-1}(\exp_p(R_i' v_i'))|} = \lim_{i\to\infty} \frac{\exp_q^{-1}(\phi(p_i))}{|\exp_q^{-1}(\phi(p_i))|}$$

for all $q \in M$. Take $q = \phi(p)$ and use the identity $\exp_{\phi(p)}^{-1} \circ \phi = d\phi(p) \circ \exp_p^{-1}$ and equation (7.5.13) to obtain

$$F_{\phi(p),p}(v) = \lim_{i\to\infty} \frac{d\phi(p)\exp_p^{-1}(p_i)}{|d\phi(p)\exp_p^{-1}(p_i)|} = d\phi(p) \lim_{i\to\infty} v_i = d\phi(p)v.$$

Hence $\phi_*[p, v] = [p, v]$ for all $\phi \in \mathcal{I}$ and this proves Lemma 7.5.7. □

The next step in the proof of Theorem 7.5.6 is to examine the gradient flow of the convex function $f : M \to \mathbb{R}$. The flow equation has the form

$$\dot{\gamma}(t) = -\nabla f(\gamma(t)), \tag{7.5.14}$$

where the gradient vector field is defined by $\langle \nabla f(p), v \rangle := df(p)v$ for $p \in M$ and $v \in T_p M$. An important consequence of convexity is that the distance between any two solutions of the gradient flow equation is nonincreasing. This is the content of the next lemma (see [17, Lemma 6]).

Lemma 7.5.8 *Let $f : M \to \mathbb{R}$ be a smooth function. Then f is convex if and only if it satisfies the condition*

$$\langle \nabla_v \nabla f(p), v \rangle \geq 0 \tag{7.5.15}$$

for all $p \in M$ and all $v \in T_p M$. If f is convex, then the following holds.

(i)   *Equation (7.5.14) has a solution $\gamma : [0, \infty) \to M$ on the entire positive real axis for every initial condition $\gamma(0) = p_0$.*

(ii) *Let $\gamma_0 : [0, \infty) \to M$ and $\gamma_1 : [0, \infty) \to M$ be two solutions of (7.5.14). Then the function $[0, \infty) \to \mathbb{R} : t \mapsto d(t) := d(\gamma_0(t), \gamma_1(t))$ is nonincreasing.*

Proof Let $\gamma : \mathbb{R} \to M$ be a geodesic. Then $f \circ \gamma : \mathbb{R} \to \mathbb{R}$ is convex if and only if

$$0 \leq \frac{d^2}{dt^2} f(\gamma(t)) = \frac{d}{dt} \langle \nabla f(\gamma(t)), \dot{\gamma}(t) \rangle = \langle \nabla_{\dot{\gamma}(t)} \nabla f(\gamma(t)), \dot{\gamma}(t) \rangle$$

for all t. This holds for all geodesics if and only if f satisfies (7.5.15). In the remainder of the proof we assume that f is convex.

Let $p_0 \in M$ and for $T > 0$ define

$$K_T := \{p \in M \mid d(p_0, p) \le T|\nabla f(p_0)|\}.$$

This set is compact because M is complete. Now let $\gamma : [0, T) \to M$ be the solution of (7.5.14) with $\gamma(0) = p_0$ on some time interval $[0, T)$. Then

$$\frac{d}{dt}|\nabla f(\gamma)|^2 = 2\langle\nabla_{\dot\gamma}\nabla f(\gamma), \nabla f(\gamma)\rangle = -2\langle\nabla_{\dot\gamma}\nabla f(\gamma), \dot\gamma\rangle \le 0$$

by (7.5.15). Thus the function $t \mapsto |\nabla f(\gamma(t))| = |\dot\gamma(t)|$ is nonincreasing and so $\gamma(t) \in K_T$ for $0 \le t < T$. Since K_T is a compact subset of M, the solution γ extends to a longer time interval $[0, T + \delta)$ for some $\delta > 0$ by Corollary 2.4.15. Since $T > 0$ was chosen arbitrary, this proves (i).

We prove part (ii). Assume without loss of generality that $\gamma_0(0) \ne \gamma_1(0)$ and so $\gamma_0(t) \ne \gamma_1(t)$ for all t. For $t \ge 0$ let $[0, 1] \to M : s \mapsto \gamma(s, t)$ be the unique geodesic that satisfies $\gamma(0, t) = \gamma_0(t)$ and $\gamma(1, t) = \gamma_1(t)$. Then $d(t) = d(\gamma_0(t), \gamma_1(t)) = L(\gamma(\cdot, t)) = |\partial_s\gamma(s, t)|$ for all s, t and hence

$$\dot d(t) = \int_0^1 \frac{\langle\partial_s\gamma(s, t), \nabla_t\partial_s\gamma(s, t)\rangle}{|\partial_s\gamma(s, t)|}\,ds = \frac{1}{d(t)}\int_0^1 \frac{\partial}{\partial s}\langle\partial_s\gamma(s, t), \partial_t\gamma(s, t)\rangle\,ds$$

$$= -\frac{1}{d(t)}\Big(\langle\partial_s\gamma(1, t), \nabla f(\gamma(1, t))\rangle - \langle\partial_s\gamma(0, t), \nabla f(\gamma(0, t))\rangle\Big)$$

$$= -\frac{1}{d(t)}\int_0^1 \langle\partial_s\gamma(s, t), \nabla_{\partial_s\gamma(s,t)}f(\gamma(s, t))\rangle\,ds \le 0$$

by (7.5.15). This proves (ii) and Lemma 7.5.8. $\qquad\square$

Lemma 7.5.9 *Let $f : M \to \mathbb{R}$ be a convex function that has a critical point p_∞. Then $f(p) \ge f(p_\infty) =: c$ for all $p \in M$, and the set $C_f := f^{-1}(c)$ of minima of f is geodesically convex.*

Proof Let $p \in M$ and let $\gamma : [0, 1] \to M$ be the unique geodesic with the endpoints $\gamma(0) = p_\infty$ and $\gamma(1) = p$. Then $\beta := f \circ \gamma : [0, 1] \to \mathbb{R}$ is a convex function satisfying $\beta(0) = f(p_\infty) = c$ and $\dot\beta(0) = 0$, hence $\beta(t) \ge c$ for all t, and so $f(p) = \beta(1) \ge c$. Thus f attains its minimum at p_∞.

Now let $p_0, p_1 \in C_f$ and let $\gamma : [0, 1] \to M$ be the unique geodesic with the endpoints $\gamma(0) = p_0$ and $\gamma(1) = p_1$. Then the function $\beta := f \circ \gamma : [0, 1] \to \mathbb{R}$ is convex, satisfies $\beta(0) = \beta(1) = c$, and takes values in the interval $[c, \infty)$. Hence $\beta \equiv c$ and so $\gamma(t) \in C_f$ for all t. This proves Lemma 7.5.9. $\qquad\square$

Proof of Theorem 7.5.6 Choose an element $p_0 \in M$ and let $\gamma : [0, \infty) \to M$ be
the unique solution of equation (7.5.14) that satisfies the initial condition $\gamma(0) = p_0$.
Assume first that

$$\sup_{t \geq 0} d(p_0, \gamma(t)) = \infty \qquad (7.5.16)$$

and choose a sequence $t_i \to \infty$ such that $\lim_{i \to \infty} d(p_0, \gamma(t_i)) = \infty$. Now let
$g \in G$. Since $\phi_g : M \to M$ is an isometry and $f \circ \phi_g = f$, it follows that
$d\phi_g(p) \nabla f(p) = \nabla f(\phi_g(p))$ for all $p \in M$ (**Exercise:** Prove this.) Hence the
curve $[0, \infty) \to M : t \mapsto \phi_g(\gamma(t))$ is another solution of equation (7.5.14) and
hence $d(\gamma(t_i), \phi_g(\gamma(t_i))) \leq d(p_0, \phi_g(p_0))$ for all i and all $g \in G$ by part (ii) of
Lemma 7.5.8. Hence Lemma 7.5.7 asserts that there exists a $(p, v) \in SM$ such that
$d\phi_g(p)v = F_{\phi_g(p),p}(v)$ for all $g \in G$, in contracdiction to our assumption that the
G-action has no fixed point at infinity.

   This shows that our assumption (7.5.16) must have been wrong. Thus

$$\sup_{t \geq 0} d(p_0, \gamma(t)) =: R < \infty,$$

and so our solution $\gamma : [0, \infty) \to M$ of (7.5.14) takes values in the compact set
$B := \{p \in M \mid d(p_0, p) \leq R\}$. Since the function $t \mapsto f(\gamma(t))$ is nonincreasing,
this implies that the limit

$$c := \lim_{t \to \infty} f(\gamma(t)) \geq \min_{p \in B} f(p) \qquad (7.5.17)$$

exists and is a real number (and not $-\infty$). Since $\frac{d}{dt} f(\gamma(t)) = -|\nabla f(\gamma(t))|^2$ and the
function $t \mapsto f(\gamma(t))$ is bounded below by c, there must exist a sequence $t_i \to \infty$
such that

$$\lim_{i \to \infty} \nabla f(\gamma(t_i)) = 0. \qquad (7.5.18)$$

Since $\gamma(t_i) \in B$ for all i, we may also assume that the limit

$$p_\infty := \lim_{i \to \infty} \gamma(t_i) \qquad (7.5.19)$$

exists (after passing to a subsequence, if necessary). This limit is a critical point
of f by (7.5.18), and $f(p_\infty) = c$ by (7.5.17). Hence, by Lemma 7.5.9, f attains its
minimum at p_∞ and the set $C_f := \{p \in M \mid f(p) = c\}$ of minima of f is geodesically convex. We must find an element of C_f that is fixed under the action of K. By
Cartan's Fixed Point Theorem 6.5.6 there exists a $q \in M$ such that $\phi_u(q) = q$ for
all $u \in K$. Since M is complete and C_f is a nonempty closed subset of M, there
exists an element $p_0 \in C_f$ such that

$$d(q, p_0) = \inf_{p \in C_f} d(q, p) =: \delta. \qquad (7.5.20)$$

We claim that $\phi_u(p_0) = p_0$ for all $u \in K$. To see this, fix an element $u \in K$, let $\gamma : [0, 1] \to M$ be the geodesic joining $\gamma(0) = p_0$ to $\gamma(1) = \phi_u(p_0)$, and denote by $m := \gamma(1/2)$ the midpoint of this geodesic. Then Lemma 6.5.7 asserts that

$$2d(q, m)^2 + \frac{d(p_0, \phi_u(p_0))^2}{2} \leq d(q, p_0)^2 + d(q, \phi_u(p_0))^2. \qquad (7.5.21)$$

Since $\phi_u(q) = q$ and ϕ_u is an isometry, we have $d(q, \phi_u(p_0)) = d(q, p_0) = \delta$, and since C_f is geodesically convex, we have $m \in C_f$ and so $d(q, m) \geq \delta$. Hence it follows from (7.5.21) that $p_0 = \phi_u(p_0)$. This shows that $p_0 \in C_f$ is a fixed point for the action of K on M and proves Theorem 7.5.6. $\qquad \square$

Example 7.5.10   To illustrate the argument in the proof of Theorem 7.5.6, take $M = \mathbb{C}$ with the standard flat metric. Then the equivalence relation on the sphere bundle $SM = \mathbb{C} \times S^1$ is given by translation in $\mathbb{C}$ and so the sphere at infinity is $S_\infty(M) = S^1$. The orthogonal group $G = O(2)$ acts by isometries on M and has no fixed point at infinity. The subgroup $K = \mathbb{Z}/2\mathbb{Z}$ acts by complex conjugation and its fixed point set is the real axis. Choose a smooth convex function $h : \mathbb{R} \to \mathbb{R}$ that vanishes on the interval $[-1, 1]$ and is positive elsewhere. Then the function $f(z) := h(|z|)$ is convex and G-invariant and the set C_f of minima of f is the closed unit disc in $\mathbb{C}$. If $q = 2$ is the fixed point of the K-action chosen in the proof, then $p_0 = 1 \in C_f$. If $G = K = \mathbb{Z}/2\mathbb{Z}$, then ± 1 are the fixed points at infinity and $f(z) := e^{\mathrm{Re}(z)}$ is convex and G-invariant, but does not take on its infimum.

Example 7.5.11 ([17])   Consider the case where $M = \mathbb{D}^m$ is the Poincaré model of hyperbolic space (Exercise 6.4.22). Then the sphere at infinity is the boundary $\partial \mathbb{D}^m = S^{m-1}$ (Exercise 6.4.24). If the convex function f extends continuously to the closed ball and does not take on its minimum in M, then it attains its minimum at a unique point on the boundary, because any two boundary points are the asymptotic limits of a geodesic in M. Hence the minimum on the boundary is fixed under the action of any Lie group on M by isometries, that leave f invariant. This is reminiscent of the Kempf Uniqueness Theorem in GIT (see [37] and [20, Theorem 10.2]), where $M = G/K$ is associated to the complexification G of a compact Lie group K and f is the Kempf–Ness function (see [38, 53] and [20, §4]).

7.5.2   Inner Products and Weighted Flags

We will now turn to a specific example, where the Hadamard manifold is the space of positive definite symmetric matrices with determinant one (Sect. 6.5.3). Following [17], we choose a finite-dimensional real vector space V equipped with a fixed inner product $\langle \cdot, \cdot \rangle$. Then every inner product on V has the form $\langle v, v' \rangle_P := \langle v, P^{-1}v' \rangle$ for some self-adjoint positive definite automorphism P. Denote the set of such automorphisms with determinant one by

$$\mathcal{P}_0(V) := \{P \in \text{End}(V) \mid P^* = P > 0, \det(P) = 1\}. \tag{7.5.22}$$

Here $P^* \in \text{End}(V)$ is defined by $\langle v, P^* v' \rangle := \langle P v, v' \rangle$ for $v, v' \in V$, and the notation "$P > 0$" means $\langle v, P v \rangle > 0$ for all $v \in V \setminus \{0\}$. Thus $\mathcal{P}_0(V)$ is a codimension-1 submanifold of the space of self-adjoint endomorphisms of V. Its tangent space at $P \in \mathcal{P}_0(V)$ is given by

$$T_P \mathcal{P}_0(V) := \{\widehat{P} \in \text{End}(V) \mid \widehat{P} = \widehat{P}^*, \text{trace}(\widehat{P} P^{-1}) = 0\}. \tag{7.5.23}$$

The Riemannian metric on $\mathcal{P}_0(V)$ is defined by

$$|\widehat{P}|_P := \sqrt{\text{trace}(\widehat{P} P^{-1} \widehat{P} P^{-1})} \tag{7.5.24}$$

for $P \in \mathcal{P}_0(V)$ and $\widehat{P} \in T_P \mathcal{P}_0(V)$ as in Sect. 6.5.3, and so $\mathcal{P}_0(V)$ is a Hadamard manifold by Theorem 6.5.10. For $\widehat{P} \in T_P \mathcal{P}_0(V)$ the endomorphism $\widehat{P} P^{-1}$ is self-adjoint with respect to the inner product $\langle \cdot, P^{-1} \cdot \rangle$ on V.

The group $\text{SL}(V) \subset \text{GL}(V)$ of automorphisms of V with determinant one acts on $\mathcal{P}_0(V)$ by the isometries $\phi_g(P) := g P g^*$ for $g \in \text{SL}(V)$. The isotropy subgroup of $\mathbb{1} \in \mathcal{P}_0(V)$ is the special orthogonal group $\text{SO}(V)$. The action of a subgroup $G \subset \text{SL}(V)$ on V is called **irreducible** iff there does not exist a linear subspace $E \subset V$, other than $E = \{0\}$ and $E = V$, such that $gE = E$ for all $g \in G$. This notion can be used to carry over the general existence theorem in Sect. 7.5.1 for critical points of a convex function to the present setting (see [17, Theorem 3]).

Theorem 7.5.12 (Donaldson) *Let $G \subset \text{SL}(V)$ be a Lie subgroup such that the action of G on V is irreducible. Let $K \subset G$ be a compact subgroup and let $f : \mathcal{P}_0(V) \to \mathbb{R}$ be a convex function such that $f(g P g^*) = f(P)$ for all $g \in G$ and all $P \in \mathcal{P}_0(V)$. Then there exists a $P_0 \in \mathcal{P}_0(V)$ such that*

$$f(P_0) = \inf_{P \in \mathcal{P}_0(V)} f(P), \qquad u P_0 u^* = P_0 \quad \text{for all } u \in K. \tag{7.5.25}$$

The goal will be to deduce Theorem 7.5.12 from Theorem 7.5.6. Thus we must understand the sphere at infinity of the space $\mathcal{P}_0(V)$. This will be accomplished with the help of the following definition.

Definition 7.5.13 A **weighted flag** in V is a pair (F, μ), where F is a finite sequence of linear suspaces

$$\{0\} = F_0 \subset F_1 \subset F_2 \subset \cdots \subset F_r = V$$

such that $n_i := \dim(F_i)/\dim(F_{i-1}) > 0$ for $i = 1, \ldots, r$, and μ is a finite sequence of real numbers $\mu_1 > \mu_2 > \cdots > \mu_r$ satisfying the conditions

$$\sum_{i=1}^{r} n_i \mu_i = 0, \qquad \sum_{i=1}^{r} n_i \mu_i^2 = 1. \tag{7.5.26}$$

Let $\mathcal{F} = \mathcal{F}(V)$ be the set of weighted flags. The group $\mathrm{SL}(V)$ acts on $\mathcal{F}(V)$ by $g \cdot (F, \mu) := (gF_i, \mu_i)_i$ for $(F, \mu) = (F_i, \mu_i)_i \in \mathcal{F}(V)$ and $g \in \mathrm{SL}(V)$.

For $P \in \mathcal{P}_0(V)$ the unit sphere in the tangent space $T_P\mathcal{P}_0(V)$ is the set

$$S_P := \left\{ \widehat{P} = \widehat{P}^* \in \mathrm{End}(V) \,\middle|\, \mathrm{trace}\!\left(\widehat{P} P^{-1}\right) = 0, \ \mathrm{trace}\!\left(\widehat{P} P^{-1}\widehat{P} P^{-1}\right) = 1 \right\}.$$

Let $P \in \mathcal{P}_0(V)$ and $\widehat{P} \in S_P$. Then the endomorphism $\widehat{P} P^{-1}$ is self-adjoint with respect to the inner product $\langle \cdot, P^{-1}\cdot\rangle$ and hence has only real eigenvalues $\mu_1 > \mu_2 > \cdots > \mu_r$. For each i let $E_i \subset V$ be the eigenspace for μ_i and define $n_i := \dim(E_i)$. Since $\mathrm{trace}\!\left(\widehat{P} P^{-1}\right) = 0$ and $\mathrm{trace}\!\left(\widehat{P} P^{-1}\widehat{P} P^{-1}\right) = 1$, the μ_i, n_i satisfy (7.5.26). The **weighted flag of** $(P, \widehat{P})$ is defined by

$$\iota_P(\widehat{P}) := (F, \mu) = \big((F_1, \mu_1), \dots, (F_r, \mu_r)\big) \in \mathcal{F}(V), \tag{7.5.27}$$

where $F_i := E_1 \oplus E_2 \oplus \cdots \oplus E_i$ for $i = 1, \dots, r$. For each $P \in \mathcal{P}_0(V)$ the map $\iota_P : S_P \to \mathcal{F}(V)$ defined by (7.5.27) is bijective, and thus induces a (compact, metrizable) topology on $\mathcal{F}(V)$. This topology is independent of P as the next lemma shows. The lemma also shows that the space of weighted flags is the sphere at infinity (see [17, Lemma 4]).

Lemma 7.5.14 *The equivalence relation in Definition 7.5.1 on the unit sphere bundle $S\mathcal{P}_0(V)$ is given by*

$$(P, \widehat{P}) \sim (Q, \widehat{Q}) \qquad \Longleftrightarrow \qquad \iota_P(\widehat{P}) = \iota_Q(\widehat{Q}) \tag{7.5.28}$$

for $P, Q \in \mathcal{P}_0(V)$, $\widehat{P} \in S_P$, and $\widehat{Q} \in S_Q$. Thus the map $\mathcal{F}_{Q,P} : S_P \to S_Q$ as defined in Lemma 7.5.2 is given by $\mathcal{F}_{Q,P} = \iota_Q \circ \iota_P^{-1}$. Moreover, the map $S\mathcal{P}_0(V) \to \mathcal{F}(V) : (P, \widehat{P}) \mapsto \iota_P(\widehat{P})$ is $\mathrm{SL}(V)$-equivariant, i.e.

$$\iota_{gPg^*}(g\widehat{P}g^*) = g \cdot \iota_P(\widehat{P}) \tag{7.5.29}$$

for all $P \in \mathcal{P}_0(V)$, all $\widehat{P} \in S_P$, and all $g \in \mathrm{SL}(V)$.

Proof Since $(g\widehat{P}g^*)(gPg^*)^{-1} = g(\widehat{P}P^{-1})g^{-1}$ for all $\widehat{P} \in S_P$ and $g \in \mathrm{SL}(V)$, equation (7.5.29) follows directly from the definitions. The proof that the equivalence relation satisfies (7.5.28) rests on the following claims.

Claim 1 *Let $S \in S_1$, let $(F_i, \mu_i)_{i=1}^r = \iota_1(S)$ be the flag associated to S, and let $h \in \mathrm{SL}(V)$ be an automorphism of V with determinant one such that*

$$hF_i = F_i \qquad \text{for } i = 1, \dots, r. \tag{7.5.30}$$

Then $(1, S) \sim (hh^*, hSh^*)$.

Claim 2 *Let $g \in SL(V)$ and fix any flag $\{0\} = F_0 \subsetneq F_1 \subsetneq \cdots \subsetneq F_r = V$. Then there exist elements $h \in SL(V)$ and $u \in SO(V)$ such that h satisfies (7.5.30) and $g = hu$.*

We prove, that these two claims imply (7.5.28). Fix any element $S \in \mathcal{S}_1$, let $(F_i, \mu_i)_{i=1}^r = \iota_1(S)$ be the weighted flag of S, and let $P \in \mathcal{P}_0(V)$. Choose an element $g \in SL(V)$ such that $gg^* = P$, and choose u and h as in Claim 2. Then $hh^* = P$, and the pairs $(P, \widehat{P}) := (hh^*, hSh^*)$ and $(1, S)$ have the same flag by (7.5.30) and are equivalent by Claim 1. Hence any pair $(P, \widehat{P})$ is equivalent to $(1, S)$ if and only if it has the same flag. By transitivity of the equivalence relation we deduce that (7.5.28) holds for all $P, Q \in \mathcal{P}_0(V)$.

We prove Claim 2. Let $F_i' := g^{-1} F_i$ and $m_i := \dim(F_i)$ for $i = 1, \ldots, r$. Then choose orthonormal bases $e_1, \ldots, e_m$ and $e_1', \ldots, e_m'$ of V such that, for each i, the vectors $e_1, \ldots, e_{m_i}$ form a basis of F_i and the vectors $e_1', \ldots, e_{m_i}'$ form a basis of F_i'. Define the orthogonal transformation u by $ue_i' := e_i$ for $i = 1, \ldots, m$. It satisfies $F_i' = u^{-1} F_i$ and hence $gu^{-1} F_i = F_i$ for all i. Thus $h := gu^{-1}$ satisfies the requirements of Claim 2.

We prove Claim 1, following [17, Lemma 4]. Define the geodesics γ_0, γ_1 in $\mathcal{P}_0(V)$ by $\gamma_0(t) = \exp(tS)$ and $\gamma_1(t) = h \exp(tS)h^*$ (see Lemma 6.5.18). By equation (6.5.12) the square of their distance is given by

$$\rho(t) := d(\exp(tS), h \exp(tS)h^*)^2 = \mathrm{trace}\left(\left(\log(M(t)M(t)^*)\right)^2\right),$$
$$M(t) := \exp(-tS/2)h \exp(tS/2). \tag{7.5.31}$$

In the eigenspace decomposition $V = E_1 \oplus \cdots \oplus E_r$ of the self-adjoint endomorphism S the automorphisms h and $\exp(tS/2)$ have the form

$$h = \begin{pmatrix} h_{11} & h_{12} & \cdots & h_{1r} \\ 0 & h_{22} & \ddots & \vdots \\ \vdots & \ddots & \ddots & h_{r-1,r} \\ 0 & \cdots & 0 & h_{rr} \end{pmatrix}, \tag{7.5.32}$$

$$\exp(tS/2) = \mathrm{diag}\left(e^{t\mu_1/2}1_{E_1}, e^{t\mu_2/2}1_{E_2}, \ldots, e^{t\mu_r/2}1_{E_r}\right).$$

Here the upper triangular form of h follows from (7.5.30). Hence

$$M(t) = \begin{pmatrix} h_{11} & h_{12}(t) & \cdots & h_{1r}(t) \\ 0 & h_{22} & \ddots & \vdots \\ \vdots & \ddots & \ddots & h_{r-1,r}(t) \\ 0 & \cdots & 0 & h_{rr} \end{pmatrix}, \tag{7.5.33}$$

where $h_{ij}(t) := e^{-t(\mu_i - \mu_j)/2} h_{ij}$ for $1 \le i < j \le r$. Since $\mu_i > \mu_j$ for $i < j$, it follows that the limit $M_\infty := \lim_{t \to \infty} M(t) = \mathrm{diag}(h_{11}, \ldots, h_{rr})$ exists and is an invertible endomorphism of V. Hence the function $\rho : [0, \infty) \to \mathbb{R}$ in (7.5.31) is bounded. This proves Claim 1 and Lemma 7.5.14. □

Proof of Theorem 7.5.12   Let $G \subset SL(V)$ be a Lie subgroup which acts irreducibly on V. Then G acts on $\mathcal{P}_0(V)$ by isometries. By Lemma 7.5.14 the induced action on the sphere at infinity $S_\infty(\mathcal{P}_0(V)) \cong \mathcal{F}(V)$ is given by $G \times \mathcal{F}(V) \to \mathcal{F}(V) : (g, (F_i, \mu_i)_{i=1,\dots,r}) \mapsto (gF_i, \mu_i)_{i=1,\dots,r}$. This action has no fixed points because the action of G on V is irreducible and $r \geq 2$ for each weighted flag $(F_i, \mu_i)_{i=1,\dots,r} \in \mathcal{F}(V)$. Hence all the assertions of Theorem 7.5.12 follow directly from Theorem 7.5.6 with $M = \mathcal{P}_0(V)$. □

7.5.3   Lengths of Vectors

The material in this section goes back to ideas in geometric invariant theory developed by Kempf–Ness [38], Ness [53], and Kirwan [40] in the complex setting and by Richardson–Slodowy [59] and Marian [48] in the real setting. We assume throughout that V, W are finite-dimensional real vector spaces and $\rho : SL(V) \to SL(W)$ is a Lie group homomorphism. Note that every Lie group homomorphism from $GL(V)$ to $GL(W)$ restricts to a Lie group homomorphism from $SL(V)$ to $SL(W)$, because every Lie group homomorphism from $GL(V)$ to the multiplicative group of nonzero real numbers is some power of the determinant.

Fix a nonzero vector $w \in W$ and denote by $G_w \subset SL(V)$ the connected component of the identity in the isotropy subgroup of w, i.e.

$$G_w := \left\{ g \in SL(V) \,\middle|\, \begin{array}{l} \exists \text{ a smooth path } \gamma : [0, 1] \to SL(V) \\ \text{such that } \gamma(0) = \mathbb{1}, \ \gamma(1) = g, \text{ and} \\ \rho(\gamma(t))w = w \text{ for } 0 \leq t \leq 1 \end{array} \right\}. \qquad (7.5.34)$$

By Theorem 2.5.27 this is a Lie subgroup of $SL(V)$ with the Lie algebra

$$\mathfrak{g}_w := \left\{ \xi \in \mathfrak{sl}(V) \,\middle|\, \dot\rho(\xi)w = 0 \right\}.$$

There are many examples of this setup that are related to interesting questions in geometry. The vector space W can be the space of all symmetric bilinear forms on V and w can be an inner product, in which case G_w is the special orthogonal group associated to the inner product, or w can be the quadratic form (6.4.9), in which case G_w is the identity component of the isometry group of hyperbolic space. Or W can be the space of skew-symmetric bilinear forms on V and w a symplectic form, in which case G_w is the symplectic linear group. Or W can be the space of skew-symmetric bilinear maps on V with values in V. Then w can be a cross product in dimension three or seven, or w can be the Lie bracket of a Lie algebra $\mathfrak{g} = V$ and then G_w is the identity component in the group of automorphisms of $\mathfrak{g}$. The latter example will be examined in detail in Sect. 7.6.2. Of particular interest are the cases where the group G_w is noncompact.

Definition 7.5.15 An inner product $\langle \cdot, \cdot \rangle$ on V is called (ρ, w)-**symmetric** iff the Lie subalgebra $\mathfrak{g}_w \subset \mathfrak{sl}(V)$ is invariant under the involution $A \mapsto A^*$, defined by $\langle v, A^* v' \rangle := \langle Av, v' \rangle$ for $v, v' \in V$.

Exercise 7.5.16 Let $\langle \cdot, \cdot \rangle$ be a (ρ, w)-symmetric inner product on V. Prove that $g \in G_w$ implies $g^* \in G_w$. **Hint:** Choose a smooth path $g : [0, 1] \to G_w$ with the endpoints $g(0) = \mathbb{1}$ and $g(1) = g$, and define $\xi(t) := g(t)^{-1} \dot{g}(t)$. Show that the initial value problem $\dot{h}(t) = \xi(t)^* h(t)$, $h(0) = 1$, has a unique solution $h : [0, 1] \to G_w$ (Exercise 2.5.36) and that $h(t) = g(t)^*$ for all t.

The following theorem asserts the existence of a (ρ, w)-symmetric inner product on V under an irreducibility assumption (see [17, Theorem 2]).

Theorem 7.5.17 *Assume that the group G_w in (7.5.34) acts irreducibly on V. Then there exists a (ρ, w)-symmetric inner product on V with the following properties. The subgroup*

$$\mathrm{K}_w := \mathrm{G}_w \cap \mathrm{SO}(V)$$

is connected and is a maximal compact subgroup of G_w. Moreover, every compact subgroup of G_w is conjugate in G_w to a Lie subgroup of K_w. Thus, if K is any maximal compact subgroup of G_w, there exists an element $h \in \mathrm{G}_w$ such that $\mathrm{K} = h \mathrm{K}_w h^{-1}$.

Proof See Lemma 7.5.23. $\square$

Example 7.5.18 The hypothesis that the group G_w acts irreducibly on V cannot be removed in Theorem 7.5.17. Consider the case where $W = V$ has dimension at least two, the homomorphism $\rho : \mathrm{SL}(V) \to \mathrm{SL}(V)$ is the identity, and $w \in V$ is any nonzero vector. Then the one-dimensional linear subspace $\mathbb{R}w \subset V$ is evidently invariant under the action of G_w, and there does not exist any (ρ, w)-symmetric inner product on V.

To begin with, we will fix any inner product $\langle \cdot, \cdot \rangle_V$ on V and define the space $\mathcal{P}_0(V)$ of self-adjoint positive definite automorphisms of V with determinant one in terms of this fixed inner product. We will then use Theorem 7.5.12 to find an element $P \in \mathcal{P}_0(V)$ such that the inner product

$$\langle v, v' \rangle_{V,P} := \langle v, P^{-1} v' \rangle_V \tag{7.5.35}$$

on V satisfies the requirements of Theorem 7.5.17. The proof is based on three lemmas. The fourth lemma restates the theorem in a modified form.

Lemma 7.5.19 *There exists an inner product $\langle \cdot, \cdot \rangle_w$ on W such that $\rho(\mathrm{SO}(V)) \subset \mathrm{SO}(W)$ and $\dot{\rho}(A^*) = \dot{\rho}(A)^*$ for all $A \in \mathfrak{sl}(V)$.*

Proof   The proof follows the argument in [17, §3]. Assume without loss of generality that $V = \mathbb{R}^m$ is equipped with the standard inner product, and that $W = \mathbb{R}^n$ and that $\rho : \mathrm{SL}(m, \mathbb{R}) \to \mathrm{SL}(n, \mathbb{R})$ is a Lie group homomorphism. We prove first that there exists a unique Lie group homomorphism $\rho^c : \mathrm{SL}(m, \mathbb{C}) \to \mathrm{SL}(n, \mathbb{C})$ (the **complexification of** ρ) such that

$$\rho^c|_{\mathrm{SL}(m,\mathbb{R})} = \rho, \qquad \dot\rho^c(A + iB) = \dot\rho(A) + i\dot\rho(B) \qquad (7.5.36)$$

for all $A, B \in \mathfrak{sl}(m, \mathbb{R})$. Since $\mathrm{SL}(m, \mathbb{C})$ is connected, we can define $\rho^c(g)$ for $g \in \mathrm{GL}(m, \mathbb{C})$ by choosing a smooth path $\alpha : [0, 1] \to \mathrm{SL}(m, \mathbb{C})$ with the endpoints $\alpha(0) = \mathbb{1}_m$ and $\alpha(1) = g$, and taking

$$\rho^c(g) := \beta(1), \qquad \beta(s)^{-1}\dot\beta(s) = \dot\rho^c\big(\alpha(s)^{-1}\dot\alpha(s)\big), \qquad \beta(0) = \mathbb{1}_n. \qquad (7.5.37)$$

To verify that $\beta(1)$ is independent of the choice of α, one can choose a smooth map $[0, 1]^2 \to \mathrm{SL}(m, \mathbb{C}) : (s, t) \mapsto \alpha(s, t)$ satisfying $\alpha(0, t) = \mathbb{1}_m$, $\alpha(1, t) = g$, define $S := \alpha^{-1}\partial_s\alpha$ and $T := \alpha^{-1}\partial_t\alpha$, and define $\beta : [0, 1]^2 \to \mathrm{SL}(n, \mathbb{C})$ as the solution of the initial value problem $\beta^{-1}\partial_s\beta = \dot\rho^c(S)$, $\beta(0, t) = \mathbb{1}_n$. Since $\dot\rho^c$ is a Lie algebra homomorphism and $\partial_t S - \partial_s T = [S, T]$, it follows that $\beta^{-1}\partial_t\beta = \dot\rho^c(T)$ and so $\beta(1, t)$ is independent of t. Moreover, $\mathrm{SL}(m, \mathbb{C})$ retracts onto $\mathrm{SU}(m)$ by polar decomposition and so is simply connected by a standard homotopy argument. That the map $\rho^c : \mathrm{SL}(m, \mathbb{C}) \to \mathrm{SL}(n, \mathbb{C})$ thus defined is smooth follows from the smooth dependence of solutions on the parameter in a smooth family of differential equations. That it is a group homomorphism follows by catenation of paths, and that it satisfies (7.5.36) follows directly from the definition.

Now consider the action of the compact subgroup $\mathrm{SU}(m) \subset \mathrm{SL}(m, \mathbb{C})$ on the Hadamard manifold $\mathcal{Q}_0$ of positive definite Hermitian $n \times n$-matrices Q of determinant one (Remark 6.5.21) by $(g, Q) \mapsto \rho^c(g)Q\rho^c(g)^*$. This action is by isometries and hence, by Cartan's Fixed Point Theorem 6.5.6, there exists an element $Q_0 \in \mathcal{Q}_0$ such that $\rho^c(g)Q_0\rho^c(g)^* = Q_0$ for all $g \in \mathrm{SU}(m)$. Differentiate this equation at $g = \mathbb{1}_m$ to obtain $\dot\rho^c(B)Q_0 + Q_0\dot\rho^c(B)^* = 0$ for every skew-Hermitian matrix $B = -B^* \in \mathfrak{sl}(m, \mathbb{C})$ with trace zero. Now let $A \in \mathfrak{sl}(m, \mathbb{R})$, define $R := \frac{1}{2}(A - A^\mathsf{T})$, $S := \frac{1}{2}(A + A^\mathsf{T})$, and take $B = R$ and $B = iS$. Then $\dot\rho(R)Q_0 + Q_0\dot\rho(R)^\mathsf{T} = 0$ and $\dot\rho(S)Q_0 = Q_0\dot\rho(S)^\mathsf{T}$. Hence the positive definite symmetric $n \times n$-matrix $Q := \mathrm{Re}(Q_0)$ satisfies

$$\dot\rho(A^\mathsf{T}) = \dot\rho(-R + S) = Q\dot\rho(R + S)^\mathsf{T}Q^{-1} = Q\dot\rho(A)^\mathsf{T}Q^{-1}$$

for all $A \in \mathfrak{sl}(m, \mathbb{R})$. This shows that the inner product $\langle w, w'\rangle := w^\mathsf{T}Q^{-1}w'$ on $\mathbb{R}^n$ satisfies the requirements of Lemma 7.5.19. $\square$

In the remainder of this subsection we will fix an inner product on W as in Lemma 7.5.19. Recall also that we have already chosen a nonzero vector $w \in W$. The norm squared of this vector determines a function on the space of inner products on V (see [17, Lemma 1]).

Lemma 7.5.20 *Define the function $f_w : \mathcal{P}_0(V) \to \mathbb{R}$ by*

$$f_w(P) = \langle w, \rho(P^{-1})w \rangle_W \tag{7.5.38}$$

for $P \in \mathcal{P}_0(V)$. Then an element $P \in \mathcal{P}_0(V)$ is a critical point of f_w if and only if $\langle \dot{\rho}(A)w, \rho(P^{-1})w \rangle_W = 0$ for all $A \in \mathfrak{sl}(V)$. Moreover, if P is a critical point of f_w, then the inner product (7.5.35) on V is (ρ, w)-symmetric.

Proof Fix an element $P \in \mathcal{P}_0(V)$ and a tangent vector $\widehat{P} \in T_p\mathcal{P}_0(V)$. Then it follows from Lemma 7.5.19 that

$$df_w(P)\widehat{P} = -\left\langle w, \dot{\rho}(P^{-1}\widehat{P})\rho(P^{-1})w \right\rangle_W$$
$$= -\left\langle \dot{\rho}(\widehat{P}P^{-1})w, \rho(P^{-1})w \right\rangle_W.$$

Now let $A \in \mathfrak{sl}(V)$ and take $\widehat{P} := AP + PA^*$ to obtain

$$df_w(P)(AP + PA^*) = -\langle \dot{\rho}(A + PA^*P^{-1})w, \rho(P^{-1})w \rangle_W$$
$$= -2\langle \dot{\rho}(A)w, \rho(P^{-1})w \rangle_W. \tag{7.5.39}$$

Thus P is a critical point of f_w if and only if the right hand side of (7.5.39) vanishes for all $A \in \mathfrak{sl}(V)$. To prove the last assertion, define the norm

$$|w'|_{W,P} := \sqrt{\langle w', \rho(P^{-1})w' \rangle_W}$$

for $w' \in W$. Now let $P \in \mathcal{P}_0(\mathfrak{g})$ be a critical point of f_w, let $\xi \in \mathfrak{sl}(V)$, and take $A := [\xi, P\xi^*P^{-1}] \in \mathfrak{sl}(V)$ in (7.5.39). Then

$$0 = \langle \dot{\rho}([\xi, P\xi^*P^{-1}])w, \rho(P^{-1})w \rangle_W = |\dot{\rho}(P\xi^*P^{-1})w|^2_{W,P} - |\dot{\rho}(\xi)w|^2_{W,P}.$$

If $\xi \in \mathfrak{g}_w$, then $\dot{\rho}(\xi)w = 0$, hence $\dot{\rho}(P\xi^*P^{-1})w = 0$, and hence $P\xi^*P^{-1} \in \mathfrak{g}_w$. But the endomorphism $P\xi^*P^{-1}$ is the adjoint of ξ with respect to the inner product (7.5.35) on V and this proves Lemma 7.5.20. □

Example 7.5.21 This example shows that the (ρ, w)-symmetry of the inner product (7.5.35) does not imply that P is a critical point of f_w. Take $W = V \times V$ and let $\rho : \mathrm{SL}(V) \to \mathrm{SL}(W)$ be the diagonal action. Assume $\dim(V) = 2$ and choose $w = (u, v) \in W$ such that $u, v \in V$ are linearly independent. Then $G_w = \{1\}$ and so every inner product on V is (ρ, w)-symmetric (and the assertions of Theorem 7.5.17 are satisfied), however, the function $f_w(P) = \langle u, P^{-1}u \rangle_V + \langle v, P^{-1}v \rangle_V$ does not have any critical point.

Let $M_w \subset \mathcal{P}_0(V)$ be the set of minima of the function $f_w : \mathcal{P}_0(V) \to \mathbb{R}$ in Lemma 7.5.20 and let $\mathrm{Crit}(f_w) \subset \mathcal{P}_0(V)$ be its set of critical points. The next result shows that f_w is convex and that G_w acts transitively on M_w whenever this set is nonempty (see [17, Lemma 2]).

Lemma 7.5.22 *The function* f_w *has the following properties.*

(i) f_w *is convex and* G_w*-invariant, and thus* $M_w = \text{Crit}(f_w)$.
(ii) *Fix two elements* $P_0 \in M_w$ *and* $P \in \mathcal{P}_0(V)$. *Then* $P \in M_w$ *if and only if there exists an element* $\eta \in \mathfrak{g}_w$ *such that* $\eta = P_0\eta^*P_0^{-1}$ *and* $\exp(\eta) = PP_0^{-1}$, *or equivalently* $\rho(PP_0^{-1})w = w$.
(iii) *The group* G_w *acts transitively on* M_w.

Proof We prove part (i). Let $\gamma : \mathbb{R} \to \mathcal{P}_0(V)$ be a geodesic and define

$$P := \gamma(0), \qquad \widehat{P} := \dot\gamma(0), \qquad S := P^{-1/2}\widehat{P}P^{-1/2}.$$

Then, by Lemma 6.5.18, $\gamma(t) = P^{1/2}\exp(tS)P^{1/2}$ and hence

$$f_w(\gamma(t)) = \langle \rho(P^{-1/2})w, \exp(-t\dot\rho(S))\rho(P^{-1/2})w\rangle_W.$$

This implies

$$\frac{d^2}{dt^2}f_w(\gamma(t)) = \langle \dot\rho(S)\rho(P^{-1/2})w, \exp(-t\dot\rho(S))\dot\rho(S)\rho(P^{-1/2})w\rangle_W \geq 0$$

for all t and hence f_w is convex. Hence $M_w = \text{Crit}(f_w)$ by Lemma 7.5.9. That f_w is G_w-invariant follows directly from the definition. This proves (i).

We prove part (ii). If $\eta \in \mathfrak{g}_w$ satisfies $\eta = P_0\eta^*P_0^{-1}$, $\exp(\eta) = PP_0^{-1}$, then $\rho(PP_0^{-1})w = w$. If $\rho(PP_0^{-1})w = w$, then $\rho(P^{-1})w = \rho(P_0^{-1})w$, hence

$$\langle \dot\rho(A)w, \rho(P^{-1})w\rangle_W = \langle \dot\rho(A)w, \rho(P_0^{-1})w\rangle_W = 0$$

for all $A \in \mathfrak{sl}(V)$, and hence $P \in M_w$ by Lemma 7.5.20 and part (i). Now assume $P \in M_w$ and let $\gamma : [0,1] \to \mathcal{P}_0(V)$ be the unique geodesic with the endpoints $\gamma(0) = P_0$ and $\gamma(1) = P \in M_w$. Then $\gamma(t) \in M_w$ for all t by Lemma 7.5.9. Hence $\langle \dot\rho(\eta)w, \rho(\gamma(t)^{-1})w\rangle_W = 0$ for all t and all $\eta \in \mathfrak{sl}(V)$, by Lemma 7.5.20. Differentiate this equation at $t = 0$ to obtain

$$\langle \dot\rho(\eta)w, \rho(P_0^{-1})\dot\rho(\widehat{P}P_0^{-1})w\rangle_W = 0, \qquad \widehat{P} := \dot\gamma(0) \in T_{P_0}\mathcal{P}_0(V).$$

Take $\eta := \widehat{P}P_0^{-1}$ to obtain $\dot\rho(\eta)w = 0$, and thus $\eta \in \mathfrak{g}_w$ and $\eta = P_0\eta^*P_0^{-1}$. By Lemma 6.5.18 we also have $P = \gamma(1) = \exp(\widehat{P}P_0^{-1})P_0 = \exp(\eta)P_0$ and this proves (ii).

We prove part (iii). Let $P_0, P \in M_w$, choose an element $\eta \in \mathfrak{g}_w$ as in (ii) so that $\eta P_0 = P_0\eta^*$ and $P = \exp(\eta)P_0$, and define $h := \exp(\eta/2) \in G_w$ to obtain $P = hP_0h^*$. This proves (iii) and Lemma 7.5.22. $\square$

With these preparations we are ready to prove Theorem 7.5.17.

Lemma 7.5.23

(i)   *If G_w acts irreducibly on V, then $M_w \neq \emptyset$.*
(ii) *If $P \in M_w$, then the inner product $\langle \cdot, P^{-1} \cdot \rangle_V$ on V satisfies all the requirements of Theorem 7.5.17.*

Proof The function f_w is convex and G_w-invariant by Lemma 7.5.22. Hence part (i) follows from Theorem 7.5.12. To prove part (ii), assume that P is a critical point of f_w. Then, by Lemma 7.5.20, the inner product $\langle \cdot, P^{-1} \cdot \rangle_V$ is (ρ, w)-symmetric. We prove the remaining assertions in four steps. Define

$$K_P := G_w \cap SO(V, \langle \cdot, P^{-1} \cdot \rangle_V) = \left\{ u \in G_w \mid u P u^* P^{-1} = \mathbb{1} \right\}.$$

Step 1 *Let $K \subset G_w$ be any compact subgroup. Then there exists an element $h \in G_w$ such that $h^{-1} K h \subset K_P$.*

By Theorem 7.5.12, there exists a $P_0 \in \mathcal{P}_0(V)$ such that $f_w(P_0) = \inf f_w$ and $u P_0 u^* = P_0$ for all $u \in K$. By Lemma 7.5.22, there exists an element $h \in G_w$ such that $P_0 = h P h^*$. Hence $u h P h^* u^* = h P h^*$ for all $u \in K$, and hence the automorphism $h^{-1} u h$ is orthogonal with respect to the inner product $\langle \cdot, P^{-1} \cdot \rangle_V$ for every $u \in K$.

Step 2 K_P *is a compact connected subgroup of G_w.*

By the Closed Subgroup Theorem 2.5.27 K_P is a closed, and hence compact, subgroup of $SO(V, \langle \cdot, P^{-1} \cdot \rangle_V)$ and so is a compact Lie subgroup of G_w. We prove that K_P is connected. Let $u \in K_P \subset G_w$. Since G_w is connected, there exists a smooth path $g : [0, 1] \to G_w$ such that $g(0) = \mathbb{1}$ and $g(1) = u$. Thus, by Exercise 7.5.16, we have $g(t) P g(t)^* P^{-1} \in G_w$ for all t. Hence, by part (ii) of Lemma 7.5.22, there exists a smooth path $\eta : [0, 1] \to \mathfrak{g}_w$ such that $\eta(t) = P \eta(t)^* P^{-1}$, $\exp(\eta(t)) = g(t) P g(t)^* P^{-1}$, and $\eta(0) = \eta(1) = 0$. Hence $u(t) := \exp(-\eta(t)/2) g(t)$ is a path in K_P joining $u(0) = \mathbb{1}$ to $u(1) = u$.

Step 3 K_P *is a maximal compact subgroup of G_w.*

Let $K \subset G_w$ be a compact subgroup containing K_P. Then by Step 1 there exists an $h \in G_w$ such that $h^{-1} K h \subset K_P$. Hence $K_P \subset K \subset h K_P h^{-1}$ and so $\mathrm{Lie}(K_P) \subset \mathrm{Lie}(K) \subset h \mathrm{Lie}(K_P) h^{-1}$. Since $\mathrm{Lie}(K_P)$ and $h \mathrm{Lie}(K_P) h^{-1}$ have the same dimension, this implies $\mathrm{Lie}(K_P) = h \mathrm{Lie}(K_P) h^{-1}$. Since K_P and $h K_P h^{-1}$ are connected, this implies $K_P = h K_P h^{-1}$ and so $K_P = K$.

Step 4 *Let K be a maximal compact subgroup of G_w. Then there exists an element $h \in G_w$ such that $K = h K_P h^{-1}$.*

By Step 1 there exists an $h \in G_w$ such that $h^{-1}Kh \subset K_P$, thus $K \subset hK_P h^{-1}$ and so $K = hK_P h^{-1}$, because K is a maximal compact subgroup of G_w. This proves Step 4, Lemma 7.5.23, and Theorem 7.5.17.   □

Lemma 7.5.23 shows that all minima of the function f_w in Lemma 7.5.20 give rise to inner products that satisfy the requirements of Theorem 7.5.17. The next lemma shows that the set of minima of f_w is a Hadamard manifold.

Lemma 7.5.24 *The set M_w is a geodesically convex and totally geodesic submanifold of $\mathcal{P}_0(V)$. Hence, if it is nonempty, it is a Hadamard manifold and a symmetric space.*

Proof By Lemma 7.5.22 the function f_w is convex, and so M_w is geodesically convex by Lemma 7.5.9. Now assume M_w is nonempty and fix any element $P_0 \in M_w$. Then the exponential map

$$T_{P_0}\mathcal{P}_0(V) \to \mathcal{P}_0(V) : \widehat{P} \mapsto \exp(\widehat{P} P_0^{-1})P_0$$

is a diffeomorphism (Theorem 6.5.10 and Lemma 6.5.18) and M_w is the image of the linear subspace $\{\widehat{P} \in T_{P_0}\mathcal{P}_0(\mathfrak{g}) \mid \widehat{P} P_0^{-1} \in \mathfrak{g}_w\}$ under this diffeomorphism (Lemma 7.5.22). Since P_0 can be chosen to be any element of M_w, this shows that M_w is a totally geodesic submanifold of $\mathcal{P}_0(V)$. Moreover, the isometry $\mathcal{P}_0(V) \to \mathcal{P}_0(V) : P \mapsto \phi_0(P) = P_0 P^{-1} P_0$ in Step 2 of the proof of Theorem 6.5.10 satisfies

$$\phi_0\left(\exp(\widehat{P} P_0^{-1})P_0\right) = \exp(-\widehat{P} P_0^{-1})P_0$$

for all $\widehat{P} \in T_P\mathcal{P}_0(V)$ and so restricts to an isometry of M_w. Hence M_w is a symmetric space and this proves Lemma 7.5.24.   □

We emphasize that Lemma 7.5.24 does not require the hypothesis that the group G_w acts irreducibly on V. This hypothesis was only used to prove that the space M_w is nonempty. The next example shows that f_w can have critical points in cases where G_w acts reducibly on V.

Example 7.5.25 Let $V = \mathbb{R}^2$, let $W = \mathcal{S} \subset \mathbb{R}^{2 \times 2}$ be the space of symmetric matrices, equipped with the standard inner product and the standard action $S \mapsto gSg^\mathsf{T}$ of $SL(2, \mathbb{R})$, and let $w = S := \mathrm{diag}(1, -1) \in \mathcal{S}$ so that

$$G_S = \left\{ \begin{pmatrix} a & b \\ b & a \end{pmatrix} \middle| a > 0, \, a^2 - b^2 = 1 \right\}.$$

The action of G_S on $\mathbb{R}^2$ is reducible, because the diagonal in $\mathbb{R}^2$ is G_S-invariant, however, the function $f_S(P) = \mathrm{trace}(SP^{-1}SP^{-1})$ attains its minimum on the set

$M_S = G_S \subset \mathcal{P}_0(\mathbb{R}^2)$, corresponding to the symmetric inner products on $\mathbb{R}^2$. If one modifies this example by taking $W = \mathcal{S} \times \mathcal{S}$ and $w = (\mathbb{1}, S)$, then $f_w(P) = \text{trace}(P^{-2} + SP^{-1}SP^{-1})$ has a unique critical point at $P = \mathbb{1}$, the group $G_w = \{\mathbb{1}\}$ acts reducibly on V, and the assertions of Theorem 7.5.17 are trivially satisfied for every inner product on V.

Example 7.5.21 and Example 7.5.25 show that, in general, one cannot expect there to be a one-to-one correspondence between the minima of f_w and the (ρ, w)-symmetric inner products on V. However, we shall see below that such a one-to-one correspondence does exist in many cases.

Remark 7.5.26 Once it is known, that the function $f_w : \mathcal{P}_0(V) \to \mathbb{R}$ has a critical point $P_0 \in \mathcal{P}_0(V)$, we can modify the entire setup as follows. Replace the inner product on V by $\langle \cdot, \cdot \rangle_{V,0} := \langle \cdot, P_0^{-1} \cdot \rangle_V$ and the inner product on W by $\langle \cdot, \cdot \rangle_{W,0} := \langle \cdot, \rho(P_0^{-1}) \cdot \rangle_W$. This pair of inner products satisfies the requirements of Lemma 7.5.19. Let $\mathcal{P}_{00}(V)$ be the space of self-adjoint positive definite endomorphisms with respect to the new inner product and define the function $f_{w,0} : \mathcal{P}_{00}(V) \to \mathbb{R}$ by the analogous formula. Then $P \in \mathcal{P}_0(V)$ if and only if $PP_0^{-1} \in \mathcal{P}_{00}(V)$ and $f_{w,0}(PP_0^{-1}) = f_w(P)$ for all $P \in \mathcal{P}_0(V)$. Thus $f_{w,0}$ attains its minimum at $P = \mathbb{1}$.

In the next corollary we do not assume that G_w acts irreducibly on V.

Corollary 7.5.27 (Cartan Decomposition) *Assume the function f_w in Lemma 7.5.20 has a critical point at $P_0 = \mathbb{1}$ and define*

$$K_w := G_w \cap \text{SO}(V), \qquad \mathfrak{p}_w := \{\eta \in \mathfrak{g}_w \mid \eta = \eta^*\}. \tag{7.5.40}$$

Then the map

$$K_w \times \mathfrak{p}_w \to G_w : (u, \eta) \mapsto \exp(\eta)u =: \phi_w(u, \eta) \tag{7.5.41}$$

is a diffeomorphism. Hence the map $G_w \to \mathcal{P}_0(V) : g \mapsto \sqrt{gg^*}$ descends to a diffeomorphism from the quotient space G_w / K_w to $M_w = G_w \cap \mathcal{P}_0(V)$.

Proof By part (ii) of Lemma 7.5.22 with $P_0 = \mathbb{1}$ the function f_w attains its minimum on the set $M_w = G_w \cap \mathcal{P}_0(\mathfrak{g})$. Now define the map

$$\psi_w : G_w \to K_w \times \mathfrak{p}_w$$

by $\psi_w(g) := (u, \eta)$, where

$$\eta := \tfrac{1}{2} \exp^{-1}(gg^*) \in \mathfrak{p}_w, \qquad u := \exp(-\eta)g \in K_w$$

for $g \in G_w$. This map is well defined and smooth because, for every $g \in G_w$, we have $gg^* \in G_w \cap \mathcal{P}_0(V) = M_w$ (see Exercise 7.5.16), and the exponential map descends to a diffeomorphism $\exp : \mathfrak{p}_w \to M_w$ (see Lemma 7.5.22). Since $\phi_w \circ \psi_w = \text{id}$ and $\psi_w \circ \phi_w = \text{id}$, it follows that ϕ_w is a diffeomorphism. This proves Corollary 7.5.27. $\qquad \square$

As a warmup for the main application of Theorem 7.5.17 in Sect. 7.6 it may be useful to consider the following two examples.

Exercise 7.5.28

(i)   Let V be an m-dimensional real vector space and

$$W := S^2 V^*$$

be the vector space of all symmetric bilinear forms $Q : V \times V \to \mathbb{R}$. Define the homomorphism $\rho : \mathrm{SL}(V) \to \mathrm{SL}(W)$ by the standard action of the group $\mathrm{SL}(V)$ on W, i.e.

$$(\rho(g)Q)(v, v') := Q(g^{-1}v, g^{-1}v')$$

for $g \in \mathrm{SL}(V)$, $Q \in S^2 V^*$, and $v, v' \in V$. Assume that V is equipped with an inner product and an orthonormal basis $e_1, \ldots, e_m$, and define an inner product on W by $\langle Q, Q' \rangle := \sum_{i,j} Q(e_i, e_j)Q'(e_i, e_j)$ for $Q, Q' \in S^2 V^*$. Show that this inner product is independent of the choice of the orthonormal basis and satisfies the requirements of Lemma 7.5.19.

(ii)   The inner product on V is an element $Q_0 \in W$ whose isotropy subgroup is the special orthogonal group $\mathrm{SO}(V)$. Show that the function f_{Q_0} in (7.5.38) is given by $f_{Q_0}(P) = \mathrm{trace}(P^2)$ for $P \in \mathcal{P}_0(V)$ and that it has a unique critical point at $P = \mathbb{1}$.

(iii)   Examine the case where $V = \mathbb{R}^{m+1}$ is equipped with the standard inner product and $Q \in W$ is the quadratic form in (6.4.9). Relate this example to the isometry group of hyperbolic space (Sect. 6.4.4). Find a maximal compact subgroup of the identity component of $\mathrm{O}(m, 1)$.

Exercise 7.5.29

(i)   Let V be a $2n$-dimensional real vector space and

$$W = \Lambda^2 V^*$$

be the vector space of all skew-symmetric bilinear forms $\tau : V \times V \to \mathbb{R}$. Define the homomorphism $\rho : \mathrm{SL}(V) \to \mathrm{SL}(W)$ by the standard action of the group $\mathrm{SL}(V)$ on W, i.e.

$$(\rho(g)\tau)(v, v') := (g_*\tau)(v, v') := \tau(g^{-1}v, g^{-1}v')$$

for $g \in \mathrm{SL}(V)$, $\tau \in \Lambda^2 V^*$, and $v, v' \in V$. Assume that V is equipped with an inner product and an orthonormal basis $e_1, \ldots, e_{2n}$, and define an inner product on W by $\langle \tau, \tau' \rangle := \sum_{i,j} \tau(e_i, e_j)\tau'(e_i, e_j)$ for $\tau, \tau' \in \Lambda^2 V^*$. Show that this inner product is independent of the choice of the orthonormal basis and satisfies the requirements of Lemma 7.5.19.

(ii) Let $\omega : V \times V \to \mathbb{R}$ be a nondegenerate skew-symmetric bilinear form. Then the pair (V, ω) is called a **symplectic vector space** and ω is called a **symplectic form on** V. The isotropy subgroup of ω in $GL(V)$ is called the **symplectic linear group**. Denote this group and its Lie algebra by

$$Sp(V, \omega) := \{g \in GL(V) \,|\, \omega(g\cdot, g\cdot) = \omega\},$$
$$\mathfrak{sp}(V, \omega) := \mathrm{Lie}(Sp(V, \omega)) = \{A \in \mathrm{End}(V) \,|\, \omega(A\cdot, \cdot) + \omega(\cdot, A\cdot) = 0\}.$$

The group $Sp(V, \omega)$ is connected and contained in $SL(V)$ (see [49]). An automorphism $J : V \to V$ is called a **linear complex structure** iff $J^2 = -\mathbb{1}$. A linear complex structure J is called **compatible with** ω iff the bilinear form $\omega(\cdot, J\cdot)$ is an inner product on V. An inner product $\langle \cdot, \cdot \rangle$ on V is called **compatible with** ω iff there exists a linear complex structure J such that $\omega(\cdot, J\cdot) = \langle \cdot, \cdot \rangle$. Prove that an inner product on V is compatible with ω if and only if it satisfies the conditions (for any basis $e_1, \dots, e_{2n}$ of V)

$$\det\bigl(\omega(e_i, e_j)\bigr) = \det\bigl(\langle e_i, e_j \rangle\bigr), \tag{7.5.42}$$
$$A \in \mathfrak{sp}(V, \omega) \quad\Longrightarrow\quad A^* \in \mathfrak{sp}(V, \omega). \tag{7.5.43}$$

If an inner product on V is compatible with ω, prove that $g \in Sp(V, \omega)$ implies $g^* \in Sp(V, \omega)$ (without using the fact that $Sp(V, \omega)$ is connected). **Hint:** Define $J \in \mathrm{End}(V)$ by $\omega(\cdot, J\cdot) := \langle \cdot, \cdot \rangle$ and show that $J + J^* = 0$. Show that $A \in \mathfrak{sp}(V, \omega)$ if and only if $AJ + JA^* = 0$. Use (7.5.43) to prove that J^2 commutes with every self-adjoint endomorphism of V and hence satisfies $J^2 = \lambda \mathbb{1}$ for some $\lambda \in \mathbb{R}$. Use (7.5.42) to conclude that $\lambda = -1$.

(iii) Fix an inner product on V and a symplectic form $\omega : V \times V \to \mathbb{R}$ that satisfies (7.5.42). Then the function f_ω in (7.5.38) is given by

$$f_\omega(P) = \sum_{i,j} \omega(e_i, e_j)\omega(Pe_i, Pe_j), \qquad P \in \mathcal{P}_0(V), \tag{7.5.44}$$

where $e_1, \dots, e_{2n}$ is an orthonormal basis of V. Prove that this is the norm squared of ω with respect to the inner product $\langle \cdot, P^{-1}\cdot \rangle$. Prove that P is a critical point of f_ω if and only if the inner product $\langle \cdot, P^{-1}\cdot \rangle$ is compatible with ω. Prove that the space $\mathcal{J}(V, \omega)$ of ω-compatible linear complex structures is a Hadamard manifold and a symmetric space (Exercise 6.5.24). Prove that, if $J \in \mathcal{J}(V, \omega)$, then the unitary group $U(V, \omega, J) := \{g \in Sp(V, \omega) \,|\, gJg^{-1} = J\}$ is a maximal compact subgroup of $Sp(V, \omega)$ and that every compact subgroup of $Sp(V, \omega)$ is conjugate to a subgroup of $U(V, \omega, J)$. All this is of course well known, but this exercise shows how these results can be derived from Theorem 7.5.17. Moreover, it is not necessary to assume that $Sp(V, \omega)$ is connected. One can start with the identity component of $Sp(V, \omega)$, prove that it is contained in $SL(V)$, and deduce the connectivity of $Sp(V, \omega)$ from that of the unitary group.

7.6 Semisimple Lie Algebras*

This section discusses applications of the results in Sect. 7.5.3 to Lie algebra theory, following the work of Donaldson [17]. It examines symmetric inner products on Lie algebras (Sect. 7.6.1), establishes their existence on simple Lie algebras (Sect. 7.6.2), and derives as consequences several standard results in Lie algebra theory, such as the uniqueness of maximal compact subgroups up to conjugation for semisimple Lie algebras (Sect. 7.6.3), and Cartan's theorem about the compact real form of a semisimple complex Lie algebra (Sect. 7.6.4).

Here are some basic definitions that will be used throughout this section. Let $\mathfrak{g}$ be a finite-dimensional real Lie algebra (Definition 2.4.22). A vector subspace $\mathfrak{h} \subset \mathfrak{g}$ is called an **ideal** iff $[\xi, \eta] \in \mathfrak{h}$ for every $\xi \in \mathfrak{g}$ and every $\eta \in \mathfrak{h}$. The Lie algebra $\mathfrak{g}$ is called **abelian** iff the Lie bracket vanishes. It is called **simple** iff it is not abelian and does not contain any ideal other than $\mathfrak{h} = \{0\}$ and $\mathfrak{h} = \mathfrak{g}$. Examples of ideals in any Lie algebra $\mathfrak{g}$ are the center $Z(\mathfrak{g})$ (Exercise 2.5.34) and the commutant $[\mathfrak{g}, \mathfrak{g}]$ (Exercise 5.2.24), defined by

$$Z(\mathfrak{g}) := \{\xi \in \mathfrak{g} \mid [\xi, \eta] = 0 \text{ for all } \eta \in \mathfrak{g}\},$$
$$[\mathfrak{g}, \mathfrak{g}] := \mathrm{span}\{[\xi, \eta] \mid \xi, \eta \in \mathfrak{g}\}.$$

Recall also the definition of the adjoint representation $\mathrm{ad} : \mathfrak{g} \to \mathrm{Der}(\mathfrak{g})$ in Example 2.5.23 by $\mathrm{ad}(\xi) := [\xi, \cdot]$ for $\xi \in \mathfrak{g}$ and the definition of the Killing form $\kappa : \mathfrak{g} \times \mathfrak{g} \to \mathbb{R}$ in Example 5.2.25 by $\kappa(\xi, \eta) := \mathrm{trace}(\mathrm{ad}(\xi)\mathrm{ad}(\eta))$ for $\xi, \eta \in \mathfrak{g}$. Let $\mathrm{Aut}_0(\mathfrak{g})$ be the connected component of the identity in the group $\mathrm{Aut}(\mathfrak{g})$ of automorphisms of $\mathfrak{g}$. This is a Lie subgroup of $\mathrm{GL}(\mathfrak{g})$ whose Lie algebra is the space $\mathrm{Lie}(\mathrm{Aut}_0(\mathfrak{g})) = \mathrm{Der}(\mathfrak{g})$ of derivations on $\mathfrak{g}$.

7.6.1 Symmetric Inner Products

In [17] Donaldson introduced the following notion.

Definition 7.6.1 Let $\mathfrak{g}$ be a finite-dimensional real Lie algebra. An inner product $\langle \cdot, \cdot \rangle$ on $\mathfrak{g}$ is called **symmetric** iff it satisfies the condition

$$\delta \in \mathrm{Der}(\mathfrak{g}) \qquad \Longrightarrow \qquad \delta^* \in \mathrm{Der}(\mathfrak{g}). \tag{7.6.1}$$

Here $\delta^* : \mathfrak{g} \to \mathfrak{g}$ denotes the adjoint of the endomorphism δ with respect to the inner product, i.e. it satisfies $\langle \xi, \delta^* \eta \rangle = \langle \delta \xi, \eta \rangle$ for all $\xi, \eta \in \mathfrak{g}$.

Exercise 7.6.2 Let $\mathfrak{g}$ be a finite-dimensional real Lie algebra equipped with a symmetric inner product. Prove that $g \in \mathrm{Aut}_0(\mathfrak{g}) \implies g^* \in \mathrm{Aut}_0(\mathfrak{g})$. **Hint:** See Exercise 7.5.16.

Some consequences of the existence of a symmetric inner product are derived in Lemma 7.6.8 below. To begin with we discuss some examples.

Every vector space endomorphism of a Lie algebra $\mathfrak{g}$ that takes values in the center $Z(\mathfrak{g})$ and vanishes on the commutant $[\mathfrak{g}, \mathfrak{g}]$ is a derivation. Conversely, every derivation that takes values in $Z(\mathfrak{g})$ necessarily vanishes on $[\mathfrak{g}, \mathfrak{g}]$. The space of such derivations is an ideal in $\mathrm{Der}(\mathfrak{g})$, denoted by

$$\mathrm{Der}_Z(\mathfrak{g}) := \{\delta \in \mathrm{Der}(\mathfrak{g}) \mid \mathrm{im}(\delta) \subset Z(\mathfrak{g})\}. \tag{7.6.2}$$

Example 7.6.3 Consider the abelian Lie algebra $\mathfrak{g} = \mathbb{R}^m$. The adjoint representation is trivial and $\mathrm{Der}(\mathfrak{g}) = \mathfrak{gl}(m, \mathbb{R}) = \mathbb{R}^{m \times m}$ is the Lie algebra of all vector space endomorphisms of $\mathfrak{g}$. Thus $\mathrm{Der}(\mathfrak{g}) = \mathrm{Der}_Z(\mathfrak{g})$, the Killing form on $\mathfrak{g}$ vanishes, and every inner product on $\mathfrak{g}$ is symmetric.

Example 7.6.4 Consider the Lie algebra $\mathfrak{g} = \mathfrak{gl}(m, \mathbb{R})$ with $[\mathfrak{g}, \mathfrak{g}] = \mathfrak{sl}(m, \mathbb{R})$ and $Z(\mathfrak{g}) = \mathbb{R}\mathbb{1}$. It satisfies $\mathfrak{g} = [\mathfrak{g}, \mathfrak{g}] \oplus Z(\mathfrak{g})$ and $\mathrm{Der}(\mathfrak{g}) = \mathrm{ad}(\mathfrak{g}) \oplus \mathrm{Der}_Z(\mathfrak{g})$, where $\mathrm{Der}_Z(\mathfrak{g}) \subset \mathrm{Der}(\mathfrak{g})$ is the one-dimensional subspace generated by the derivation $\delta_Z(A) = \mathrm{trace}(A)\mathbb{1}$ (whose trace is m). Moreover, the standard inner product $\langle A, A' \rangle = \mathrm{trace}(A^\mathsf{T} A')$ on $\mathfrak{g}$ is symmetric and the kernel of the Killing form $\kappa(A, A') = 2m\,\mathrm{trace}(AA') - 2\mathrm{trace}(A)\mathrm{trace}(A')$ is $Z(\mathfrak{g})$.

Example 7.6.5 Consider the Lie algebra $\mathfrak{g} = \mathfrak{gl}(m, \mathbb{R}) \times \mathbb{R}^m$ with the Lie bracket $[(A, v), (A', v')] := ([A, A'], Av' - A'v)$. This Lie algebra can be identified with the space of all affine vector fields on $\mathbb{R}^m$. It has a trivial center and the commutant $[\mathfrak{g}, \mathfrak{g}] = \mathfrak{sl}(m, \mathbb{R}) \times \mathbb{R}^m$ has codimension one. Moreover, $\mathrm{trace}(\mathrm{ad}(A, v)) = \mathrm{trace}(A)$ for every $(A, v) \in \mathfrak{g}$, the kernel of the Killing form $\kappa((A, v), (A', v')) = (2m + 1)\mathrm{trace}(AA') - 2\mathrm{trace}(A)\mathrm{trace}(A')$ is the abelian ideal $\{0\} \times \mathbb{R}^m$, and the adjoint representation $\mathrm{ad} : \mathfrak{g} \to \mathrm{Der}(\mathfrak{g})$ is a Lie algebra isomorphism.

Example 7.6.6 Consider the Heisenberg algebra $\mathfrak{h} = V \times \mathbb{R}$ of a symplectic vector space (V, ω) with the Lie bracket $[(v, t), (v', t')] = (0, \omega(v, v'))$ (see Exercise 2.5.15). It satisfies $Z(\mathfrak{h}) = [\mathfrak{h}, \mathfrak{h}] = \{0\} \times \mathbb{R}$ and the Killing form vanishes. Every derivation on $\mathfrak{h}$ has the form $\delta(v, t) = (Av + \lambda v, \Lambda(v) + 2\lambda t)$, where $\lambda \in \mathbb{R}$, $\Lambda \in V^*$, and $A \in \mathfrak{sp}(V, \omega)$. The subspace $\mathrm{ad}(\mathfrak{h}) = \mathrm{Der}_Z(\mathfrak{h})$ consists of all derivations of the form $\delta(v, t) = (0, \Lambda(v))$.

Example 7.6.7 The Heisenberg algebra of a symplectic vector space (V, ω) extends to a Lie algebra $\mathfrak{g} = \mathfrak{sp}(V, \omega) \times V \times \mathbb{R}$ with the Lie bracket

$$[(A, v, t), (A', v', t')] = ([A, A'], Av' - A'v, \omega(v, v')).$$

It satisfies $[\mathfrak{g}, \mathfrak{g}] = \mathfrak{g}$ and has a one-dimensional center $Z(\mathfrak{g}) = \{0\} \times \{0\} \times \mathbb{R}$. The kernel of the Killing form is the Heisenberg algebra and the adjoint representation $\mathrm{ad} : \mathfrak{g} \to \mathrm{Der}(\mathfrak{g})$ is surjective.

The next lemma shows that the Lie algebras in Examples 7.6.5, 7.6.6, and 7.6.7 do not admit symmetric inner products.

Lemma 7.6.8 *Let $\mathfrak{g}$ be a finite-dimensional real Lie algebra equipped with a symmetric inner product. Then*

$$\mathrm{Der}(\mathfrak{g}) = \mathrm{ad}(\mathfrak{g}) \oplus \mathrm{Der}_Z(\mathfrak{g}), \qquad \mathfrak{g} = [\mathfrak{g}, \mathfrak{g}] \oplus Z(\mathfrak{g}), \qquad (7.6.3)$$

the kernel of the Killing form $\kappa : \mathfrak{g} \times \mathfrak{g} \to \mathbb{R}$ is the center of $\mathfrak{g}$, and there exists an involution $\mathfrak{g} \to \mathfrak{g} : \xi \mapsto \xi^*$ such that, for all $\xi, \eta \in \mathfrak{g}$,

$$\mathrm{ad}(\xi^*) = \mathrm{ad}(\xi)^*, \qquad [\xi, \eta]^* = [\eta^*, \xi^*]. \qquad (7.6.4)$$

Proof Consider the orthogonal decomposition

$$\mathrm{Der}(\mathfrak{g}) = \mathcal{A} \oplus \mathcal{B}, \qquad \mathcal{A} := \mathrm{ad}(\mathfrak{g}), \qquad \mathcal{B} := \mathcal{A}^\perp, \qquad (7.6.5)$$

with respect to the inner product $\langle \delta, \delta' \rangle = \mathrm{trace}(\delta^* \delta')$ for $\delta, \delta' \in \mathrm{Der}(\mathfrak{g})$. Since $[\delta, \mathrm{ad}(\xi)] = \mathrm{ad}(\delta\xi)$ for $\delta \in \mathrm{Der}(\mathfrak{g})$ and $\xi \in \mathfrak{g}$, the subspace $\mathcal{A}$ is an ideal in $\mathrm{Der}(\mathfrak{g})$. Moreover, if $\delta \in \mathcal{B}$ and $\varepsilon \in \mathrm{Der}(\mathfrak{g})$, then $\varepsilon^* \in \mathrm{Der}(\mathfrak{g})$, hence $\mathrm{trace}([\varepsilon, \delta]^* \mathrm{ad}(\xi)) = \mathrm{trace}(\delta^* [\varepsilon^*, \mathrm{ad}(\xi)]) = \mathrm{trace}(\delta^* \mathrm{ad}(\varepsilon^* \xi)) = 0$ for all $\xi \in \mathfrak{g}$, and hence $[\varepsilon, \delta] \in \mathcal{B}$. Thus $\mathcal{B}$ is also an ideal in $\mathrm{Der}(\mathfrak{g})$.

Next define $\mathrm{Der}_Z(\mathfrak{g})^* := \{\delta^* \mid \delta \in \mathrm{Der}_Z(\mathfrak{g})\}$. We prove that

$$\mathcal{B} = \mathrm{Der}_Z(\mathfrak{g}) = \mathrm{Der}_Z(\mathfrak{g})^*. \qquad (7.6.6)$$

Since $\mathcal{A}$ and $\mathcal{B}$ are ideals we have $[\delta, \delta'] = 0$ for all $\delta \in \mathcal{B}$ and $\delta' \in \mathcal{A}$. Thus $\mathrm{ad}(\delta\xi) = [\delta, \mathrm{ad}(\xi)] = 0$ for all $\delta \in \mathcal{B}$ and $\xi \in \mathfrak{g}$, hence $\mathrm{im}(\delta) \subset Z(\mathfrak{g})$ for all $\delta \in \mathcal{B}$, and so $\mathcal{B} \subset \mathrm{Der}_Z(\mathfrak{g})$. Now let $\delta \in \mathrm{Der}_Z(\mathfrak{g})^*$. Then $\delta^* \in \mathrm{Der}_Z(\mathfrak{g})$, hence $[\mathfrak{g}, \mathfrak{g}] \subset \ker(\delta^*)$, hence $\mathrm{trace}(\delta^* \mathrm{ad}(\xi)) = 0$ for all $\xi \in \mathfrak{g}$, and so $\delta \in \mathcal{B}$. Thus $\mathrm{Der}_Z(\mathfrak{g})^* \subset \mathcal{B} \subset \mathrm{Der}_Z(\mathfrak{g})$ and so (7.6.6) holds for dimensional reasons. The first equation in (7.6.3) follows directly from (7.6.5) and (7.6.6). It follows also from (7.6.6) that $\mathrm{trace}(\delta^* \mathrm{ad}(\xi)^*) = \mathrm{trace}(\mathrm{ad}(\xi)\delta) = 0$ for all $\xi \in \mathfrak{g}$ and all $\delta \in \mathcal{B}$, and so $\mathrm{ad}(\xi)^* \in \mathcal{B}^\perp = \mathcal{A}$ for all $\xi \in \mathfrak{g}$. Thus

$$\delta \in \mathcal{A} \qquad \Longrightarrow \qquad \delta^* \in \mathcal{A}. \qquad (7.6.7)$$

By (7.6.7), an element $\zeta \in \mathfrak{g}$ belongs to $Z(\mathfrak{g})$ if and only if $\mathrm{ad}(\xi)^* \zeta = 0$ for all $\xi \in \mathfrak{g}$ if and only if $\langle \zeta, \mathrm{ad}(\xi)\eta \rangle = 0$ for all $\xi, \eta \in \mathfrak{g}$, if and only if $\zeta \in [\mathfrak{g}, \mathfrak{g}]^\perp$. Thus $[\mathfrak{g}, \mathfrak{g}]^\perp = Z(\mathfrak{g})$ and this proves the second equation in (7.6.3).

By (7.6.3), the map $\mathrm{ad} : \mathfrak{g} \to \mathrm{Der}(\mathfrak{g})$ restricts to a Lie algebra isomorphism from $[\mathfrak{g}, \mathfrak{g}]$ to $\mathcal{A}$. Hence, by (7.6.7) there exists a unique involution $[\mathfrak{g}, \mathfrak{g}] \to [\mathfrak{g}, \mathfrak{g}] : \xi \mapsto \xi^*$ that satisfies (7.6.4) for all $\xi, \eta \in [\mathfrak{g}, \mathfrak{g}]$. By (7.6.3) this involution extends uniquely to an involution $\mathfrak{g} \to \mathfrak{g} : \xi \mapsto \xi^*$ such that $\zeta^* = \zeta$ for all $\zeta \in Z(\mathfrak{g})$, and the extended involution satisfies (7.6.4) for all $\xi, \eta \in \mathfrak{g}$.

Now let $\zeta \in \mathfrak{g}$ belong to the kernel of the Killing form, i.e. $\kappa(\zeta, \xi) = 0$ for all $\xi \in \mathfrak{g}$. Then $|\mathrm{ad}(\zeta)|^2 = \kappa(\zeta, \zeta^*) = 0$ and hence $\zeta \in Z(\mathfrak{g})$. Conversely, it follows directly from the definitions that $Z(\mathfrak{g})$ is contained in the kernel of the Killing form and this proves Lemma 7.6.8. $\qquad \square$

7.6.2 Simple Lie Algebras

The goal of this section is to establish the existence of symmetric inner products on simple Lie algebras. First, the following lemma derives some immediate consequences of the definition of a simple Lie algebra.

Lemma 7.6.9 *Let $\mathfrak{g}$ be a finite-dimensional simple real Lie algebra. Then the center of $\mathfrak{g}$ is trivial, the adjoint representation* ad $: \mathfrak{g} \to \mathrm{Der}(\mathfrak{g})$ *is injective, the commutant is* $[\mathfrak{g}, \mathfrak{g}] = \mathfrak{g}$, *and* $\mathrm{trace}(\mathrm{ad}(\xi)) = 0$ *for all* $\xi \in \mathfrak{g}$.

Proof The center $Z(\mathfrak{g})$ is an ideal in $\mathfrak{g}$. It is not equal to $\mathfrak{g}$ because $\mathfrak{g}$ is not abelian, and hence $Z(\mathfrak{g}) = \{0\}$ because $\mathfrak{g}$ is simple. The subspace $[\mathfrak{g}, \mathfrak{g}]$ is also an ideal in $\mathfrak{g}$. It is nonzero because $\mathfrak{g}$ is not abelian, and hence $[\mathfrak{g}, \mathfrak{g}] = \mathfrak{g}$ because $\mathfrak{g}$ is simple. The adjoint representation ad $: \mathfrak{g} \to \mathrm{Der}(\mathfrak{g})$ is injective because its kernel is the center of $\mathfrak{g}$. The last assertion follows from the fact that $[\mathfrak{g}, \mathfrak{g}] = \mathfrak{g}$ and $\mathrm{trace}\big(\mathrm{ad}([\xi, \eta])\big) = \mathrm{trace}\big([\mathrm{ad}(\xi), \mathrm{ad}(\eta)]\big) = 0$ for all $\xi, \eta \in \mathfrak{g}$. This proves Lemma 7.6.9. $\square$

That the Killing form of a simple Lie algebra is nondegenerate is a deeper result that does not follow directly from the definition. In [17, Theorem 1] Donaldson deduces the existence of symmetric inner products on simple Lie algebras from Theorem 7.5.17 and derives as corollaries various standard results in Lie algebra theory, including nondegeneracy of the Killing form.

Theorem 7.6.10 *Let $\mathfrak{g}$ be a finite-dimensional simple real Lie algebra. Then the Killing form on $\mathfrak{g}$ is nondegenerate, the adjoint representation* ad $: \mathfrak{g} \to \mathrm{Der}(\mathfrak{g})$ *is bijective, every derivation $\delta : \mathfrak{g} \to \mathfrak{g}$ has trace zero, and every automorphism in the identity component* $\mathrm{Aut}_0(\mathfrak{g})$ *has determinant one.*

In particular, Theorem 7.6.10 establishes for every simple Lie algebra $\mathfrak{g}$ the existence of a connected Lie group $\mathrm{Aut}_0(\mathfrak{g}) \subset \mathrm{SL}(\mathfrak{g})$ whose Lie algebra is isomorphic to $\mathfrak{g}$.

Theorem 7.6.11 (Donaldson) *Every finite-dimensional simple real Lie algebra admits a symmetric inner product. Moreover, if $\mathrm{SO}(\mathfrak{g})$ is the special orthogonal group associated to a symmetric inner product on $\mathfrak{g}$, then*

$$K := \mathrm{Aut}_0(\mathfrak{g}) \cap \mathrm{SO}(\mathfrak{g}) \tag{7.6.8}$$

is connected and is a maximal compact subgroup of $\mathrm{Aut}_0(\mathfrak{g})$, every compact subgroup of $\mathrm{Aut}_0(\mathfrak{g})$ is conjugate to a Lie subgroup of K, and every maximal compact subgroup of $\mathrm{Aut}_0(\mathfrak{g})$ is conjugate to K.

Given an inner product $\langle\cdot,\cdot\rangle$ on any m-dimensional Lie algebra $\mathfrak{g}$ and an orthonormal basis $e_1,\ldots,e_m$ of $\mathfrak{g}$, define the function $f_\mathfrak{g} : \mathcal{P}_0(\mathfrak{g}) \to \mathbb{R}$ by

$$f_\mathfrak{g}(P) := \sum_{i,j}\langle [e_i,e_j], P^{-1}[Pe_i,Pe_j]\rangle \tag{7.6.9}$$

for $P \in \mathcal{P}_0(\mathfrak{g})$. The right hand side of (7.6.9) is independent of the choice of the orthonormal basis and is the norm squared of the Lie bracket with respect to the inner product $\langle\cdot, P^{-1}\cdot\rangle$ on $\mathfrak{g}$.

Theorem 7.6.12 (Donaldson)   *Let $\mathfrak{g}$ be a finite-dimensional simple real Lie algebra equipped with a symmetric inner product. Then the set*

$$\begin{aligned} M_\mathfrak{g} &:= \big\{P \in \mathcal{P}_0(\mathfrak{g}) \,\big|\, df_\mathfrak{g}(P) = 0\big\} = \big\{P \in \mathcal{P}_0(\mathfrak{g}) \,\big|\, f_\mathfrak{g}(P) = \inf f_\mathfrak{g}\big\} \\ &= \mathcal{P}_0(\mathfrak{g}) \cap \mathrm{Aut}(\mathfrak{g}) = \big\{\exp(\delta) \,\big|\, \delta \in \mathrm{Der}(\mathfrak{g}),\, \delta = \delta^*\big\} \\ &= \big\{P \in \mathcal{P}_0(\mathfrak{g}) \,\big|\, \text{the inner product } \langle\cdot, P^{-1}\cdot\rangle \text{ is symmetric}\big\} \end{aligned} \tag{7.6.10}$$

of critical points of $f_\mathfrak{g}$ is a geodesically convex and totally geodesic submanifold of $\mathcal{P}_0(\mathfrak{g})$. Hence it is a Hadamard manifold and a symmetric space.

The proofs require three preparatory lemmas. We do not assume that every derivation has trace zero. Thus it is necessary as an intermediate step to introduce the subspace $\mathrm{Der}_0(\mathfrak{g}) := \{\delta \in \mathrm{Der}(\mathfrak{g}) \,|\, \mathrm{trace}(\delta) = 0\}$. We will consider inner products on $\mathfrak{g}$ that satisfy the condition

$$\delta \in \mathrm{Der}_0(\mathfrak{g}) \quad\Longrightarrow\quad \delta^* \in \mathrm{Der}_0(\mathfrak{g}). \tag{7.6.11}$$

Lemma 7.6.13   *Let $\mathfrak{g}$ be a finite-dimensional real Lie algebra satisfying the conditions $Z(\mathfrak{g}) = \{0\}$ and $[\mathfrak{g},\mathfrak{g}] = \mathfrak{g}$, and fix an inner product $\langle\cdot,\cdot\rangle$ on $\mathfrak{g}$. Then the inner product is symmetric if and only if it satsfies (7.6.11). Moreover, if such an inner product exists, then $\mathrm{Der}(\mathfrak{g}) = \mathrm{ad}(\mathfrak{g}) = \mathrm{Der}_0(\mathfrak{g})$.*

Proof   Assume (7.6.11) and consider the decomposition $\mathrm{Der}_0(\mathfrak{g}) = \mathcal{A}_0 \oplus \mathcal{B}_0$, where $\mathcal{A}_0 := \mathrm{ad}(\mathfrak{g}) \subset \mathrm{Der}_0(\mathfrak{g})$ (because $[\mathfrak{g},\mathfrak{g}] = \mathfrak{g}$) and $\mathcal{B}_0 := \mathcal{A}_0^\perp$ are ideals in $\mathrm{Der}_0(\mathfrak{g})$ as in the proof of Lemma 7.6.8. Then $\mathrm{ad}(\delta\xi) = [\delta, \mathrm{ad}(\xi)] = 0$ for all $\delta \in \mathcal{B}_0$ and all $\xi \in \mathfrak{g}$. Since $Z(\mathfrak{g}) = \{0\}$, this implies $\mathcal{B}_0 = 0$, and hence the adjoint representation $\mathrm{ad} : \mathfrak{g} \to \mathrm{Der}_0(\mathfrak{g})$ is bijective. By (7.6.11), this implies the existence of an involution $\xi \mapsto \xi^*$ such that $\mathrm{ad}(\xi^*) = \mathrm{ad}(\xi)^*$ for all $\xi \in \mathfrak{g}$. Since the adjoint representation $\mathrm{ad} : \mathfrak{g} \to \mathrm{Der}(\mathfrak{g})$ is injective, this in turn implies that the Killing form is nondegenerate and so $\mathrm{Der}(\mathfrak{g}) = \mathrm{ad}(\mathfrak{g}) = \mathrm{Der}_0(\mathfrak{g})$ by Lemma 7.4.3. Thus the inner product on $\mathfrak{g}$ is symmetric. Conversely, if the inner product on $\mathfrak{g}$ is symmetric, then it satisfies (7.6.11) because δ and δ^* have the same trace. This proves Lemma 7.6.13. $\square$

Lemma 7.6.14 *Let $\mathfrak{g}$ be a finite-dimensional simple real Lie algebra with a symmetric inner product, and let $\mathfrak{g} \to \mathfrak{g} : \xi \mapsto \xi^*$ be the unique involution that satisfies (7.6.4). Then there exists a constant $c > 0$ such that*

$$\kappa(\xi^*, \eta) = c\langle \xi, \eta \rangle \qquad \text{for all } \xi, \eta \in \mathfrak{g}. \tag{7.6.12}$$

Proof By Lemma 7.6.9 the adjoint representation $\mathrm{ad} : \mathfrak{g} \to \mathrm{Der}(\mathfrak{g})$ is injective. Hence the map $\mathfrak{g} \times \mathfrak{g} \to \mathbb{R} : (\xi, \eta) \mapsto \kappa(\xi^*, \eta) = \mathrm{trace}(\mathrm{ad}(\xi)^*\mathrm{ad}(\eta))$ is an inner product on $\mathfrak{g}$. Thus there exists a self-adjoint positive definite vector space isomorphism $A : \mathfrak{g} \to \mathfrak{g}$ such that $\kappa(\xi^*, \eta) = \langle A\xi, \eta \rangle$ for all $\xi, \eta \in \mathfrak{g}$. Let $c > 0$ be an eigenvalue of A and define $\mathfrak{h} := \{\eta \in \mathfrak{g} \mid A\eta = c\eta\}$. We prove that $\mathfrak{h}$ is an ideal in $\mathfrak{g}$. Let $\xi \in \mathfrak{g}$ and $\eta \in \mathfrak{h}$. Then, for all $\zeta \in \mathfrak{g}$,

$$\langle A[\xi, \eta], \zeta \rangle = \kappa([\xi, \eta]^*, \zeta) = \kappa([\eta^*, \xi^*], \zeta) = \kappa(\eta^*, [\xi^*, \zeta]) = \langle A\eta, [\xi^*, \zeta] \rangle$$
$$= \langle A\eta, \mathrm{ad}(\xi^*)\zeta \rangle = \langle \mathrm{ad}(\xi)A\eta, \zeta \rangle = \langle [\xi, A\eta], \zeta \rangle = c\langle [\xi, \eta], \zeta \rangle.$$

Hence $A[\xi, \eta] = c[\xi, \eta]$ and so $[\xi, \eta] \in \mathfrak{h}$. This shows that $\mathfrak{h}$ is a nonzero ideal and hence $\mathfrak{h} = \mathfrak{g}$. Thus $A = c\mathbb{1}$ and this proves Lemma 7.6.14.   □

Given any inner product on $\mathfrak{g}$, call an element $P \in \mathcal{P}_0(\mathfrak{g})$ **symmetric** iff the inner product $\langle \cdot, P^{-1} \cdot \rangle$ on $\mathfrak{g}$ is symmetric. Thus every symmetric element $P \in \mathcal{P}_0(\mathfrak{g})$ determines an involution $\tau_P : \mathrm{Der}(\mathfrak{g}) \to \mathrm{Der}(\mathfrak{g})$ given by $\tau_P(\delta) := P\delta^* P^{-1}$. Denote its determinant by $\varepsilon_P := \det(\tau_P) \in \{-1, +1\}$.

Lemma 7.6.15 *Let $\mathfrak{g}$ be a finite-dimensional simple real Lie algebra, choose any inner product on $\mathfrak{g}$, and let $P, P_0 \in \mathcal{P}_0(\mathfrak{g})$ be symmetric elements. Then $PP_0^{-1} \in \mathrm{Aut}(\mathfrak{g})$.*

Proof The composition of the involutions τ_P and τ_{P_0} is the Lie algebra automorphism $\tau_P \circ \tau_{P_0}(\delta) = P\left(P_0\delta^* P_0^{-1}\right)^* P^{-1} = PP_0^{-1}\delta P_0 P^{-1}$ for $\delta \in \mathrm{Der}(\mathfrak{g})$. By Lemma 7.6.8 and Lemma 7.6.9 the very existence of a symmetric inner product on $\mathfrak{g}$ implies that the adjoint representation $\mathrm{ad} : \mathfrak{g} \to \mathrm{Der}(\mathfrak{g})$ is a Lie algebra isomorphism. Hence there exists a $g \in \mathrm{Aut}(\mathfrak{g})$ such that

$$\mathrm{ad}(g\xi) = PP_0^{-1}\mathrm{ad}(\xi)P_0 P^{-1} \qquad \text{for all } \xi \in \mathfrak{g}. \tag{7.6.13}$$

Since $\mathrm{ad}(g\xi) = g\mathrm{ad}(\xi)g^{-1}$, this automorphism satisfies the equations

$$g^{-1}PP_0^{-1}[\xi, \eta] = [\xi, g^{-1}PP_0^{-1}\eta] \qquad \text{for all } \xi, \eta \in \mathfrak{g}, \tag{7.6.14}$$

$$\det(g) = \varepsilon_P\varepsilon_{P_0} \in \{-1, +1\}. \tag{7.6.15}$$

Since P_0 is symmetric, it follows from Lemma 7.6.8 and Lemma 7.6.14 that there exists an involution $\mathfrak{g} \to \mathfrak{g} : \xi \mapsto \xi^*$ and a constant $c > 0$ such that

$$\mathrm{ad}(\xi^*) = P_0\mathrm{ad}(\xi)^* P_0^{-1}, \qquad \kappa(\xi^*, \eta) = c\langle \xi, P_0^{-1}\eta \rangle \tag{7.6.16}$$

for all $\xi, \eta \in \mathfrak{g}$. By (7.6.13) and (7.6.16) we have

$$
\begin{aligned}
\kappa((g\xi)^*, \eta) &= \text{trace}\left(P_0 \text{ad}(g\xi)^* P_0^{-1} \text{ad}(\eta)\right) \\
&= \text{trace}\left(\left(P P_0^{-1} \text{ad}(\xi) P_0 P^{-1}\right)^* P_0^{-1} \text{ad}(\eta) P_0\right) \\
&= \text{trace}\left(P^{-1} P_0 \text{ad}(\xi)^* P_0^{-1} P P_0^{-1} \text{ad}(\eta) P_0\right) \\
&= \text{trace}\left(\text{ad}(\xi^*) P P_0^{-1} \text{ad}(\eta) P_0 P^{-1}\right) \\
&= \text{trace}\left(\text{ad}(\xi^*) \text{ad}(g\eta)\right) \\
&= \kappa(\xi^*, g\eta).
\end{aligned}
$$

This shows that g is self-adjoint and positive definite with respect to the inner product $\langle \cdot, P_0^{-1} \cdot \rangle$, and so $\det(g) = 1$ by (7.6.15). Since $P P_0^{-1}$ is self-adjoint and positive definite with respect to the same inner product, the vector space isomorphism $g^{-1} P P_0^{-1} = g^{-1/2}(g^{-1/2} P P_0^{-1} g^{-1/2}) g^{1/2}$ has only positive real eigenvalues. Let $\lambda > 0$ be one such eigenvalue. Then the eigenspace $\ker(\lambda \mathbb{1} - g^{-1} P P_0^{-1})$ is a nonzero ideal in $\mathfrak{g}$ by (7.6.14), and so is equal to $\mathfrak{g}$, because $\mathfrak{g}$ is simple. Thus $g^{-1} P P_0^{-1} = \lambda \mathbb{1}$, and since g, P, P_0 all have determinant one, it follows that $\lambda = 1$ and so $P P_0^{-1} = g \in \text{Aut}(\mathfrak{g})$. This proves Lemma 7.6.15. $\qquad\square$

Proof of Theorems 7.6.10, 7.6.11, and 7.6.12 We use the results of Sect. 7.5.3 in the situation where $V := \mathfrak{g}$ is the Lie algebra itself and $W := \Lambda^2 \mathfrak{g}^* \otimes \mathfrak{g}$ is the space of all skew-symmetric bilinear maps $\tau : \mathfrak{g} \times \mathfrak{g} \to \mathfrak{g}$. The Lie group homomorphism $\rho : \text{SL}(\mathfrak{g}) \to \text{SL}(W)$ is given by the standard action of the group $\text{SL}(\mathfrak{g})$ on W, i.e.

$$
(\rho(g)\tau)(\xi, \eta) := g\tau(g^{-1}\xi, g^{-1}\eta)
$$

for $g \in \text{SL}(\mathfrak{g})$, $\tau \in W$, and $\xi, \eta \in \mathfrak{g}$. Fix an inner product $\langle \cdot, \cdot \rangle$ on $\mathfrak{g}$ and an orthonormal basis $e_1, \ldots, e_m$ of $\mathfrak{g}$, and define

$$
\langle \sigma, \tau \rangle_W := \sum_{i,j} \langle \sigma(e_i, e_j), \tau(e_i, e_j) \rangle \tag{7.6.17}
$$

for $\sigma, \tau \in W$. This inner product satisfies the requirements of Lemma 7.5.19, i.e. $\rho(A^*) = \rho(A)^*$ for all $A \in \mathfrak{sl}(\mathfrak{g})$. The vector $w := [\cdot, \cdot] \in W$ is chosen to be the Lie bracket. This vector is nonzero because $\mathfrak{g}$ is not abelian. The isotropy subgroup of w is the group $\text{Aut}(\mathfrak{g}) \cap \text{SL}(\mathfrak{g})$ of all automorphisms of determinant one. Denote its identity component by $\text{G} \subset \text{Aut}_0(\mathfrak{g}) \cap \text{SL}(\mathfrak{g})$. This is a Lie subgroup of $\text{SL}(\mathfrak{g})$ with the Lie algebra $\text{Lie}(\text{G}) = \text{Der}_0(\mathfrak{g})$.

Claim 1 G *acts irreducibly on* $\mathfrak{g}$.

Let $\mathfrak{h} \subset \mathfrak{g}$ be a subspace that is invariant under the action of G. Then $\delta \mathfrak{h} \subset \mathfrak{h}$ for every $\delta \in \text{Der}_0(\mathfrak{g})$. By Lemma 7.6.9 this implies $\text{ad}(\xi)\mathfrak{h} \subset \mathfrak{h}$ for all $\xi \in \mathfrak{g}$, so $\mathfrak{h}$ is an ideal. Thus $\mathfrak{h} = \{0\}$ or $\mathfrak{h} = \mathfrak{g}$ because $\mathfrak{g}$ is simple.

Claim 2 $f_{\mathfrak{g}}$ *is convex and G-invariant and has a critical point. Moreover, every critical point* $P \in \mathcal{P}_0(\mathfrak{g})$ *is symmetric, and* $\mathrm{Der}_0(\mathfrak{g}) = \mathrm{Der}(\mathfrak{g})$.

In the present setting the function f_w in Lemma 7.5.20 agrees with the function $f_{\mathfrak{g}}$ in (7.6.9) and $G_w = G$. Hence $f_{\mathfrak{g}}$ is convex and G-invariant by Lemma 7.5.22, and G acts irreducibly on $\mathfrak{g}$ by Claim 1. Thus Lemma 7.5.23 asserts that $f_{\mathfrak{g}}$ has a critical point. Let $P \in \mathcal{P}_0(\mathfrak{g})$ be a critical point of $f_{\mathfrak{g}}$. Then by Lemma 7.5.23 the inner product $\langle \cdot, P^{-1} \cdot \rangle$ satisfies (7.6.11) and so, by Lemma 7.6.13, this inner product is symmetric and $\mathrm{Der}(\mathfrak{g}) = \mathrm{Der}_0(\mathfrak{g})$.

Claim 3 *Fix a critical point* $P_0 \in \mathcal{P}_0(\mathfrak{g})$ *of* $f_{\mathfrak{g}}$ *and any element* $P \in \mathcal{P}_0(\mathfrak{g})$. *Then the following are equivalent.*

(a) *P is a critical point of* $f_{\mathfrak{g}}$.
(b) $f_{\mathfrak{g}}(P) = \inf f_{\mathfrak{g}}$.
(c) $PP_0^{-1} \in \mathrm{Aut}(\mathfrak{g})$.
(d) *There exists a* $\delta \in \mathrm{Der}(g)$ *such that* $\delta = P_0 \delta^* P_0^{-1}$, $\exp(\delta) = PP_0^{-1}$.
(e) *The inner product* $\langle \cdot, P^{-1} \cdot \rangle$ *on* $\mathfrak{g}$ *is symmeric.*

Since $f_{\mathfrak{g}}$ is convex by Claim 2, the equivalence of (a) and (b) follows from Lemma 7.5.9. Since $\mathrm{Der}(\mathfrak{g}) = \mathrm{Der}_0(\mathfrak{g})$ by Claim 2, the equivalence of (b), (c), and (d) follows from part (ii) of Lemma 7.5.22. Moreover, (a) implies (e) by Claim 2, and (e) implies (c) by Lemma 7.6.15. This proves Claim 3.

The existence of a symmetric inner product on $\mathfrak{g}$ was proved in Claim 2. Thus the nondegeneracy of the Killing form follows from Lemma 7.6.8. The remaining assertions of Theorem 7.6.10 are direct consequences of the nondegeneracy of the Killing form. In particular, the adjoint representation $\mathrm{ad} : \mathfrak{g} \to \mathrm{Der}(\mathfrak{g})$ is bijective and $\mathrm{Der}(\mathfrak{g}) \subset \mathfrak{sl}(\mathfrak{g})$ by Lemma 7.4.3. Hence $\mathrm{Aut}_0(\mathfrak{g}) \subset \mathrm{SL}(\mathfrak{g})$ and so $G_w = G = \mathrm{Aut}_0(\mathfrak{g})$ in the notation of Sect. 7.5.3.

Now fix a symmetric inner product on $\mathfrak{g}$. Then $P_0 = \mathbb{1}$ is a critical point of $f_{\mathfrak{g}}$ by "(e) $\Longrightarrow$ (a)" in Claim 3. Hence the assertions about the subgroup $K = \mathrm{Aut}_0(\mathfrak{g}) \cap \mathrm{SO}(\mathfrak{g})$ in Theorem 7.6.11 follow from Lemma 7.5.23. The equalities in (7.6.10) follow from the equivalence of (a), (b), (c), (d), (e) in Claim 3 with $P_0 = \mathbb{1}$, and the remaining assertions about the space $M_{\mathfrak{g}}$ in Theorem 7.6.12 follow from Lemma 7.5.24. This completes the proof of Theorems 7.6.10, 7.6.11, and 7.6.12.   □

7.6.3 Semisimple Lie Algebras

The following theorem characterizes semisimple Lie algebras in terms of symmetric inner products. First observe that, if $\mathfrak{h}$ is an ideal in a finite-dimensional real Lie algebra $\mathfrak{g}$, then the subspace

$$\mathfrak{h}' := \{\eta' \in \mathfrak{g} \mid \kappa(\eta, \eta') = 0 \text{ for all } \eta \in \mathfrak{h}\} \tag{7.6.18}$$

is also an ideal, because every $\eta' \in \mathfrak{h}'$ satisfies $\kappa(\eta, [\xi, \eta']) = \kappa([\eta, \xi], \eta') = 0$ for all $\xi \in \mathfrak{g}$ and all $\eta \in \mathfrak{h}$, and so $[\xi, \eta'] \in \mathfrak{h}'$ for all $\xi \in \mathfrak{g}$.

Theorem 7.6.16 (Semisimple Lie algebras) *Let $\mathfrak{g}$ be a finite-dimensional real Lie algebra. Then the following are equivalent.*

(i)   *$\mathfrak{g}$ has a trivial center and admits a symmetric inner product.*
(ii) *The Killing form $\kappa : \mathfrak{g} \times \mathfrak{g} \to \mathbb{R}$ is nondegenerate.*
(iii) *If $\mathfrak{h} \subset \mathfrak{g}$ is an ideal, then $\mathfrak{g} = \mathfrak{h} \oplus \mathfrak{h}'$.*
(iv) *$\mathfrak{g}$ is a direct sum of simple ideals.*
(v)   *There exists an inner product $\langle \cdot, \cdot \rangle$ on $\mathfrak{g}$, an involution $\mathfrak{g} \to \mathfrak{g} : \xi \mapsto \xi^*$, and a constant $c > 0$ such that, for all $\xi, \eta \in \mathfrak{g}$,*

$$\mathrm{ad}(\xi^*) = \mathrm{ad}(\xi)^*, \qquad [\xi, \eta]^* = [\eta^*, \xi^*], \qquad \kappa(\xi^*, \eta) = c \langle \xi, \eta \rangle. \quad (7.6.19)$$

Definition 7.6.17 A real Lie algebra $\mathfrak{g}$ is called **semisimple** iff it is finite-dimensional and satisfies the equivalent conditions in Theorem 7.6.16.

Proof of Theorem 7.6.16 That (i) implies (ii) was shown in Lemma 7.6.8.

We prove that (ii) is equivalent to (iii). Assume first that the Killing form is nondegenerate and let $\mathfrak{h} \subset \mathfrak{g}$ be an ideal. Then

$$\eta \in \mathfrak{h}, \quad \eta' \in \mathfrak{h}' \quad \Longrightarrow \quad [\eta, \eta'] = 0, \quad (7.6.20)$$

because $\kappa([\eta, \eta'], \xi) = \kappa(\eta, [\eta', \xi]) = 0$ for all $\xi \in \mathfrak{g}$, all $\eta \in \mathfrak{h}$, and all $\eta' \in \mathfrak{h}'$. Now let $\eta \in \mathfrak{h} \cap \mathfrak{h}'$. Then $\eta' := \mathrm{ad}(\xi)\mathrm{ad}(\eta)\zeta = [\xi, [\eta, \zeta]] \in \mathfrak{h}'$ for all $\xi, \zeta \in \mathfrak{g}$. Hence, by (7.6.20) we have $(\mathrm{ad}(\xi)\mathrm{ad}(\eta))^2\zeta = [\xi, [\eta, \eta']] = 0$ for all $\xi, \zeta \in \mathfrak{g}$. Thus $(\mathrm{ad}(\xi)\mathrm{ad}(\eta))^2 = 0$ and so $\kappa(\xi, \eta) = \mathrm{trace}(\mathrm{ad}(\xi)\mathrm{ad}(\eta)) = 0$ for all $\xi \in \mathfrak{g}$. This implies $\eta = 0$ by nondegeneracy of the Killing form. Thus $\mathfrak{h} \cap \mathfrak{h}' = \{0\}$ and so $\mathfrak{g} = \mathfrak{h} \oplus \mathfrak{h}'$ because $\dim(\mathfrak{h}) + \dim(\mathfrak{h}') = \dim(\mathfrak{g})$. This shows that (ii) implies (iii). To prove the converse, take $\mathfrak{h} = \mathfrak{g}$ so that $\mathfrak{h}' = \{0\}$ is the kernel of the Killing form.

It follows from (ii) and (iii) that, for every ideal $\mathfrak{h} \subset \mathfrak{g}$, the Killing forms of $\mathfrak{h}$ and $\mathfrak{h}'$ are both nondegenerate. Hence an induction argument shows that (iii) implies (iv).

We prove that (iv) implies (v). Assume that $\mathfrak{g} = \mathfrak{g}_1 \oplus \cdots \oplus \mathfrak{g}_r$ is a direct sum of simple ideals $\mathfrak{g}_j \subset \mathfrak{g}$. Fix an index $j \in \{1, \ldots, r\}$ and denote by $\kappa_j : \mathfrak{g}_j \times \mathfrak{g}_j \to \mathbb{R}$ the Killing form of $\mathfrak{g}_j$. By Theorem 7.6.11 the Lie algebra $\mathfrak{g}_j$ admits a symmetric inner product $\langle \cdot, \cdot \rangle_j$. Hence, by Lemma 7.6.8 there exists an involution $\mathfrak{g}_j \to \mathfrak{g}_j : \xi \mapsto \xi^*$ that satisfies (7.6.4), and by Lemma 7.6.14 there exists a constant $c_j > 0$ such that $\kappa_j(\xi^*, \eta) = c_j \langle \xi, \eta \rangle_j$ for all $\xi, \eta \in \mathfrak{g}_j$. Thus the inner product $\langle \xi, \eta \rangle := \sum_j c_j \langle \xi_j, \eta_j \rangle_j$ for $\xi_j, \eta_j \in \mathfrak{g}_j$ and $\xi = \xi_1 + \cdots + \xi_r, \eta = \eta_1 + \cdots + \eta_r$ satisfies the requirements of part (v) with $c = 1$ and $\xi^* := \xi_1^* + \cdots + \xi_r^*$.

If (v) holds, then the Killing form is nondegenerate, hence $Z(\mathfrak{g}) = 0$ and $\mathrm{Der}(\mathfrak{g}) = \mathrm{ad}(\mathfrak{g})$ by Lemma 7.4.3, and hence the inner product in (v) is symmetric. Thus (v) implies (i) and this proves Theorem 7.6.16. $\square$

Corollary 7.6.18 (Cartan Involution) *Let $\mathfrak{g}$ be a nonzero semisimple real Lie algebra equipped with a symmetric inner product and let $\xi \mapsto \xi^*$ be the involution in Lemma 7.6.8. Then the map*

$$\mathfrak{g} \to \mathfrak{g} : \xi \mapsto -\xi^* \tag{7.6.21}$$

*is a Lie algebra homomorphism (called a **Cartan involution**). The Cartan involution (7.6.21) gives rise to a splitting*

$$\mathfrak{g} = \mathfrak{k} \oplus \mathfrak{p}, \quad \mathfrak{k} := \{\xi \in \mathfrak{g} \,|\, \xi + \xi^* = 0\}, \quad \mathfrak{p} := \{\eta \in \mathfrak{g} \,|\, \eta = \eta^*\}, \tag{7.6.22}$$

such that

$$[\mathfrak{k}, \mathfrak{k}] \subset \mathfrak{k}, \qquad [\mathfrak{k}, \mathfrak{p}] \subset \mathfrak{p}, \qquad [\mathfrak{p}, \mathfrak{p}] \subset \mathfrak{k}. \tag{7.6.23}$$

Moreover, $\mathfrak{k}$ is nontrivial, the Killing form $\kappa : \mathfrak{g} \times \mathfrak{g} \to \mathbb{R}$ is negative definite on $\mathfrak{k}$ and positive definite on $\mathfrak{p}$, and $\kappa(\xi, \eta) = 0$ for all $\xi \in \mathfrak{k}$ and all $\eta \in \mathfrak{p}$.

Proof That the map (7.6.21) is a Lie algebra homomorphism follows directly from (7.6.4). It follows also from (7.6.4) that the subspaces $\mathfrak{k}, \mathfrak{p} \subset \mathfrak{g}$ in (7.6.22) satisfy (7.6.23). That $\mathfrak{g}$ is the direct sum of these subspaces, follows from the identity $\zeta^{**} = \zeta$ for $\zeta \in \mathfrak{g}$, which implies

$$\zeta = \xi + \eta, \qquad \xi := \tfrac{1}{2}(\zeta - \zeta^*) \in \mathfrak{k}, \qquad \eta := \tfrac{1}{2}(\zeta + \zeta^*) \in \mathfrak{p}.$$

By (7.6.23) the summand $\mathfrak{k}$ must be nontrivial, because $\mathfrak{g}$ is nonzero and hence is not abelian.

Next observe that the formula $\langle A, B \rangle := \mathrm{trace}(A^* B)$ defines an inner product on $\mathrm{End}(\mathfrak{g})$ with the norm $|A| := \sqrt{\mathrm{trace}(A^* A)}$, and that

$$\kappa(\xi, \eta^*) = \kappa(\xi^*, \eta) = \mathrm{trace}\big(\mathrm{ad}(\xi)^* \mathrm{ad}(\eta)\big) = \langle \mathrm{ad}(\xi), \mathrm{ad}(\eta) \rangle.$$

For $\xi \in \mathfrak{k}$ and $\eta \in \mathfrak{p}$ this implies $\kappa(\xi, \xi) = -|\mathrm{ad}(\xi)|^2$, $\kappa(\eta, \eta) = |\mathrm{ad}(\eta)|^2$, and $\kappa(\xi, \eta) = 0$. This proves Corollary 7.6.18. □

Lemma 7.6.19 *Let $\mathfrak{g}$ be a semisimple real Lie algebra equipped with a symmetric inner product, let $\xi \mapsto \xi^*$ be the involution in Lemma 7.6.8, let $\mathfrak{g} = \mathfrak{g}_1 \oplus \cdots \oplus \mathfrak{g}_r$ be a decomposition into simple ideals $\mathfrak{g}_j$, and let $\mathfrak{k}, \mathfrak{p} \subset \mathfrak{g}$ be as in Corollary 7.6.18. Then the following holds.*

(i)   *The simple ideals $\mathfrak{g}_j \subset \mathfrak{g}$ are pairwise orthogonal.*
(ii)   *Each ideal $\mathfrak{g}_j$ is invariant under the involution $\xi \mapsto \xi^*$.*
(iii) *$\mathfrak{p}$ is the orthogonal complement of $\mathfrak{k}$.*
(iv)   *For each j the restriction of the inner product to $\mathfrak{g}_j$ is symmetric and $\mathfrak{g}_j = \mathfrak{k}_j \oplus \mathfrak{p}_j$, where $\mathfrak{k}_j = \mathfrak{k} \cap \mathfrak{g}_j$, $\mathfrak{p}_j = \mathfrak{p} \cap \mathfrak{g}_j$ are as in Corollary 7.6.18.*

Proof   We prove part (i). For each i the orthogonal complement $\mathfrak{g}_i^\perp$ is an ideal, because $\langle[\xi,\eta],\zeta\rangle = \langle\eta,[\xi^*,\zeta]\rangle = 0$ for all $\xi \in \mathfrak{g}$, $\eta \in \mathfrak{g}_i^\perp$, $\zeta \in \mathfrak{g}_i$, and so $[\xi,\eta] \in \mathfrak{g}_i^\perp$ for all $\xi \in \mathfrak{g}$, $\eta \in \mathfrak{g}_i^\perp$. This implies that $\mathfrak{h}_j := \bigcap_{i \neq j} \mathfrak{g}_i^\perp$ is an ideal, and so is the subspace $\mathfrak{h}_j \cap \mathfrak{g}_j$. So either $\mathfrak{h}_j \cap \mathfrak{g}_j = \{0\}$ or $\mathfrak{h}_j = \mathfrak{g}_j$, because $\mathfrak{g}_j$ is simple. If $\mathfrak{h}_j \cap \mathfrak{g}_j = \{0\}$, then $[\xi,\eta] = 0$ for all $\xi \in \mathfrak{g}_j$ and all $\eta \in \mathfrak{h}_j$, hence $\mathfrak{g}_j \subset Z(\mathfrak{g})$, and this is impossible because the center of $\mathfrak{g}$ is trivial. Thus $\mathfrak{h}_j = \mathfrak{g}_j$ for all j and this proves (i).

We prove part (ii). The subspace $\mathfrak{g}_j^* := \{\eta^* \mid \eta \in \mathfrak{g}_j\}$ is an ideal, because $[\xi,\eta^*] = [\eta,\xi^*]^* \in \mathfrak{g}_j^*$ for all $\xi \in \mathfrak{g}$ and all $\eta \in \mathfrak{g}_j$. By part (i) we have $\langle[\xi,\eta^*],\zeta\rangle = -\langle\xi,[\eta,\zeta]\rangle = 0$ for all $\xi \in \mathfrak{g}_i$ and $\eta \in \mathfrak{g}_j$ with $i \neq j$, and all $\zeta \in \mathfrak{g}$. Hence $[\mathfrak{g}_i,\mathfrak{g}_j^*] = 0$ for $i \neq j$. Hence the ideal $\mathfrak{g}_j^* \cap \mathfrak{g}_j$ cannot be zero, because otherwise $[\mathfrak{g}_j,\mathfrak{g}_j^*] = 0$ and so $\mathfrak{g}_j^* \subset Z(\mathfrak{g})$. Hence $\mathfrak{g}_j^* = \mathfrak{g}_j$, because $\mathfrak{g}_j$ is simple. This proves (ii).

We prove part (iii). By (ii) and Lemma 7.6.14 the involution $\xi \mapsto \xi^*$ preserves the inner product on $\mathfrak{g}_j$, and so by (i) it preserves the inner product on all of $\mathfrak{g}$. Hence $\langle\xi,\eta\rangle = \langle\xi^*,\eta^*\rangle = -\langle\xi,\eta\rangle$ for all $\xi \in \mathfrak{k}$ and all $\eta \in \mathfrak{p}$. Thus $\mathfrak{k}$ and $\mathfrak{p}$ are orthogonal to each other and this proves (iii).

Part (iv) follows directly from (ii) and this proves Lemma 7.6.19. $\square$

Theorem 7.6.20   *Let $\mathfrak{g}$ be an m-dimensional real Lie algebra that is not abelian and fix an inner product on $\mathfrak{g}$ and an orthonormal basis $e_1,\ldots,e_m$ of $\mathfrak{g}$. Then the following are equivalent.*

(i)   *$P = \mathbb{1}$ is a critical point of $f_\mathfrak{g}$.*

(ii)   *There exists a real number c such that*

$$\sum_{i=1}^m \Big(2\mathrm{ad}(e_i)^*\mathrm{ad}(e_i) - \mathrm{ad}(e_i)\mathrm{ad}(e_i)^*\Big) = c\mathbb{1}. \tag{7.6.24}$$

(iii)   *$\mathfrak{g}$ is semisimple, the inner product is symmetric, and there exists an involution $\mathfrak{g} \to \mathfrak{g} : \xi \mapsto \xi^*$ and a constant $c > 0$ such that (7.6.19) holds.*

Proof   By Lemma 7.5.20 the element $P = \mathbb{1} \in \mathcal{P}_0(\mathfrak{g})$ is a critical point of the function $f_\mathfrak{g}$ in (7.6.9) if and only if, for all $A \in \mathfrak{sl}(\mathfrak{g})$,

$$0 = \sum_{i,j=1}^m \langle[e_i,e_j], A[e_i,e_j] - [Ae_i,e_j] - [e_i,Ae_j]\rangle$$

$$= \sum_{i=1}^m \mathrm{trace}\Big(\mathrm{ad}(e_i)^* A\,\mathrm{ad}(e_i) - 2\mathrm{ad}(e_i)^*\mathrm{ad}(e_i)A\Big).$$

This holds if and only if there exists a constant $c \in \mathbb{R}$ that satisfies (7.6.24). Thus we have proved that (i) is equivalent to (ii).

We prove that (i) and (ii) imply (iii). Since $P = \mathbb{1}$ is a critical point of $f_\mathfrak{g}$, Lemma 7.5.20 asserts that the inner product on $\mathfrak{g}$ is symmetric. Thus by

Lemma 7.6.8 there exists an involution $\mathfrak{g} \to \mathfrak{g} : \xi \mapsto \xi^*$ that satisfies (7.6.4). Hence by (ii) there exists a real number c such that

$$Q_{\mathfrak{g}} := \sum_{i=1}^{m} \left(2\mathrm{ad}(e_i^*)\mathrm{ad}(e_i) - \mathrm{ad}(e_i)\mathrm{ad}(e_i^*) \right) = c\mathbb{1}. \qquad (7.6.25)$$

Since $\mathfrak{g}$ is not abelian, the endomorphism $Q_{\mathfrak{g}}$ has a positive trace, so $c > 0$. This implies that the center of $\mathfrak{g}$ is trivial, because $Z(\mathfrak{g}) \subset \ker(Q_{\mathfrak{g}})$. Hence, by Lemma 7.6.8, the Killing form on $\mathfrak{g}$ is nondegenerate. By Lemma 7.6.19 this implies that the decomposition $\mathfrak{g} = \mathfrak{k} \oplus \mathfrak{p}$ in Corollary 7.6.18 is orthogonal. Hence the orthonormal basis $e_1, \ldots, e_m$ of $\mathfrak{g}$ can be chosen such that $e_1, \ldots, e_k$ is a basis of $\mathfrak{k}$ and $e_{k+1}, \ldots, e_m$ is a basis of $\mathfrak{p}$. Thus $e_i^* = -e_i$ for $i \leq k$ and $e_i^* = e_i$ for $i > k$, and so it follows from (7.6.25) that

$$\sum_{i=1}^{m} \mathrm{ad}(e_i^*)\mathrm{ad}(e_i) = c\mathbb{1}. \qquad (7.6.26)$$

Hence $\kappa(\xi^*, \eta) = \sum_i \langle e_i, \mathrm{ad}(\xi^*)\mathrm{ad}(\eta)e_i \rangle = \sum_i \langle \xi, \mathrm{ad}(e_i^*)\mathrm{ad}(e_i)\eta \rangle = c\langle \xi, \eta \rangle$ for all $\xi, \eta \in \mathfrak{g}$. This proves (iii).

That (iii) implies (ii) follows by reversing this argument. By (iii) the Killing form is nondegenerate and the inner product is symmetric and is preserved by the involution $\xi \mapsto \xi^*$. Thus the splitting $\mathfrak{g} = \mathfrak{k} \oplus \mathfrak{p}$ is orthogonal, and hence the orthonormal basis $e_1, \ldots, e_m$ of $\mathfrak{g}$ can be chosen such that $e_1, \ldots, e_k$ is a basis of $\mathfrak{k}$ and $e_{k+1}, \ldots, e_m$ is a basis of $\mathfrak{p}$. Moreover,

$$\sum_{i=1}^{m} \langle \xi, \mathrm{ad}(e_i^*)\mathrm{ad}(e_i)\eta \rangle = \sum_{i=1}^{m} \langle e_i, \mathrm{ad}(\xi^*)\mathrm{ad}(\eta)e_i \rangle = \kappa(\xi^*, \eta) = c\langle \xi, \eta \rangle$$

for all $\xi, \eta \in \mathfrak{g}$ by (7.6.19). This implies (7.6.26). Since $e_i^* = \pm e_i$ for all i, equation (7.6.24) follows from (7.6.26). This proves Theorem 7.6.20. $\qquad \square$

Corollary 7.6.21 (Cartan Decomposition) *Let $\mathfrak{g}$ be a semisimple real Lie algebra equipped with a symmetric inner product, let $\mathfrak{g} \to \mathfrak{g} : \xi \mapsto \xi^*$ be the involution in Lemma 7.6.8, and define*

$$K := \mathrm{Aut}_0(\mathfrak{g}) \cap \mathrm{SO}(\mathfrak{g}), \qquad \mathrm{Der}^+(\mathfrak{g}) := \{\delta \in \mathrm{Der}(\mathfrak{g}) \mid \delta = \delta^*\}.$$

Then the following holds.

(i) *K is connected and is a maximal compact subgroup of $\mathrm{Aut}_0(\mathfrak{g})$, every compact subgroup of $\mathrm{Aut}_0(\mathfrak{g})$ is conjugate in $\mathrm{Aut}_0(\mathfrak{g})$ to a Lie subgroup of K, and every maximal compact subgroup of $\mathrm{Aut}_0(\mathfrak{g})$ is conjugate to K.*

(ii) *The map*

$$K \times \mathrm{Der}^+(\mathfrak{g}) \to \mathrm{Aut}_0(\mathfrak{g}) : (u, \delta) \mapsto \exp(\delta)u$$

is a diffeomorphism.

(iii) *If there exists a $c > 0$ such that $\kappa(\xi^*, \eta) = c\langle\xi, \eta\rangle$ for all $\xi, \eta \in \mathfrak{g}$, then*

$$M_\mathfrak{g} := \mathrm{Crit}(f_\mathfrak{g}) = \mathcal{P}_0(\mathfrak{g}) \cap \mathrm{Aut}(\mathfrak{g}) = \{\exp(\delta) \,|\, \delta \in \mathrm{Der}^+(\mathfrak{g})\}$$

is a totally geodesic and geodesically convex submanifold of $\mathcal{P}_0(\mathfrak{g})$ and so is a Hadamard manifold and a symmetric space, and the map

$$\mathrm{Aut}_0(\mathfrak{g}) \to \mathcal{P}_0(\mathfrak{g}) : g \mapsto \sqrt{gg^*}$$

descends to a diffeomorphism from the quotient space $\mathrm{Aut}_0(\mathfrak{g})/\mathrm{K}$ to $M_\mathfrak{g}$.

Proof By Theorem 7.6.16 there exists a splitting $\mathfrak{g} = \mathfrak{g}_1 \oplus \cdots \oplus \mathfrak{g}_r$ into simple ideals, this splitting is preserved by every derivation of $\mathfrak{g}$, and by Lemma 7.6.19 it is also preserved by the involution $\xi \mapsto \xi^*$. Thus $\mathrm{Aut}_0(\mathfrak{g})$ is isomorphic to the product of the groups $\mathrm{Aut}_0(\mathfrak{g}_j)$, and $\mathrm{K} = \mathrm{Aut}_0(\mathfrak{g}) \cap \mathrm{SO}(\mathfrak{g})$ is isomorphic to the product of the subgroups $\mathrm{K}_j := \mathrm{Aut}_0(\mathfrak{g}_j) \cap \mathrm{SO}(\mathfrak{g}_j)$. Hence part (i) follows from Theorem 7.6.11. Moreover, by Lemma 7.6.19, we have $\mathfrak{p} = \mathfrak{p}_1 \oplus \cdots \oplus \mathfrak{p}_r$, and so part (ii) follows from Corollary 7.5.27.

Under the assumptions of (iii) Theorem 7.6.20 asserts that $P_0 = \mathbb{1}$ is a critical point of $f_\mathfrak{g}$, so (iii) follows from Lemma 7.5.22, Lemma 7.5.24, and Corollary 7.5.27. This proves Corollary 7.6.21. $\qquad\square$

Remark 7.6.22 The Lie algebra of the group $\mathrm{K} = \mathrm{Aut}_0(\mathfrak{g}) \cap \mathrm{SO}(\mathfrak{g})$ in Corollary 7.6.21 is given by $\mathrm{Lie}(\mathrm{K}) = \{\mathrm{ad}(\xi) \,|\, \xi \in \mathfrak{k}\}$ (see Corollary 7.6.18). If the summand $\mathfrak{p}$ in (7.6.22) is trivial, then $\mathrm{Aut}_0(\mathfrak{g}) = \mathrm{K}$ is a compact Lie group. If $\mathfrak{p}$ is nontrivial, then the quotient space $\mathrm{Aut}_0(\mathfrak{g})/\mathrm{K}$ is a nontrivial Hadamard manifold diffeomorphic to $\mathfrak{p}$.

Remark 7.6.23 One can replace the space $\mathcal{P}_0(\mathfrak{g})$ of positive definite self-adjoint vector space isomorphisms $P : \mathfrak{g} \to \mathfrak{g}$ of determinant one by the space $\mathcal{H}_\mathfrak{g}$ of inner products on $\mathfrak{g}$ with a fixed determinant (Remark 6.5.11). This eliminates the dependence on the background inner product and there is then only one function $f_\mathfrak{g} : \mathcal{H}_\mathfrak{g} \to \mathbb{R}$ whose set of minima is the totally geodesic submanifold $\mathcal{M}_\mathfrak{g} \subset \mathcal{H}_\mathfrak{g}$ of all symmetric inner products on $\mathfrak{g}$ with a fixed determinant that satisfy part (iii) of Theorem 7.6.20. The main result asserts that, when $\mathfrak{g}$ is not abelian, the space $\mathcal{M}_\mathfrak{g}$ is nonempty if and only if $\mathfrak{g}$ is semisimple (Theorem 7.6.16 and Theorem 7.6.20).

7.6.4 Complex Lie Algebras

A **complex Lie algebra** is a complex vector space $\mathfrak{g}$ equipped with a Lie bracket $\mathfrak{g} \times \mathfrak{g} \to \mathfrak{g} : (\xi, \eta) \mapsto [\xi, \eta]$ that is complex bilinear, i.e. it is a skew-symmetric bilinear map that satisfies the Jacobi identity and

$$[\mathrm{i}\xi, \eta] = [\xi, \mathrm{i}\eta] = \mathrm{i}[\xi, \eta]$$

for all $\xi, \eta \in \mathfrak{g}$. Thus every complex Lie algebra is also a real Lie algebra.

Let $\mathfrak{g}$ be a finite-dimensional complex Lie algebra. A **complex ideal** in $\mathfrak{g}$ is a complex linear subspace $\mathfrak{h} \subset \mathfrak{g}$ that satisfies $[\xi, \eta] \in \mathfrak{h}$ for all $\xi \in \mathfrak{g}$ and all $\eta \in \mathfrak{h}$. The complex Lie algebra $\mathfrak{g}$ is called **simple** iff it is not abelian and has no complex ideals other than $\mathfrak{h} = \{0\}$ and $\mathfrak{h} = \mathfrak{g}$. It is called **semisimple** iff it is finite-dimensional and the **complex Killing form** $\kappa^c : \mathfrak{g} \times \mathfrak{g} \to \mathbb{C}$, defined by $\kappa^c(\xi, \eta) :=$ trace$^c(\mathrm{ad}(\xi)\mathrm{ad}(\eta))$ for $\xi, \eta \in \mathfrak{g}$, is nondegenerate.

Since $\kappa = 2\mathrm{Re}\kappa^c$, a complex Lie algebra is semisimple if and only if it is semisimple as a real Lie algebra. The next lemma shows that the analogous assertion holds for simple complex Lie algebras (see [17, Lemma 7]).

Lemma 7.6.24 *A finite-dimensional complex Lie algebra $\mathfrak{g}$ is simple if and only if it is simple as a real Lie algebra.*

Proof Let $\mathfrak{g}$ be a simple complex Lie algebra of complex dimension n and let $\mathfrak{h} \subset \mathfrak{g}$ be a real linear subspace of $\mathfrak{g}$ that satisfies $[\xi, \eta] \in \mathfrak{h}$ for all $\xi \in \mathfrak{g}$ and all $\eta \in \mathfrak{h}$. Then the subspaces

$$\mathfrak{h} \cap i\mathfrak{h}, \qquad \mathfrak{h} + i\mathfrak{h}$$

are complex ideals in $\mathfrak{g}$, and their real dimensions satisfy the equation

$$\dim^{\mathbb{R}}(\mathfrak{h} \cap i\mathfrak{h}) + \dim^{\mathbb{R}}(\mathfrak{h} + i\mathfrak{h}) = 2\dim^{\mathbb{R}}(\mathfrak{h}).$$

Since both summands on the left are either 0 or $2n$, the real dimension of $\mathfrak{h}$ is either 0, n, or $2n$. We claim that the dimension cannot be n.

Assume, by contradiction, that $\dim^{\mathbb{R}}(\mathfrak{h}) = n$. Then

$$\mathfrak{h} \cap i\mathfrak{h} = \{0\}, \qquad \mathfrak{h} + i\mathfrak{h} = \mathfrak{g}.$$

Hence, for all $\zeta, \zeta' \in \mathfrak{g}$ there exist $\xi, \eta \in \mathfrak{h}$ such that $\zeta = \xi + i\eta$, and so

$$[\zeta, \zeta'] = [\xi, \zeta'] + [\eta, i\zeta'] = i([\xi, -i\zeta'] + [\eta, \zeta']) \in \mathfrak{h} \cap i\mathfrak{h} = \{0\}.$$

This contradicts the fact that $\mathfrak{g}$ is not abelian. Thus $\mathfrak{h} = \{0\}$ or $\mathfrak{h} = \mathfrak{g}$, and this proves Lemma 7.6.24. $\qquad\square$

Lemma 7.6.25 *Let $\mathfrak{g}$ be a semisimple complex Lie algebra. Then $\mathfrak{g}$ is a direct sum of simple complex ideals.*

Proof Let $\mathfrak{h} \subset \mathfrak{g}$ be a real ideal and let $\mathfrak{h}' \subset \mathfrak{g}$ be as in (7.6.18). Then it follows from "(ii) $\Longrightarrow$ (iii)" in Theorem 7.6.16 that $[\mathfrak{h} + i\mathfrak{h}, \mathfrak{h}'] = \{0\}$. Hence $\mathfrak{h} + i\mathfrak{h} \subset \mathfrak{h}'' = \mathfrak{h}$ and so $i\mathfrak{h} = \mathfrak{h}$. Thus every real ideal in $\mathfrak{g}$ is a complex ideal. Hence the assertion follows from "(ii) $\Longrightarrow$ (iv)" in Theorem 7.6.16. $\qquad\square$

The next result is a theorem of Cartan [13] which asserts that every semisimple complex Lie algebra has a compact real form. The proof given here is due to Donaldson [17, Lemma 8].

Theorem 7.6.26 (Cartan) *Let $\mathfrak{g}$ be a nonzero semisimple complex Lie algebra equipped with a symmetric inner product. Then*

$$\mathfrak{p} = i\mathfrak{k}, \qquad \mathfrak{g} = \mathfrak{k} \oplus i\mathfrak{k}, \tag{7.6.27}$$

where $\mathfrak{k}, \mathfrak{p}$ are as in Corollary 7.6.18. Moreover, the group $\mathrm{Aut}_0(\mathfrak{g})$ is the complexification of the maximal compact subgroup $K = \mathrm{Aut}_0(\mathfrak{g}) \cap \mathrm{SO}(\mathfrak{g})$, i.e.

$$\mathrm{Aut}_0(\mathfrak{g}) = \{\exp(i\delta)u \mid u \in K, \ \delta \in \mathrm{Lie}(K)\} \tag{7.6.28}$$

and the map $K \times \mathrm{Lie}(K) \to \mathrm{Aut}_0(\mathfrak{g}) : (u, \delta) \mapsto \exp(i\delta)u$ is a diffeomorphism.

Proof Assume first that $\mathfrak{g}$ is simple. Then $\mathfrak{g}$ is simple as a real Lie algebra (Lemma 7.6.24), and so has a nondegenerate Killing form (Theorem 7.6.10). By the inclusions in (7.6.23) the subspace

$$\mathfrak{h} := (\mathfrak{k} \cap i\mathfrak{p}) + (\mathfrak{p} \cap i\mathfrak{k}) = (\mathfrak{k} \cap i\mathfrak{p}) + i(\mathfrak{k} \cap i\mathfrak{p})$$

is a complex ideal in $\mathfrak{g}$. Hence it is either $\{0\}$ or $\mathfrak{g}$. If $\mathfrak{h} = \mathfrak{g}$, then it follows from (7.6.22) that $\mathfrak{p} = i\mathfrak{k}$. Assume, by contradiction, that $\mathfrak{h} = \{0\}$. Then

$$\mathfrak{k} \cap i\mathfrak{p} = \{0\}, \qquad \mathfrak{g} = \mathfrak{k} \oplus i\mathfrak{p}.$$

Hence the map $\sigma : \mathfrak{g} \to \mathfrak{g}$, defined by

$$\sigma(\xi + i\eta) := \xi - i\eta$$

for $\xi \in \mathfrak{k}$ and $\eta \in \mathfrak{p}$ is a Lie algebra homomorphism and an involution. This implies $\mathrm{ad}(\sigma(\zeta)) = \sigma\mathrm{ad}(\zeta)\sigma$ and so $\kappa(\sigma(\zeta), \sigma(\zeta')) = \kappa(\zeta, \zeta')$ for all $\zeta, \zeta' \in \mathfrak{g}$. Hence $\kappa(\xi, i\eta) = 0$ for all $\xi \in \mathfrak{k}$ and all $\eta \in \mathfrak{p}$. Thus $i\mathfrak{p}$ is the orthogonal complement of $\mathfrak{k}$ with respect to the Killing form. Thus, by Corollary 7.6.18,

$$\mathfrak{p} = i\mathfrak{p}.$$

However, for all $\eta \in \mathfrak{p}$ we have $\kappa(i\eta, i\eta) = -\kappa(\eta, \eta) = -|\mathrm{ad}(\eta)|^2$. Hence the Killing form is negative definite on $i\mathfrak{p}$ and positive definite on $\mathfrak{p}$, and hence $\mathfrak{p} = \{0\}$. This implies $\mathfrak{k} = i\mathfrak{k}$. Since the Killing form is positive definite on $i\mathfrak{k}$ and negative definite on $\mathfrak{k}$, this is a contradiction. This contradiction shows that our assumption $\mathfrak{h} = \{0\}$ must have been wrong. Thus $\mathfrak{h} = \mathfrak{g}$ and hence $\mathfrak{p} = i\mathfrak{k}$. This completes the proof of (7.6.27) in the simple case.

For general semisimple complex Lie algebras, the proof of the equations in (7.6.27) reduces to the simple case by Lemma 7.6.19 and Lemma 7.6.25. Equation (7.6.28) follows directly from (7.6.27) and Corollary 7.6.21. This proves Theorem 7.6.26. □

Remark 7.6.27 Let $\mathfrak{g}$ be a semisimple complex Lie algebra. Then $\mathrm{Aut}(\mathfrak{g})$ is a **complex Lie group**, i.e. it admits the structure of a complex manifold such that the structure maps

$$G \times G \to G : (h, g) \mapsto hg, \qquad G \to G : g \mapsto g^{-1}$$

are holomorphic. The proof uses the fact that $\mathrm{Der}(\mathfrak{g})$ is isomorphic to $\mathfrak{g}$ and that the resulting almost complex structure on $\mathrm{Aut}(\mathfrak{g})$ is integrable (as it is preserved by the torsion-free connection $g^{-1}\nabla\widehat{g} = \frac{d}{dt}(g^{-1}\widehat{g}) + [g^{-1}\dot{g}, g^{-1}\widehat{g}]$). Theorem 7.6.26 asserts that the identity component $G = \mathrm{Aut}_0(\mathfrak{g})$ of the group of automorphisms of $\mathfrak{g}$ is the **complexification** of the maximal compact subgroup $K = \mathrm{Aut}_0(\mathfrak{g}) \cap \mathrm{SO}(\mathfrak{g})$, i.e. its Lie algebra $\mathrm{Der}(\mathfrak{g}) = \mathrm{ad}(\mathfrak{g})$ is the complexification of the Lie algebra $\mathrm{Lie}(K) = \mathrm{ad}(\mathfrak{k})$ and the quotient space G/K is contractible. These conditions imply the universality property that every Lie group homomorphism $\rho : K \to \mathcal{G}$ with values in a complex Lie group $\mathcal{G}$ extends to a unique holomorphic Lie group homomorphism $\rho^c : G \to \mathcal{G}$ such that $\rho^c|_K = \rho$. Such a complexification exists for every compact Lie group K, whether or not it is semisimple. (For an exposition see [20, Appendix B].)

The following exercise is inspired by a remark in [17, §5.1] concerning a positive curvature manifold that is dual to $M_\mathfrak{g}$.

Exercise 7.6.28 Let $\mathfrak{g}$ be a semisimple real Lie algebra equipped with a symmetric inner product that satisfies condition (iii) in Theorem 7.6.20. Consider the complexified Lie algebra

$$\mathfrak{g}^c := \mathfrak{g} \oplus i\mathfrak{g}$$

with the Lie bracket

$$[\zeta, \zeta'] := [\xi, \xi'] - [\eta, \eta'] + i([\xi, \eta'] + [\eta, \xi']) \tag{7.6.29}$$

and the Hermitian form

$$\langle \zeta, \zeta' \rangle^c := \langle \xi, \xi' \rangle + \langle \eta, \eta' \rangle + i(\langle \xi, \eta' \rangle - \langle \eta, \xi' \rangle) \tag{7.6.30}$$

for $\zeta = \xi + i\eta \in \mathfrak{g}^c$ and $\zeta' = \xi' + i\eta' \in \mathfrak{g}^c$. With this convention the Hermitian form (7.6.30) is complex anti-linear in the first variable and complex linear in the second variable. Prove the following.

(a) $\mathfrak{g}^c$ is semisimple. **Hint:** $\mathfrak{g}^c$ has a trivial center and the real part of (7.6.30) is a symmetric inner product on $\mathfrak{g}^c$.
(b) If $\mathfrak{g}$ is simple, then $\mathfrak{g}^c$ is simple. **Hint:** If $\mathfrak{h}^c$ is a complex ideal in $\mathfrak{g}^c$, then the linear subspace

$$\mathfrak{h} := \{\mathrm{Re}(\zeta) \mid \zeta \in \mathfrak{h}^c\}$$

is an ideal in $\mathfrak{g}$.

(c) Every real linear derivation on $\mathfrak{g}^c$ is complex linear and has complex trace zero.
(d) The identity component $\mathrm{Aut}_0(\mathfrak{g}^c)$ of the group of real linear Lie algebra auto-morphisms of $\mathfrak{g}^c$ consists of complex linear automorphisms of complex deter-minant one. (Complex conjugation is a Lie algebra automorphism of $\mathfrak{g}^c$ not in the identity component.)
(e) The subgroup

$$K^c := \mathrm{Aut}_0(\mathfrak{g}^c) \cap \mathrm{SU}(\mathfrak{g}^c)$$

is connected and is a maximal compact subgroup of $\mathrm{Aut}_0(\mathfrak{g}^c)$. Its Lie algebra

$$\mathrm{Lie}(K^c) \cong \mathfrak{k} + i\mathfrak{p}$$

is the compact real form of $\mathfrak{g}^c$ (Corollary 7.6.18 and Theorem 7.6.26).
(f) Let $\Phi + i\Psi : \mathfrak{g} \to \mathfrak{g}^c$ be a Lie algebra homomorphism, i.e. for all $\xi, \eta \in \mathfrak{g}$,

$$\Phi[\xi, \eta] = [\Phi\xi, \Phi\eta] - [\Psi\xi, \Psi\eta], \qquad \Psi[\xi, \eta] = [\Phi\xi, \Psi\eta] + [\Psi\xi, \Phi\eta].$$

Assume $\Phi + i\Psi$ is injective and denote its image by

$$\mathfrak{l} := \{\Phi\xi + i\Psi\xi \,|\, \xi \in \mathfrak{g}\}.$$

Then the following are equivalent.
(I)   The real part of (7.6.30) restricts to a symmetric inner product on $\mathfrak{l}$.
(II) $\Phi^*\Phi + \Psi^*\Psi \in \mathrm{Aut}(\mathfrak{g})$.
(III) $\mathfrak{l}$ is a Lagrangian subspace of $\mathfrak{g}^c$ with respect to the imaginary part of (7.6.30), i.e. $\Phi^*\Psi - \Psi^*\Phi = 0$.
Hint: If $\Phi^*\Phi + \Psi^*\Psi \in \mathrm{Aut}(\mathfrak{g})$, then there exists a derivation $\alpha \in \mathrm{Der}(\mathfrak{g})$ such that $\alpha = \alpha^*$ and $\exp(2\alpha) = \Phi^*\Phi + \Psi^*\Psi$ (Corollary 7.6.21). Prove that

$$\delta := \exp(-\alpha)(\Phi^*\Psi - \Psi^*\Phi)\exp(-\alpha)$$

is a derivation whose image is abelian. Prove that

$$\kappa(\delta\xi, \delta\eta) = 0$$

for all $\xi, \eta \in \mathfrak{g}$ and deduce that $\delta^*\delta = -\delta^2 = 0$.
(g) The space of oriented Lagrangian Lie subalgebras $\mathfrak{l} \subset \mathfrak{g}^c$ isomorphic to $\mathfrak{g}$ (that can be joined to $\mathfrak{g}$ by a path of such subspaces) is diffeomorphic to the quotient space

$$L_\mathfrak{g} := K^c/K,$$

where $K^c := \mathrm{Aut}_0(\mathfrak{g}^c) \cap \mathrm{SU}(\mathfrak{g}^c)$ and $K := \mathrm{Aut}_0(\mathfrak{g}) \cap \mathrm{SO}(\mathfrak{g})$. **Hint:** Choose the embedding $\Phi + i\Psi$ in (f) such that

$$\Phi^*\Phi + \Psi^*\Psi = \mathbb{1}, \qquad \Phi^*\Psi - \Psi^*\Phi = 0,$$

and extend it to a unitary automorphism of $\mathfrak{g}^c$.

(h) The space $L_{\mathfrak{g}}$ in part (g) embeds as a totally geodesic submanifold of dimension $\dim(L_{\mathfrak{g}}) = \dim(\mathfrak{p})$ into the symmetric space $U(\mathfrak{g}^c)/SO(\mathfrak{g})$ of all oriented Lagrangian subspaces of $\mathfrak{g}^c$. Hence $L_{\mathfrak{g}}$ has nonnegative sectional curvature. (**Hint:** Example 5.2.18.) One can think of the positive curvature manifold $L_{\mathfrak{g}}$ of all Lagrangian Lie subalgebras of $\mathfrak{g}^c$ that are isomorphic to $\mathfrak{g}$ as dual to the negative curvature manifold $M_{\mathfrak{g}}$ of all symmetric inner products on $\mathfrak{g}$ that satisfy condition (iii) in Theorem 7.6.20.

Remark 7.6.29 The idea of minimizing the norm of the Lie bracket was the approach to the existence of a compact real form of a simple complex Lie algebra suggested by Cartan [14] and carried out by Richardson [58]. One significant difference in the method of Donaldson [17], which we follow in the section, is that there is no need to assume that the Killing form is nondegenerate, but that this result emerges as a byproduct of the proof (Theorem 7.6.10). There is also no need to use the structure theory of Lie algebras as in the work of Weyl [77]. Instead one can use the existence of a symmetric inner product as a starting point to develop the structure theory of Lie algebras.

Remark 7.6.30 As pointed out by Donaldson [17], a more direct approach to Theorem 7.6.26 would be to carry over the entire program in the present section and Sect. 7.5 to the complex setting, starting in Sect. 7.5.2 with convex functions on the Hadamard manifold $M = \mathcal{Q}_0(V) \cong SL(V)/SU(V)$ of positive definite Hermitian automorphisms with determinant one of a complex vector space V equipped with a Hermitian inner product (Remark 6.5.20).

In the complex Lie algebra setting with $\mathcal{Q}_0(\mathfrak{g}) \cong SL(\mathfrak{g}, \mathbb{C})/SU(\mathfrak{g}, \mathbb{C})$ the logarithm of the function $f_{\mathfrak{g}} : \mathcal{Q}_0(\mathfrak{g}) \to \mathbb{R}$ is the log-norm function of Kempf and Ness in geometric invariant theory [20, 38]. Thus the existence of a critical point of $f_{\mathfrak{g}}$ is the polystability condition in GIT. This approach was developed by Lauret [44] and he proved that the polystable points are precisely the semisimple Lie algebras. This is the content of Theorem 7.6.20 in the complex setting. Lauret's proof uses Cartan's theorem about the compact real form of a semisimple complex Lie algebra.

One can also deduce the theorems in the real setting from those in the complex setting by complexifying the relevant real inner product space V to obtain a complex vector space $V^c = V \oplus iV$ with a Hermitian inner product, and embedding the space $\mathcal{P}_0(V) \cong SL(V)/SO(V)$ as a totally geodesic submanifold into $\mathcal{Q}_0(V^c) \cong SL(V^c)/SU(V^c)$.

Appendix A: Notes

This appendix explains some notations and standard results from first year analysis that are used throughout this book.

A.1 Maps and Functions

The notation $f : X \to Y$ means that f is a function which assigns to every point x in the set X a point $f(x)$ in the set Y. When $Y = \mathbb{R}$ we express this by saying that f is a *real valued* function defined on the set X and, if Y is a vector space, we may say that f is a *vector valued* function. However in general it is better to say that f is a **map** from X to Y and call the set X the **source** of the map and the set Y its **target**. The **graph** of f is the set

$$\operatorname{graph}(f) := \{(x, y) \in X \times Y \mid y = f(x)\}.$$

We always distinguish two maps with the same graph when their targets are different.

A map $f : X \to Y$ is said to be

$$
\left.\begin{array}{l}
\textbf{injective} \\
\textbf{surjective} \\
\textbf{bijective}
\end{array}\right\} \quad \text{iff} \quad
\left\{\begin{array}{l}
f(x_1) = f(x_2) \implies x_1 = x_2 \\
\forall y \in Y \, \exists x \in X \text{ s.t. } y = f(x) \\
\text{it is both injective and surjective.}
\end{array}\right\}
$$

Then

(a) f is injective $\iff$ it has a left inverse $g : Y \to X$ (i.e. $g \circ f = \operatorname{id}_X$);
(b) f is surjective $\iff$ it has a right inverse $g : Y \to X$ (i.e. $f \circ g = \operatorname{id}_Y$);
(c) f is bijective $\iff$ it has a two sided inverse $f^{-1} : Y \to X$.

(Item (b) is the *Axiom of Choice*.)

© The Editor(s) (if applicable) and The Author(s), under exclusive license to   403
Springer-Verlag GmbH, DE, part of Springer Nature 2022
J.W. Robbin, D.A. Salamon, *Introduction to Differential Geometry*,
Springer Studium Mathematik (Master), https://doi.org/10.1007/978-3-662-64340-2

The analogous principle holds for linear maps: if $A \in \mathbb{R}^{m \times n}$, then the linear map $\mathbb{R}^n \to \mathbb{R}^m : x \mapsto Ax$ is

(a) injective $\iff$ $BA = \mathbb{1}_n$ for some $B \in \mathbb{R}^{n \times m}$;
(b) surjective $\iff$ $AB = \mathbb{1}_m$ for some $B \in \mathbb{R}^{n \times m}$;
(c) bijective $\iff$ A is invertible (i.e. $m = n$ and $\det(A) \neq 0$).

(Here $\mathbb{1}_k$ is the $k \times k$ identity matrix.) However, this principle fails completely for continuous maps: the map $f : [0, 2\pi) \to S^1$ defined by $f(\theta) = (\cos\theta, \sin\theta)$ is continuous and bijective but its inverse is not continuous. (Here $S^1 \subset \mathbb{R}^2$ is the unit circle $x^2 + y^2 = 1$.)

A.2 Normal Forms

The *Fundamental Idea of Differential Calculus* is that near a point $x_0 \in U$ a smooth map $f : U \to V$ behaves like its linear approximation, i.e.

$$f(x) \approx f(x_0) + df(x_0)(x - x_0).$$

The *Normal Form Theorem from Linear Algebra* says that if $A \in \mathbb{R}^{m \times n}$ has rank r, then there are invertible matrices $P \in \mathbb{R}^{m \times m}$ and $Q \in \mathbb{R}^{n \times n}$ such that

$$P^{-1}AQ = \begin{pmatrix} \mathbb{1}_r & 0_{r \times (n-r)} \\ 0_{(m-r) \times r} & 0_{(m-r) \times (n-r)} \end{pmatrix}.$$

By the Fundamental Idea we can expect an analogous theorem for smooth maps.

Theorem A.2.1 (Local Normal Form for Smooth Maps) *Let $U \subset \mathbb{R}^n$ and $V \subset \mathbb{R}^m$ be open, $x_0 \in U$, and $f : U \to V$ be smooth. Assume that the derivative $df(x_0) \in \mathbb{R}^{m \times n}$ has rank r. Then there is an open neighborhood U_0 of x_0 in U, an open neighborhood V_0 of $f(x_0)$ in V, a diffeomorphism $\phi : U_1 \times U_2 \subset \mathbb{R}^r \times \mathbb{R}^{n-r}$, a diffeomorphism $\psi : V_0 \to U_1 \times V_2 \subset \mathbb{R}^r \times \mathbb{R}^{m-r}$, such that $\phi(x_0) = (0, 0)$, $\psi(f(x_0)) = (0, 0)$, and*

$$\psi^{-1} \circ f \circ \phi(x, y) = (x, g(x, y)) \quad \text{and} \quad dg(0, 0) = 0$$

for $(x, y) \in U_1 \times U_2$.

The Local Normal Form Theorem is an easy consequence of the Inverse Function Theorem.

Theorem A.2.2 (Inverse Function Theorem)   *Let $U \subset \mathbb{R}^n$, $V \subset \mathbb{R}^m$, $x_0 \in U$ and $f : U \to V$ be a smooth map. If $df(x_0)$ is invertible, then ($m = n$ and) there are neighborhoods U_0 of x_0 in U and V_0 of $f(x_0)$ in V so that the restriction $f_{|U_0} : U_0 \to V_0$ is a diffeomorphism.*

Here follow some other consequences of the Inverse Function Theorem. The terms *submersion* and *immersion* are defined in Sect. 2.6.1 and Definition 2.3.2 of Sect. 2.3.

Corollary A.2.3 (Submersion Theorem)   *When $r = m$ the diffeomorphisms ϕ and ψ in Theorem A.2.1 may be chosen so that the local normal form is*

$$\psi^{-1} \circ f \circ \phi(x, y) = x.$$

Corollary A.2.4 (Immersion Theorem)   *When $r = n$ the diffeomorphisms ϕ and ψ in Theorem A.2.1 may be chosen so that the local normal form is*

$$\psi^{-1} \circ f \circ \phi(x) = (x, 0).$$

Corollary A.2.5 (Rank Theorem)   *If the rank of $df(x) = r$ for all $x \in U$, then for every $x_0 \in U$ the diffeomorphisms ϕ and ψ in Theorem A.2.1 may be chosen so that the local normal form is*

$$\psi^{-1} \circ f \circ \phi(x) = (x_1, \ldots, x_r, 0, \ldots, 0).$$

Corollary A.2.6 (Implicit Function Theorem)   *Let $U \subset \mathbb{R}^m \times \mathbb{R}^n$ be an open set, let $F : U \to \mathbb{R}^n$ be smooth, and let $(x_0, y_0) \in U$ with $x_0 \in \mathbb{R}^m$ and $y_0 \in \mathbb{R}^n$. Define the partial derivative $d_2 F(x_0, y_0) \in \mathbb{R}^{n \times n}$ by*

$$d_2 F(x_0, y_0)v := \frac{d}{dt}\bigg|_{t=0} F(x_0, y_0 + tv)$$

for $v \in \mathbb{R}^n$. Assume that $F(x_0, y_0) = 0$ and that $d_2 F(x_0, y_0)$ is invertible. Then there exist neighborhoods U_0 of x_0 in $\mathbb{R}^m$ and V_0 of y_0 in $\mathbb{R}^n$ and a smooth map $g : U_0 \to V_0$ such that

$$U_0 \times V_0 \subset U, \qquad g(x_0) = y_0$$

and

$$F(x, y) = 0 \qquad \Longleftrightarrow \qquad y = g(x)$$

for $x \in U_0$ and $y \in V_0$.

A.3   Euclidean Spaces

This is the arena of Euclidean geometry; i.e. every figure which is studied in Euclidean geometry is a subset of Euclidean space. To define it one could proceed axiomatically as Euclid did; one would then verify that the axioms characterized Euclidean space by constructing "Cartesian Coordinate Systems" which identify the n-dimensional Euclidean space E^n with the n-dimensional numerical space $\mathbb{R}^n$. This program was carried out rigorously by Hilbert. We shall adopt the mathematically simpler but philosophically less satisfying course of taking the characterization as the definition.

We shall use three closely related spaces: n-dimensional Euclidean affine space E^n, n-dimensional Euclidean vector space $\mathbf{E}^n$, and the space $\mathbb{R}^n$ of all n-tuples of real numbers. The distinction among them is a bit pedantic, especially if one views as the purpose of geometry the interpretation of calculations on $\mathbb{R}^n$. The purpose for distinguishing these three spaces is the same as in elementary vector calculus; it aids geometric intuition. Here is the precise definition.

Definition A.3.1   An n-dimensional **Euclidean vector space** is a real n-dimensional vector space $\mathbf{E}^n$ equipped with a (real valued symmetric positive definite) inner product $\mathbf{E}^n \times \mathbf{E}^n \to \mathbb{R} : (v, w) \mapsto \langle v, w \rangle$. An n-dimensional *Euclidean affine space* consists of a set E^n and an n-dimensional Euclidean vector space $\mathbf{E}^n$ and maps

$$E^n \times E^n \to \mathbf{E}^n : (p, q) \mapsto p - q,$$
$$E^n \times \mathbf{E}^n \to E^n : (p, v) \mapsto p + v$$

satisfying the axioms

$$p + 0 = p, \qquad p + (v + w) = (p + v) + w, \qquad q + (p - q) = p$$

for all $p, q \in E^n$ and all $v, w \in \mathbf{E}^n$. The vector $p - q \in \mathbf{E}^n$ is called the **vector** from q to p and the point $p + v$ is called the **translate** of p by v. It follows easily that each choice of a point $o \in E^n$ determines a bijection $v \mapsto o + v$ from $\mathbf{E}^n$ onto E^n. The inner product on $\mathbf{E}^n$ equips the space E^n with a metric via the formula

$$|p - q| = \sqrt{\langle p - q, p - q \rangle}, \qquad p, q \in \mathbf{E}^n.$$

The **standard Euclidean space** of dimension n is $E^n = \mathbf{E}^n = \mathbb{R}^n$ with the usual matrix algebra operations $(x \pm y)^i = x^i \pm y^i$, $\langle x, y \rangle = \sum_i x^i y^i$.

Lemma A.3.2   *Any choice of an origin $o \in E^n$ and an orthonormal basis $e_1, \ldots, e_n$ for $\mathbf{E}^n$ determines an isometric bijection:*

$$\mathbb{R}^n \to E^n : (x^1, \ldots, x^n) \mapsto o + \sum_{i=1}^n x^i e_i$$

*(the inverse of which is called a **Cartesian coordinate system** on E^n).*

Lemma A.3.3 *If $E^n \to \mathbb{R}^n : p \mapsto (x^1, \ldots, x^n)$, $(y^1, \ldots, y^n)$ are two Cartesian coordinate systems, the **change of coordinates map** has the form*

$$y^j(p) = \sum_{j=1}^{n} a_i^j x^i(p) + v^i,$$

where the matrix $a = (a_i^j) \in \mathbb{R}^{n \times n}$ is an orthogonal matrix and $v \in \mathbb{R}^n$.

Example A.3.4 Any n-dimensional affine subspace of some numerical space $\mathbb{R}^k$ (with $k > n$) is an example of a Euclidean space. The corresponding vector space $\mathbf{E}^n$ is the unique vector subspace of $\mathbb{R}^k$ for which:

$$E^n = o + \mathbf{E}^n$$

for $o \in E^n$. This subspace is independent of the choice of $o \in E^n$. Note that $\mathbf{E}^n$ contains the "preferred" point 0 while E^n has no preferred point. Such spaces E^n and $\mathbf{E}^n$ would arise in linear algebra by taking E^n to be the space of solutions of $k - n$ independent inhomogeneous linear equations in k unknowns while $\mathbf{E}^n$ is the space of solutions of the corresponding homogeneous equations. The correspondence between E^n and $\mathbf{E}^n$ illustrates the mantra

> *The general solution of an inhomogeneous system of linear equations is a particular solution plus the general solution of the corresponding homogeneous linear system.*

This discussion shows that a Euclidean space E^n is an n-dimensional manifold with its Cartesian coordinate systems whose tangent space at each point is naturally isomorphic to $\mathbf{E}^n$. Thus it is natural to introduce submanifolds of Euclidean space as submanifolds of E^n whose tangent spaces are then linear subspaces of the associated vector space $\mathbf{E}^n$. Instead we have chosen in this book for simplicity of the exposition to describe manifolds as subsets of the vector space $\mathbb{R}^n$ equipped with its standard inner product.

References

1. Ralph Herman Abraham & Joel William Robbin, *Transversal Mappings and Flows*. Benjamin Press, 1967.
2. Lars Valerian Ahlfors & Leo Sario, *Riemann Surfaces*. Princeton University Press, 1960.
3. Aleksandr Danilovich Alexandrov, Über eine Verallgemeinerung der Riemannschen Geometrie. *Schriftreihe des Forschinstituts für Mathematik* **1** (1957), 33–84.
4. Michael Francis Atiyah & Nigel James Hitchin & Isodore Manuel Singer, Self-duality in four-dimensional Riemannian geometry. *Proceedings of the Royal Society of London, Series A* **362** (1978), 425–461.
5. Thierry Émilien Flavien Aubin, Équations différentielles non linéaires et problème de Yamabe concernant la courbure scalaire. *Journal de Mathématiques Pures et Appliquées* **55** (1976), 269–296.
6. Marcel Berger, Les variétés Riemanniennes 1/4-pincées, *Annali della Scuola Normale Superiore di Pisa* **14** (1960) , 161–170.
7. Richard Lawrence Bishop & Barrett O'Neill, Manifolds of negative curvature. *Transactions of the American Mathematical Society* **145** (1969), 1–49.
8. Simon Brendle, *Ricci Flow and the Sphere Theorem*. Graduate Studies in Mathematics **111**. AMS, Providence, RI, 2010.
9. Simon Brendle & Richard Melvin Schoen, Manifolds with 1/4-pinched curvature are space forms. *Journal of the American Mathematical Society* **22** (2009), 287–307. https://arxiv.org/pdf/0705.0766.pdf
10. Simon Brendle & Richard Melvin Schoen, Curvature, Sphere Theorems, and the Ricci Flow. *Bulletin of the American Mathematical Society* **48** (2011), 1–32. https://arxiv.org/pdf/1001.2278.pdf
11. Martin Bridson, & André Haefliger, *Metric spaces of non-positive curvature*. Springer, 1999. https://www.math.bgu.ac.il/~barakw/rigidity/bh.pdf
12. Theo Bühler & Dietmar Arno Salamon, *Functional Analysis*. Graduate Studies in Mathematics **191**, American Mathematical Society, Providence, Rhode Island, 2018.
13. Élie Joseph Cartan, Les groupes réels simples, finis et continus. *Annales scientifiques de l'École Normale Sup'erieure* **31** (1914), 263–355.
14. Élie Joseph Cartan, Groupes simples clos et ouverts et géométrie Riemannienne. *Journal de Mathématiques Pures et Appliquées* **8** (1929) 1–33.
15. Élie Joseph Cartan, La théorie des groupes finis et continus et l'Analysis Situs. *Mémorial des Sciences Mathématiques* **42** (1930), 1–61.
16. Bill Casselmann, Symmetric spaces of semi-simple groups. *Essays on representations of real groups*, August, 2012. http://www.math.ubc.ca/~cass/research/pdf/Cartan.pdf

17. Simon Kirwan Donaldson, Lie algebra theory without algebra. In *"Algebra, Arithmetic, and Geometry, In honour of Yu. I. Manin"*, edited by Yuri Tschinkel and Yuri Zarhin, Progress in Mathematics **269**, Birkhäuser, 2009, pp. 249–266. https://arxiv.org/pdf/math/0702016.pdf

18. Simon Kirwan Donaldson, Towards enumerative theories for structures on 4-manifolds. Seminar, 15 September 2020. https://www.youtube.com/watch?v=44hwQKQggbA

19. David Bernard Alper Epstein, Foliations with all leaves compact. *Annales de l'institute Fourier* **26** (1976), 265–282.

20. Valentina Georgoulas & Joel William Robbin & Dietmar Arno Salamon, *The Moment-Weight Inequality and the Hilbert–Mumford Criterion*. Lecture Notes in Mathematics **2297**, Springer, 2021.

21. Mikhael Leonidovich Gromov, Pseudoholomorphic curves in symplectic manifolds. *Inventiones Mathematicae* **82** (1985), 307–347.

22. Michael Leonidovich Gromov & Herbert Blaine Lawson, The classification of simply connected manifolds of positive scalar curvature, *Annals of Mathematics* **111** (1980), 423–434.

23. Victor Guillemin & Alan Pollack, *Differential Topology*. Prentice-Hall, 1974.

24. Allen Hatcher, *Algebraic Topology*, Cambridge, 2002.

25. Sigurour Helgason, *Differential Geometry, Lie Groups, and Symmetric Spaces*. Graduate Studies in Mathematics **35**, AMS, 2001.

26. Noel Justin Hicks, *Notes on Differential Geometry*, Van Nostrand Rheinhold Mathematical Studies, 1965.

27. David Hilbert & Stephan Cohn-Vossen, *Geometry and the Imagination*. Chelsea Publishing Company, 1952.

28. Joachim Hilgert & Karl-Hermann Neeb, *Structure and Geometry of Lie groups*. Springer Monographs in Mathematics, Springer, 2012.

29. Morris William Hirsch, *Differential Topology*. Springer, 1994.

30. Nigel James Hitchin, Harmonic spinors. *Advances in Mathematics* **14** (1974), 1–55.

31. Heinz Hopf, Zum Clifford–Kleinschen Raumproblem. *Mathematische Annalen* **95** (1926), 313–339.

32. Heinz Hopf, Differentialgeometrie und topologische Gestalt. *Jahresbericht der Deutschen Mathematiker-Vereinigung* **41** (1932), 209–229.

33. Ralph Howard, Kuiper's theorem on conformally flat manifolds. Lecture Notes, University of South Caroline, Columbia, July 1996. http://ralphhoward.github.io/SemNotes/Notes/conformal.pdf

34. Mustafa Kalafat, Locally conformally flat and self-dual structures on simple 4-manifolds. *Proceedings of the Gökova Geometry-Topology Conference 2012*, International Press, Somerville, MA, 2013, pp 111–122. http://gokovagt.org/proceedings/2012/ggt12-kalafat.pdf

35. Tosio Kato, *Perturbation Theory for Linear Operators*. Grundlehren der Mathematische Wissenschaften **132**, Springer, 1976.

36. John Leroy Kelley, *General Topology*. Graduate Texts in Mathematics, **27**. Springer 1975.

37. George Kempf, Instability in invariant theory. *Annals of Mathematics* **108** (1978), 299–317.

38. George Kempf & Linda Ness, The lengths ov vectors in representation spaces. *Springer Lecture Notes* **732** (1978), 233–242.

39. Wilhelm Karl Joseph Killing, Über die Grundlagen der Geometrie. *Journal für die reine und angewandte Mathematik* **109** (1892), 121–186.

40. Frances Kirwan, *Cohomology of Quotients in Symplectic and Algebraic Geometry*. Princeton University Press, 1984.

41. Wilhelm Klingenberg, Über Riemannsche Mannigfaltigkeiten mit positiver Krümmung. *Commentarii Mathematici Helvetici* **35** (1961), 47–54.

42. Nicolaas Hendrik Kuiper, On conformally flat spaces in the large. *Annals of Mathematics* **50** (1949), 916–924.

43. Wolfgang Kühnel, *Differential geometry: Curves-surfaces-manifolds*, Second Edition. Student Mathematical Library, 16, AMS, Providence, RI, 2006.

44. Jorge Lauret, On the moment map for the variety of Lie algebras. *Journal of Functional Analysis* **202** (2003), 392–423.
45. Claude LeBrun, On the topology of self-dual 4-manifolds. *Proceedings of the American Mathematical Society* **98** (1986), 637–640.
46. André Lichnerowicz, Spineurs harmoniques, *Comptes rendus de l'Académie des Sciences, Série A-B* **257** (1963), 7–9.
47. Anatolij Ivanovich Malcev, On the theory of the Lie groups in the large. *Matematichevskii Sbornik N. S.* **16** (1945), 163–189.
    Corrections to my paper "On the theory of Lie groups in the large". *Matematichevskii Sbornik N. S.* **19** (1946), 523–524.
48. Alina Marian, On the real moment map. *Mathematical Research Letters* **8** (2001), 779–788.
49. Dusa McDuff & Dietmar Arno Salamon, *Introduction to Symplectic Topology, Third Edition.* Oxford University Press, 2017.
50. John Willard Milnor, *Topology from the Differentiable Viewpoint.* The University Press of Virginia 1969.
51. James Raymond Munkres, *Topology*, Second Edition. Prentice Hall, 2000.
52. Sumner Byron Myers & Norman Earl Steenrod, The group of isometries of a Riemannian manifold. *Annals of Mathematics* **40** (1939), 400–416.
53. Linda Ness, A stratification of the null cone by the moment map. *American Journal of Mathematics* **106** (1984), 128–132.
54. John von Neumann, Über die analytischen Eigenschaften von Gruppen linearer Transformationen und ihrer Darstellungen. *Mathematische Zeitschrift* **30** (1929), 3–42.
55. Takushiro Ochiai & Tsunero Takahashi, The group of isometries of a left invariant Riemannian metric on a Lie group. *Mathematische Annalen* **223** (1976), 91–96.
56. Roger Penrose, Nonlinear gravitons and curved twistor theory. The riddle of gravitation – on the occasion of the 60th birthday of Peter G. Bergmann (Proc. Conf., Syracuse Univ., Syracuse, NY., 1975), *General Relativity and Gravitation* **7** (1976), 31–52.
57. Harry Earnest Rauch, A contribution to differential geometry in the large. *Annals of Mathematics* **54** (1951), 38–55.
58. Roger Richardson, Compact real forms of a complex semisimple Lie algebra. *Journal of Differential Geometry* **2** (1968), 411–419.
59. Roger Richardson & Peter Slodowy, Minimal vectors for real reductive group actions. *Journal of the London Mathematical Society* **42** (1990), 409–429.
60. Joel William Robbin, *Extrinsic Differential Geometry.* 1982.
    http://www.math.wisc.edu/~robbin/0geom.pdf
61. Joel William Robbin & Dietmar Arno Salamon, The spectral flow and the Maslov index. Bulletin of the London Mathematical Society **27** (1995), 1–33. https://people.math.ethz.ch/~salamond/PREPRINTS/spec.pdf
62. Mary Ellen Rudin, A new proof that metric spaces are paracompact. *Proceedings of the American Mathematical Society* **20** (1969), 603.
63. Dietmar Arno Salamon, *Analysis I.* Lecture Course, ETHZ, HS 2016.
64. Dietmar Arno Salamon, *Analysis II.* Lecture Course, ETHZ, FS 2017.
    https://people.math.ethz.ch/~salamon/PREPRINTS/analysis2.pdf
65. Dietmar Arno Salamon, *Spin Geometry and Seiberg–Witten invariants.* Unpublished Manuscript, 1999. https://people.math.ethz.ch/~salamond/PREPRINTS/witsei.pdf
66. Dietmar Arno Salamon, *Measure and Integration.* European Mathematical Society, Textbooks in Mathematics, 2016. https://people.math.ethz.ch/~salamond/PREPRINTS/measure.pdf
67. Dietmar Arno Salamon, Uniqueness of symplectic structures. *Acta Mathematica Vietnamica* **38** (2013), 123–144.
68. Dietmar Arno Salamon & Thomas Walpuski, Notes on the Octonions. *Proceedings of the 23rd Gökova Geometry-Topology Conference 2016*, edited by S. Akbulut, D. Auroux, and T. Önder, International Press, Somerville, Massachusetts, 2017, pp 1–85. https://people.math.ethz.ch/~salamon/PREPRINTS/Octonions.pdf

69. Richard Melvin Schoen, Conformal deformation of a Riemannian metric to constant scalar curvature. *Journal of Differerential Geometry* **20** (1984), 479–495.
70. Richard Melvin Schoen & Shing-Tung Yau, On the structure of manifolds with positive scalar curvature. *Manuscripta Mathematica* **28** (1979), 159–183.
71. Carl Ludwig Siegel, *Symplectic Geometry*. Academic Press, 1964.
72. Stephen Smale, Diffeomorphisms of the 2-sphere. *Proceedings of the American Mathematical Society* **10** (1959), 621–626.
73. Clifford Henry Taubes, Self-dual Yang-Mills connections on non-self-dual 4-manifolds. *Journal of Differential Geometry* **17** (1982), 139–170.
74. Clifford Henry Taubes, SW $\Longrightarrow$ Gr: From the Seiberg–Witten equations to pseudoholomorphic curves. *Journal of the American Mathematical Society* **9** (1996), 845–918.
75. Neil Sidney Trudinger, Remarks concerning the conformal deformation of Riemannian structures on compact manifolds. *Annali della Scuola Normale Superiore di Pisa* **22** (1968), 265–274.
76. Joa Weber, Perturbed closed geodesics are periodic orbits: Index and Transversality. *Mathematische Zeitschrift* **241** (2002), 45–81.
77. Hermann Weyl, Theorie der Darstellung kontinuierlicher halbeinfacher Gruppen durch lineare Transformationen, I, II, III; Nachtrag zu der Arbeit III. *Mathematische Zeitschrift* **23** (1925), 271–309; **24** (1926), 328–376, 377–395, 789–791.
78. Hidehiko Yamabe, On a deformation of Riemannian structures on compact manifolds. *Osaka Journal of Mathematics* **12** (1960), 21–37.

Index

Printed in the United States
by Baker & Taylor Publisher Services